D1158214

BETWEEN HUMANITIES AND THE DIGITAL

BETWEEN HUMANITIES AND THE DIGITAL

edited by Patrik Svensson and David Theo Goldberg

The MIT Press
Cambridge, Massachusetts
London, England

MIT Press books may be purchased at special quantity discounts for business or sales promotional use. For information, please email special_sales@mitpress.mit.edu.

This book was set in Gentium Plus by Toppan Best-set Premedia Limited. Printed and bound in the United States of America.

Library of Congress Cataloging-in-Publication Data is available.

ISBN: 978-0-262-02868-4

10 9 8 7 6 5 4 3 2 1

CONTENTS

CONTRIBUTORS

Ian Bogost is the Ivan Allen College Distinguished Chair in Media Studies and Professor of Interactive Computing at the Georgia Institute of Technology.

Anne Cong-Huyen is the Digital Scholar at Whittier College's Digital Liberal Arts center and a former Mellon Postdoctoral Fellow in the Humanities and Visiting Assistant Professor of Asian American Studies at the University of California, Los Angeles.

Mats Dahlström is an Associate Professor at the University of Borås.

Cathy N. Davidson is Distinguished Professor and Director of The Futures Initiative at the Graduate Center, City University of New York.

Johanna Drucker is Breslauer Professor of Bibliographical Studies in the Department of Information Studies, University of California, Los Angeles.

Amy E. Earhart is Associate Professor, Department of English, Texas A&M University.

Kathleen Fitzpatrick is Director of Scholarly Communication of the Modern Language Association and Visiting Research Professor of English at New York University.

Maurizio Forte is William and Sue Gross Professor of Classical Studies Art, Art History, and Visual Studies at Duke University.

Zephyr Frank is Associate Professor of Latin American History and Director of the Center for Spatial and Textual Analysis at Stanford University.

David Theo Goldberg is Professor of Comparative Literature and Anthropology at the University of California, Irvine, and the Director of the UC Humanities Research Institute and of the Digital Media and Learning Research Hub.

Jennifer González is Professor of History of Art and Visual Culture, Contemporary Art, and Race and Representation at the University of California, Santa Cruz.

Jo Guldi is Hans Rothfels Assistant Professor of History at Brown University.

N. Katherine Hayles is Professor of Literature at Duke University.

Geraldine Heng is Perceval Fellow and Associate Professor of English and Comparative Literature at the University of Texas, Austin.

Larissa Hjorth is Associate Professor in the Games Programs and Co-Director of the Digital Ethnography Research Centre, RMIT.

Tim Hutchings is a William Leech Research Fellow at the CODEC Research Centre for Digital Theology at Durham University.

Henry Jenkins is the Provost's Professor of Communication, Journalism, and Cinematic Arts at the University of Southern California.

Matthew G. Kirschenbaum is Associate Professor in the Department of English at the University of Maryland and Associate Director of the Maryland Institute for Technology in the Humanities.

Cecilia Lindhé is a Senior Lecturer in Comparative Literature and Director of the HUMlab at Umeå University.

Alan Liu is Professor in the English Department at the University of California, Santa Barbara, and an affiliated faculty member of the Media Arts and Technology graduate program, University of California, Santa Barbara. He was Chair of his department in 2008 to 2012.

Elizabeth Losh is Associate Teaching Professor, Sixth College, and Director, Culture, Art, and Technology Program, University of California, San Diego.

Tara McPherson is Associate Professor of Critical Studies at the School of Cinematic Arts, University of Southern California.

Nick Montfort is Associate Professor of Digital Media at the Massachusetts Institute of Technology.

Chandra Mukerji is Professor of Communication and Science Studies at the University of California, San Diego.

Jenna Ng is Anniversary Research Lecturer in Film and Interactive Media at the University of York.

Bethany Nowviskie is Special Advisor to the Provost at the University of Virginia and Director of the Scholars' Lab and Department of Digital Research and Scholarship at the UVa Library. She is also a Distinguished Presidential Fellow at CLIR, the Council for Library and Information Resources.

Jennie Olofsson is a Senior Assistant Lecturer at HUMlab at Umeå University.

Lisa Parks is Professor, Department of Film and Media Studies at the University of California, Santa Barbara.

Natalie Phillips is Assistant Professor of English at Michigan State University.

Todd Presner is Professor of Germanic Languages, Comparative Literature, and Jewish Studies at the University of California, Los Angeles.

Steve Rachman is Associate Chair for Graduate Studies in the department of English and Director of the American Studies Program and Co-Director of the Digital Humanities Literary Cognition Laboratory at Michigan State University.

Jentery Sayers is Assistant Professor of English at the University of Victoria.

Patricia Seed is Professor of History at University of California, Irvine.

Nishant Shah is a professor at the Institute of Culture and Aesthetics of Digital Media at Leuphana University, Germany and the co-founder of the Centre for Internet and Society, India.

Ray Siemens is Canada Research Chair in Humanities Computing and Distinguished Professor in the Faculty of Humanities in English and Computer Science at the University of Victoria.

Jonathan Sterne is Professor and James McGill Chair in Culture and Technology in the Department of Art History and Communication Studies at McGill University and maintains sterneworks.org.

Patrik Svensson is Professor and Chair in the Humanities and Information Technology at HUMlab at Umeå University.

William G. Thomas III is the Chair of the Department of History and the John and Catherine Angle Chair in the Humanities and Professor of History at the University of Nebraska, Lincoln.

Whitney Anne Trettien is a PhD candidate in English at Duke University.

Sherry Turkle is Abby Rockefeller Mauzé Professor of the Social Studies of Science and Technology in the Program in Science, Technology, and Society at the Massachusetts Institute of Technology.

Michael Widner is the Academic Technology Specialist for the Division of Literatures, Cultures, and Languages at Stanford University.

ACKNOWLEDGMENTS

We would like to thank our research assistants—Anna Finn, Maia Krause, and Emma Ewadotter—for bringing this volume to fruition. The book would not have been realized without their tireless collective efforts. We would also like to acknowledge the important support from the Wallenberg Foundation and the University of California Office of the President (UC Humanities Initiative Grant ID 142835). Finally, we would like to thank our institutions: HUMlab, Umeå University and University of California Humanities Research Institute for supporting us in this work.

INTRODUCTION

Patrik Svensson and David Theo Goldberg

Between Humanities and the Digital seeks to offer a comprehensive vision of a quickly evolving, contested and exciting field, as well as of its future directions. The volume describes the breadth and depth of how the humanities engage with the digital and information technology, including discipline-specific studies and perspectives, research infrastructure, innovative tools, and creative expression. We are committed in and through the volume to articulate an expansive and large-scale vision of the field, of the humanities at large, and so also of their constitutive relations to each other. This vision is anchored in several traditions: those invested in the humanities and their constituent disciplines, in the digital, in the digital's key contributions to the humanities, and in the humanistic considerations regarding the digital. Respect for these is vital to the book project as well as to a strong engagement with the future of the humanities. This volume aspires to be field-defining in an opening and open-ended sorts of ways, to be inspirational and bridge building through suggesting and manifesting a "large" and capacious sense of the digital, of the humanities, of their relation, and, in the final analysis, of what is marked by the name "digital humanities." It is all this that we intend by the "between" in the title.

The humanities have been undergoing profound changes regarding research and pedagogical practices, funding structures, the role of creative expression, infrastructural bases, reward systems, and interdisciplinary sentiment and structures. At the deepest level, these considerations have been bound up with the transforming and contested roles of the humanities within the changing structure of the university and the place and structure of higher education in contemporary political economy, and so also with a crisis in self-confidence. These shifts have been mediated at the same time by the emergence of a deeply networked humanities both in relation to the processes of knowledge production and its products. Humanities scholars, as scholars more generally, have turned increasingly to using and exploring information technology as both scholarly tool and cultural object in need of analysis.

Currently, there is a cumulative set of experiences, practices, and models flourishing in what digital humanities is invoked to reference. In historical terms, the humanities first engaged with computing technologies in the early 1950s as a tool to make concordances and carry out textual analysis. Despite several other early examples of humanities engagement with computation, for instance, in language learning, in rhetoric and composition, and in the speculative analysis about the future of intelligence and analytic systems and categories, it is the former tradition that mainly came to be identified with humanities engagement with technology, under the auspices of "humanities computing." This latter umbrella was principally associated with the development of such first-generation tools as text mining and searching, concordances, encoding, and early humanities archives. In the past decade the discursive shift from humanities computing to what is now being termed the digital humanities has concerned shifts in institutional, disciplinary, and social organization. This renaming, attendant as it has been to new modes of technological production and practice in the humanities, has brought with it a set of new conceptions concerning information technology, knowledge production and institutional arrangements and practices, predicated as they have been on a range of epistemic traditions.

There are multiple possible futures for what is today marked as digital humanities. One possibility would be to create a technologically and methodologically focused platform, somewhat similar to what humanities computing used to be. This would be a perfectly valuable part of the humanities, but it would be a much smaller project than the current scope and footprint of the digital humanities. However valuable such a trajectory, it would seem to be counterproductive not to build on the current situation with a field that is engaging with the humanities as never before. The established platform of the field, the influx of new scholars and technology experts to the field, and the larger intellectual and institutional agenda make the digital humanities a place where some of the most interesting emerging humanities scholarship and building can be done. Some of this work will be produced in what would be called the core digital humanities community, but much of it will be carried out across the humanities more broadly but still be part of the field. A main argument of this volume is that the digital humanities need to engage with the humanities multifariously and deeply. But equally, that which lies between the humanities and the digital will help shape the future of the field of digital humanities.

Digital humanities has also come to serve as a touchstone and laboratory for thinking about the current state and future of the humanities at large. In the view of some traditionalists, the digital has come to represent an abandonment of what the humanities are supposed to stand for, and by extension, the representative expression of the supposed crisis of the humanities; moreover, in the call of some digitalists, the technological has been taken presumptuously to save the humanities from the insularities of its traditionalisms. Contributors to this volume reveal the poverty of both positions, exemplifying the most probing thinking precisely at the interface. It is in this spirit of rethinking, now made self-reflexive to include the digital humanities itself, that we are concerned in this volume to offer a comprehensive framework for exploring the spaces of engagement, intellectually, materially, and institutionally, between humanities and the digital both generally and generously conceived.

At the risk of caricature, existing work around questions of humanities and informational technology can be divided at the outer edges roughly between the technological and methodological imperatives represented by work under the rubric of humanities computing, on one hand, and critical studies of the role, structures and infrastructures, interventions, and impacts of the range of work falling under new media, on the other. *Between Humanities and the Digital* is concerned to address the entangled interfaces of new media technology with the humanities in all their complexity. The volume maps the range of interactive engagements between humanities and digital technologies and the transformative impacts as a consequence on both.

Between Humanities and the Digital brings together a multifaceted view of digital humanities and its intellectual culture, of the entangled interfaces between humanities and the digital, as well as the interactive engagements between their practicing scholars. It speaks accordingly also to the future of the humanities, to their institutional-level and strategic perspectives. *Between Humanities and the Digital* thus sets out to:

- articulate digital humanities both as a large-scale and multiplicitous, heterogeneous humanities project, or really a set of projects;
- build bridges between different traditions invested in the intersection, at the interface;
- engage critically and productively with high-quality current humanities research and current research challenges within and outside the disciplines;
- extend the engagement with the digital in the humanities, between the humanistic and informational technology;
- consider digital humanities as one way among others of reconfiguring the humanities and the academy;
- consider critical work and "making" as intertwined practices;
- operate on three levels (coinciding more or less with the volume's three parts): institutional, scholarly, and infrastructural;
- complement other work on digital humanities, and on the interface between humanities and the digital.

Structure and Composition

The volume is divided into three complementary parts each with their own introduction. The first part addresses the current state and future of the field of digital humanities through bringing together scholars and leaders from different traditions and perspectives. As a whole, an expansive and intersectional vision of digital humanities is forwarded. The second part includes digitally inflected and enabled work in the humanities from a range of fields and disciplines. The range of work presented demonstrates the rich quality of work at this interface as well as how the digital intersects with significant humanities-based research challenges and emerging areas of inquiry. The final part explores how the digital pressures models for knowledge production and infrastructure inside and outside the humanities, suggests alternative modalities for academic expression, and challenges traditional models of learning.

The thirty-five contributions reflect the ethos in the volume in the sense of drawing on different traditions and modalities. For instance, a few of the chapters originated as blog entries and were chosen because they add importantly to the narrative. Three chapters are based on previous publications and selected for the same reason. All the other chapters were solicited from the authors for the direct purpose of this volume. The book includes contributions set across a span of disciplines and areas including cinema studies, humanities computing, English, archeology, media studies, science and technology studies, history, art theory, library science, religious studies, media history, gender studies, computer science, ethnic studies, and comparative literature.

Katherine Hayles closes the volume with a productive provocation. While she notes the reach and importance of the work presented, she also finds much of this work set in a revisionist rather than revolutionary framework. From the point of view of the central aspiration of this volume—exploring, building, and interrogating what is between humanities and the digital—Hayles's observation may indeed be seen as a confirmation of engagement with that liminal space. This does not preclude truly transformational work, of course, but it is likely that much of the most important work will be placed within or in relation to existing structures and lead to long-term renewal (to use that term) rather than immediate revolution and transformation.

In all, then, *Between Humanities and the Digital* reveals that a new turn—perhaps a new temporal chapter—has emerged in the relational engagement between humanities and the digital. The most interesting and innovative work today in what has come to be referenced by digital humanities is less in its discrete self-formation as a self-encapsulating, self-informed, and self-referencing discipline so much as in the ways the digital has been taken up to push the boundaries of the sorts of questions and challenges the humanities have long addressed within, across, and beyond their own disciplinary formations. It is in the "between" that the most interesting, creative, and provocative work of the digital and the humanistic is today being done. In this, the digital has not only prompted the humanities to open up to their own beyond, their own horizons of possibility; the humanities have likewise pushed the digital to become more than techne, more than a narrowly technological application. This liminal position is simultaneously precious, productive, and precarious. In taking seriously the interactive, relational, and interfacing challenges of the ampersand (and)—the *between* of the digital and humanities—lies the possibilities of their mutually engaged and creative transposition, their re-vision and re-formation.

A Visioning Statement

What today falls under the designation of "digital humanities" both builds on and challenges what has been characterized by contrast as "traditional humanities," in conception and application. On one hand, while digital humanities has sought discrete status, it accordingly is not a distinct and separate discipline, although it may seek institutional integrity to be productive. It looks to engage with, build on, and connect with pretty much any and all humanities disciplines, while looking at once to push the traditional and conventional to new insights and newly productive ways of making knowledge. It seeks persistently to touch the heart of humanities disciplines and interdisciplines, to advance them individually and interconnectedly, but also to look outside the humanities proper for relational possibilities of opening up new insights, new ways of thinking about subject matters, indeed new subject matters. It looks to provide productive resources to address traditional research questions but equally to pose old questions anew and new questions for long established or more recent fields. Digital humanities should be sufficiently open to incorporate data heavy projects, encoding methodologies, the textual with the visual and the sonic as well as critically and theoretically based analyses. And indeed to have its principal and driving premises challenged by the latter. One of the major challenges at the intersection of humanities and the digital is the interweaving of intellectual, exploratory and technological modalities. Johanna Drucker describes this challenge in terms of creating "intellectual middleware."

"Big" digital humanities—laboratory-like initiatives that incorporate numerous projects—draws on multiple epistemic traditions, seeking to find common and sometimes contesting ground and language while being open to productive tensions, critical interventions, and new directions. Digital humanities puts into play computational and interpretive analytics, textual and visual rhetorics, logical systematicity, and metaphoricity. It seeks to combine the technics of enhanced searchability, archiving, and mapping capacity with provocative new (counter-) possibilities in composability, curation, and curatability, and analytic as well as representational experimentation.

Digital humanities thus offers to the humanities a capacity to reach beyond itself in four key and sometimes unique senses. First, it puts the humanities into serious play with technical disciplines such as engineering and computing sciences in terms both of engaging with the latter's conceptual and instrumental capabilities and of engaging its epistemological assumptions as objects of (critical) analysis and comprehension. As such, it opens up challenges to engineering and computational thinking to engage more humanistically, whether by way of addressing more conventionally humanistic subject matters or by finding more humanistically disposed processes in their own applications and practices. Second, it provides the humanities with productive possibilities to reach new publics and well-disposed intellectuals in new ways. The reach of digital media provides the potential to engage a far broader range of publics, while in turn challenging humanists to become more self-reflective about how to represent their own work in more publicly accessible ways without necessarily compromising the quality or criticality of their work. Third, it challenges conventional modes of knowledge production and encourages exploration of multimodal expressions. Humanistic engagement has long predicated itself dominantly on more or less narrowly construed textual media, and digital technology has made multimodal production far more readily available. This, in turn, pressures humanists to think both more creatively and provocatively about the possibilities of multimediating composition, curation, and argumentation. And fourth, digital humanities enables the humanities to question in far more robust ways the practices, place, and role of the humanities in the twenty-first century. Digital technology has come to magnify the range and applications of meaning, value, and significance, the possibilities and actual expressions of interactive and relational cultural expression across broad divides hitherto less interactive. It has made possible new modes of translatability, with all the attendant challenges regarding power divides, interpretive presumption and failures, and translational hubris and misdirection thus posed. These concerns with meaning, value, significance, and textual, representational, and broadly cultural translation are issues long central to humanistic concern and focus.

At the same time, new media technology, as with any initiative depending on more expensive infrastructure and equipment, is prone to re-inscribing existing resource differentials and inequalities of power. As so far elaborated, digital humanities is little different. But digital technologies and social media have proven to offer up some unexpected opportunities to the resourceful. First, the technologies and media have been taken up in unpredictable and innovative ways to open up analytic, political, and social possibilities otherwise less readily or altogether unavailable. And second, the relative lack of resources has led to creative workarounds using available technologies in boundary breaking ways. We think here of the creative use of mobile technologies in societies with less reliable digital infrastructure, often outstripping better-resourced societies. Digital humanities is challenged both to attend to these developments at once as a check on its own sense of self-declared "progress" and as experiences from which valuable lessons are to be learned as much about the humanistic as about productive uses of the digital.

Commitment to such an expansive digital humanities accordingly comes with responsibilities and cannot be exclusively mapped onto individual traditions in a convincing way. In enabling the humanities to reach beyond itself both disciplinarily and in terms of long established and reproduced cultural and national boundaries, the digital challenges the humanities to be concerned with more than self-reproducing professionalization, to be fully mindful to whom the humanities speak and how. But it also entails that digital humanities be more than a self-reproducing technicist discipline, that it live up to its promise to marry productive rigor, instrumental usability, rhetorical openness and clarity, along with cultural humility and sensitivity. The work of the digital humanities must be intellectually driven in such a way that it challenges, stimulates, and provokes humanist scholars from the traditional disciplines. Through engaging with technology, including infrastructure shaped by humanistic questions, digital humanities can offer innovative platforms for exploring materials, expressions and research questions.

Yet digital humanities must resist a consuming romance with the newest technology without questioning its productive value while allowing appreciation of and experimentation with new (and old) technologies. The concern that digital humanities will eat up scarce resources in the humanities would best be offset by engaging the technological capacities to advance multimodal composition, to facilitate searchability and translatability, linguistic and cultural, to push new modes of expression and cartographic possibility while posing novel questions, to challenge fixed ways of knowledge-making, and to encourage collaborative engagements as the most effective means to advance knowledge about human being, sociality, and culture in and across heterogeneous environments. Importantly, we need to move beyond a zero-sum mentality and seek an expanded and reinvigorated place for the humanities within and across both the academy and public sphere(s).

Digital humanities accordingly has the power to stimulate visionary and transformative thinking, to be a site for innovation, reconfiguration, and exploration. This capacity derives from its broad and intersectional reach, from being situated at the periphery and challenging established structures at both the humanistic and digital boundaries, and so too from its abiding nondisciplinary status even in the face of drives to disciplinization. The digital serves as a potent point of shaping for this transformative sentiment. As a consequence digital humanities can become a proxy site where the digital, analog, and hybrid humanities can be discussed, contested, negotiated, and projected. Digital humanities requires material and technological grounding in order to facilitate the often intertwined practical, expressive, and critical work associated with it.

Today, even those traditional humanists who might openly deplore "digital humanities" as too much technology and too little humanities very likely use digital technology on a daily basis, for searching and archival consultation, for reading and composition. How this use transforms what each of us does as and in the name of humanistic endeavor too is a critical self-reflective question. Digital media and technology thus must be engaged deeply, sensitively, indeed critically in order to shape our own means of knowledge production. A properly humanities-based notion of research infrastructure requires humanists to think carefully through the intellectual challenges and ideational underpinnings of the humanities generally as well as of their individual fields. But it challenges us also critically to explore the technologies and methodologies the now established but also constantly morphing endeavors at the interface of the digital and the humanistic enable and produce.

I THE FIELD OF DIGITAL HUMANITIES

This is not a volume about digital humanities, conventionally understood; rather, it is one that engages with the interface between humanities and the digital. Digital humanities has become the primary institutional actor dominating this interfacing space. By contrast, part I addresses some perspectives important to the formation and understanding of what the digital brings to the humanities, sometimes in the name of "digital humanities" and sometimes unmarked. The foundational narrative of digital humanities as a field dates to the late 1940s, and much of the early humanities-based engagement with computational technologies was related to linguistic and literary analysis of textual materials. A strong tradition of computational, textual, and instrumental engagement was forged under the rubric "humanities computing," notably from the 1960s through the 1990s. And yet the first volume of the journal *Computers and the Humanities* (1966–1967) included articles on a range of humanistic engagements including "Art, art history, and the computer" and "Musicology and the computer in New Orleans" (informatik.uni-trier.de). Further there are other longstanding traditions, such as computational or digital archeology (from the late 1950s onward), that are normally not included in the story of digital humanities (Huggett 2012).

In the early 2000s "humanities computing" gradually began to be rebranded as "digital humanities." Funding agencies and other institutional bodies started to become more interested in the field. These movements coincided with the development of a renewed interest in research infrastructure under rubrics such as cyberinfrastructure and e-science. The new name and the associated discourse came with an expectation of a more broadly conceived field, both from institutional actors and from incoming, often junior scholars from other fields. For instance, the idea of "big-tent digital humanities" (used for the subheading of the 2011 Digital Humanities conference at Stanford University) can be read as a response to, or consequence of, this expanded notion of the digital humanities.

Nevertheless, it seems that at least implicitly, the epistemic tradition of humanities computing is still presumptively quite prevalent in digital humanities as institutionalized by the Alliance of Digital Humanities Organizations (ADHO). This means, for instance, that there tends to be a primary focus on methodology, tools, and encoding, and the starting point is often not research issues focused on content, but rather methodology, technology, and instrumentality. The idea of a methodological commons as a mechanism to hold the field together is strongly entrenched (McCarty 2005). This does not mean that research questions are not relevant, but that there is a tendency to foreground methodology and its uses in specific projects. Subsequently the realization has begun to emerge that there is a need to move the focus from the output of digital analysis to interpretation, evidenced notably in the description of a 2014 MLA session:

> The panelists' brief talks will offer interpretations of texts, language, literature and/or literary history that definitely began with a digital approach. But—and this is crucial—we have asked our presenters to focus not on their methods but instead on the interpretations they have reached as result of their digital praxis.

It seems clear that even though the session was an attempt to move away from data and methodology as a central focus, starting out with a digital approach and a set of methods is standard practice. These lead to an interpretation, but somehow the actual research questions are not given much centrality in the process even as framed in this session description. Our argument is not that they are not there, rather, that they are not necessarily the driving force in the framing of the projects. The digital humanities have also been critiqued for not engaging strongly with critical perspectives—such as power, the environment, the postcolonial, and gender considerations—central to key areas across the humanities today (Koh and Risam 2013). The focus on methodology signals a weak connection between the digital humanities and the core of the humanities. And although the field is much more diverse now, the notion of a methodological core is even maintained by groups critiquing the field for the lack of critical engagement. The Postcolonial Digital Humanities website describes the field in this way: "For our purposes, our working definition of the digital humanities is a set of methodologies engaged by humanists to use, produce, teach, and analyze culture and technology" (Koh and Risam "Mission Statement").

With the increased interest in the field, it has become clear that the digital humanities has also increasingly come to be seen as a vehicle for promoting, discussing, and developing the humanities at large. This is not a development coming primarily from the tradition of humanities computing. It is associated rather with incoming scholars (many junior) from other fields along with institutional platforms such as HASTAC. While linked to many of the concerns that have been important in the history of digital humanities, such as tenure systems and assessing digital products, this development is much broader, more capacious, even more "revolutionary." There is sometimes a tension between this kind of "speak," which often tends to be institutional, visionary, and high powered, and the discourse of traditional digital humanities. Not everyone in the digital humanities thinks that it is their task to take on the future of the humanities and the academy. Even in milder forms, as in the following quote from the much debated book *Digital_Humanities*, this kind of discourse is very different from more traditional engagements of the field:

> Digital Humanities represents a major expansion of the purview of the humanities, precisely because it brings the values, representational and interpretive practices, meaning-making strategies, complexities, and ambiguities of being human into every realm of experience and knowledge of the world. It is a global, trans-historical, and transmedia approach to knowledge and meaning-making.

The contrast is stark if we compare this with the definition of digital humanities offered by the Digital.humanities formation at Oxford University: "By digital humanities, we mean research that uses information technology as a central part of its methodology, for creating and/or processing data" (digital.humanities@oxford).

This contrast between a narrow, even technicist, focus and a more expansive, even bloated and questionably accurate sense of digital humanities can be quite productive, but it is also precarious. The authors of *Digital_Humanities* have been involved in the digital humanities (as humanities computing) for a long time, so to some extent they exist in both worlds. The expansion of the field has also led to an influx of researchers from humanities disciplines that do not necessarily have this grounding. These disciplines have their own epistemic traditions, which at least exist in some contrast to digital humanities. For instance, as Tara McPherson points out, a discipline such as cinema studies has not really had a strong engagement with production and making (McPherson 2009a). In such instances, the experiences and legitimacy of the digital humanities can help to renew a particular discipline. For negotiations and meetings across epistemic traditions to be productive, there has to be openness and willingness to engage critically and technologically without losing the sharpness or the disciplinary anchoring.

In chapter 1, Jonathan Sterne opens the volume with a timely provocation. Situated between humanities and the digital, Sterne is not normally associated with the digital humanities, at least in some more narrow sense, as a field. In his contribution here, he engages with the digital humanities while maintaining a certain friendly distance. Sterne recognizes the infrastructural achievements of the digital humanities, new textual forms and experiments, but questions whether the field has fully delivered at the level of ideas and cautions against the revolutionary rhetoric surrounding digital media. He is moved consequently to posit an "analogue humanities" similarly broad in range and outset to the digital humanities. Looking at the establishment of infrastructures and materials in a pre-digital world, he points to the material engagement in what is often seen as purely theoretical work, and how the humanities have always been multimodal and engaging with new sets of technologies. He relates this earlier uptake of technologies to low-key integration of tools in his own area of sound studies, arguing that good scholarship requires some degree of situated transcendence, causing us sufficiently to lose track of our tools so that we can get lost in a world of ideas. After all, he maintains, the humanities are about ideas. Sterne's thoughtful opening intervention thus throws down a challenge (or two) for digital humanities.

In chapter 2, Alan Liu and William G. Thomas III add a strong institutional perspective to the volume. Focusing on the United States, they relate the digital turn to the pressured situation for the humanities. The digital is seen as something strategically meaningful on the university level, as exemplified by MOOCs, online universities and, indeed, the recent interest in the digital humanities. In a sense it is a call to action; the authors argue that either those in the humanities can take systemic action to shape its long-term digital future or they will cede the opportunity and have their digital infrastructure built for them. The former alternative is a reframing project for both education and research that will require leadership, imagination and experimentation. As Liu and Thomas argue, this reframing is a matter of responsibility, in the end, to the humanities themselves. There is a sense of urgency in Liu and Thomas's organizational vision of the digital humanities that contrasts with Sterne's longer term and less urgent perspective.

The revolutionary and institutional power of technology that Liu and Thomas address, and which is important in the discourse of digital humanities, can be traced back to Renaissance technological utopianism. Chandra Mukerji (chapter 3) engages with the organizational and logistical power manifested through technologies and logistical traditions. Her earlier work on the gardens of Versailles as a system of territorial governance and impersonal rule serves as a basis for reinvestigating the gardens, labyrinths, and mazes of Versailles as an immersive environment. She stresses the materiality of both the gardens and digital culture, and she claims that the digital revolution, like the logistical tradition, has restructured ideas, global relations of power, and social identities through material innovation. In this way the history of technology becomes important to the digital humanities, and Mukerji suggests that we can rethink the history of technology through asking different and new questions about technology and humanist values. In this she is urging that we rethink digital culture through asking questions about how it engages with cultural traditions more broadly. Ultimately both sets of questions are equally critical to the humanities and digital humanities. Mukerji's is one of numerous contributions to the volume to exemplify how productive the engagement at the interface between the digital and the humanistic can indeed be.

The interface between the humanistic and the digital is also the topic of Todd Presner's (chapter 4) investigation of how critical theory and the digital humanities connect. He outlines a cultural-critical praxis that can potentially be central to the construction of twenty-first century cultural critique. While Presner concurs with the contemporary critique of digital humanities as not sufficiently critically engaged, he also locates some of the critical work of the field in the discursively suppressed or forgotten "mangle of practice." Finished projects, stable systems, and factual archives do not demonstrate the ambiguity, imaginary power and experimental modalities of mangled and "dirty" practice. There is a material sensitivity here that is both reminiscent of and different from Mukerji's work. Presner mainly focuses on the materiality and critical potential of digital interfaces, invoking Web-based projects as case studies. He traces the constructive and performative lineage of critical theory and connects it to the making practices of the digital humanities as well as the speculative and participatory core of the field. On this reading, digital humanities is productively rendered socially engaged, participatory, and critically aware.

Henry Jenkins (chapter 5) represents a humanities that is public, critically aware, and strongly connected to actual cultural phenomena and people. His primary focus is not technology or informations systems. Jenkins's practice includes extensive blogging (not per se about the digital humanities, but about specific research issues and popular culture), and some of this work is collaborative and curatorial. His contribution to the volume is based on an interview with Sherry Turkle originally published as a blog entry. In many ways their conversation seems like one core to digital humanities about humanistic engagement with technology. They both relate to what it means to be a humanist embedded in a technological setting. Important questions concern the intimacy of technologies, identity negotiation, companionship, and different ways of engaging with technological development. Turkle has been depicted as anti-technological after her most recent book *Alone Together*, but in the interview she makes it clear that she is interested in an empowering and critical engagement with technology. Technology is a means to enable work and life, open to testing and challenge, not to be pursued for its own sake.

Like Turkle, Johanna Drucker (chapter 6) has an interest in the relation between individuality and networked collectivity. Drucker explores the currency of concepts such as individual voice and authoritative text in a time of mass collaboration, machine readings, collective identity, and crowd psychology. She considers texts as iterative and interim objects that may be the result of search queries and algorithmic processing as well as textual forms resulting from aggregation and mediation. These objects and forms contrast with the notion of individual subjectivity as a central idea in humanistic thinking. Drucker considers installations such as *We Feel Fine* to detect emergent, aggregate voices arising, in this case from "feelings" expressed online through social networking sites. She connects the development of Renaissance humanism and the emergence of an individuated perspective represented in the voice as developed in modernism to human expression grounded in emergent systems and processes that connect to individual experience. Drucker's ways of addressing new textual conditions and forms in relation to this development are highly relevant to the digital humanities, demonstrating how the digital can reach deeply into disciplinary and trans-disciplinary core issues and challenges.

Nishant Shah (chapter 7) speaks from a very different position, even though Drucker invokes his work on putting political and social media in perspective. Instead of approaching long disciplinary and terminological trajectories, Shah uses "digital humanities" as a tool to imagine the future of higher education and educational infrastructure in India. He argues that there have been few "takers" for digital humanities in South Asia, although there have been practices that can be identified as belonging to the field. In looking at educational infrastructure under the auspices of digital humanities, he presents an alternative genealogy of the digital humanities that contrasts with the main narrative of the field in the global north. He regards digital humanities as a key opportunity to contest and question how language, labor and life are changing in relation to the unequal and diverse adaptation of infrastructure. Under this view, the field can both re-humanize the state-orchestrated conception of infrastructure and provide a critical perspective that can help us keep a humanistic sensibility and individuality in a networked society.

Anne Cong-Huyen also suggests an alternative genealogy in her chapter 8 on the intersection of Asian American Studies and digital humanities. She envisions a common future pathway for the two fields, but her work also entails a rewriting of what digital humanities is and could be. Importantly, Cong-Huyen argues that we need to move beyond the binaries to a spectrum of experiences, locations, and problematics. She stresses that inclusiveness and belonging are not necessarily experienced equally, inside and outside the digital humanities. In her analysis, Cong-Huyen points to many of the benefits of an open, intersectional model for digital humanities as she describes the overlaps and complementarities of the field. The focus on situated knowledge, historical legacy, as well as on social injustice in relation to race, class, and gender in Asian American Studies can be instrumental in developing and critically situating digital humanities. Cong-Huyen sees digital humanities as a place of engagement with technology, social justice, and praxis. She powerfully represents a large group of junior scholars seeing digital humanities as a significant opportunity and source of hope for the humanities and the academy.

Ian Bogost (chapter 9) argues strongly for a more active and worldly engaged humanities. In a provocative piece, invoking the Kantian notion of lower and higher faculties, he asserts that the humanities actually have a disdain for humans and a serious disconnect with the world. According to Bogost, the humanities need to engage with the world at large, and he looks at digital humanities as a serious candidate for supporting this process. At the same time he is cautious about the revolutionary and hopeful sentiment often associated with the field. He argues that reforming the humanities and establishing digital humanities broadly can be seen as internally focused institution building rather than world building. This view contrasts strongly with that of the integrative and aspirational vision articulated in Cong-Huyen's chapter 8. Bogost focuses on the digital as a tool, and argues that this is an area where the humanities are lagging behind and where there is considerable, if more low-key potential. Moving in this direction requires humanities to actually build tools to support the work done in its name and not just let existing technologies shape the conditions for humanistic work.

In chapter 10, Cathy Davidson's idea of the digital humanities, in part contrast to Bogost's, is clearly expansive. Essentially she argues that the digital humanities can and should have a key role in redesigning higher education for a digital age. Unlike Bogost, she does not make a distinction between institution building and world building, but rather argues that in order to manage digital knowledge and contemporary knowledge production, we need to challenge siloed university structures. The (digital) humanities can provide leadership in a world facing major global challenges, which require working with the best people across fields and organizing open and dynamic structures. Davidson uses the much debated binary between thinking ("yack") and making ("hack") to show how accepting such challenges requires us to be richly collaborative and heterogeneously configured, to be respectful of other competencies, to combine thinking and making, and to see Web-making as a new form of literacy. Digital humanists need to be public intellectuals, assuming a leadership roles within the academy and advocating publicly for the humanities.

In contrast with most of the previous chapters in this section, Ray Siemens and Jentery Sayers (chapter 11) engage with the long history of digital humanities as humanities computing and argue that understanding the past of the field is critical to its future development. They point to a number of examples from this genealogy and position humanities computing as quite different from the current media attention and hype associated with digital humanities. There is a tension here between humanities computing, embedded in a pragmatic and down-to-earth context, and the transformative approaches suggested by Cathy Davidson and others. Importantly, the problem-based model suggested by Siemens and Sayers is inclusive and speaks to several traditions at the same time. Their commitment to this model is shown through serious engagement with examples from a number of traditions. According to the authors, significant factors for moving in a problem-based direction include large-scale collaboration, linked data, nonempirical enquiry (e.g., play), computational cultural studies and physical computing. Under this model, computation has to be tied to culture, ideology and social justice in a way that has not previously been the case in digital humanities. This in turn means that there has to be a much more direct engagement with public humanities. Metadata considerations have to be integrated into scholarly workflows and there will be an increased emphasis on real-time analysis and predication. Furthermore the material engagement evidenced by physical computing points to the need of increased engagement with interaction design and demonstrates how digital humanities increasingly overlap with the lived spaces of everyday life and research.

In the concluding chapter 12 for part I, David Theo Goldberg explores the relation between the humanities and the digital as manifested inside and outside what has come to be ordinarily characterized as "digital humanities." He notes how the field in some ways has come to shape humanistic pursuit, act as a model for the humanities, and serve as a beacon of institutional hope. Goldberg's starting point is the humanities at large rather than digital humanities and he reminds us about the critical and interpretative drives that are central to any type of humanities. For Goldberg, digital humanities is alien and marginal as well as dynamic and the bearer of agency. Goldberg argues that digital humanities as commonly conceived comes with a number of assumptions about the humanities and that uncritically taking digital humanities as a discipline is highly problematic. Indeed, if the digital provokes deep humanistic questions, why would not the humanities be digitally engaged? And how could a digital discipline primarily invested in instrumentalization invoke the humanities? At the same time Goldberg asserts that the most compelling and transformative work is likely to emerge at the intersection of the interpretative-critical and the instrumental. His argument is concerned both to prioritize the humanities in digital humanities and to emphasize the contributions the digital can make to the humanities if they are to be productively reconceived. The driving question, in Goldberg's words, is "what kind of humanists we choose to be in and for our times."

1 THE EXAMPLE: SOME HISTORICAL CONSIDERATIONS

Jonathan Sterne

Far more rare are considerations of how knowledge in the humanities comes about.

—Lisa Gitelman (2006: 153)

Dreamed that I was in a used bookstore, flipping through a tattered, old hardback, checking out the videos on the touch screens embedded in some of its crumbling pages. In the world of the dream this was completely unremarkable.

—Mark Katz, Facebook Status Update, July 17, 2012

As humanists, our work often rests on the kind of forgetting or mystification dramatized in Mark Katz's dream. We depend on elaborate technical accomplishments to do our work; we adapt them to our own purposes, then we talk about the work as if it's just a set of ideas, separate from the material forms in which those ideas circulate. Katz dreamed about the pleasures of used book shopping, so the books' multimodality was "unremarkable" in the world of his dream. But in the world of the digital humanities, and especially in the world of commentary on the digital humanities (such as in this book), we treat multimodality as quite remarkable. On this front, commentaries on the digital humanities have a lot in common with current trends in media studies: both claim a certain kind of materialism, and insist that we attend to the materiality of scholarly productions. This essay mostly takes that tack. From the outset, however, I want to acknowledge the limits of that approach as a solution to the intellectual challenges facing digital humanists.[1] Sure, we must better understand the articulations of language and technology that make up the humanist stratum (if not the whole human stratum). But I make the point in the shadow of a bigger one: great scholarship also requires a kind of situated transcendence, where we lose track of our tools enough (or become *good enough at them*) to get lost in the world of ideas. We know this from the study of technology in other domains. A musician who thinks too much about fingerings on the instrument loses the flow of the song. If I think too much about the word processor, screen and keyboard, I will lose the flow of this sentence. And if you think too much about the texture of the page, the font of the print, the backlighting of your screen, or the quality of the text-to-speech program you are using (however you are reading this text), you will lose track of the argument.

Introducing (for the Very First Time): The Analog Humanities

If the term *digital humanities* applies a technical modifier to a broad mode of inquiry in the human sciences, we must be careful to avoid the well-documented mistakes of other digital utopians who confuse digitization with technologization. Digital technologies are certainly technological, but they are not necessarily more technological than what came before them. They are simply more apparent as technologies because of their relative novelty. Crotchety colleagues have sometimes responded to the millenarian rhetoric of digital humanities by positing "analog humanities" as an alternative, even borrowing from vinyl record collectors to call for a revival. I wish to imply neither nostalgia nor revivalism, but the term *analog humanities* is a wonderful heuristic for understanding what humanists used to do, and mostly still do.

For this essay, I will use the term *analog humanities* in a matter of fact fashion to refer to humanists' uses of analog media technologies—and the analog components of digital technologies—in academic settings and in print, in exactly the same expansive way that *digital humanities* refers to a complex of technologies and engagements without specifying any particular discipline. Katherine Hayles (2012: 27) defines the digital humanities as "a diverse field of practices associated with computational techniques and reaching beyond print in its modes of inquiry, research, publication and dissemination." Although Hayles says this definition allows her to "understand the digital humanities as broadly as possible," it actually excludes a wide swath of work on and with digital media in so-called new media studies. Her definition is about the media humanists work with, rather than the media they may study. Other writers echo this distinction. Matthew Kirschenbaum (2010: 56) writes that the digital humanities are defined by a "methodological outlook" and a "social undertaking"; Patrik Svensson (2012: para. 21) writes that digital humanities scholarship "operates across all of the humanities." While there is some inevitable overlap in Hayles's implied distinction (e.g., she is read in both fields), it has the advantage of not collapsing the study of technology into the use of technology, just as we would separate a field such as creative writing from literary criticism, poetics, or the history of the book. As Tara McPherson argued in 2009, while there is much potential for the interesting fusion of digital humanities work and scholarship in media studies (and new media studies), it currently exists more as a promise and potential (2009b). McPherson suggests that this synthesis might best happen through multimodal scholarship. In this chapter I want to argue for recognizing that humanities scholarship has *always* been multimodal. As Lisa Gitelman (2006: 153) writes, "Media history offers access to the epistemologies and interpretive practices of the humanities at a vernacular as well as scholarly or academic level . . . media aren't instruments of scholarship in the humanities; they are the instruments of humanism at large . . ." If we recognize the long-term multimodality and mediality of humanistic scholarship, then it is perhaps not radical at all to incorporate new technological modes into our work. It is simply a matter of doing the best job that we can.

For the purpose of my argument, I will extend Hayles's, Kirschenbaum's, and Svensson's reasoning backward.[2] The *analog humanities* refers to a nexus of methodological, technological, and institutional conditions across the humanities that have only come into clear focus in retrospect. They refer to the cultural and material infrastructures on which humanists depended and still depend. They were (and are) not uniform across fields. Just as "there is no single vision of the digital humanities, nor can a single vision even be possible" (Svensson: para. 127), we could say the same for the analog formations of humanistic scholarship. Borrowing a periodization from the history of sound recording, I will refer to the "analog era" as the period dominated by analog technologies of reproduction for print, images, video, and sound and the "digital era" as our present moment.

"Dominated by" is not the same as "exclusive"—just as the odd digital reverb started appearing in recording studios in the 1970s, humanists also had engagement with digital technologies in their work long before it was thematized, for instance, every time they made a long-distance phone call. *Analog era* is also strictly heuristic and rhetorical for the purposes of this essay. Just as the retrospective term "analog recording" includes massive shifts from cylinders to disc to tape, there were many important changes in humanistic interpretation and its connections with institutional-technical formations across human history such that a monolithic "analog period" makes sense only as a point of relief, a *rhetorical before*, to accompany the digital. This point is well documented in classic technological histories like those of Harold Innis (1991) and Lewis Mumford (1934), as well as new work like Andrew Piper's (2009) that shows how romanticism could not have emerged without the technical, literary, and practical effects that "generated bookish thought." As I argue elsewhere (Sterne, forthcoming), *analog* used to refer to very specific kinds of technologies; it is only with the advent of widespread computing, and a journalistic field that heavily promoted it, that people came to understand *the analog* as "that which is not digital." For the purposes of an essay in a digital humanities volume, I am allowing myself to sustain that historical anachronism. But please understand that every time I use the word *analog*, I am making a retrospective move, and *not* describing the past as it actually was for the people who lived there.

As with the recording metaphor, analog technologies and analog components of digital technologies are still essential. Just as recording engineers and musicians still make use of mechanical or analog electronic instruments, microphones, speakers, rooms, and room treatment, digital humanists also operate with technologies that are in no way digital (Massumi 2002). But just as the institutional shape of education is changing—due to a mix of changes in hardware and a series of political economic shifts—so too are our orientations to our tools, and through our tools, the texts and artifacts we study and the ideas we advance. As John Durham Peters (2003: para. 7) has written, the fundamental insight of media studies is that "texts cannot be interpreted apart from an interpretation of the processes that produced them." Let us consider some of the processes behind the production of humanists' texts.

The ability to represent and manipulate texts, images, and sounds—in publications, in classrooms, at conferences—facilitates conversation both within and across humanities fields. For all the mileage scholars have gotten praising abstraction, we do a great deal of our work through indexicality—through pointing and showing. The changing options for representing images and sounds suggest a different future in the age of digital humanities. By understanding the uses of technologies now deemed passé by digital savants, we can get a better sense of how the ubiquity of computational power might lend itself to real conceptual advances and transformations in humanistic fields. Below, I will pull a few old books from my shelf and refer to some canonic uses of examples to show how this logic works in practice and then consider the changes happening in the digital humanist universe. In the process, I will argue for a less revolutionary and more workaday sensibility for digital humanists' relationship to technology. The revolutionary rhetoric surrounding new media is a commercially driven discourse; we import it uncritically at our own great peril.

Insofar as digital humanist work helps us construct new kinds of objects that lead to new insights, it is useful, but it cannot get there through methodologism or technological innovation without also passing through pedagogy and, if we don't monumentalize the term, theory (or perhaps, more modestly, concept formation). At the risk of excessive generalization: after philology, almost all of the great humanist work has done one of two things. It has either helped its readers interpret the world in new ways, producing new understandings of phenomena that matter to those who attend to culture. Or, it has itself been a masterful interpretation of some cultural artifact (here I follow Hayden White's prescriptions for historians in "The Burden of History"; see White 1978). Ian Bogost's recent call for relevance—"we must want to be of the world, rather than hidden from it" (quoted in Svensson 2012: para. 36)—restates a century-long aspiration of humanists, from W. E. B. Du Bois ([1903] 1999) turn to African-American spirituals, to Hayden White's advocacy of presentism in historiographic thought, to Stuart Hall's (1981) famous line that popular culture matters because it is a terrain of struggle; "otherwise, I don't give a damn about it."[3] There is no reason to think that digital humanities work is a radical break with the substance of questions that have spanned across centuries and generations of technical media simply because the tools are different.

Humanists, Examples, and Media Systems

If you take apart any so-called digital technology, you will quickly discover that most of its components are (by the retrospective definition) analog, and that some of the most important innovations that make it work affect its analog components, not its microprocessor, firmware, or software. Lev Manovich (2001: 64–65) may be right that the Graphical User Interface "shapes how the user conceives of the computer itself." But the prevalence of repetitive stress injuries among humanists, not to mention recurrent conversations about ergonomic chairs and desk options, suggests that embodied knowledge of the machine comes from a place other than the GUI.

Consider the source of those injuries: the keyboard interface. The very form of the QWERTY keyboard was originally designed to impose slowness on the user of a mechanical typewriter (so the keys wouldn't jam). Liberated from both the need and the mechanics, at first glance the keyboard seems to be a creature of the same kind of vestigial history as the hat tip: "a man who raises his hat in greeting is unwittingly reactivating a conventional sign inherited from the Middle Ages, when, as Erwin Panofsky reminds us, armed men used to take off their helmets to make clear their peaceful intentions" (Bourdieu 1981: 305). Somewhere between a historical tradition and a vestige, the QWERTY keyboard links computers with a long history of middle-class office practice. The keyboard on early computers helped make them intelligible or scrutable to early users, just as Robert Moog's choice to put a piano keyboard on his analog synthesizer made it accessible to a broader swath of musicians than synthesizers that did not have piano keyboards (Pinch and Trocco 2002). But, of course, as any player or typist knows, a Moog keyboard is not a piano keyboard, and a computer keyboard is not a mechanical or electric typewriter. It carries the form and spatial relations. It allows for the transfer of embodied knowledge from one apparatus to another (like the hat-tip). But in providing some embodied continuity, it affords the possibility of accommodating other kinds of change. The mechanical and electric typewriter was connected to a special set of tasks, one that was often closely articulated to women's bodies, as both Friedrich Kittler (1999) and Lochlann Jain (2006) remind us. The personal computer keyboard comes with the generalization of typing to more genders, more social stations, and more bureaucratic spaces. Today nobody even talks about their typing skills, except, of course, on new interfaces like smartphones, which add new challenges and difficulties.

Attention to the keyboard highlights something we know intuitively but too often leave aside in discussions of what comes next in the digital humanities: texts in the humanities have always presupposed and referred directly to extensive articulations of people, practices, and media inside the classroom and out. Long before the digital humanities, humanists depended on a whole social and technical assemblage for their production, one that included typing and typists; notecards, card catalogs, and file cabinets; buildings, bookshelves, locks, and lights; tables, chairs, and writing implements; publishers, printers, editors, copyeditors, and proofreaders (the last three being in sadly short supply today); classrooms, desks, tables, blackboards, screens, and overhead projectors; cameras, copiers, mimeographs, record players, headphones, speakers, VCRs, DVD players, cathode ray tubes, eventually computers, then laptops, then dongles, and on and on. Humanist texts were never just isolated containers of words, divorced from other media systems. They depended on elaborate configurations and articulations (Slack 1996; Slack and Wise 2006) of materials, relations, and knowledges. They still do.

One last tarry with the computer keyboard will illustrate my point, through an example of an example. Consider Foucault's famous use of the French keyboard sequence A, Z, E, R, T (which was adapted from the American QWERTY at the turn of the twentieth century). He takes it up to illustrate his concept of "statement" in the *Archaeology of Knowledge* (not a proposition or a phrase, not language, not technology ...), to show that thought is inseparable from *techné*, but that *techné* is not simply technology.

> The letters marked on a keyboard are not statements: at most they are tools with which one can write statements. On the other hand, what are the letters that I write down haphazardly on to a sheet of paper, just as they come to mind, and to show that they cannot, in their disordered state, constitute a statement? What figure do they form? Are they not a table of letters chosen in a contingent way, the statement of an alphabetical series governed by other laws than those of chance? (Foucault 1972: 85)

A statement is "a transmission of particular elements distributed in a corresponding space," writes Deleuze (1988: 3) in his commentary on this passage. AZERT is a statement in a typewriting manual, not on a typewriter. Here was an example that every French intellectual knew, but that was also native to its historical moment: when male academics might well try their luck at typing for themselves, and then publicly reflect on that fact.

The keyboard example depends on common knowledge, the availability of shared experience, and a particular moment and place in the history of technology: men typing and personal typewriters, the existence of typing manuals and the readiness of the typewriter layout to everyday consciousness. Foucault's theory of the statement, instructive for those of us tempted to move too quickly from what Madeleine Akrich (1992) calls the "script" to the software and hardware itself, also illustrates the dependency of humanist reasoning on technocultural formations. Thus even an article or book like *Archaeology of Knowledge,* one with no illustrations, could assume that its reader had access to certain kinds of textual, visual, or audiovisual material, or in the case of a computer keyboard, a shared set of experiences.

The analog humanities weren't just bound up in media systems; they sought to create, develop, and propagate them, or worked with allies in libraries and university administration to do so. In the process they produced a shared set of experiences as the basis for a collective project of knowledge. Analog media systems also helped define the relationship between publication and pedagogy or intellectual practice. Students learned to study images and texts by looking at them. Looking at them assumed a system of availability, dissemination, and utility. Consider Laura Mulvey's use of Hitchcock in her classic essay "Visual Pleasure and Narrative Cinema." Hitchcock may only have been in the process of canonization in the 1970s, but he was certainly known to film scholars. In a passage referring to *Marnie, Rear Window*, and *Vertigo*, Mulvey (1975: 15) writes, "Hitchcock's skillful use of identification processes and liberal use of subjective camera from the point of view of the male protagonist draw the spectators deeply into his position, making them share his uneasy gaze. The audience is absorbed into a voyeuristic situation within the screen scene and diegesis which parodies his own in the cinema." She goes on to describe scenes in detail enough that the reader will understand her argument about the function of the gaze and identification with a male viewing position, but the essay ultimately assumes that readers might, could, should, go out and experience the films for themselves.

Read functionally, this passage places an expectation upon the reader: you should know these films, and if not, you should be able to find a way to get to know them. To put it another way, the problem of canons in the humanities has always been a media problem, and specifically a media resources problem. The imperative "should" implies both a moral imperative and a reasonable expectation of an institutionalized media situation. I realize it is also a blatant display of a kind of cultural capital, in the very well-trod Bourdieuan sense, just as I am asking readers to know names like Mulvey, Foucault, Bourdieu, and so on. But the point is, however we read the class politics, that there is a pedagogy of texts that animates and supports humanistic reasoning and argument. That is why we teach our students to have bibliographies, and we ourselves go out and read the texts in our favorite authors' bibliographies. For texts other than print, there were also sets of expectations deemed reasonable in terms of getting one's hands on them. The library would have the novel or play, the slide library would have the painting, the music library would have the score or record, the map library would have the map. The analog humanities depended on an institutional bargain with libraries and a whole set of academic media industries. They also depended on consistent enough pedagogy that students would know both canons of texts and the accepted protocols for interacting with them (ones that were often shaped by the various availabilities and rarities built into the analog humanities' media system).

This was the basis of fields like literary criticism, art history, musicology, and cinema studies. Arguments were made not just with respect to written texts, images, or audiovisual texts, but to specific features of materials that were either immediately available to the senses or assumed to have been available to students. This kind of knowledge is *referential* in character but it is not *semantico-referential* in an abstract sense of the word "tree" having lexical reference to the sound image of a tree in a person's mind (Saussure). Rather, it implies a specific experience of the object of study for the reader—whether or not that experience actually happened. If you were going to write about an image, you would want people to be able to see the image, either inside your text or by going out and finding it without too much trouble. As Robert Nelson (2000: 422) has argued, slides in art history lectures *became* the objects of study, so that "arguments based upon slides alone are persuasive, even if the evidence only exists within the rhetorical/technological parameters of the lecture itself." The same logic can be found in film studies. Lengthy descriptions of plot, formal aspects of a film, or scenes in a movie would accompany formal analysis in film study, but they assumed that their readers would have to audioview the film somehow to truly understand the critique (even if this didn't always happen in practice).

Any work of interpretation in the analog humanities was subtended by various assemblages of print, chalkboards, screens, seating, libraries, overhead projection, slide projection and cinematic projection, and sound and image reproduction, all of which were designed to allow for certain kinds of discussions to happen. Art history departments would maintain vast slide libraries and have classrooms specially fitted with slide projectors. Film departments would have projection equipment and film collections. Musicologists depended mightily on sheet music, pianos in classrooms, and later—though still with surprising selectivity—record players and tape decks. To be clear, it was not the technologies themselves that inherently shaped intellectual practice but rather their institutional configurations and the practices that grew up around their particular affordances. It is true that an overhead only projects transparent images and cannot reproduce sounds or send emails. At the same time an overhead projector is not a substantially different technology from a magic lantern, even though a classroom lecture with overheads is substantially different from a magic lantern show. The nature of the spectacle produced by the professor—up to and including the lecturer accidentally spilling coffee on the overhead to produce a psychedelic, brown bubbling effect—was fundamentally different from a nineteenth-century magic lantern show when we consider the space of projection and what the audience was doing with its bodies. Consider the note-taking and poses of attention, and the (pardon the anachronism) pre-Facebook technologies of boredom like newspapers, doodling, note-passing, gossip, and textbooks for other classes. When we get into slide projection, film viewing (and here I mean movies and projectors, not bringing up a YouTube clip on the laptop), and record listening, there is a substantial range of institutional resources, technologies, and cultural practices that are implicated in the mere project of placing an example before students in a class for the purposes of analysis and discussion. The same resources were necessary for scholars to do work in their offices, home offices, libraries, or dedicated screening rooms.

The need to manage examples helps us to understand the real utility of canons for previous generations of humanists. A canon guaranteed a manageable core of shared knowledge and experience for the purposes of criticism and analysis. This was exceptionally important in a condition of artificial scarcity of images, texts, and recordings. If professors and students all had access to a shared repertoire of works, the analytical task became much easier; it was easier for people to talk with one another across different examples. Where these shared canons were absent, other modes were prized, such as archival research, because they involved access to primary texts. Archival fetishism in historical research also emanates from this condition of rarity (Steedman 2001).[4]

Another prized move was to expand techniques used on one familiar set of texts to another, less familiar set. Standard humanistic practices of citation, interpretation, and criticism follow this practice. Dick Hebdige's *Subculture: The Meaning of Style* (1979) was considered heretical for its application of semiotics—traditionally a mode of analysis reserved for high culture—to punk. But that also made punk intelligible to a wide range of humanists who might otherwise not have had the tools to make sense of it. John Berger's *Ways of Seeing* (1977) juxtaposed well-known oil painting to modern advertisements, advancing his "men act, women appear" thesis through a juxtaposition of different kinds of images. Indeed the book worked so well for so long as a pedagogical tool precisely because it combined simple, polemical prose with dozens of images placed in dynamic tension with one another. There was no canon of advertising images, so the book made its own corpus, which it displayed for the purposes of analysis and critique. I wonder today if any publisher would have the courage to print that many commercial images without either paying out steep royalties or placing lots of words around each picture—the image credits at the end of *Ways of Seeing* seem solely geared toward art galleries.

For Mulvey, for Berger, for Foucault, for our humanist critic in the classroom, a major argumentative and pedagogical gesture might be described as pointing, like Peirce's (1955: 109) weathervane: "look, over there, how that process I just described *works*." The writer, student, or teacher zooms in, as it were, on the detail: a turned head in a painting, the relationship between a look and a person looked at; the iconographic similarity between aspects of two images; the rhythm or semiotic richness of a piece of prose. Or, consider a Socratic moment in the classroom, "open your book to page 109. What does Peirce's discussion of the weathervane tell us about the nature of indexical signs?" This command assumes that the students will have the example (in this case, Peirce's *Philosophical Writings)* ready to hand, that they will all read the passage, and then they can discuss their interpretations. Again, a common base of examples is hugely helpful in each of these situations. Common textual experience is one of the main reasons that humanities classes *work* at all, and a central means through which humanities disciplines reproduce themselves, since, once there is a common experience, various interpretive operations can be taught and experimented with.

To say that we understand images or texts so well (if we do in fact understand them well) because of abstract language thus completely misses the point of how knowledge is conveyed, how students in the humanities learn through experiencing material, and how we actually make our arguments to one another. To be sure, some aspects of that conveyance are usefully mystified for the purposes of classroom discussion. John Corbett brings this point home in imagining the experience of record players in the music library:

> Imagine several partitioned cubicles, each of which contains a headphoned student who faces an amplifier and a turntable: on each platter spins a record of Beethoven's Ninth Symphony. One student lifts his needle to run to the bathroom; another listens twenty times to a difficult passage; a third is frustrated by a skip in the record and proceeds directly to the next movement of the symphony; at the same time another finds it difficult to concentrate due to the volume of her neighbor's headphones. Even as they do these things that are made possible only by the technology of recording, these students are required to develop a historico-theoretical interpretation as if the technical means through which the music is accessed—right there, starting them in the face—are of no significance whatsoever. (Corbett 1994: 36)

Corbett was making a point about recording, but his example shows the importance not only of the record players and headphones at that moment but libraries, collection practices, furniture, and listening practices tied to music history pedagogy. His point was that music scholars ought to pay more attention to sound-reproduction technology, and of course, I agree that it's a good thing for music scholars to study. But at the same time technology is not the only thing. It is important to behold all of those things Corbett's passage holds in dynamic tension. But humanistic scholarship also requires moments of situated transcendence. Like musicians, humanists must learn their tools, care for them, and understand them, and then selectively and situationally (but never totally) forget about them as they become second nature. To put it another way, reflexivity is a necessary part of scholarship, but it need not always be displayed in the end result.

Technics and the Humanities

The need for an institutionalized, reproducible technocultural infrastructure—and its occasioned mystification—is one of the central problems that confronts digital humanists today. I don't think it makes a big difference in terms of what "kind" of digital humanists we are talking about. Whether you are interested in new modes of disseminating knowledge and new platforms for doing it, or new digitally aided approaches to inquiry, the struggles over the "how" are so monumental that the *technics for thought* take up some of the space that *thoughts about stuff* used to occupy. I'm all for discussions of technics, but there is a well-documented issue that humanists now share with musicians and cooks. You can have the best gear, and music can sound terrible and the food can taste bad. Too much focus on the gear (or its techniques) tends to take time away from actual cooking and music. If digital humanists want to serve up a new cuisine, it's well and good to get into the finer points of molecular gastronomy and culinary chemistry. But sooner or later, someone's got to be served something tasty.

This point is aptly made by Patrick Juola (2008) in his essay "Killer Applications in the Digital Humanities." Using the language of computer science (or perhaps more accurately, venture capital and the Silicon Valley tech industry), Juola defines a "killer application" as a device or application that identifies "a need of the user group that can be filled, and by filling it, create an acceptance of that tool and the supporting methods/results" (5). Juola gives some historical examples, like photocopying and scanning, portable computing, word processing, networked communication (and software for it like browsers and email programs), and search engines (7–9). His examples are striking because none of these tools were initially developed by or for humanists. Rather, each was adopted by or adapted for humanist use. Somewhat perversely, technologies developed by and for humanists are not necessarily or automatically more useful than technologies adapted from other fields. He discusses a digital edition of *Clotel*, which he calls a "technical masterpiece," but which does not connect with any stated demands of scholars. "It is not clear who among Clotel scholars may be interested in using this capacity or this edition. . . . The Clotel edition is essentially a service offered to the broader research community in the hope that it will be used, and runs a great risk of becoming simply yet another tool developed by the DH specialists to be ignored" (75). The work of digital annotation is actually much closer to an older, philological model of the humanities, and in areas where manuscripts are still relatively rare and hard to access, massive projects of digitization and annotation may be incredibly useful. New digital editions of canonic texts may also be useful in unpredictable ways, just as monographs may languish on library shelves for years or decades and suddenly be plucked off to be read by scholars of a new generation who have completely different interests in the material.

Juola's proposed killer apps are based around needs that he can see and identify—annotation tools, resource exploration, automatic essay grading. Six years later we can say progress has been made on all of these fronts. But that is only one measure of success. As fields mature, we expect that scholars within them will develop works that impact not only their own members but start to have influence on adjacent fields. By that measure, digital humanists do not yet have their *Being and Time, The Savage Mind,* "The Traffic in Women," *History of Sexuality, Postmodernism, Gender Trouble* or *Orientalism*—a work that is intellectually transformative and inspirational for scholars across fields. What defines and sets apart each of these works is that they ask fundamental questions in new ways and develop methods to set about answering them that, in the process, open up other areas of inquiry. They inspire scholars by teaching them to think differently about a range of subjects. They understand that the "summum of the art" is the "construction of the object" of research—breaking with both commonsense and prior scholarly notions (see Bourdieu, Chamboredon, and Passeron 1991: 33–55).

Juola's historical examples are thus interesting for a quality they share with the works I cite immediately above. Despite his definition, they aren't based on the identification and filling of a need. Rather, killer apps go beyond filling needs to provide inspiring tools that expand the sense of the possible. Word processing works once humanists become their own typists (a locution that barely makes sense anymore since *who has a typist? We barely have typesetters!*). I am not saying a shiny new word processor (or later, word processing application) is inspiring in the same way as Butler's critique of the sex-gender binary or Jameson getting lost in a Los Angeles hotel and writing about it. But digital humanists should take a cue that comes from epiphanies gained from reading very good books. If you don't share my tastes in theory or my education, fill in your own epiphanic moments as a reader. Truly transformative humanist work stimulates thought and opens out possibilities. There is no reason why it must appear in book or journal article form as it has for the last generation, but the fundamental intellectual tasks have not changed all that much.

In the digital era we have been very good at creating new tools and new modes of dissemination. The list of options for the aspiring digital humanist is mind-bending. But there is another step. Presumably most people who enter the humanities do so because of a passion for ideas. This can manifest itself in several ways. The great philologists and aesthetes of previous generations tracked progress in terms of discoveries, descriptions, judgments, and interpretations. There are many new digital archives, databases, and annotation tools that fit the old philological model very well, while systematically undermining the structural condition of rarity that went with it. For people who came of age in the cultural studies era, progress is marked somewhat differently. We seek out conceptual transformations that offer new ways of describing the world, as well as major political upheavals in the organization of knowledge as the signposts. Each of these notions of progress is experienced at the level of ideas, but also exists crucially at the level of institutions, materials, and technologies. New objects, modes of evidence, modes of argument, and ways of working with evidence require resources and infrastructure. All these modes of knowledge depend crucially on argument and the employment of evidence. As a scholar, I value ideas much more than I value methods, or rather I do not see the creation of a method as an end in itself. I am more interested in what we can do with it (I'd say the same for theories). By this measure, the digital humanities have produced a lot of infrastructure and resources, a host of experiments and new textual forms. But they have not fully delivered at the level of ideas. How might the delivery arrive?

At one level, those of us who came up at the tail end of the analog era could just sit back and wait for some brilliant young scholars, people who came of age with sets of digital tools, to encounter some enduring question and transform it in some unexpected way. Or maybe in a generation, once all these archives are actually digitized and online, and all this data-crunching software is old hat, people will use the material to transform how we think about cultures and their artifacts. But there is now more than one generation of scholars who have experimented with digital tools in an attempt to produce new modes of knowledge, and as of yet, no digital humanist I know of has produced that signature work that forces people outside the digital humanities to stand up and take notice.

A savvy reader will object that in pointing to works that are today considered monumental, I am unfairly monumentalizing them. They are, after all, only books, produced incrementally and over time with a great deal of patience by their authors, who may or may not have had a sense that they were writing anything convincingly monumental. Each work built on a body of texts that went before it, and recombined a set of influences that alone did not add up to the synthesis that the author could provide. Each work also depended on vast networks of intellectual collaboration and interdependency that disappeared under the sign of the single author.[5] Juola's killer apps follow this model. They are adaptations of existing applications, just as much of the greatness in my list of monumental books may be found in adapting one set of ideas for another purpose.

Today there is much to adapt. Where the analog humanities invented institutions and practices to deal with a fundamental scarcity of textual material, today we are confronted with problems of abundance, rather than scarcity. Where analog humanists had to invent clever ways to get around the limits in the display of textual and audiovisual materials, we have a stunning and sometimes bewildering array of alternatives before us. Where analog humanists didn't really have to think in terms of intellectual property (remember Berger's appropriation of ads), today we are confronted with timid or greedy publishers, aggressive and venturous lawyers, and institutions more committed to protecting themselves from liability than protecting the mission of the humanities, which require access to cultural materials, the opportunity to manipulate them, and to redistribute them in transformed contexts for the purposes of analysis and criticism. The average humanist now has at her fingertips the kinds of computing power that would have made the 1980s humanities computing crowd explode with delight and anticipation. And yet the real intellectual transformations occasioned by all this digital gear may well come up more sneakily, with more subtlety, and with less fanfare than the millennial rhetoric of the digital humanities might suggest.

Here, by way of ending—and with no apology for resorting to anecdote in an age of big data—I want to turn to the recent interdisciplinary boom in sound studies as an example of mundane tool adoption, while acknowledging that a few purpose-built tools (in collaboration with librarians and "humanities infrastructure people") would help. Over a hundred books have appeared in the field since the early 2000s, so it is certainly making use of traditional analog channels. It is often also cited as a response to the interdisciplinary boom in studies of visual culture in the 1980s and 1990s, and certainly that is confirmed in my own experience. But sound scholarship has been expanding for other reasons as well. The ready availability of recorded audio, and the ready availability of playback technology in classrooms and conference spaces only arrived with their outfitting for computers. Before that, and compared with overhead projectors and VCRs, record or tape players were considered specialist equipment, often only available in music departments. When I started teaching about sound as an assistant professor in the early 2000s, I still had to bring my own speakers and portable CD player to the classroom. The availability of large libraries of audio on the Internet also allowed sound scholars to circumvent the sometimes limited collection practices of music libraries. And over the course of the 1990s, a surfeit of software and hardware for manipulating sound inside computers became cheaply available, meaning that a generation of graduate students arrived in humanities PhD programs with deep personal familiarity with sound as a tangible cultural material, rather than an ineffable, ephemeral force, dramatically different from fixed images and texts. Sound became more like the other kinds of things studied by humanists, rather than something that was mystified and dealt with behind closed doors, by experts in recording studios.

This condition also reverberates at conferences and in public presentations. Where it was once difficult and elaborate to get any kind of sound examples happening in an academic talk, sound can now be more or less seamlessly integrated. For those of us who want to move beyond linear and sequence-based slideware, software originally designed for audio artists and musicians can be repurposed for academic presentations. I now use Ableton Live—a program originally designed for electronic musicians, DJs, and sound artists—in talks where I want to work with sound examples because it allows easy editing, manipulation, and playback in an order I can control and transform on the fly. I think of it as nonlinear slideware for sound—the equivalent of PowerPoint, Prezi, or Keynote, but more flexible and more powerful. When presenting my current research on signal processing, I am able to actually present signal processing technologies to audiences while analyzing them, in exactly the fashion a film scholar might show a brief film clip and then break it down. This was not really possible ten years ago in most academic presentation settings. None of these changes were planned ahead of time, none were engineered by digital humanists, and none were touted as revolutionary before they happened. Their very banality means that they may not call much attention to themselves, but over time, they may facilitate more profound intellectual discovery, innovation and dissemination.

A growing group of new sound studies periodicals also nicely illustrates the patterns of adaptation rather than disruption. Though work on sound appeared in older paper journals like *Ethnomusicology*, *Cultural Studies*, *Organized Sound,* and in newer journals like *Senses and Society,* newer periodicals explicitly dedicated to supporting interdisciplinary scholarship on sound are almost entirely in the digital domain. *The Journal of Sonic Studies, Sound Effects*, and

Interference are all "born digital" publications, which is essential for their work. To juxtapose sound and criticism necessitates a natively electronic text where audio files can be embedded directly into the article. This is a core feature of the three journals above, all of which aspire to the traditional refereed-journal model but want to do so in a more fully multimodal environment. *Sounding Out!* dispenses with the older journal model more fully. Though it calls itself a blog (and it is more curated and edited than refereed in the traditional sense), it is more like a space for work in progress and fragments of bigger projects, proudly working across media platforms, and it is nicely skewed toward the work of junior scholars. As with the TV studies site *Flow*, one significant effect of this approach is the production of voluminous and timely material for teaching and outreach to nonacademic readers (Santo and Lucas 2009). Another is that it co-opts some of the social media aspects of community-building to a specifically academic project. It affords its readers a sense of how sound studies is happening across many different fields (though for now its coverage of academic conferences tends to emphasize North America)—it does the work of an association newsletter without the infrastructure of an association, but it goes far, far beyond that. Most of these new periodicals make use of existing software and platforms: WordPress, Open Journal Systems, Soundcloud, YouTube, and so forth. There are real limits to these options, both because of technical limitations (e.g., if you don't want your audio transcoded as part of your presentation of it to listeners) and because of real issues around scholarly use of copyrighted material. Humanist-oriented sites like Critical Commons potentially offer alternative hosting strategies, but their support for audio-only files has thus far been somewhat limited. Print publications also had paid labor relations built into their production practices that as of yet digital publications do not. To mount an electronic journal, there are fewer professional copyeditors and production professionals, and that means that academics do more of their own publication production work.[6]

As of yet, there is no "born digital" sound studies book, at least not one that has the same scholarly heft and sophistication as existing major works in the field.[7] Even the most vociferous advocates of the digital humanities are still publishing paper codex books that may also be available in a digital format (like this one). But the delay is not simply one based on a lack of hardware and software or concerns about durability and future-proofing. It's not surprising that journal articles go digital faster than books. The print codex is five hundred years old; the modern journal article seems to morph in form and tone every quarter-century or so. Advancing beyond the codex requires new protocols of reading and writing, of media making and consumption, which are only slowly emerging across large swaths of the humanities. Confronted with imperfect and unfinished word processors a generation ago, academic writers simply used other tools or worked around existing limitations (for my own undergraduate thesis, I hand numbered endnotes since my word processor had no footnote or endnote function—it sounds ridiculous today, but it was still faster than typing the whole thing). If someone was ready to write a born-digital book, it could be written. In countries where there isn't money for academics to develop custom software and apps, a motivated writer could use WordPress and host files on a private or rented server while waiting for platforms like Scalar to mature a little more.

It is more likely that the major, large-scale digital works will emerge from writers already comfortable in other, shorter, less risky-feeling digital platforms. Over time ambitious ideas will come together and be carried out and the large-scale digital humanist work will emerge, one that is a masterpiece of interpretation, a creature of postphilological scholarship. It may be explicitly collaborative, or it may be subsumed under a traditional single author. We will know that it works when the experience is as seamless as Mark Katz's dream, when humanists' digital technologies can be a point of focus *or* a banal and easily ignored part of the experience of thought and learning, whether or not they at first seem spectacular. On the way there, we should expect to conduct many failed experiments; to carry out many fights over intellectual property and open access, promotion and tenure criteria, battles with unsympathetic library executives and publishers; and further economic and political challenges to our working conditions. The irony may well turn out to be that while humanists are often accused of being nostalgic for an analog era, in the cold light of day it appears that we are adapting to digital models and methods just fine. It's just that in retrospect, the best parts of the digital humanities may turn out to be analog.

Notes

For their comments on earlier drafts, many thanks to Carrie Rentschler and Andrew Piper, as well as the volume editors. Thanks also to Dylan Mulvin for research assistance and to the Social Sciences and Humanities Research Council of Canada for financial support.

1. This chapter will consider the relationship of the tools of humanistic thought to these intellectual challenges, even as they are necessarily entangled with institutional issues ranging from university technology regimes, to publication practices, to funding priorities and practices, to accreditation, tenure, and prestige economies.

2. To be fair, Hayles gestures in this direction with her analysis of telegraph code books, though her interest in them as documents of the "first globally pervasive binary signaling system" (123) instantly refers us back to the digital.

3. Disturbingly, Bogost essentially calls for a purge of analog humanists—"we must cull." The irony of that call is that it would effect exactly the kinds of historical errors Bogost commits a few sentences above, when he wrongly attributes the revelation of a "world of things" to "computing." While computing can

certainly reveal the world differently, digital humanists are far from the first to confront materiality and far from the first to call for worldly engagement. Writers from earlier generations like Du Bois, White, and Hall might have had better and more well-thought-out political reasoning, precisely because their calls did not comfortably and unreflectively echo the bureaucratic language of funding initiatives.

4. As Hayden White has argued, historians had a vexed position between humanistic and social scientific status, in part because of anxious relations to interpretation in their own field, and in part because the practice of historical research, though textual in orientation, may have had more in common with practices like ethnography in its go-out-and-retrieve-and-describe gestures.

5. I also don't want to fetishize the single author as genius. I did not include any collaborative work in my list, though one easily could, like J. K. Gibson-Graham's still underappreciated *The End of Capitalism (as We Knew It)*. Part of this has to do with the fact that until recently, humanists didn't acknowledge or formalize collaboration all that much. I think many of the important works going forward will be collaboratively produced, or at least there will be opportunities for that to happen.

6. The same can be said for electronic books, companion websites, and other kinds of online academic publication. Compared with a print book, the author does *much* more of the production work, which is a major problem in an age where institutions are pressing more and more labor responsibilities onto faculty, both full time and sessional.

7. There is a large body of multimodal work undertaken in Canada and some European countries under the flag of "research creation," but it tends to align itself with the expressive practices of art (and be tied to art and production programs)—and quite often to confrontational aesthetics of the avant-garde—rather than the more didactic practices of humanistic textual production.

2 HUMANITIES IN THE DIGITAL AGE

Alan Liu and William G. Thomas III

Since the 2008 Great Recession, American higher education has experienced a new round of uncertainties and reductions—especially, but not only, in public institutions. British academics refer to the current season of top-down austerity as "the cuts," but in the United States, we might speak of *lingchi*, "death by a thousand cuts." Faculty lines slashed, programs eliminated, course seats lowered, graduate student aid reduced—the decentralized US higher education system has struggled to maintain quality across the disciplines. Humanities programs, in particular, have appeared threatened.

Yet we are now also in the first phase of a digital revolution in higher education. Much of the teaching and learning apparatus has moved online. Computational technologies and methodologies have transformed research practices in every discipline, leading to exciting discoveries and tools. New interdisciplinary initiatives exploiting the digital, such as bioinformatics, human cognition, and digital humanities, are bringing faculty members together in ways never before attempted.

For the humanities, the threat of diminished resources has ironically been simultaneous with this digital turn. Recent events at the University of Virginia, for example, demonstrate just how influential the digital paradigm has become, but also how unevenly applied its pressures can be across the disciplines. The university's board members seemed to be swayed by the model of massive open online courses (MOOCs) under development at the Massachusetts Institute of Technology and Stanford University, among other institutions, many of the key instances of which have been in the science, technology, engineering, and mathematics (STEM) fields. Some board members proposed to eliminate classics and German to save money in the face of the university's massive structural budget deficit. They apparently did not realize how many students actually take these subjects (a lot) or that the subjects have been required in state codes chartering the university.

As humanities chairs with a long involvement in digital issues, we have seen clearly that top-down budget cuts are often justified with arguments about how digital technologies are driving change in higher education. Just as the MOOC course model played a signature role in the University of Virginia case, for instance, so the all-digital UC campus proposed in 2009 by Berkeley Law School Dean Christopher Edley Jr., stirred up controversy in the University of California system at the onset of the epic California budget crisis.

We believe that humanities faculty members, chairs, and administrators now have a choice. One option is to take no systematic action on the digital humanities front and thus let the long-term digital future be built for them. By taking "no systematic action," we mean the present practices of many of us in the humanities who automatically denounce university ambitions for digital education without looking into the issues, allow digital humanities to be just the special province of "power users," and treat digital humanities as a discretionary field. The results of this course of inaction have included settling the responsibility for leading the humanities into the digital age on adjunct faculty or library staff, ignoring the need to find appropriate ways to assess digital humanities projects that do not accord with established ways of measuring academic performance, and restricting the scope of digital humanities to siloed projects. In addition the humanities too often outsource digital humanities to a special center on campus or tiptoe into digital humanities by advertising for faculty in traditional fields and merely adding that "digital proficiency is a plus."

The other option for humanities faculty, chairs, and administrators is to plan how to integrate the digital humanities systematically through our departments—that is, to infuse our departments with digital technologies and practices that create models of organically interrelated humanities digital research, teaching, administration, and staff work. Of course, we have no proof that systematically integrating digital humanities in this way will "save the humanities," a goal we share but that we fear is counterproductive when posed as an all-or-nothing proposition. Good strategy requires picking some point on the line to apply leverage. The leverage point in the policies now shaping the future university is the digital, and we feel that it is crucial that the humanities try for well-conceived, humanities-friendly digital models that are institutionally cohesive enough to influence university policy.

How can we change the dynamic and create new structures for the humanities to flourish in the digital age? We recommend the following four principles for faculty members, department chairs, and administrators to follow in integrating the digital humanities in the humanities.

Think Departmentally

It all starts with where scholars live and work natively: in their departments (or similar units). Currently, digital initiatives in universities are predominantly institutionalized in campus units, library annex programs, or interdisciplinary entities; whereas in departments they spring up accidentally like weeds around particular faculty members, areas of research, or projects. We propose an organic strategy for integrating digital initiatives in core departmental research, teaching, administration, and staff work. Departments could help spread digital methods and tools across the curriculum, for example, by sponsoring graduate students to study digital pedagogies and then encouraging their implementation or by engaging students and faculty in the process of building websites. Departments could also cultivate digital humanities among a larger range of faculty members and fields by alternating between two kinds of job searches: those that prioritize traditional fields, with digital expertise considered a plus, and those that instead prioritize digital expertise, with attachment to an established field considered supplementary.

Chairs and faculty should adopt guidelines for tenure and promotion reviews that value such activities as writing grant proposals, collaborating on projects, creating digital archives, building cyber infrastructures, or contributing influential nonrefereed articles or blog posts (starting with steps as simple as standardizing categories for these activities in CVs). In our own departments at University of California, Santa Barbara, and University of Nebraska-Lincoln, we have worked to define digital scholarship expectations when recruiting digital humanists, to train graduate students in digital humanities (e.g., through introductory digital humanities courses), and to improve administrative and clerical support of research and teaching through digital methods that meet campus standards, where they exist, of accessibility, preservation, privacy, and security.

Think Collaboratively (across Departments and Divisions)

In our experience, the digital humanities act not just as a particular field in a department but as a shared conduit between university programs. Digital technologies and media typically require a broad set of working methods and skills to carry out—as in the case of computational and archival projects requiring the combined expertise of computer scientists, social scientists, artists, and humanists. Digital methods can thus serve as a common link between departments or divisions, bringing them together to collaborate on shared grants, research projects, and curricular initiatives. Such collaboration can strengthen the humanities, partnering them intellectually and practically with other fields and also making them magnets for cost-share and other funding. We ourselves have benefited from collaborating with other departments and divisions on digital projects on our campuses, and we have seen impressive results in our university administrations' cost-share support.

In teaching, the need for partnership is especially acute. For example, the humanities could play an important role in helping to develop innovative digital alternatives to the standard thrice-weekly 50- or 75-minute large lecture course. Such alternatives could better serve a university's students than supersized MOOC courses distributed worldwide to unknown masses. In general, humanities departments could expand their collaborative reach by meeting with other departments (and deans) to explore how multiple departments can coordinate a digital course, project, or administrative tool; providing incentives to faculty members to try for collaborative grants (e.g., by offering course release for grant writing that would eventually repay the lost teaching through curriculum development work or a course buyout); and creating lecture series and workshops that enable faculty to learn about digital research and pedagogy carried out elsewhere on campus.

Think Computationally

Humanities departments need intensive computing power to keep up their research in today's era of large-scale text and data sets, distributed archival resources, and multi-modal (visual, aural, cartographic, etc.) materials. Yet they often lag in both simple and complex technology. This has spillover effects on teaching as well. Though universities and colleges often furnish classroom technology through central campus agencies, we believe that boosting department-specific technology for the humanities could lead to curricular gains.

The fact is that the latest technology can improve humanities research and teaching in tandem, affecting the way faculty members shape their classes by interweaving research and teaching to the benefit of both activities. For example, the National Endowment for the Humanities funded English Broadside Ballad Archive (EBBA) project in the University of California at Santa Barbara's English department and the Digital History Project at the University of Nebraska-Lincoln's history department have driven the adoption of higher grades of department technology (workstations, servers, backup systems, remote conferencing tools, text-encoding, and image handling tools), all of which has created a thriving digital environment (and busy shared physical space) where undergraduate and graduate students work directly on the project as part of their learning in courses. In general, the humanities are now at a point where they cannot settle for the minimal provision of one aging workstation in each faculty member's office plus a computer with digital projector in each classroom.

Departments and chairs should seek larger start-up packages for all new hires (and larger retention packages for faculty with offers elsewhere) so as to encourage the adoption of powerful computational technologies; initiate a replacement cycle for faculty computers; explore creating a shared department computational research facility (or at least a grouped set of research workstations) if none exists; provide at least one departmentally controlled server for project development or collaborative experimentation that would not be possible on mission-critical university servers; boost large-scale faculty data storage and backup facilities; create remote conferencing facilities to accommodate the increasing number of online meetings and job interviews; and sponsor workshops to keep faculty current on new technologies and methods.

Think Societywide

The humanities have enjoyed immense cultural authority and interest in every state and community. Yet they have made little attempt to maintain, renew, and update their cultural influence in the digital age. Such an effort is now vital as austerity measures make some leaders discount the value of the humanities on the basis of misinformed cost-benefit calculations.

The coin of the realm in the digital age, we predict, will be service to society. On the one hand, crowdsourcing and other partnerships with "citizen scholars" will increasingly contribute to humanities scholarship. On the other hand, the humanities must continue to develop their expertise as differentiated from the new, networked public knowledge. The trick will be to evolve the roles of the humanities both in, and distinct from, digital public knowledge so that they will be valued as a necessary public resource.

While the established humanities model of research followed by presentation of finished results in scholarly lectures and publications will continue to be important, that model can no longer stand alone. Digital technologies enable humanities scholars to engage in open discourse about ongoing research and also to explore a wider range of forums accessible to the public.

In this regard humanities faculty members, chairs, and administrators could start by reviewing and renewing their collective understanding of what is meant by "service," which traditionally denoted committee work supplemented by ill-defined "community" or "other" work. The goal of recommitting to service and adapting that concept to today's needs is not to take faculty time away from research and teaching, as if academic work were a zero-sum game, but instead to explore ways to integrate service with research and teaching so as to benefit all aspects of scholarly activity. We know from our own experience as digital scholars, for example, that the simple act of taking the extra step to ensure that the webpage for a project addresses the public has the effect not only of boosting the public service value of the project but of enriching our understanding of the project's research and teaching potential.

In effect, digital technologies could be a catalyst for change in the relation of the humanities to society. While digital humanists have already been exploring methods for publishing in open, crowd-reviewed, blog-based ways, there is incentive for them to go even further in assisting the humanities in reengaging the public. For example, the digital humanities could help create next-generation scholarly platforms that integrate public engagements seamlessly with core research and teaching. Online journals might employ text-mining, topic-modeling, linked-data, visualization, and other tools to create on-demand summaries or "WorldCat

Identities"-like pages—to be used directly by the public as well as by scholars for easy import into public websites or course pages.

Humanities departments can also take such initial, imaginative steps as conducting all-faculty exercises in revising the departmental website. Tomorrow's humanities departmental sites must go beyond presenting people, courses, and events just one level deep. They could expose to public view some of the real content and activity of humanities scholars and activities--for example, by publishing interviews with faculty, creating showcases of student projects, and presenting excerpts of faculty lectures and articles. Other initial steps might be organizing online events that allow faculty and students to share their research with alumni or the community or creating a new service role in the department for an annual "public faculty member" charged with cultivating public engagement, agreeing to meet with members of the community, working on collaboration with local public libraries and museums, or keeping a blog and creating an online showcase for all these activities.

We are aware that there are valid concerns by many of our colleagues that signing on to the digital revolution in higher education in any systematic way is tantamount to undermining some of the core principles and strengths of the humanities. After all, leading philanthropists have suggested that the World Wide Web will soon eclipse all "place-based institutions" of higher education, and enormous sums of venture capital funding have moved into "for profit" and MOOC-based higher ed. Faculty could reasonably conclude that the digital project means participating in the demise of their institution of learning, ceding ever more influence to an oligarchy of elite, private universities with the resources and cachet to start online course consortiums, detracting from the humanistic ideal of close inquiry carried out in intimate conversation, and—it must also be said—eroding the need for as many faculty and instructor positions as now exist.

But what the current interest in the digital also tells us is that we have an opportunity and a responsibility to reframe the humanities for the digital age. In our respective universities we also see many administrators, colleagues, students, and the public eager to assist. The questions and concerns of the humanities continue to speak to and inspire these constituencies, and we should work to enlist them in a collective foray into new digital methods. The reframing project that humanities leaders face will require imagination, leadership, and experimentation. The work we propose is to adopt the necessary level of organizational vision to systematically harness the digital age for the humanities.

Note

This chapter originally appeared as Alan Liu and William G. Thomas III, "Humanities in the Digital Age," *Inside Higher Ed*—Views, 1 October 2012. It is here republished in revised form.

3 ME? A DIGITAL HUMANIST?

Chandra Mukerji

As a historian of early modern technology, I don't often think of myself as a digital humanist. Scholars of new media and digitalization often treat the digital age as a break with the past—the product of a chaotic process, spiraling off in new directions. Their work implies that the pre-digital world I study is irrelevant to the digital humanities. But during the Renaissance and into the Enlightenment, there was a comparable burst of technological innovation and social change. Classical texts provided Europeans with new knowledge of nature, material practices, and engineering that seemed to hold promise of radical transformation. Early modern attempts to imitate and build on these precedents helped establish in the West a tradition of seeking and wielding power by material means that has been tapped once again in the digital revolution.

The logistical tradition, as I call it, began in the Renaissance, and was characterized by experiments in designing new (sometimes utopian) social worlds with innovative forms of art and technical control of land. Early modern governments built ideal cities, new infrastructure, fortresses, and arts programs not only to further their glory but to also extend their power (Appuhn 2008; Konvitz 1978; Langins 2004; Rosenau 1972; Scott 1998; Smith 1993; Vérin 1993). These material practices of government changed the political landscape—literally—but also created a cultural contradiction between humanist ideas of personal agency and the emerging impersonal practices of power. Ironically, the cultural sites that best addressed this contradiction—the great landscape gardens—have rarely been analyzed in these terms. These immersive environments dwarfed individuals with displays of territorial control, expressing the threat to human agency of material means of exercising power.

Perhaps surprisingly, some of the cultural devices used in these gardens have been re-appropriated in digital media where they have been used to address comparable issues of agency and identity raised by new technologies today. The digital revolution, like the logistical revolution of the early modern period, has restructured selves, social identities, and global relations of power through material innovation. Digital technologies may restructure communication rather than relation to land, but they still are tools of impersonal rule that raise questions about human agency. Doing a genealogy of technological forms found in early modern gardens and still used in digital technologies today, we can consider the historical and social problematics that link the digital world to late Renaissance culture. We can recognize the legacy of Renaissance technological utopianism in the contemporary world, and see how continuing anxiety about agency and identity in the face of impersonal governance shapes cultures of technology.

The anxieties about technologies of impersonal rule developed in the early modern period when fortresses, infrastructures, and art were found to be effective in regulating life far from the purview of political elites. Roads and canals could limit the range of social possibilities around them just as surely as agents of the state, and art could define identities through the cultivation of tastes. They did not dictate how people would live or see themselves, but they changed the physical context of life in ways that affected its meaning and conduct. Things were given powers in politics that had real effects, and the result was the movement in the West toward patterns of impersonal rule (Mukerji 2009).

In a similar way the design and use of digital objects has led to a massive outsourcing of human capabilities to artifacts, and the cultivation of new forms of logistical power. iPhones now talk; robots do manufacturing jobs; and drones fight our wars, all using attributes that were once seen as distinctly human. We let machines think for us, fight for us, talk to us, and help us sense objects in our environment. This has produced a new regime of impersonal power, just as threatening to ideas of personal agency as the territorial practices of the early modern period (Magnet and Gates 2009; Turner 2006) It has also created a Deleuzean (1994; Dosse 2010) collapse of time that makes the earlier history seem passé. So we can learn something different about digital culture by recovering the tradition of Western logistical governance and anxiety about techniques of impersonal rule.

I will make the case in this chapter that while the West has proliferated nonhuman agents, artists and engineers have also culturally explored the contradictions between impersonal rule and humanist ideals. They have given people material means for pursuing agency within nonhuman environments, adapting themselves to logistical regimes by doing so.

Within this perspective, digital media cannot be understood as gateways to an ethereal domain of freedom where people—like electrons—can move between matter and energy, liberating themselves from constraints of the body or the physical environment. Digital media—computers, cameras, game systems, and smart phones—are powerful *material* actors that shape social relations. Their nonhuman properties—their physical affordances—are what allow new media to be agential (see Latour 1987). They are artifacts used in the Western tradition of engineering social relations of power.

The desire to break with the analog world using digital media is itself part of the longstanding logistical tradition in the West, exploiting technological means for gaining social advantage. The contemporary technical "revolution" is part of a heritage of Renaissance technical utopianism that has been sustained because logistical power has proved to be so consequential for modern governments.

In the sixteenth and seventeenth centuries, European leaders in Italy, France, Holland, and Sweden, among other places, took the idea of material power seriously, and built new military and commercial infrastructures to empower their regimes (Appuhn 2009; Cook 2007; Drayton 2000; Masters 1998; Mukerji 2009). Administrations copied Rome and its uses of logistical power (Mukerji 2009, 2012). Their successes encouraged other states to cultivate technologies as well, and led corporations to follow this route to power (Harrison and Johnson 2009). Digital culture, as heir to this tradition, continues to use material means to restructure institutional forms of power, while also addressing the problems that material governance has posed for human agency and identity (Turner 2006).

I will try here to explore some genealogical links between digital culture and early modern logistical culture by focusing on two kinds of immersive environments found in early modern gardens and digital media: (1) memory p(a)laces and (2) mazes or labyrinths. Memory palaces are immersive environments that are filled with real or virtual artifacts, representing past thoughts or actions that can be used to (per)form an identity. Their social contexts help determine the significance of actions, affecting the formation of the self as actor. Memory palaces affect what G. H. Mead (1962) would call the formation of social identity through the history of the "I."

Mazes or labyrinths, in contrast, are immersive environments that test personal agency against an organized system of constraints that people must navigate. Their structures lie (mainly) outside the control and understanding of the people who enter them, challenging "players" to analyze their structures to make effective decisions. Inside mazes or labyrinths, people find themselves in and struggle against a system of unfreedom, trying to regain their sense of free will. Mazes and labyrinths often offer tricks or opportunities that people must understand to navigate wisely. To become free requires being smart, but becoming smart about a maze requires subservience to the system of control.

Both kinds of immersive environments—memory palaces and labyrinths—trouble assumptions about human agency, pointing to the power of impersonal systems of logistical or material governance. They illustrate how deeply people can be affected by the environments they inhabit, and they query ideas about free will, intelligence and dominion that lie at the heart of the Western Judeo-Christian tradition. They ask what it means for people to act as agents when they live in highly engineered and materially constrained social worlds. Can people sustain independent identities if they must submit themselves to the material world they also rely upon? Precisely because such questions have never have been or could be fully answered, they continue to reappear in the Western culture and are addressed in its cultural forms.

To look at the genealogical links between early modern gardens and contemporary digital culture as cultures of impersonal rule, I will compare memory palaces and mazes in digital media and the gardens of Versailles. Both are showy sites of technological display, showcasing the latest techniques of impersonal rule (Rosental 2004). And both offer fictive spaces where people can play with their identities, using precisely the technologies designed to constrain them in order to assert their continued agency and will.

Versailles as an Immersive Environment

As Oliver Grau (2003) has argued, immersive environments are ancient forms that have taken on many shapes and uses. Such spaces convey a reality that is both imaginary and real. Some are microcosms, standing for macrocosms. But all display evidence of a fundamental order that lies partly outside human control—and sometimes even human understanding.

The gardens of Versailles constituted an immersive environment dedicated to the power of the Sun King that was demonstrated in a vast system of territorial governance and impersonal rule. Louis XIV's park at Versailles was beautiful but also structured with military and hydraulic engineering, presenting a model of territorial order that seemed to make nature more perfect and the earth more divine. The royal park had a complex geometrical structure that was more or less bilaterally symmetrical, itself following rules of design. The space was huge and the walkways intersected in complicated ways, making it easy for people to get lost. The geometrical forms were always palpable, however, suggesting to visitors that they were navigating a world governed by a larger design.

The gardens at Versailles were a microcosm of France in which people could witness the territorial powers of the state. The park had fortress-style walls, large metal statues cast at the armory, and an infrastructure of walks, roads, and canals. The royal park illustrated uses of logistical power that were also being applied elsewhere in France, so it stood as a model of the new technical regime. But Versailles was also a collector's garden full of classical statues or imitations of them, so it was also a memory palace, an Olympus for courtiers revolving around Apollo, the Sun King (Apostelidès 1981; Goldstein 2008). At the same time it was a play space with a labyrinth, a kind of maze that stood for the complex world of power revolving around the king. In these ways the park was designed to demonstrate the capacity of Louis XIV to use techniques of impersonal rule and play with these rules.

Versailles as a Memory Palace

The gardens of Versailles were an immersive environment that served as a memory palace filled with elements of classical culture to associate France with Rome. Artworks and fountains invoking Roman classical art and engineering silently asserted a new political identity for France as heir to Rome and its imperial destiny. The flowerbeds and walkways may seem innocently cultural in their mute materiality, but the history they suggest was radical. It displaced France's medieval historical identity with a history of classical inheritance, forcing nobles to inhabit a dream world of cultural descent in which medieval forms were erased. This new political imaginary did not question patrimonialism per se. But it devalued the cultural foundations of noble and clerical authority by making them seem at odds with France's imperial destiny.

Ambassadors and courtiers strolling through the gardens of Louis XIV entered an Olympus, a land of gods and heroes, where they could imagine themselves as semi-deities in a program of Roman revival. The dizzying displays of waterworks and neoclassical statuary did not so much celebrate Rome as bring its culture back to life. Rome was not a lost past, but a living tradition in this memory palace (Mukerji 2012).

The memory palace was a concept developed by classical writers and elaborated by later scholars as a virtual space used for memory and identity. It was a space (remembered or just imaged) that orators would use to increase their capacity to deliver speeches from memory. To organize their words in perfect detail, they were urged to "place" ideas or sentences in identifiable locations within specific rooms. They had only to imagine revisiting those rooms to restore those memories (Carruthers 1990: 33–37, 71–79).

Quintillian (1774: 292–93) tried to explain the origins of the memory palace by repeating a famous story from the Greeks:

> The first person to discover an art of memory is said to have been Simonides, of whom the following well-known story is told. He had written an ode of the kind usually composed in honour of victorious athletes, to celebrate the achievement of one who had gained the crown for boxing.... [A] great banquet was given in honour of the boxer's success, [but] Simonides was summoned forth from the feast.... [He] had scarcely crossed the threshold on his way out, when the banqueting hall fell in upon the heads of the guests and wrought such havoc among them that the relatives of the dead who came to seek the bodies for burial were unable to distinguish not merely the faces but even the limbs of the dead. Then it is said, Simonides, who remembered the order in which the guests had been sitting, succeeded in restoring to each man his own dead....

Based on this story, the Greeks believed that they could use imaginary spaces to locate memories in an equally robust way. Memory palaces were meant to be tools for perfecting performances of the self, or honing a social identity through feats of memory. Orators gained their reputations as speakers for their effective use of language. The memory palace was not just a place to "find" words or ideas for a speech, then, but a tool for crafting identities for people as authors and agents of their own lives.

When the ancients argued that you could remember events better by thinking about the spaces in which the events took place, they were implicitly making the case that people were sensitive to the physical world around them. Qunitillian made this clear when he argued, "... when we return to a place after considerable absence, we not merely recognise [sic] the place itself, but remember things that we did there, and recall the persons whom we met and even the unuttered thoughts which passed through our minds when we were there before" (Quintillian 293). Memory palaces then were designed as human constructs that could do the same thing, stir people's memories in ways that affected their actions and senses of self.

The gardens of Versailles were designed to employ this power for political effect. It was a place to remember filled with objects invoking Rome. It was a memory palace for reorganizing noble identities around the classical past, using a site structured to suit the will of the king. It was a place to experience Louis XIV, as Apollo, the Sun King, and to see him surrounded by gods and heroes (Mukerji 2012; Apostolidès 1981).

Visitors to the gardens of Versailles entered a virtual world of classically inspired technology too. The park with its terraces of fortress-style walls, sophisticated hydraulic system, and cast bronze statues from the arsenal mutely and effectively testified France's engineering capacity to take up the Roman heritage. The canal was even outfitted with reduced-sized warships to underscore the point. Versailles was a wonderland of Roman revival, a virtual reality of territorial power and classical inheritance, where gods in artworks and courtiers in fine attire mixed as equals.

To control how people experienced this memory palace, the king wrote itineraries for his visitors for promenades. He did not explain the stories embedded in the garden art. He simply asked them to obey his rules of conduct, forgoing their individual agency to enter a wonderland of classical memory and territorial power. In royal festivals or divertissements, nobles acted out classical stories or scripts by Molière, taking on roles as gods and heroes. They used their agency to bring the classical past to life (Mukerji 1997).

This landscape of art and engineering seemed apolitical because it was mute. But as an immersive environment, it defined a new political reality for France by placing it in a line of descent from Rome. The classical past and present were collapsed, setting the stage (literally) for the pursuit of modern, impersonal forms of territorial politics.

Immersive Digital Environments and Memory Palaces

Digital culture is full of immersive environments and memory palaces around which people shape their identities and exercise agency. The Web too is a kind of immersive environment, a technical world that people can inhabit, entertaining new social possibilities and political transformation. It is a memory palace as well, with websites serving as places where information is stored and can be retrieved. Each person can construct his or her own "personal truth" by finding unique ways to navigate this world of information. But in fact most people use search engines to look for information on the Internet, retrieving what Google suggests they want, and building personal knowledge through the routes suggested for them (see chapter 18 by Olofsson in this volume). Like Versailles, the Web is a designed site, but rather than celebrating and controlling the classical tradition, it celebrates and controls access to information.

Precisely because the Web is an impersonal source of information, it treats information as detached from human learning. In this sense the Web goes against humanist assumptions about the mind and thought that treat true knowledge as a matter of human judgment. In response, many people try to curate or create critical tools for navigating Web-based information, reasserting human judgment over it. Wikipedia is particularly notable as a curatorial effort to define what is worth placing in a shared memory palace. It celebrates humanist ideas about knowledge as something made possible by persons, reacting against the free flow of impersonal information. Wikipedia presents itself as a memory palace of the humanist self, defending the importance of true knowledge over belief or rumor. For this reason Wikipedia is the most obvious, if not the only, site on the Web that struggles against the assault on humanist ideas of agency structured into the Internet itself (Reagle 2010).

Perhaps the quintessential online memory palace, however, is Facebook. Clearly, many users get lost in it for hours, checking on friends, family, allies, and adversaries, taking it as their social milieu. They look for new people to cultivate or shun in this space, and post new evidence of their lives to serve as memories of their actions. Facebook pages are clearly designed as sites for attaching identity to online activity. They are full of memorabilia, mainly pictures, music, and quotes, that define persons according to cultural categories of taste and distinction. The "timelines" now organized on Facebook explicitly call on users to create a history of the "I," and to know each other and themselves as historical beings that continually invent themselves through online postings. It is also a place where people recount their lives through software that consults them on the important components of an (online) identity.

If Facebook is a memory palace used to enhance and construct performances of the self, there are other memory palaces that explore more deeply what it means to be a self. These are, for example, the Virtual Reality Therapies studied by Marisa Brandt (2012). Her example is Virtual Iraq, a program of virtual reality (VR) therapy for soldiers suffering from posttraumatic stress disorder or PTSD. Virtual Iraq is a technological means of addressing a political problem, much like the gardens of Versailles. Just like courtiers at Versailles were meant to rethink their identities around artifacts invoking Rome, soldiers suffering from PTSD are meant to remediate their psychological experiences by entering a virtual world of war in Iraq.

Virtual Iraq is like a first-person shooter game, but without a gun. Patients are told to navigate their way through a virtual world of places that are meant to be familiar to soldiers who have served in Iraq. In Virtual Iraq, generic spaces are meant to help veterans recall what happened to them when they served as soldiers, using the evocative powers of places that Quintillian described. Real memories of war are supposed to be conjured up by association with the generic ones. They do not precisely represent Iraq but simply call out the feelings of being there (Brandt 2012).

Traumatized soldiers or veterans are given memory palaces to enter because with PTSD, they cannot control their memories. They can experience flashbacks of Iraq when they don't expect it and can become disoriented by the power of their memories. Soldiers whose experiences of war seem to have taken on a life of their own are the ones for whom these virtual worlds of military memory are constructed. They have been overwhelmed by the impersonal world of war. Virtual Iraq is meant to allow them to refashion their memories and narratives of self, using an immersive environment, or to desensitize themselves to triggers to their memories of Iraq. In either case, VR therapy for PTSD begins from the assumption that impersonal forces like bombs and bullets can threaten the sense of personal agency in soldiers, and that providing them with a new "memory palace" of Iraq may give them a way to gain some control over that experience and constitute a new and more viable identity (Brandt 2012).

The Labyrinth at Versailles

Mazes and labyrinths even more than memory palaces are immersive environments that address questions of agency in engineered worlds. Mazes are generally geometrically organized spaces with complex sets of paths and dead ends where people get lost before they learn how to navigate the space and exit from it. Mazes are pure sites for exploring impersonal constraint and the struggle for agency. Labyrinths are slightly different in that they do not require a strict geometry, and they can be unicursal, having only one path rather than multiple choices and dead ends. They are sites of physical dangers, however, that make completion of the highly constrained course difficult. Both mazes and labyrinths are then complicated environments of routes and unforeseeable dangers (e.g., dead ends) through which people try to travel to gain their freedom or at least new levels of action. The point of entering a labyrinth or maze is to exit it, and the point of the exercise is to complete a very difficult journey. Simply put, mazes or labyrinths are sites of unfreedom, and places to struggle to regain agency. But achieving this goal requires subordination to the rules of the space imposed from the outside. So each effort to assert agency is also a moment of subordination to an impersonal order (Thacker 1979).

One of the most beloved spaces in the gardens at Versailles was the labyrinth where the problem of gaining and losing control of one's life was made a game. It was designed by Charles Perrault, author of fairy tales including Cinderella (Perrault 1697; Perrault and Aesops 1982). He was also the secretary of the propaganda machine of Louis XIV (Mukerji 2012). Perrault's labyrinth was a play space in which visitors tested their agency in a world of tall hedges and small fountains that had meandering paths and a core "garden room." Visitors had to act as agents to move through the labyrinth, but they surrendered their agency when they got lost. Negotiating the labyrinth was a humbling lesson in subordination to an impersonal order they could not control. They had to "learn" the labyrinth to escape from it. So, as much as it was a site of playful heroic accomplishment, it was also a place to learn humility and obedience.

The meaning of labyrinth in Western culture gained significance from the classical story of the Minotaur, the dangerous beast who lay at the heart of the labyrinth at Crete. The Minotaur was a wild beast, half man, half bull, kept within a labyrinth designed by Daedalus and Icarus for King Minos. Minos demonstrated his authority by sacrificing young Athenians to the beast in an annual offering. Theseus, an Athenian, sought to stop this cruel practice, and with the help of Ariadne, Minos' daughter, he did. The hero was able both to slay the Minotaur and find his way out of the labyrinth by learning its form. Minos's labyrinth was a material tool of political control, a means for exercising power, and a test of powerful men and beasts—even a test of Theseus's sexual powers over Ariadne. It was a model of logistical power and impersonal control that true heroes could overcome. The labyrinth at Versailles, in this sense, was a place for heroes and for connecting political forms of impersonal rule with the classical past.

Importantly for the political life of the maze, medieval Christians started to use mazes as pedagogical tools. Churches from Ravenna to Chartres were built with large floor mazes that believers would walk to experience the mystery of God's plan. They were supposed to use their confusion in navigating the maze to experience the limits of human powers and to highlight the hubris of those who mistook human abilities for godlike powers. At Versailles, the labyrinth was a place to experience the godlike powers of the Sun King over his lands. As an immersive environment with high hedge walls, it served as a virtual world of territorial domination, teaching visitors the power of impersonal rule.

The labyrinth was made a memory palace too, with fountains depicting Aesop's fables. The lessons taught here were meant to be eternal ones, although the stories that were illustrated had strong connections to the culture of the French court. The fountain of the owl and the birds conveyed the danger of silly creatures not listening to those with wisdom. The fountain of the cock, dog, and fox demonstrated that deceptions would only be met with other deceptions. Similarly the fountain of the fox and stork taught the golden rule: do unto others what you would have others do unto you. These and other fables illustrated in Perrault's fountains provided lessons in character and conduct for courtiers that explained how to flourish under the rule of the king (Perrault and Aesops 1982).

The entrance to the labyrinth was flanked by sculptures of Aesop and Cupid, presenting alternative guides to the labyrinth of life. Cupid stood for the thread of love that one could follow through one's mortal journey, and Aesop stood for the thread of wisdom. The fountains embodied Aesop's choice of wisdom as the way for navigating the maze (Perrault and Aesops 1982). It thus asserted the importance of human agency and intelligence even as it demonstrated how easily people could be ruled by things (Mukerji 2012). It stood for life, and the problem of navigating it in a regime of mute, material governance.

Mazes and Labyrinths in Digital Culture

Mazes or labyrinths did not die out with the decline of formal gardens in the West but rather became the backbone for many modern games, including digital games. They have remained fun to play in because they continue to address questions of freedom and control. Mazes provide immersive experiences of struggle, confusion, improvisation, and efficacy. They constitute play worlds separate from ordinary reality where people can test their skills or compete with others in ways, Huizinga (1951) says, that can teach them real life skills. I argue that they teach players to accept their dependency on the material order (or digital world) that they must use to play.

Digital games usually require people to struggle just to live, facing the always-present danger of dying as they move through virtual labyrinths. They pose deep questions like earlier labyrinths, including the labyrinth of Minos. How can one be a hero in the sense of exercising agency in a meaningful way? How can one proceed in a dangerous world where the threats are inhuman or nonhuman? What can one do to gain some advantages for making a journey through such a space? What can one learn about the structure of the space to make its navigation possible? Maze or labyrinth games still pose the problem of how to proceed in a structured and deeply impersonal world where personal agency seems not just threatened but dwarfed by nonhuman actants.

Pac-Man was one of the first popular games to employ mazes in digital platforms, pitting Pac-Man against ghosts. The game used what looked like a traditional maze in its geometry, but it had no entrance or exit. Playing the game was not about escape but rather the struggle in life against death. The physical constraints of the game space made evident that life's choices were constrained. Players had to make use of opportunities available to them, and use their wits to survive.

In cultural terms, the game space was more of a labyrinth than a maze because there was no exit, only new levels of play. And the only way to be a hero was by surviving, nothing very dramatic. Still, in playing Pac-Man, people developed skills in managing risks, and negotiating a logistical regime beyond their control. Over time players could come to predict the actions of game figures because they had stable characteristics, and use their technical features against them. But learning these patterns invested the player in the rules of the game, and made personal efficacy a matter of learning about technology.

The Mario Brothers games did not use geometrical maze-like patterns like Pac-Man, but players were placed inside a labyrinth too—an immersive environment that was defined by walls, floors, ceilings, fire, water, and other kinds of material constraints. The players could break through some walls, altering the structures that limited their freedom, but they mainly remained caught inside the labyrinth, playing with the theme of unfreedom and agency without any real power.

The maze or labyrinth structure was also transferred to some early first-person shooter games like Maze War. Players faced more kinds of dangers in these constrained environments, where they had to move fast, and be quick-witted. They were faced with impersonal threats that were both architectural and nonhuman, but also inhuman or extra-human. This became a tradition in many first-person shooter games, even as the complexity of the environment increased to more three-dimensional possibilities with games such as Wolfenstein 3-D, the threats remained impersonal.

These labyrinths confronted players with impersonal forces that threatened their very existence, again making the games morality tales about conduct in a system of material control. Giant spiders and weird creatures from other worlds confronted human players, pitting human agency against nonhuman powers. Players could extend their agency by acquiring more tools and weapons, matching their impersonal powers with those of their enemies. Dangers of different sorts would pop up in these mazes. Even if they were called war games, the play was not just about war and violence but the threat to human agency in the digital world.

Asserting agency by gamers became more complex as the problems of agency were made more complicated. Gamers had to engage in killing in order to avoid being killed, but what constrained them most were the environments they had to enter to play at all. If the focus was on targets, the problem was visibility in a virtual world of architecture. These were often memory palaces too, where resources were stashed and could be collected by moving through the space.

Interestingly, as avatars became more like gladiators, some games seemed to associate American empire with Rome, dominating the world with logistical power. This new Rome of digital games was hardly the world of white marble constructed at Versailles but rather a dark or demonic labyrinth filled with new Minotaurs.

Digital games have continued to engage questions of agency and identity in a world of impersonal rule. The powers of nonhuman actants in digital games are more complex than those of early modern technical systems, but this only reflects the increased range of agential powers being designed into contemporary digital artifacts. Online multiplayer games have reasserted the importance of social relations to survival in a digital age, but they still are constrained spaces of play.

Conclusions

What can we make of the connection between early modern cultures of technology and digital culture? We can rethink the history of technology, asking more questions about technology and humanist values, or we can rethink digital culture, asking more questions about the cultural traditions it deploys.

Jessica Riskin (2003, 2008) has already written about technologies from the early modern period and the problem of human agency, but has studied a very different type of technology than the ones used in the gardens of Versailles. She has focused on automata, describing them as philosophical machines used to clarify the distinction between humans and nonhumans. She focuses on the automata built in the late medieval and early modern period that were used to judge human character. Her point has been to ask what Descartes meant by mechanism when he defined human beings in contrast to it (Riskin 2008), not to assess the threats to human agency posed by technology. So her work has not pointed to the tradition of Western technological utopianism and impersonal power.

Riskin also focuses on the relations between people and machines but does not engage how this line began to be troubled by the logistical revolution of the sixteenth through eighteenth centuries. Philosophical machines, Riskin writes, were important into the eighteenth century for demonstrating Cartesian ideas. They defined humanist values in contrast to mechanism, and defined human exceptionalism in terms of the difference. Automata would demonstrate what attributes could or could not be mimicked by technological means. The machines could act like living creatures, but could not engage in independent (moral) reasoning. The automata feigned life, helping to pose questions about true human nature, focusing on the distinction between humans and nonhumans and the moral foundations of human exceptionalism (Riskin 2007: 1–34; Grafton in Riskin 2007: 46–62). Clearly, machines were important enough to become philosophical agents, but the limits of their agency were continually asserted.

But in roughly the same period, logistical powers were tested against social ones, and land control seemed to prevail (Mukerji 2009). In the park at Versailles this message was forcefully conveyed. Impersonal power prevailed over people who lost themselves in the garden of the king. Water spouted, trees grew, flowers gave off scents, and gilt statues glistened in the sun. This was the land of the Sun King, a semi-deity. He ruled nature and, in this way, ruled the world. People were dwarfed by the land he ordered, and subject to an environment that only he could control. The political technologies of territorial governance were the opposite of Riskin's philosophical machines, and suggested the dystopian possibilities of impersonal rule.

Riskin's arguments are important to consider for thinking about genealogical links between logistical cultures of the early modern period and digital culture today because she contends that we still use modern media as philosophical machinery. She focuses on artificial intelligence to point to our continued fascination with making machines more human, and our desire to continually reinscribe the differences between the two (Riskin 2007: 1–34). What is interesting is that the cultural forms created and used on these machines have presented human exceptionalism—agency and intelligence—as much more problematic.

What can this genealogical analysis of impersonal governance and digital culture tell us about contemporary life? Simply put, it tells us that digital culture is not part of an unprecedented world of technological transformation. Quite the opposite. Digital culture is the current heir to the logistical tradition of power developed in early modern Europe. And it makes use of cultural forms from that period because they still make sense. They were designed to address the fate of human beings in worlds controlled by things and, if anything, contemporary life is even more clearly the product of a regime of impersonal rule.

Digital culture still tries to champion human creativity and freedom, holding out the promise of new agency in new technologies, but it continues to outsource powers to things. If artificial intelligence seeks ways to understand the exceptional capacities of human beings, defining this exceptionalism as it goes, digital culture is left to address the losses of power, identity and agency posed by the tradition of technological utopianism we have inherited from the Renaissance.

4 CRITICAL THEORY AND THE MANGLE OF DIGITAL HUMANITIES

Todd Presner

As the various fields of the digital humanities have matured and gained institutional traction, a debate has started to coalesce around the relationship between the "critical" function of the humanities and the "building" and "making" claims of the digital humanities (Ramsay, Mandell, and Liu 2011, 2012). While "making" was obviously central to the formation of the artifacts and objects of study in the humanities (whether musical compositions, films, works of art, literary or philosophical texts), the institutional and disciplinary formations of the humanities have largely focused their intellectual energies on criticism and interpretation. Although a simplification, it would not be a great exaggeration to say this has been true of humanistic fields of inquiry for centuries, ranging from Enlightenment ideals of rational subjects engaging in critique to the Marxist-inflected social and cultural criticism of the Frankfurt School, not to mention more contemporary modes of deconstructive critique in fields such as postcolonialism, feminism, critical race theory, and cultural studies. Recently, however, the digital humanities has distinguished itself as an enterprise deeply informed by design, making, and building, even developing "a materialist epistemology" (Ramsay and Rockwell 2012: 77), something that seems to place it at odds with the established notions of humanistic inquiry characterized by "reading," "interpreting," and "critiquing." To be sure, one might argue that the accent has simply shifted, to engage with and shape the material, artifactual, design, and praxis-oriented aspects of the human cultural record. But that leaves a lingering question: What is the relationship between the "critical" function of the humanities and the "building" and "making" espoused by the digital humanities?

Alan Liu has recently thrown down the gauntlet for digital humanists, arguing that "the digital humanities have been oblivious to cultural criticism," which, for him, moves the understanding of cultural, social, and economic dynamics to the foreground through, for example, approaches informed by New Criticism, Marxist social criticism, and the methods of *Kulturgeschichte* and *Kulturkritik* spawned, crudely put, since Hegel (Liu 2012: 491). Without "adequate critical awareness of the larger social, economic, and cultural issues at stake" (Liu 2011: 11), digital humanities, Liu argues, will not be able to engage seriously with the changing nature of higher education in the postindustrial state. I think that Liu is right, and I would add that without this critical awareness, the digital humanities will largely ape and extend the technological imaginary as defined by corporate needs and the bottom line through instrumentalized approaches to technology that are insufficiently aware of their cultural and social conditions of possibility, not to mention the critically transformative potential of the digital humanities to construct new models of culture and society. Such models, I will argue here, have the possibility of fundamentally rethinking the public sphere and knowledge systems by revealing the operations of structures of power and exclusion, while also imagining possibilities for heterological knowledge rooted in an ethic of participation without condition. To that end, I will discuss projects that may not (yet) be considered "canonical" to the digital humanities. However, precisely because they engage with and bring into focus these issues, I am arguing that they have a vital role to play in linking digital humanities with critical theory and, thereby, engage a broader public with methods and values specific to positioning the Humanities in a leadership position for the twenty-first century.

In a word, then, the purpose of this chapter is to concretely connect the core values, methods, and concepts of critical theory with what I will call, following Andrew Pickering, "the mangle of digital humanities." The mangle, for Pickering, is a metaphor for understanding scientific practices that are marked by a "dance of agency" played out through human, material, and social strategies of resistance and accommodation (Pickering 1995: 22–23). As such, digital humanities is a practice and performance of making that is conditioned by human, social, and material contingencies, all of which have the potential to engage in transformative praxis. I do not think, however, that the digital humanities have been completely oblivious to cultural criticism. There are in fact a number of compelling initiatives and projects that either explicitly build these bridges or, in the instantiation of the project, perform such a cultural-critical function. To be sure, there is much bridge-building that still needs to happen. I will thus try to articulate a network of salient, conceptual sites of contact between critical theory and digital humanities in a speculative mode that moves beyond any pure factuality or givenness of either cultural artifacts or technologies.

While the intellectual origins of critical theory stretch back to embrace elements of the Kantian critiques of reason, ethics, and aesthetics, as well as, perhaps most saliently, Marxist critiques of political economy, we can situate the flourishing of critical theory in the 1930s and 1940s with the Frankfurt Institute for Social Research. More than a worldview or cosmology, critical theory was a method of dialectical critique for the analysis and transformation of society (Buck-Morss 1977; Jay 1973). Plenty of comprehensive accounts of the Frankfurt School intellectuals exist, which place them within the cultural-historical context of Germany during the rise of Nazism, American exile in the 1940s and 1950s, and the resurgence and application of aspects of critical theory within a wide array of disciplinary arenas from debates within postmodernism to possibilities for global democracy (Wolin 2006). I will not rehearse that history in this short essay but merely point to some of the key concepts and problematics that, I believe, should inform the cultural-critical function of the "making" in the digital humanities.

In its most programmatic formulations by Herbert Marcuse and Max Horkheimer in 1937, critical theory was a method for engaging in productive social critique; it did not merely mimic, register, or cite the social or cultural conditions that existed but sought to expose their conditions of possibility, their will to truth, and, perhaps most important, effect a transformative change. As Marcuse wrote (1969: 143): "[The] constructive quality of critical theory . . . came from the force with which it spoke against the facts and confronted bad facticity with its better potentialities." In this regard it is both a critical view on society and a constructive practice, one that is future-oriented and engaged in the imagination of a "coming society," even one that has unrealized and untapped utopian elements (Marcuse 1969: 146). For Marcuse, explicitly citing the concluding questions in Kant's first critique, the task of critical theory is not to simply register "what is or was" (questions of fact) but also pose speculative questions of "what could or might be" as well as ethical questions of "what should or ought" I to do (Marcuse 1969: 146). Reformulating Marcuse's critique of traditional philosophy and its ways of investigating the world and creating knowledge, critical theory might be able to expose "the specific social conditions at the root of [digital humanities'] inability to pose the problem in a more comprehensive way" (Marcuse 1969: 149–50). A "more comprehensive way," I suggest, takes us out of the domain of facticity, objective knowledge, and technology tools ready-for-use and into the domain of social practices, the speculative, the future-oriented, and the ethical.

For Horkheimer ([1937] 2002: 197), "traditional theory"—whether that of philosophy or science—was about the pursuit of factual knowledge and the technological mastery of the world isolated from its social and material conditions, whereas "critical theory" was always aware of and engaged with the social and material conditions of both the researcher and object of study. Horkheimer denounces the seemingly objective intellectual pursuits of scholars, which largely accept the world as it is and pursue knowledge outside of or regardless of social, material, and cultural conditions. Instead, like Marcuse and Adorno, he privileged notions of "negation" or "negative dialectics," because it is here that he sees the power to interrogate and undo totalizing systems, expose immanent knowledge claims, salvage the heteronymous or nonidentical, and, in the words of Ernst Bloch, recognize traces of the "not-yet-existing" (see also Buck-Morss 1977: 76). Particularly for Bloch, Adorno, and Benjamin, the notion of futurity, especially the utopian or messianic idea, was a crucial part of the transformative possibilities that they imagined for critical theory. In what follows, I will begin with social practices—particularly, "the mangle of digital humanities"—before turning to the critical-theoretical, speculative, and ethical dimensions of digital humanities.

Countless critiques of "traditional theory" (Horkheimer) and "scientific objectivity" (Marcuse 1969: 156) in the pursuit of knowledge have come in the wake of the Frankfurt School.[1] One need only look to the work of Georges Canguilhem and his student, Michel Foucault, to understand, via Nietzschean antifoundationalism, the nexus of knowledge and power in distinguishing the normative from the pathological, or more recently, the burgeoning field of science, technology, and society studies, marked by the critical work of thinkers such as Bruno Latour, Barbara Herrnstein-Smith, Ian Hacking, and Andrew Pickering. For Pickering (1995: 22), in particular, science is a resolutely performative practice in which the material, temporal, social, and cultural dimensions of "doing" or "making" are characterized by "the mangle," which, in his formulation, is a "dialectic of resistance and accommodation." In practice, this means resistance and failure (material, conceptual, etc.) and the active human response to this failure through various kinds of accommodations, work-arounds, and revisions. In effect, what Pickering does—not unlike the critical theory articulated by Marcuse—is consider "the social dimensions of scientific culture . . . in the plane of practice and, as always, in principle, subject to mangling there, just like and together with the material and conceptual dimensions" (Pickering 1995: 61).

With Marcuse and Pickering in mind, we might ask: Where is critical theory in digital humanities practices? How do we characterize the mangle of digital humanities—that is to say, the performative dimension of resistance and accommodation that characterizes the "doing" of digital humanities through various kinds of agency: human, material, computational, conceptual, and disciplinary? As someone who has conceptualized, developed, and helped build a number of digital humanities projects, I would argue that they are all marked by the mangle of practice: sometimes things work out one way and sometimes they do not; sometimes funding exists to do one thing, but half way there you need to change directions to accommodate something else; sometimes particular kinds of expertise exist and sometimes they do not (and sometimes you don't even know what kind of expertise you need); sometimes conceptual models map onto practice and sometimes the practice, or the software, or the storage system, or the data push forward new conceptual models; sometimes the work-around, the quick-and-dirty, the hack, or the kludge is all there is to move a project forward, not a set of systematic code and design principles followed in a replicable, logical order. Some digital projects iterate; some digital projects fail; some engender whole new fields of investigation, while others close in on themselves. Digital humanities is experimental, dirty, and completely suffused by social and material dialectics of resistance and accommodation, failure and revision, hacks and protocols. This is the mangle of digital humanities, and it is certainly part of current practice, although often erased, forgotten, or variously overcome.

Who, after all, wants to betray the kludge at the core of their practice? Who wants to expose the mangle of practice—the contingencies, resistances, accommodations, constraints, failures, agencies, and revisions—that make up every digital humanities project? Instead (and often for good reason), we encounter most digital humanities projects in their "un-mangled" state: seemingly objective knowledge, stable systems and platforms, and (more or less) complete archives of factuality, which adhere to well-established meta-data and encoding standards to facilitate robust search, discovery, and use. Where is the mangle in such digital archives? Do such projects look more like "traditional theory" and "factual knowledge"? What is at stake in exposing and documenting this mangle of practices, performances, constructions, social relations, and disciplinary powers at the core of both the technological factuality ("the technical tools") and content-based factuality ("the cultural archive")? Of course, the two cannot and should not be separated from one another, as they are recursive elements of any knowledge system. The point is to interrogate the stakes of their presumed "givenness" and, thereby, shift attention to their conditions of possibility, their cultural and social contingency, and, finally, their transformative potentiality. As the emerging subfield of platform and critical code studies has already shown, the mangle of material, cultural, social, and conceptual forces denaturalizes the seeming objectivity and givenness of computational systems, platforms, code modules, inscription practices, and storage devices (Bogost and Montfort 2009), revealing their structuring assumptions, protocols, and even ideologies of power (Chun 2011). But there is still much more work to be done in both critical and speculative modes in order to bring the mangle of practice together with the transformative dimensions of critical theory.

To do so, one would have to add the knowledge practices and rules for the control of discourse analyzed, for example, by Michel Foucault in his 1970 inaugural address to the Collège de France. Here, Foucault points to the rules of exclusion, the establishment of the difference between reason and folly, and the will to truth supported by a whole strata of practices and institutional sites, ranging from pedagogy, learned societies, and laboratories to libraries, the book publishing system, and, we might add, grant and funding agencies (Foucault 1972: 219ff). Disciplines function, according to Foucault, as anonymous systems to regulate discourse through various kinds of practices of rarefaction, rituals, doctrines of truth, and social appropriations. Foucault looks to the conditions of possibility of discourse and, thereby, focuses attention on "notions of chance, discontinuity, and materiality" (Foucault 1972: 231), bringing both a critical and a genealogical principle to unmask the unification, normalization, and diffusion of discourse. Ultimately, his challenge is to "reestablish contact with the non-philosophical" (236), the so-called noise of the madman or the utterances of bare life, which are not even "within the true" (224) or part of public discourse because such utterances are not recognized as knowledge worthy of the distinction between true and false, right and wrong, important and insignificant.

Foucault's investment in reestablishing contact with the nonphilosophical accords, in many ways, with the critical theory of Benjamin and Adorno, both of whom articulated a philosophy of negative dialectics rooted in the preservation of the particular, the nonidentical, and the heterological. For Benjamin, the task of the historical materialist was to "brush history against the grain," revealing the "barbarism" lurking within every "cultural treasure" of civilization (Benjamin 1968: 256–57). Writing after the Second World War and the Holocaust, Adorno (1973) considered genocide to be "the absolute integration" (362), the imposition of an identity principle in which the other was made to perish. Negative dialectics contravened the principle of synthesis (of historical processes, of knowledge systems), which Adorno called "the definition of the difference that perished" (157). Opposed to totalizing systems of knowledge, universal histories, and final syntheses, critical theory privileges the heteronymous, the particular, the voice of the other, and the fragment—all of which underscore openness, unfinishedness, and the refusal of closure in any knowledge system.

As such, the first challenge for digital humanities is to develop both critical and genealogical principles for exposing its own discursive structures and knowledge formations at every level of practice, from the materiality of platforms, the textuality of the code, and the development of content objects to the systems of inclusion and exclusion, truth and falsehood governing its disciplinary rituals, doctrines, and social systems. This is what I earlier termed the critical-theoretical dimension of the digital humanities. But equally important is the speculative dimension of the digital humanities because it is here that one engages with the possible, the future, the not-yet, with that which might or could be. Here, I am especially interested in the subjunctive nature of speculative making, as the creation of a possible future, for this is what links digital humanities to the vaguely utopian dimensions of critical theory.

Nowadays utopian ideas have a bad rap because they appear hopelessly naïve or programmatically prescriptive; however, without an idea of change for the better, there can be no constructive social critique. For the digital humanities, I believe that there is a utopian idea at its core: participation without condition.[2] To be sure, "participatory" is a foundational concept of many digital humanities projects insofar as they create conditions for engagement with communities and individuals not traditionally involved with humanities research and the documentation of the human cultural record.[3] Participatory is arrayed against the rules of prohibition and exclusion, the rarefaction principles and fellowships of discourse that create knowledge hierarchies and closed communities of practitioners; instead, participatory culture is, in its best sense, open-ended, nonhierarchical, and transmigratory, aimed at reestablishing contact with the nonphilosophical. "Participation without condition" is not a principle that can be willed into place, but rather an ideal to build toward through imaginative speculation and ethically informed engagement, one that promises—in the Derridean sense of the *arrivant*—to go beyond the limits and boundaries erected by prior formations of the humanities (Derrida 2002), many of which were deeply exclusionary and remain stratified in countless ways today.

Digital humanities scholarship has begun to render the walls of the university porous by engaging with significantly broader publics in the design, creation, and dissemination of knowledge. By conceiving of scholarship in ways that foundationally involve community partners, cultural institutions, the private sector, nonprofits, government agencies, and slices of the general public, digital humanities expands both the notion of scholarship and the public sphere in order to create new sites and nodes of engagement, documentation, and collaboration. With such an expanded definition of scholarship, digital humanists are able to place questions of justice, social responsibility, and civic engagement, for example, front-and-center; they are able to revitalize the cultural record in ways that involve citizens in the academic enterprise and bring the academy into the expanded public sphere. The result is a form of scholarship that is, by definition, applied: it applies the knowledge and methods of the humanities to pose new questions, to design new possibilities, and to create citizen-scholars who value the complexity, ambiguity, and differences that comprise our cultural record as a species. I will now discuss several digital humanities projects that I think tarry with this notion of participation without condition. While they may not be considered canonical to current definitions of digital humanities, these projects present possibilities for expanding the field of what counts as "digital humanities" precisely by bringing facets of critical theory into the mangle of practice.

The "Jan25 Voices" and "Feb 17 Voices" documentary projects are compelling examples of how social technologies like Twitter can be used to give voice to people who were silenced in the 2011 revolutions in Egypt and Libya. Started by John Scott-Railton, then a graduate student at UCLA, the projects used Twitter to disseminate suppressed messages from protestors to the world. Scott-Railton began the Jan25 Voices project when Egypt effectively "turned off" the Internet between January 28th and February 2nd of 2011. Relying primarily on landlines, Scott-Railton began calling friends in Egypt who knew protesters and could provide highly localized and accurate accounts of what was happening on the ground. He assembled a network of trusted informants who agreed to have their phone calls recorded and published to the world on AudioBoo, an audio hosting service. Scott-Railton simultaneously posted messages to twitter, often with links to audio files and other media reports that would help the world "see" and "hear" what was going on in Egypt in real-time. In effect the digital portal became a global public sphere, however fragile and endangered, that was fundamentally linked to the deeply embodied and precisely located events on the ground. The thousands of voices are now part of a living Web archive and documentary memorial (Scott-Railton 2012).

While the "role" of social media has been feverishly debated in fomenting, planning, and sustaining revolutions since Twitter was first hailed—quite exaggeratedly—as a revolutionary technology in Moldova in 2009 (Mungiu-Pippidi and Munteanu 2009; Morozov 2011) and YouTube became a people's archive for election protests in Tehran during the summer of that same year,[4] it seems incontestable that the "public images" of broadcast media (often singular, uni-directional, and hierarchical) are being supplanted by decentralized, multi-directional "public utterances" that are changing the way in which events unfold, become represented, and are disseminated (almost instantaneously) on a global scale. Two decades ago, Paul Virilio (1993) thought that the physical space of the public sphere had become replaced by the media of the "public image" (9); but now we are seeing the reassertion of the public sphere through radically non-Cartesian geographies enabled by what I would call the Web's "contiguity of the non-contiguous."[5] What this means is a massive contraction and alignment of the event (an embodied and location specific phenomenon), the representation of the event (through Twitter messages, Facebook posts, cell phone video and photographs, etc.), and the dissemination of the event (through Web-based social networks and information channels). The result is a significantly more adaptable, amorphous, global, but also ephemeral and multi-mediated public sphere, one that may, for example, be constituted as a thinly contiguous space connecting Westwood, California, and, simultaneously, Tahrir Square in Cairo or Benghazi, Libya.

At the same time we need to be critical and suspicious of any evidentiary function of social media, not only because they can be easily manipulated and are hard (although not impossible) to verify but also because social media can be used by anyone, in the service of both democratic and authoritarian ends. And even with the best intentions, social media can amplify misinformation on a global scale, creating an echo chamber of falsehoods that are easily accepted as truths by virtue of their sheer repetition. There is no clarity of meaning or channels of truth to be found here, only multiple levels of mediation within complex and ever-shifting dynamics of power and participation.

Around the same time that Scott-Railton began the Jan25 Voices project, the HyperCities team at UCLA created a mash-up for live streaming and archiving twitter feeds from Egypt and visualizing them on a Google Map. The project, "HyperCities Now," made live calls to the standard Twitter search API for tweets originating within 200 miles of Cairo's city center and containing hashtags such as #jan25, #tahrir, or #egypt.[6] Over the course of several weeks, about 450,000 tweets from Egypt were archived, with the most tweets (nearly 25,000) occurring in the hours preceding Mubarak's resignation on February 11th, 2011. In all, the project archived messages from more than 40,000 distinct Twitter user handles documenting the events of the

Egyptian Revolution. Slightly more than half of these voices began "tweeting" for the first time after February 10th, a figure that represents a significant uptake in Twitter usage over the course of the Revolution. While this number is significant, we need to bear in mind that only about 5 percent of Egyptians actually used any social media during the revolution, and Twitter users (less than a fraction of one percent) were part of a fairly homogeneous group in terms of education level, class, and generation (Srinivasan 2011, 2012).

In addition to Egypt, the team mapped and archived Twitter feeds from Libya as well as Sendai, Japan, following the earthquake and tsunami. Spearheaded by a team of volunteers from GISCorps and CrisisCommons, the latter project mapped more than 650,000 social media feeds onto GIS data (including flood zones, evacuation centers, traffic, and public phone locations) so that real-time decisions for coordinating disaster relief could be carried out. By harnessing and repurposing the affordances of existing tools and technologies, the digital humanities team, led by Yoh Kawano, played a decisively interventionist, even public role in responding to and documenting the disaster. This work continues today through a partnership between the UCLA team and Niigata University to measure airborne radiation contamination levels within the evacuation zone around the Fukushima nuclear power plant, to map and visualize the results over time, and to make that data publicly accessible and useful.

Unlike the algorithmically aggregated, displayed, and archived data of "HyperCities Now," Railton-Scott's work was possible because of ever-expanding, although deeply fragile, networks of human witnesses who trusted him to steward and relay what they saw and experienced on the ground. In this sense the project accords with an earlier project of curation created by Xárene Eskandar, also at the time a graduate student at UCLA, documenting, day-by-day and often hour-by-hour and sometimes even minute-by-minute, the election protests in Tehran during the summer and fall of 2009. Utilizing the HyperCities platform, Eskandar painstakingly documented gunfire, protest sites, beatings, blocked streets, sites of safety, clashes, and killings through photographs, Twitter messages, and YouTube videos, creating a geo-chronology for hundreds of reports and media objects. In essence it is a map of events, people, and voices no longer on the map. This project, like Scott-Railton's, was profoundly connected to the original etymology of the term curation, meaning "care of souls" or, in some cases, "stewardship of the dead." Both sought to curate—care for, preserve, document, and archive—the lives, experiences, and actions of the protesters for a global audience, despite (or perhaps because of) the precarious material, social, and technical conditions of possibility for the very stories. At every moment various agencies, resistances, accommodations, revisions, contingencies, and even failures constituted the "mangle" of practice of these digital humanities projects.

These projects connect with critical theory insofar as they expand the concept of the public sphere through an ethic of participation, community collaboration, and socially engaged praxis. They were each deployed extremely quickly—in a matter of days—as experiments or prototypes to intervene in an event that was still unfolding and unbounded. Far from complete or total archives documenting "the whole history" of the revolution or the disaster, they are motivated by several principles that I think accord with critical theory as a socially engaged praxis: a respect for multiplicity and difference through the creation of trusted social bonds, an approach to historical documentation that builds from the fragments of participatory discourse, and a concept of archivization made possible by the contingent material technologies of communication (ranging from mobile phones, social media applications, and decentralized data centers to MySQL, PHP, and JSON scripts). Far from simply documents of the past ("what was"), these archives are spectral, in that they pose haunting questions about the possibility of a future—in the Derridean sense of "what might come"—and therefore are motivated by a responsibility or promise that remains open and undetermined (Derrida 1998: 36). As much as we may hope for a coming democracy, the future may also bring disaster, and this is something with which these projects hauntingly reckon.

While there is a growing number of compelling cultural-critical archive projects and platforms that have emerged over the past few years—including "The Real Face of White Australia," which is part of the Invisible Australians initiative, and platforms such a Mukurtu (mukurtu.org), a content management system that foregrounds cultural difference, perspective, and responsibility through differential modes of access to sensitive cultural artifacts and vulnerable communities—I will conclude by discussing two projects by Sharon Daniel and Erik Loyer, both published in the online journal *Vectors*: "Public Secrets" (2007) and "Blood Sugar" (2010). The projects are audio archives of individuals who have been precluded from participating in public discourse: "Public Secrets" gives voice to women in California state prisons, while "Blood Sugar" gives voice to heroin addicts at a needle exchange program outside of Oakland. In terms of genre, they might be considered "database documentaries," in that the segments of the stories—told by the women and men in their own words through interviews with Daniel—are linked together in a back-end database that establishes sets of relations (conceptual, semantic, contextual, and so forth) and, algorithmically, generates links to the stories in a navigable front-end as an interconnected set of documents. Designed by Loyer, the navigation interface combines a treemap and a typographic algorithm that both creates both containers for visualizing content and structures for enclosing it in various ways. As users navigate the stories, they become enmeshed deeper and deeper into the lives of those whose speech is barely recognizable as speech because it does not, in Foucault's sense, stem from "within the true." That is to say, it is speech that has been deliberately silenced, squelched, and excluded from participation and adjudication within public discourse: In the case of "Public Secrets," it is because the women are not allowed to transmit their speech beyond the prison walls and that (under most circumstances) their speech cannot even be recorded and disseminated;[7] in the case of "Blood Sugar," it is because the addicts live in a liminal zone of privation on the street, without basic civic or social services. In both projects, the speech from "inside" bears witness to the imposition of sovereign power over the other, of the transformation of a human life into bare life, of the reduction of humanity to mere biological functionality.

In discussion with N. Katherine Hayles, Daniel remarked that her projects shift attention from representations to modalities of participation in that the audio archives allow "others to provide their own representation" (qtd. in Hayles 2012: 39). Her projects are predicated on an ethic of participation in which the most imprisoning of conditions have been punctured, if only in that moment of transmission beyond the locked confines of the prison and beyond the conceptual, social, and civic walls of public society. But even more than enabling speech, the project also performs—through the dialogical interactivity of the interface itself—an ethical relationship, in which viewers/listeners are placed in the Levinasian position of responsibility vis-à-vis the voice and experiences of the other. It is a project that solicits careful and sensitive listening, of being open to the voice of the other, and therefore raises an infinite claim that haunts our own (comparatively safe and secure) social, material, and cultural circumstances of listening.

Let me now conclude. Each of these projects offers perspectives and possibilities for digital humanities to develop a cultural-critical praxis rooted in an ethic of participation and curation. But perhaps one might object: What does it mean that these projects were all built on corporate platforms and software (e.g., Google Maps, Twitter, ArcGIS, and Flash)?[8] Do they inevitably speak their language, surreptitiously mimic their worldviews, and quietly extend the dominance of the technological imaginary as put forward by corporations? Or, perhaps, might they create fissures, alternative narratives, incommensurabilities, and new moves within the existing platforms and paradigms? Such a deconstructive strategy is what Lyotard (1991) once termed "paralogy" for its "imaginative invention" of giving rise to the unknown, disturbing the order of reason, producing dissent, and imagining a new move from within the order of things (60ff). The imaginative ability "to make a new move or change the rules of the game" (52) by organizing and "arranging data in a new way" (51), for example, lies at the heart of curation in a cultural-critical mode. It teases out sites of tension and possibility that give voice to particularity and expand notions of participation; it destabilizes and de-ontologizes representational cartographies, corporate platforms, and technologies—not to mention so-called social truths and publicly accepted norms—through new modes of interactivity, memory mapping, consciousness raising, and forms of counter-mapping. One might cavalierly or cynically dismiss this as naïve, but I think it embodies the cultural-critical, *weakly* utopian possibility of the digital humanities. The purpose of this chapter has been to indicate some of the creative ways that digital humanities practices can contribute to public knowledge and even assume a leadership position in placing the core values and methods of the humanities at the heart of twenty-first century cultural critique.

The task, of course, is never finished, and as such, it demands an ever-renewed alliance between the making practices of the digital humanities and the transformative social praxis of critical theory. As Foucault (1972) writes with regard to a philosophy that seeks to reestablish contact with the nonphilosophical: "this philosophy was to examine the singularity of history, the regional rationalities of science, the depths of memory in consciousness; thus arose the notion of a philosophy that was *present, uncertain, mobile all along its lines of contact with nonphilosophy*, existing on its own, however, and revealing the meaning this non-philosophy has for us" (236; my emphasis). Digital humanities suffused with critical theory strives to mediate between and render into contact the philosophical and the nonphilosophical, the mapped and the unmapped, the global and the local, the human life and the bare life, the technologies of factuality and the fissures of the infinitely participatory. In this tension, sites for the heterological and the nonidentical may inform and open up humanities knowledge in ways that truly enable participation without condition. As both an epistemology and an ethics of materialist making, the digital humanities might become a cultural-critical praxis that engages not only with what is and was but also with what might be and what ought to be. Ultimately this is why it is deeply wed to the critical lineage of the humanities.

Notes

1. Outside the Frankfurt School (although more or less contemporaneous), we might cite Ludwig Fleck's *Genesis and Development of a Scientific Fact* (1935), which influenced thinkers such as Paul Feyerabend and Thomas Kuhn.

2. The authors of *Digital_Humanities* reference this notion once in our collaboratively authored book, and I build on that reference here (94).

3. For an excellent overview of modalities of "collaboration" and "participation" in the digital humanities, see Lisa Spiro's essay, "Computing and Communicating Knowledge: Collaborative Approaches to Digital Humanities Projects."

4. For a sobering account of the dialectical underbelly of social media technologies, see, for example, Evgeny Morozov's analysis of the "dark side" of the Twitter revolution, in which he shows how totalitarian regimes harness social media to track and detain dissidents: "Iran: Downside to the 'Twitter Revolution'" and the broader treatment in his book, *The Net Delusion*.

5. This is a play on Ernst Bloch's famous idea of the "simultaneity of the non-simultaneous" to refer to the temporal dynamics of modernity. I am reworking this phrase to emphasize the ways in which

geographically distant events and people may now be linked, via the Web, as if contiguous with one another.

6. Sensitive to the risks involved in creating an archive that precisely maps Twitter users, messages, and location, the HyperCities team truncated the exact latitude and longitude (when it was returned by Twitter's location parameter) at the hundredth decimal place, effectively placing a tweet within a two mile radius, rather than at an exact GPS location.

7. Daniel notes that as a "legal advocate" working for the nonprofit, human rights group "Justice Now," she was granted access to the facility and allowed to record the voices of the women inside, provided she adhered to the "Kafkaesque" search and surveillance procedures of the prison (2007).

8. It should be noted that they also use open standards, open source code, and free software, such as HTML5, MySQL, and PHP.

5 "DOES THIS TECHNOLOGY SERVE HUMAN PURPOSES?" A "NECESSARY CONVERSATION" WITH SHERRY TURKLE

Henry Jenkins

For me, the blog represents perhaps the most powerful tool for digital humanities. I started blogging in 2006, with no idea how important my blog, Confessions of an Aca-Fan, would become to my professional profile or how central it would become to my research process.

For starters, blogging offers academics a powerful interface with a larger public that might otherwise have very limited exposure to our ideas, given how rarely academic books cross over into mainstream bookstores, how infrequently humanistic academics appear on national television discussion programs, how few of us write op-ed pieces discussing our work, or how few opportunities many of us enjoy in speaking beyond our classrooms and formal conferences. Despite many calls for scholars to become public intellectuals, we have created surprisingly few mechanisms for achieving these goals on an ongoing basis. In a period of profound and prolonged media transition, digital humanists have a unique contribution to make in helping the public process and evaluate the changes occurring around them. To perform that role, academics need to adopt a more citizenly discourse through which to speak meaningfully to a nonspecialized public about matters of immediate public concern.

I have found that the process of blogging week in and week out has not only allowed me to strengthen my own skills and commitments as a public intellectual but also to provide a space for other academics to make this transition and to help mentor my students about the challenges of writing for a general readership. I am especially proud when I am able to use my blog to provide a larger forum for key and emerging thinkers in our field. In some occasions my blog has hosted public conversations where many different scholars could weigh in on pressing topics in our field, as occurred with large-scale discussions around "Gender and Fan Studies" in 2007 and "Aca-Fandom and Beyond" in 2011. More often I use the blog for interviews and conversations with leading figures across a range of disciplines who have had important things to say about new media literacies, transmedia storytelling, fandom, and participatory culture, among a range of other topics that are close to my heart. Some such exchanges are among friends and colleagues. I have at times tracked down authors whose work helps inform my own thinking. Whichever way, the blog interview is relational: my questions say as much about the state of my own thinking as their responses say about theirs, and often what emerges is a text that lies at the intersection between our different ways of seeing the world.

These interviews are conducted electronically, so the text begins as written prose, albeit prose that is often less formal and rigidified than traditional scholarly writing. I encourage my contributors to share something of the personal stakes and human motivations shaping their work. Disagreements seem less rigid or unresolvable, as we engage with each other on a more personal and immediate level. We are able to draw out some core themes running across multiple projects that might help readers to understand scholarship as an ongoing process. The interview can be topical, responding to current events while they are still current, in a way that would be impossible in much more slowly moving modes of publication. Often I try to push the interview subjects beyond what they have explicitly said in their work, to apply their ideas to new contexts, or to reflect back on what has happened since the book went to press. When the blog interview works, it achieves something of lasting value, but it achieves it by trying to preserve the everyday processes of scholarly communication rather than smooth over the rough edges in search of something more monumental.

All these virtues came together in the following interview I conducted with Sherry Turkle in August 2011. After having spent the bulk of my career at MIT, I had moved to Southern California just a year before. We also conducted this exchange as Sherry Turkle's book, *Alone Together: Why We Expect More from Technology and Less from Each Other* (2012) had hit the market and was being heatedly discussed. Turkle and I had conducted a public conversation at the Scratch Conference, hosted by Mitchell Resnick, a mutual friend, at the MIT Media Lab, and in many ways this interview is an extension of that earlier exchange. You can watch a video of that exchange at: http://video.mit.edu/watch/scratchmit-friday-keynote-rethinking-identity-rethinking-participation-6081/. Sherry and I had often traded guest lectures in each other's classes through the years, but this was the first time our friendly disagreements with each other had been publically discussed. The result is one of a handful of very best exchanges I've run through my blog.

After more than twenty years of living in the heart of the machine, I have concluded that there are two ways of doing humanities at MIT (perhaps anywhere): the first is entrenched and embattled, defending the traditions, from a broom closet, trying to civilize those who see virtue in the technological and who undervalue the cultural; the second is engaging, confronting the technological and demanding that it serve human needs, asking core questions about the nature of our species, and exploring how the cultural and the psychological are reasserted through those media that we make, in Marshall McLuhan's terms, into extensions of ourselves.

There is at MIT no greater advocate for humanistic engagement than Sherry Turkle, who makes technologies as "second selves," as "evocative objects," as intimate tools and "relational artifacts," the central theme of her work.

It has been my joy and honor to consider Turkle my friend for more than two decades. Our paths crossed too rarely in the years I was in Cambridge, but each time they did, I left the conversation changed by her insights about core questions that shaped both of our work. Sherry Turkle shared with me some years ago the insight that we are both victims of the public's desire for simple answers. No matter what Sherry says, which is often layered and sometimes paradoxical, about the complexity of human's relations with technology, there will be those who see her as too pessimistic and no matter what I say, people are going to see me as too celebratory. For us both, at the heart of our work is the desire to "complicate" our understanding of technological change through a focus on core human experiences.

I was reminded of Sherry's insight about the way our respective positions get simplified when I saw the response to her most recent book, *Alone Together: Why We Expect More From Technology and Less from Each Other*. Critics and supporters alike tended to read the book as a diatribe against new media, and as thus a turning of her back on the work of many at MIT who stress the ways new tools are expanding rather than constraining human potentials. Many wrote to ask me what I thought of the book, often with the expectation that we were fundamentally at odds with each other.

I should have known better, but I found myself entering the book on the defensive, looking for points of disagreement, and there are certainly some of those as the following exchange will suggest. But, as I read, I found myself struggling to answer the challenges she posed, and finding the book anything but simplistic and one-sided. She is demanding that we all enter a new phase of the "conversation," one that accepts that technological changes are fundamental and unlikely to reverse course, but one that demands that we shape technologies to core human needs and goals rather than the other way around.

This is the great theme that runs across the remarkable interview I am sharing with you here, resurfacing again and again as she presses beyond simple one-sided perspectives and forces us to address our fundamental "vulnerability" to technological shifts. Do not enter into this interview expecting to disagree with Turkle or to simply reaffirm your own comfortable and well-rehearsed arguments. Rather, use her comments to reshape your thinking and to redirect your energies to some of the core struggles of our times. What you will find throughout this discussion is a powerful intellect engaging with the shifting borders between the human and the mechanical, between psychology and technology, and between pessimism and skepticism. As always, I learn so much from reading Turkle's work, even where, or perhaps especially where, we disagree. But, again, I would stress, we disagree far less often than many, ourselves among them, might imagine.

I was struck by one of the very first sentences in the book: "Technology proposes itself as the architect of our intimacies." Can you dissect that evocative phrase a bit for me? In what forms does the proposal take and how do we signal whether or not we accept?

From my earliest days at MIT, struck by the intensity of people's emotional engagement with their objects—and especially with computational objects there were many people, and many colleagues, who were highly skeptical of my endeavor. And yet, I was inspired by Winston Churchill's words, who said, before McCluhan rephrased them: "We make our buildings, and in turn, our buildings make and shape us." We make our technologies, and our technologies make and shape us. The technologies I study, the technologies of communication, are identity technologies. I think of them as intimate machines. They are not only, as the computer has always been, mirrors of our mind; they are now the places where the shape and dimensions of our relationship are sculpted.

I think of the technological devices as having an inner history. That inner history is how they shape our relationships with them and our relationships with each other. Another way to think of this is in terms of technological affordance and human vulnerability. Technologies have certain psychological affordances; they make certain psychological offers. We are vulnerable to many of these. There is an intricate play between what technology offers and what we, vulnerable, often struggle to refuse.

There would have been a time when technology was understood as the opposite of intimate—as something cold, impersonal, mechanical, and industrial. In a sense your three books have mapped the process by which we have come to embrace technology as intimate. What factors have led to this shift in our relationships to technology?

I think there are two ways of answering your question. The first is to say that technology has never been cold, impersonal, and industrial. We simply chose to understand it that way. Technology has always had a role in shaping the inner life, the intimate life. The telephone—surely a shaping force in the making and shaping of self, and likewise, the telegram, the letter, the book.

As a teenager living in Paris in the 1960s, I remember the telephone being shunned as too "impersonal"—for significant apologies, a request for a meeting, an assignation—it was explained to me that one sent a *pneumatique*. All the post offices of Paris were connected with pneumatic tubes. One wrote a letter in a sealed envelope. It was picked up at one's apartment and brought to the post, put in the tube, sent to the post office closest to the destination's address and hand delivered. The *pneumatique* had the touch of the hand on the correspondence. This, too, was intimate technology. There was nothing cold about the letter.

Nor was there anything cold about how industrial technologies such as cars and trains shaped our sensibilities, our sense of self, of our sensuality, our possibilities. If we have succumbed to an ideology of technological neutrality that is something that needs to be studied as an independent phenomenon; it is not to be taken as a given.

But there is another way of approaching this question. And that is to say, I do believe that information technology and the digital revolution have changed something fundamental in our way of seeing the world. There is something new in our current circumstance. The computer is a mind machine, not only because it has its own very primitive psychology but also because it causes us to reflect upon our own.

From the very beginning, people saw the computer as a "second self"—an extension and reflection of self. The computer seemed much like the psychologist's inkblot test: the computer as Rorschach, a projection of personal concerns. Indeed I got the title of my first book on the computer culture from a thirteen-year-old who said, after an experience with computers: "When you work with a computer, you put a little piece of your mind into the computer's mind and you come to see yourself differently." A second self. So one might say that in a context where I believe that all technologies shape and make us, the computer takes this vocation to a higher power. Or one could say, this vocation is a centerpiece of its identity. I think of it as an intimate machine.

This vocation has been heightened in the age of always-on/always-on-you communications devices, which of course are the focus of my current work. They move from being tools, or perhaps prosthetics, to giving people the sense of being near-cyborg. The devices seem like a phantom limb, so much are they are part of us.

Your discussion of our shifting relations to robots remains focused primarily on the actual technological devices and the roles they play in our lived experience. Yet surely our shifting understanding of the robotic has also been shaped in profound ways by the cultural imagination. After all, the very term "robot" emerges from a work of science fiction—Karel Capek's *R.U.R.* (1920) and surely our relations with actual robots have been shaped by science fiction representations from Asimov's *I Robot* and Robbie the Robot and Gort to C3PO and R2D2. So what relationship might we posit between the creative imagination and our shifting relations to the robots in our physical surroundings?

This is a very important question for me. I have been tracking the flowering of a genre—there are, of course, antecedents—but now we have a flowering—of the robot who teaches people to love, and more than this, and crucially, teaches people how to be human. For me, the prototype here is *WALL-E*. The people have forgotten their sensuality, their capacity for love, their capacity for interconnectedness. It is a robot designed for industrial cleanup that rediscovers all of this, that falls in love, and that, transcendent in this capacity, is in a position to teach it to humanity. In fact he saves humanity not just in the physical sense but in the spiritual sense as well.

In *Alone Together* I talk about our having reached a "robotic moment." This is not because we have robots that are capable of loving us but because so many of the people I interviewed say that they are prepared to be loved by a robot. There is no question that imaginative literature and film have been part of this shift. We used to look to machines for physical help. Now we feel we are missing things on an emotional and spiritual dimension and we look to the machine world.

In many ways both of us have been profoundly shaped by our time among MIT students. And you wrote very explicitly about MIT hacker culture in *The Second Self*. What do you see as the strengths and limitations of MIT as a testing ground for your ideas?

I don't see MIT as a testing ground for my ideas. I would say rather that MIT is the place where my ideas are most challenged because there is a tendency at MIT to want to see human purposes and technological affordances as being one. Technology has purposes; technology is made by people. Technology and people are at one in their purpose.

From my point of view, every technology offers an opportunity for people to ask: "Does this technology serve human purposes?" and this is a wonderful thing because it enables us to ask again what these purposes are. We are well positioned to create technology whose purposes are not in our best interests. And then it is time to make the corrections.

So, from this point of view, I find that my favorite sentence in my most recent book is "Just because we grew up with the Internet, we think that the Internet is all grown up." From my point of view, seeing technology as a given is a distortion of perspective, one that is very common at MIT. From my point of view, we are in early days and it is time to make the corrections.

Perhaps the greatest ongoing difference of opinion I have had with close colleagues at MIT has been about the meaning and prospect of sociable robots. I take a very strong position in *Alone Together* that nanny-bots and elder-care bots that pretend affection are seductive. And that my research shows that we are vulnerable to them. We are alone with them, yet we feel a faux-intimacy with them.

Indeed the arc of the book is that with robots, we are alone and feel a new intimacy. In our new, mobile connectivity, we are together with each other and yet experience new solitudes.

I worked on my studies of sociable robots with colleagues at MIT who are some of the most brilliant and creative developers of sociable robotics. We had deeply felt, serious conversations about the purposes and possibilities of these machines. Some think that their ultimate significance will be profoundly humanistic. I'm listening, but I am not convinced.

Conversations with robots about love, sex, children, the arc of a life—in other words, about human meaning—to me, this has no meaning. These are things that the robot has not experienced. These are not appropriate topics for conversations with robots. So being at MIT has kept me more aware than I would ever have been about the broad differences of opinion in what the purposes of machines can be.

I took you to task, ever so mildly, in my blog a while back (http://henryjenkins.org/2010/02/killer_paragraphs_and_other_re.html) about some of your comments about MIT students and multi-tasking in the *Digital Nations* documentary.

You can see what I said here I wanted to offer you a chance to respond to my arguments.

I most often run into our disagreement about multi-tasking in the context of parents who say, "Well, is it so bad if I text while my kid is in the kitchen with me; my mom used to do the dishes while I hung around?" Or, "My dad used to read the newspaper when we watched sports on TV; what's the difference between that and my doing my email while I watch sports with my son on Sunday?"

Having interviewed the children who feel abandoned by their parents, who feel almost desperate for parental attention, has led me to do a lot of thinking about the kinds of attention that digital devices require. We don't give them the kind of attention we gave to doodling or to a newspaper, or for that matter, to cooking or watching TV. We are drawn in in quite a different way. This is made apparent when I interview teenagers who say things like "When I was little, I used to watch Sunday football with my dad and we would talk. Now, he is on his BlackBerry and he is in the 'Zone.' I can't interrupt him." Or, stories, many stories, of daughters who come into the kitchen to hang out with their mothers and find them texting and cannot make eye contact with them and who are shushed away. I observe parents and children in the playground with children desperately trying to get their parents' attention; parents are absorbed in their devices and cannot "multi-task" attention for their kids.

So I think that the narratives we use to think about our students multi-tasking in class needs to be informed by the nature of what it is to absorb oneself in digital media. Beyond this, I am persuaded by the research that suggests that when we multi-task, our performance degrades for each task that we multi-task, even as we receive a neurochemical reward for our multi-tasking. So, through no fault of our own, our biology has us feeling better and better even as we do worse and worse.

I do think that, smitten by what computers enable us to do, we have allowed multi-tasking to seem like a twenty-first century alchemy. I think that classrooms will soon be in the position of being the places where uni-tasking is taught, places where students learn to concentrate and where, additionally, they learn to cultivate the capacity for solitude.

I think that the two learning skills that are in the most jeopardy in our hyperconnected world are the ability to concentrate on one thing and the capacity for the kind of solitude that replenishes and restores.

I am going to be running a summer-long conversation on this blog about the value of the autobiographical voice in cultural criticism. You have now edited a series of books where people share autobiographical reflections on what you call evocative objects. Can you explain what you mean by evocative objects and what you think is the contributions of these kinds of reflections?

Evocative objects are objects that cause us to reflect on ourselves or on other things. Put otherwise, they give us materials that help us do this in new and richer ways. Objects can be evocative for many different reasons. Some of these reasons have been widely studied. For example, objects that are "betwixt and between" standard categories are classically evocative because they cause us to reflect on the categories themselves. This is why computational objects, standing between mind and not-mind, between the world of the animate and not animate, have been so evocative as objects-to-think-with.

Other evocative objects partake of elements of what Winnicott called "transitional objects." These are objects that blur the boundaries between self and not-self, objects that we experience as being in a special, blurred, sometimes fused relation to self. Here too computational objects have had a special role to play. From the very beginning, people experienced a kind of "mind meld" when using software, saying things such as "When I use Microsoft Word, I see my ideas form someplace between my mind and the screen." Now, in talking about always-on-them digital devices, there is an ever greater sense of the device being part of the body.

Evocative objects provide a special window onto life experience, one that is grounded and cannot avoid issues of depth psychology. Science studies, sociology, anthropology, have each in their own way welcomed the study of objects but have been hostile to depth psychology. When one pays careful attention to evocative objects, one "hears" psychodynamic issues, one "hears" family history, one "hears" a close attention to personal narrative and the texture of a life in all of its peculiarity and deeply woven interconnections with others. In science studies, studying objects and life narrative has the additional virtue of making the point, which seems to need making for every new generation of students, that technologies are not "just" tools, that our relationships with objects are profoundly interconnected to how we make meaning out of lives and think through who we are as people.

You describe both children and the elderly being drawn to robots as companions. In your discussion of social networking sites, you seem to accept the distinction between digital natives and digital immigrants, implying that generational differences matter in response to those technologies. Do these same differences matter in talking about human relations with robots?

There are, of course, important differences in how people who grew up with a given technology appropriate it in contrast to those who adopted it in adulthood. But what most fascinates me these days are common vulnerabilities of grown-ups and younger people, both in the area of communications technology and in the area of sociable robotics. I did many interviews with people in their 40s, 50s, and 60s who are willing to entertain the idea of a robot that might love them, care for them. But certainly, the sensibility of the "robotic moment," the idea that we are ready for robots that might care for us is most apparent among the young.

Their science fiction and imaginative toy and game worlds suggest to them that robots may soon be in a position to teach people how to love; they have a way of thinking about the nature of aliveness that considers objects with a new pragmatism. That is, previous generations talked about computational objects as "sort of alive" or "kind of alive." This new generation talks of computational objects as "alive enough" to do certain jobs. Robots are thus considered "alive enough" for the job of care and companionship, at the limit, alive for affection.

As you describe the many kinds of anxieties, uncertainties, disappointments, and frustrations that surround technology in everyday life, it sounds like many people are unhappy with current configurations and most have harsh judgments of the uses of new media by others in their friends and family, yet few people are breaking out of the patterns you describe. Why not?

I think that we are at a point of inflection. Our lives are enmeshed with our new technologies of connection and ever more so. We now have more experience of what this means for us as individuals, for our relationships with our families, with our parents, with our children, with our friends, with our neighbors. We are coming to a greater understanding of what this means for us politically, both in our own country and globally.

It has taken time for people to understand where life with this new kind of technology has brought them. Things came to them one gadget at a time. A phone, a navigation system, a way to listen to music, a new way to read books, books "on tape" became something else . . . and now we catch up to the idea that positioning and navigation translate into surveillance and that using social media as though it were a neutral "utility" ignores important issues about privacy and ownership of personal data.

I don't think that we grown-ups who "gave" this new communications regime to our children thought it through on several of its critical dimensions. What is intimacy without privacy? What is democracy without privacy? These are not easy questions. But they are starting to be questions that people are thinking about. They begin to have concrete meaning as people come to a new kind of life and have enough experience to take its measure.

On the simplest level, when I talk to parents who realize that it makes them anxious to walk to the corner convenience store with their child without taking their cell phone, who cannot go to the playground with their child without bringing their email enabled device, who text in the car while driving with their children in the back seat, it seems clear to me that we are not at a point of stable equilibrium. These people are not happy.

So my qualified optimism about change comes from my sense that the people with whom I have been talking are not happy and are genuinely searching for new ways of living with new technology. I hear more and more about "Internet Sabbaths" during which families disconnect for a day or a weekend. Some families modify the Sabbath and declare two hours a day as off-the-grid family time.

There is no option that we give up on our new devices. They are our partners in the human adventure. What we have to do is find a way to live with them that is healthier. A digital diet that is better for our health and the health of our families. It took a long time for Americans to learn that a diet high in sugar and processed foods was not healthy. It is going to take a long time for people to develop strategies, individually and collectively, to live with our new technologies in the most healthy way. But the stakes are high, and we can get this right.

Your book describes a world where technological demands often supersede human needs, yet you are insistent that you are not anti-technology. So what do you see as the gains that new media have brought to the culture?

In the domain of communications technology, one of the things that excites me the most is when technologies of the virtual enhance our experiences of and in the physical real. So, ironically, one of the earliest uses of the Internet as a social media—how MeetUps were used in the Howard Dean campaign in 2004—remains an inspiration to me. People "met" online for a political purpose and then "met up" in the physical world. They did not fool themselves into thinking that political action consisted of just giving money online or visiting a website and leaving a "thumbs up" sign on it.

MeetUp continues in this tradition as do many other online groups that organize in the virtual and connect in the physical. We have seen this play out on the most dramatic scale in political life where despots have been challenged by groups brought together by social networking in all of its many forms.

My criticism of social networking boils down to the necessity for us not to redefine the social as what the social network can do. The social encompasses a great deal more. This is not to put the social network down, it is simply to put it in its place. So there is no conflict between the magnificence of what the social network can do to mobilize political opinion and how it can get in the way of the development of teenagers who need to engage with each other, face to face.

On a personal note, I recently attended a reunion of my fifth-grade class. This was the fifth-grade class from PS 216 Brooklyn. This particular reunion would never have happened had it not been for Facebook. One person from the class had been connected with several others and then they each searched Facebook for a few more names they remembered. And then those people remembered a few more names. Within six months, our fifth-grade class was on the roof terrace of the Peninsula Hotel in New York. It was a small miracle. It was a gift, a profound gift. Yet in the annals of the social network, my story is banal.

You suggest that we are using new media to deal with the anxiety of separation. Is this separation anxiety itself a product of our reliance on technology or is it a reflection of, say, the increases of divorce and mobility in American culture over the past several generations? Are there ways in which the use of social media is a rational response to those social and cultural disruptions, allowing for old friends to remain in contact despite geographic distances or for separated parents to remain active parts of their children's lives?

I think it is easy to make distinctions in this domain. A parent who uses social media to keep up with a child living away from home or a child who uses social media to keep up with a parent in a different city—one recognizes and respects these cases when one sees them. My concern is with very different kinds of cases. Parents who cannot tolerate their eight-year-old not having a cell phone. Children who have developed a style of relating that I characterize as "I text therefore I am" or "I share therefore I am."

To put it too simply, things have moved from a style of relating where one thinks: "I have a feeling, I want to make a call" to "I want to have a feeling, I need to send a text." In other words, the act of sharing a nascent feeling becomes part of the constitution of the feeling.

The problem is that when we use other people in this way, as needed elements on the path toward our having our feelings, we can move toward a misuse of others. We are not relating to them as full others but as what psychologists call "part objects." We are using them as spare parts to support our fragile selves.

This takes the notion of an "other-directed" self to a higher power. Our technology supports a culture of narcissism digital style. It is a kind of self that does not tolerate being alone. And yet psychology teaches us that if you do not teach your children to be alone, they will only know how to be lonely. We are forgetting this lesson in our culture of hyperconnection. These kinds of anxieties of connection are different from the "rational responses" to staying in touch with far-flung family and friends.

In your discussion of Chatroulette, you talk about "nexting," while elsewhere, you describe "stalking." First can you explain the two concepts and then tell us what you see as the relations between them? Is the indifference to others implied by nexting the flip side of the kinds of obsessive interest in other people's business online represented by stalking?

What both nexting and stalking have in common is the objectification of people who we meet on screens. We do not consider them in their humanity. And this is one of the major themes of *Alone Together*: we are at a moment of temptation. We are tempted to treat machines as if they were people and to treat people as if they were machines.

In what ways has the persistence of information online forced you to revise earlier arguments about the potential to protean plays with identity? It seems these days, on the Internet, everyone knows you are dog and many know what dog food you eat.

Henry, this is beautifully put. My earlier enthusiasm for identity play on the Internet, for the kinds of experiences that led Amy Bruckman to call the Internet an "identity workshop," relied heavily on the work of psychologist and psychoanalyst Erik Erikson. Erikson wrote about the developmental need for a moratorium or "time out" during adolescence, a kind of play space in which one had a chance to experiment with identity. In the mid-1990s I wrote about the Internet as a space where anonymity was possible and where one could experiment with aspects of self in a safe environment.

Today adolescents grow up with a sense of wearing their online selves on their backs "like a turtle" for the rest of their lives. The Internet is forever. And anonymity on the Internet seems a dream of another century, another technology. People still use game and virtual world avatars and social network personae for identity play. But the expectation of a parallel, distinct, and anonymous virtual life is no longer a clear starting expectation. It cannot be. Many of these experiences begin by registering with a credit card.

You are skeptical of the value of the term "addiction" to describe some of the kinds of behavior you criticize in the book. What do you see as the limits of addiction as a way of understanding what's going on here?

No matter how much the metaphor of addiction may seem to fit our circumstance, we can ill afford the luxury of using it. It does not serve us well. To end addiction, you have to discard the substance. And we know that we are not going to "get rid" of the Internet. We are not going to "get rid" of social networking. We will not go "cold turkey" or forbid cell phones to our children. Using the metaphor of addiction—with its one solution that we know we won't use—makes us feel hopeless, passive.

We will find new paths, but a first step will surely be to not consider ourselves passive victims of a bad substance but to acknowledge that in our use of networked technology, we have incurred some costs that we don't want to pay. We are not in trouble because of invention but because we think it will solve everything. As we consider all this, we will not find a "solution" or a simple answer. But we cannot assume that the life technology makes easy is how we want to live. There is time to make the corrections.

You describe your book as an attempt to start a conversation. What has been your sense so far of the conversation that it has generated? What have people misunderstood about your book?

I wrote *Alone Together* to mark a time of opportunity. For example, the essence of my critique of the metaphor of Internet "addiction" is that it closes down conversation because it suggests a solution that no one is going to take. Addictive substances need to be discarded. We are not going to discard connectivity technology.

We need to form a more empowering partnership with it, one that shows (for example) greater respect for our needs for privacy, solitude, times of noninterruption. In some areas the need for empowerment has reached a state of great urgency, for example, in the area of privacy. Mark Zuckerberg, the founder and CEO of Facebook, has declared privacy to be "no longer a social norm."

In *Alone Together* I question such assumptions. Privacy may not be convenient for social networking technology, but it seems to me essential to intimacy and democracy. This is one of the conversations I wanted to contribute to. Others include conversations about child development, connectivity, autonomy, and narcissism. I think one of the most important sentences of my book is "If we don't teach our children to be alone, they will only know how to be lonely." I want people to talk about this when they give their eight-year-olds smart phones.

And yet much of the reaction to *Alone Together* criticizes me as though I have told the world to "unplug." As though I have accused technology of causing a new epidemic of mental illness. And as though I have said that technology is making us less human. I have been portrayed as an anti-technology crusader. Reviewers analyze why someone like me, someone who was once on the cover of *Wired* magazine, could now not "like" technology. Commentators talk as though technology and I were dating and I, capriciously, have decided to cheat on him.

This rhetoric points to a serious problem. Technology is not there for us to like or not like. Our job is to shape it to our human purposes. When you say a technology has problems that need to be addressed, people are quick to interpret you as saying that it offers nothing. In *Alone Together* I write of "necessary conversations" that lie ahead. I wrote the book in the hope of sparking some of them. I'm glad that people are talking. But sometimes it can be hard to know if you are in a conversation if people are shouting.

6 HUMANIST COMPUTING AT THE END OF THE INDIVIDUAL VOICE AND THE AUTHORITATIVE TEXT

Johanna Drucker

Flash mobs, Internet movements, swarm behavior, and virtual crowds are forms of mass participation driven by some unclear impulse toward collective action. Spurred by events, engineered and manipulated, or spontaneously arising, they are modes of responsive behavior but not necessarily methods of productive authorship. Meanwhile, data structures and hypertexts, code and software, have all become objects of study, but the interim textual configurations that arise from searches, queries, analyses, and processing have yet to find stable formats and sufficient status to allow for recovery and reconsideration. Both phenomena are changing, and so concepts of the individual voice and the authoritative text are being challenged in ways that fulfill a poststructuralist critical view through a dialogue with complex systems theory. The individual voice and the authoritative text are not mutually dependent; both are mythic concepts that have been central to humanistic practice. The work of establishing authoritative texts is practically at the core of humanism, and our need to accommodate new forms and textual conditions will only increase as the ways data can be processed and configured expand. As to the individual voice, its demise signals a cultural change with other implications in aesthetic, political, and critical dimensions.

Neither collaborative authorship nor notions of a hyperfluid text depend on electronic technology or networked environments, but changes to each get a boost in intensity of scale, speed, and degree of transformation under current conditions. In online projects, collaborative writing absorbs the input of a mass of individuals, not just a few, processing their streams into a single output text. And when a text is data driven, queries and other processing reconfigure a text as an interim and iterative object. Unlike the hypertrophic claims made for early hypertexts, those arising from a critical study of search results or faceted browsing as texts are still few and modest. Do these changes alter the way humanists do their work? Or will they do so in the future, by shifting the human-centered universe onto a new axis that is not merely polyvocal but a new hybrid form of mass expression speaking and writing as one voice and justifying its claims to authority in the public domain while creating untraceable, ephemeral, instantiations of textual systems?

Creative writing practices will surely hold on to individual identity and expression, to the celebrity branding that sustains the myth of uniqueness to which consumer culture is so addicted. Copyright laws and intellectual property considerations codify the stable expression that constitutes a work but cannot protect every combinatoric instance of a complex data set. More and more the collective mind-body organizes itself around nodes (event-actions) and agents who seem to be produced as an effect, rather than serving as autonomous instigator. Whether taken in its extreme form, as the flickering edge of a new-age, science-fiction, mind-meld, futuristic transformation of consciousness, or in its milder expressions of collaboration on a large scale, the phenomena of the collective text, as database writing and ephemeral work synthetically hybridized, are aspects of contemporary culture in unprecedented ways and at an unusual scale.

We can point to the tenets of poststructuralist criticism as harbingers, however simplistic it may be to equate the critical insights expressed in the "death of the author" and rise of the text with a change in broader attitudes toward aesthetic production (Foucault 1979; Barthes 1968). But the recognition that language produced subjects rather than the opposite, that we are spoken by the regimes and disciplines of knowledge, language, power, and so forth, was certainly absorbed into the innovative writing stance of language poetry and then new conceptualisms.[1] Individual voice exists, but interior life and lyric poetic form are not crucial to its content or expression. Individuality is a network node, unique and discrete, and however much inflected with affect, it is systemic and embedded in codependent relation with conditions of production. Procedural works have been a mainstay of creative practice since the mid-twentieth century, and instruction driven projects instituted a basic algorithmic procedure in the practices of Sol Lewitt, Roy Ascott, Yoko Ono, John Cage, and other major conceptual and early digitally engaged artists.[2] New media theory and that of complex systems produced a critical paradigm to match the emergent phenomena, playing catch-up to the rapid transformations wrought unwittingly by the interpenetration of human culture and networked communications. Eschewing techno-deterministic fallacies, one has still to grapple with the implications of emergent processes (Holland 1998; Mitchell 2011). Bibliographical studies have morphed into code, platform, and software studies for good reason, in an attempt to grapple with seemingly intractable issues of transient texts (criticalcodestudies.com, softwarestudies.com; Bogost 2012c).

The individual voice had its own mythologies and historical specificity. It belonged to the tradition of humanism that found its apogee in romanticism, with classical roots in the narratives of heroism and valor, courtship and love, individual subjectivity and a faith in the critical voice. A current example is the celebration of the Chinese dissident artist, Ai WeiWei. He provides a useful counterpoint to the mass spectacles of international events or to the models of collective practice (Klayman 2012). But the ideological force of Ai's image, as the paradigmatic figure of resistance, the heroic artist, is such as to make us believe he may be the last of his kind, the final gasp of a particular concept, reified in order to reinforce faith in the possibility of the existence of such a figure. And of course, he is conceived on the public stage as in opposition to a regime whose tactics of repression hark backward, to the overt systems of state control, while the mind meld writing practices of a current generation in the West are driven by consumer designs and manufactured desires of another stripe, equally manipulative and deterministic. Like other contemporary artists, Ai is engaged in the creation of collaborative, large-scale events as much as objects or artifacts, and the realization that art making is social engineering of reaction and response as much as it is the manufacture of objects has become mainstream (Kester 2011).

What is mass collaboration and how does it manifest itself? The host of a radio talk show I listen to recently announced that she'd received 25,000 postings on the blog. "We would like 50,000," she said, encouraging listeners to post. No doubt the numbers translate into revenue, subvention, or status and job security, but surely at that scale the individual posts aren't read, so what would be the motivation for writing? Communication does not seem to matter anymore. Any post is just a data point in large set. As a data point, however, its textual dimensions are available for processing, for text and trend analysis, and for any of the many re-purposings that can be contrived from such activity. This does not just mean paraphrase or restatement of results, but the reconfiguring of the initial text as a part of a sequence of textual events. In that continuum the textual event is not just a result of human reading and interpretation, but of machine readings and processing that are often strategically valuable and yet largely ephemeral. Again, I invoke the image of search results, data queries, and textual analyses that appear on our screens as texts but are not preserved despite their existence as part of a humanistic inquiry and study. This goes beyond the basic date-stamped specificity of Web-based objects, calling attention to the next level of digestion and textual iteration. We may, in theory, recapture any configured data set, but in practice, we do not have conventions for identifying, citing, curating, or conserving these intermediate textual artifacts.

The notion of collective identity plays a crucial role, as does the experiential dimension of participation (Holland, Fox, and Daro 2008). The impulse toward collective activity results in behaviors whose momentum is controlled by bonds not fully understood—though the fascination with crowd psychology has entertained a wide range of thinkers analyzing the Salem witch trials, the Reign of Terror during the French Revolution, the rise of Fascism, lynchings, and other phenomena.[3] Theories of emergent properties of complex systems combined with concepts of behavioral convergence divide sharply along lines that assign responsibility for group behavior to individuals and those that postulate a larger entity in formation. The role played by media and mass communication in creating a group identity fits both models—that in which individual subjects willingly yield their individual choice through alignment with belief and that in which a merged identity blurs the boundaries of one person's consciousness with that of a larger entity. Analogies arise with insect societies or higher level organisms whose functions are a mass of coordinated activities carried on by cellular units acting collectively, but without obvious or evident direction (Holland 1995; Wolfram 2000; Hofstadter 1979).

Mediated and aggregate thought forms have influence that can be explained without resorting to spooky explanations or systems theories of emergent entities. But the scale of collaborative and collective action combined with the real-time capacity for integrated feeds and aggregated outcomes is unprecedented. Absorption into ongoing streams of production creates modes of authorship that do not have a clear connection to an individual voice. Wiki trails may capture edits, and collaborative writing tools have their back-end log and database, but many texts read at the surface level conceal their production histories. Drilling down into the history of edits may reveal a moment-to-moment shift in tone and emphasis, content or argument of a text, but microscale changes weave together in a finished texture that is the result of the total pattern, not the image of individual elements. This has always been true in the fluid structure of transmitted texts. The transcription of a play performed becomes the basis of another authorial intervention, and many versions and voices can be traced in texts with long histories of use or broad constituencies of users. But the celebration of the individual voice remains a feature of common mythology in the arts and entertainment industries. Publishers and managers take the brand of an individual seriously, even if the product promoted is the result of many hands and minds contributing expertise and talent to create an image of a unique talent.

Slate magazine, among other journalistic venues, has been attentive to the appearance of collective writing practices—distinct from collaborations, which are usually between or among a small group of participants. In July 2004 Clive Thompson posted a piece in Slate titled "Art Mobs" that posed the question: "Can an online crowd create a poem, novel, or a painting?" Citing the still new Wiki as an instance of collaborative authorship enabled by online platforms, Thompson contrasted group creativity with other mob behavior—such as bringing down governments through popular uprisings. While recent events testify to the power of social networking to galvanize action and coordinate real-time strategic interventions and activities to an unprecedented degree, the distance between the body politic and the aesthetic domain remains marked by specific conventions and substantive differences. We still cling to a notion of art in ways that bear the hallmark of long traditions. Or do we? What if we are indeed at the end of the era of the individual voice, and works of art that come out of conceptual traditions as well as those that are sustained by online platforms that aggregate and synthesize contributions into a mass collectivity are the bellwethers of that change? What would that mean for the ways we might think about humanism and humanities computing? For our work and the place from which we speak?

One of the tenets of humanism is that individual subjectivity, the place and space of individuated experience, has some value in the social order as well as for the solipsistic pleasures of a unique consciousness. Critical subjectivity is a tenet of political philosophy, one of the terms on which the possibility of effective action is conceived, since it permits distinctions between self and other essential to the social contract and individual responsibility. Such lofty notions may seem far from the art mob scene. But the work of Nishant Shah, and others concerned with the impact of the Internet on society, argues forcefully against the mythologies of "revolution" and radical claims for citizen action organized through networked communication, showing that absorption into larger systems of state power and market forces are often at work in what are self-identified "resistance" movements (Shah 2012). In other words, at a systems level the myth of resistance tied to individual activism breaks down.

Critics have approached new poetic practices from another perspective. In her study of current writing in the un-creative vein, *Unoriginal Genius*, Marjorie Perloff begins with a discussion of the reception of T.S. Eliot's *The Waste Land* (Perloff 2010). Disdained as a collection of citations, of fragments of found, exhumed, recombined, and referenced language, the poem was dismissed as what we would now call a "mash-up"—anathema to the cult of the lyric voice. Eliot's essay, "Tradition and the Individual Talent," made clear the grounds on which poetic composition necessarily drew on the knowledge of the work of others, and though he did not go so far as current conceptual writers, his modern tendencies exposed the double-sided myth of originality—that it was generative and combinatoric, but not autonomous or independent of sources and influence. In measured careful tones, Eliot remarks "our tendency [is] to insist, when we praise a poet, upon those aspects of his work in which he least resembles anyone else." Where in fact, "we shall often find that not only the best, but the most individual parts of his work may be those in which the dead poets, his ancestors, assert their immortality most vigorously" (Eliot 1920).

The online edition of *The Waste Land* is a fetish object, of course, one that invests heavily in the individual author as its birthright even as the palimpsestic layering and editorial trails should show how intimately Eliot's work is entwined with the discourses prevailing in his time. Configured and constrained, a work comes into being within a field of discourse, not outside of it, holding its shape and aesthetic distinction as the very definition of poetic form. But language is always borrowed, appropriated, recycled, and reused. In that sense the individual voice is both a miracle (that it exists at all) and a banality (the distinction between system and use, langue and parole, language and any/every instantiation in expression).

Un-creative writing is an extreme extension of modern collage, if we can take Kenny Goldmith's "No need to write anything else, just manage the language we already have" as a restatement of the tenets of the modernist's essay.[4] But to get to that place with justification that goes beyond mere glib maneuvering, we need a bit of backup, and the process-based work of conceptualism attaches a great deal of currency to authorship and authorial identity. The individual may or may not embrace the notion of talent—or originality, as Charles Bernstein so wittily remarks, "I like originality so much I keep imitating it"—but that the notable practitioners of conceptual work identify as branded, named authors is unmistakable. So Goldsmith can take the text of a day's *New York Times* as the basis of *Day* and Caroline Bergvall can find 48 variant translations of the opening lines of Dante's *Inferno* and assemble them into a poetic text without composing a single phrase. But the authorial intent of their compositional acts not only attaches to their names and identities but provides the same leverage for professional stature as other practices in other times or veins (Goldsmith 2003; Bergvall 2005). New conceptualism's un-originality and un-creativity erase individual expression in one high art zone whose lineage tracks to Dada and other early avant-garde modernisms at the same moment in time that the leveling of individuality has become conspicuous within collective practices. What to make of this coincidence or perhaps a symptom of a larger systemic shift and an uncanny harmonic convergence within a poetic ecology? In the immediate post-WWII period, an impulse toward collective artistic actions prompted CoBrA, early Situationists, and others to postulate an anti-heroic stance as a contrast to the Übermensch that served a perverse purpose in a very different vision of collective identification as a body with a single leader at its head (cobra-museum.nl). The Danish, Dutch, Scandinavian, and European artists engaged in collectivity saw it as egalitarian, a formation that could not be forced into alignment because of the distributed nature of decision-making in the group, and because authorship was absorbed into community, it undermined heroic individualism, at least in principle. Like many utopian projects, it was fraught with contradictions and unsustainable ideals, but it serves as a touchstone for reflection.

The phenomenon of collective art production has received a major boost with networked technology. *One Million Monkeys Typing* is a site that describes its activities as "part choose your own adventure, part exquisite corpse" in a fine combination of children's diversion and surrealistic activity. Less interactive than a MOO or MUD, it allows participants to add branches to existing tales and build on each other's narratives. The reference to the combinatoric gamble for production of Shakespeare by the massive number of simian inputs is an apt rhyme with the procedural work of new conceptualism, even if the vector of creativity runs the opposite direction. (If conceptualism empties creative work of any impulse toward originality, the million monkeys paradigm suggests that enough collaborative composition, however random, might add up to a masterwork).

But if the "authors" in the monkey writing mill remain identifiable, their stories intertwined in a primary school space of pseudo-egalitarian we-all-share-nicely mode, other collaborative spaces eliminate the signs of authorship behind avatars and screen names, devices of fiction and illusion, as in the case of the multiplayer online novel creation, *Orion's Arm* (orionsarm.com). This galactica-metallica-sci-fi-space opera fiction has much in common with the game world, but the emergent whole, as in those environments, is a product of collective collaboration. The world is what emerges, not a single player or author's view or vision, and individual expression can't be extracted from the whole.

Other experiments in group art production in a graphic mode provided an interface for shared decision making. The aggregation of choices made by individuals was only exhibited in the changes in the image on display, so the formal expression of the work was the result of the will of the flock. The projects on *Art Othernet* (**art.othernet.com [now defunct]**) proceed by aggregation of individual units of choice and contribution that add up to a whole.

True aggregate work absorbs individual contributions and either effaces their distinctions through processing the input into a single data stream or by turning the unique element into a piece among so many that its identity is overwhelmed, as in the mosaic of breast cancer patients that comprise the Venus survivor portrait collage (Sniderman 2012). As in the case of the vast AIDS quilt, or other works of witness or testimonial, the presence of each individual face, name, or mark has value within the whole. The value resides in the individuation of experience to which the unique point of identification provides a link—that, there, that one, that photograph, that star in the firmament is mine. The individual voice gains from its place within the whole, is part of something larger, more forceful, and reassuring by virtue of the communal experience.

This concept of the shared, the collective, the communal is the thin edge of the hive mind showing its peculiar and unfamiliar human face. The scientists who study emergent behaviors of hives and swarms note the peculiar properties of the colony in its coordinated but un-directed actions. No single part of the "superorganism" is in control, but the group acts in consort and a whole emerges that has an identity and capacity for action independent of the individual parts. Precisely how this works or why it happens is unclear, and theories of complex systems bootstrapping an emergent sentience have a cultish allure, but may or may not be adequate explanations. What we observe in the insect and animal kingdoms may reflect some chemical processes of communication and coordination that humans might be subject to as well under certain circumstances, or not. Human motivation can be explained without resorting to these constructs, and the *One Million Masterpiece* seems to be an expression of the desire for distraction as much as an instinct for hive behavior. Still, the time for such speculation may be upon us (millionmasterpiece.com).

T(w)eensgosocial begins with the premise that social production is the only form of authorship (**teensgosocial.wordpress.com**). Social networking as a phenomenon is now well beyond the initial stages of recognition. The promotion of collective and collaborative writing practices within school environments is evidence of the extent to which such shared upload activities are perceived as a norm rather than an anomaly. The question that presses is whether the rate and volume of networked communication increases is merely bringing about changes of scale or whether there is some substantive shift in our character and identity. If I look at younger, highly wired individuals I sense that a changed condition of human-ness is coming about, and that group mind and hive behaviors are actually replacing individual identity and interior life.

A site like *Dumpster,* with its chocolate-toned graphics and sweet script reminiscent of valentines and love letters, is a perfect demonstration of this aggregation engine at the service of individual life (Levan, Nigam, and Feinberg 2006). The shared experience being dumped, let go, dismissed by a sweetheart is thrown into this huge collective trash bin of the punned name of the site, and the display of heartaches, bouncing like so many pong balls in a hot lottery soup, turns individual sadness into collective play. *Typophile* claims that the "collective consciousness" is trying to grope toward a group design of a font (typophile.com). The result is rather better than that of collaborative drawing or writing, since the level of processing is higher. Instead of individual efforts toward a single common goal, as in the case of this shared type design project that optimized at 5000 participants but tended to go astray above that number (how many cooks is too many?), the *Dumpster* site's algorithmic mode absorbs all individuals into a common soup, rather than pitting them against each other like so many electrons yoked to a single wagon trying to make headway along a set course. When the properties of the project are engineered to maximize the possibilities of emergence, rather than simply to combine elements or aggregate a whole from individual parts, the systemic quality of group identity takes on different properties.

Such emergence is not the same as other forms of collaboration. If we cast these observations backward, imagining that in fact "We are Homer," we recognize that biblical and classical texts have their authorial attributes, and we may trace these as if they are so many identity isotopes tracked through stylometric analysis or linguistic study. But the convergence of authorial intention in these works is more incidental and ad hoc than deliberate (wearehomer.blogspot.com). Only as the mode of group think becomes enabled within networked environments—electronic being the most obvious, but perhaps only symptomatic of the larger effects of a noosphere and its systematicity—do we begin to see an emergent pattern in which the individual voice is absorbed without protest into a larger whole. I am of course reminded of Arthur Clarke's (1953) still remarkable *Childhood's End* and its vivid images of such absorption, an image strikingly distinct from that of fascist culture or totalitarian subjugation and instead evolving toward a positive ecology of culture.

Is the coming of the hive mind a genuine phenomenon? And does it signal the death of our longstanding romantic ideology and all that it bequeathed to modernism (McGann 1983)? Is the neo-post-human not so much a cyborg human–machine hybrid as a new social being whose existence is part of a complex system of connections and relations in which differentiating impulses have begun to evaporate? Gerald Bruns (2011), writing in *In Ceasing to be Human*, describes a condition of liberation from the pressures of differentiation, a lowering of the defensive boundary mechanisms that police our otherness to ourselves. For Bruns this is all positive, a way to feel a one-ness with the world that corresponds to the discussions of new materialisms that posit their post-humanism in terms of an un-privileging of human consciousness and sentience in favor of a larger view of complex systems. In that critical frame, our long-standing attachment to a Cartesian cogito was both symptom and instrument of a conviction that sentience was separate from matter, that consciousness was other than the supposedly inert stuff with measure, mass, extent that was counted as the inanimate part of the world.

The modernist ideology drew directly on romanticism, even as it differentiated itself from lyric imagination of Romantic poetry. The work of T.S. Eliot, Ezra Pound, Gertrude Stein and others, far as they may be from the imagery of Caspar David Friedrich or the writings of John Keats, are emblematically stamped with the image of the One, the Individual, whose unique and original experience is uttered as a cry into the overwhelming sublimity of nature or against the supposedly crushing banality of culture. As much the result of what Jerome McGann called "the romantic ideology"—the product of criticism and literary mythmaking as of romanticism itself—this "heroicization" of the individual voice has been taken apart repeatedly (1983). Edgar Allan Poe's 1846 essay, "Philosophy of Composition," was a systematic exposure of the measured and calculating work of creative writing, a clear-headed rebuttal of inspiration and imagination, but not of individual authorship.

The notion of individual voice may have been shown to be mythic and has, equally mythically, been taken apart by Barthes, Foucault, and others, as already stated above. The idea of the author-as-effect may be a post-modern standard, in existence long before David Shields produced his all-citation *Reality Hunger* and talked about it on Stephen Colbert's show "The Colbert Report," saying "All art is theft" while Colbert sliced pages with his own citations out of the bound volume. If so, then we are still left with the paradox that named identity and individual revenue streams are linked to intellectual profiles as property (Shields 2011). The artist-icon remains a persistent gold standard for individuation in modern times, from the flagrant bad-boy fly in the face of convention actions of the noble Lord Byron or the equally upper-crust diva technics of Lady GaGa, or Ai WeiWei, mentioned above.

But some current works of art demonstrate an alternative approach to the "individual," posing the terms of a new paradigm of collectivity, aggregation, synthesis, and emergent voice. Here the "I" engages, the slippage notwithstanding, with the "We" while identification and cathexis manage to perform—not just an object of collective experience but as a product of group mind, swarm think, hive speak. In *We Feel Fine* emotions are culled and tracked in the real time processing of live feeds from publicly available social networking sites (Harris and Kamvar 2009). Every phrase or sentence beginning "I feel" or "I am feeling" is captured and added to the site, which can be searched and sorted by mood, feeling, weather, gender, location, and age to get a read on a mass sensation.

As these phenomena appear, they coincide with the emergence of concepts of new materialisms rooted in advanced systems theory, a radical re-conceptualization of ways of thinking about sentience and being at a fundamental level (Coole and Frost 2010). Such thoughts have their roots in long traditions—the works of Heraclitus, pre-Socratics, and in more modern form, the work of Spinoza, while also resonating with non-Western approaches to the nature of the world. The concept of post-humanism in this context is not linked to cyborg existence or the hybridization of humans and machines but rather to an eclipse of the centrality of human beings in our own perception of the world, a recognition that our "otherness" may not be so unique and distinct as we have imagined.

In coming to a close here, I want to suggest another aspect to this argument. Humanism, the humanism of the Renaissance, was something very different from later modern ideologies. The humanism of Leonardo, in which man is the measure, promotes a notion of perception situated within the human experiential frame, one that shifts the reference point of culture and individual experience from a theocentric to an anthropocentric one. This humanism remains valid, for unlike the disappearance of the ego-centric individual voice, the persistence of the individuated position from which one perceives, rather than gives voice. Humanism was not wedded to the individual talent of romantic ideology, but conceived as a set of intellectual, aesthetic, epistemological observations about situatedness, thoughtfulness, being, scheming, calculating, and knowing that shifted the ground from a theocentric worldview to one grounded in human values, concerns, and reference frames.

The illusions of this humanism are also all evident to us: the blindness of the Cartesian cogito, the limits of rationality, the justifications of imperialism, strivings to control nature, to put ourselves above the other beasts by imagining ourselves outside of the very ecologies that produce us. But at its core, that humanistic sensibility offers to us a compensation for the absorption of individual expression into collectivity through the recognition of the value of individuated experience of the phenomena of the world and culture, passion and nature, self and others. Humanism was also premised on bibliographical work and scholarly attention to the recovery of texts, and in that sense, it was a massive editing project aimed at establishing the authoritative version of crucial biblical and classical texts. Future editorial work will require new methodological premises aimed at defining the identity of interim and iterative conditional texts.

I'll end with Mark Hansen and Ben Rubin's (2006) elegant piece, *Listening Post*, which might be my single favorite example of art after the individual voice. The piece channels live data, in a manner similar to that of *We Feel Fine*, sorts and transforms it into LED displays and a synthetically voiced audio. The machine speak is completely neutral and genderless, always the same in pitch, tone, and inflection, and yet manages to be a poignant performance of our existence in the precise moment when we hover on the edge of that loss of individual voice while remaining witnesses to its expression.

Humanist ideology introduced a perspective rooted in individuated perspective. Modernism consolidated its ideology around the concept of the individual voice. Even as I observe what feels like the beginnings of a substantive, qualitative change in human identity, I feel compelled to continue my call for the integration of humanistic methods into the design of networked platforms for scholarship and creative work. The attempt to re-conceptualize our human-ness in light of theoretical premises and observed phenomena and to try to press for humanistic methods in the automated algorithmic procedural world of computation might seem at cross purposes. In fact they stem from similar beliefs, namely, that we are as yet unable to fully realize forms of human expression rooted in complex systems and emergent processes that also link directly to individual experience. Are changes in scale in social media transforming individuation and identity, works of art and artistic practices, the identity of what can even be considered a textual object? Looking around, I see a startling range of curious phenomena that, though they do not align or converge in any simple, systematic matter, suggest a change in the way the notion of the individual voice and the authoritative text is being understood. This is not as romantic constructs but as individuated spaces within networks of communication and as temporary, ephemeral configurations of emergent systems.

Notes

1. See Rob Fitterman and Vanessa Place, *Notes on Conceptualisms*, (Ugly Duckling Press, 2009), Marjorie Perloff, *UnOriginal Genius: Poetry by Other Means in the 21st Century* (Chicago: University of Chicago Press, 2010); see also **www.ubuweb.com** for Craig Dworkin's "Conceptual Writing" and other major related resources.

2. For work on conceptual art, including Sol Lewitt and Yoko Ono, see Ursula Meyer, *Conceptual Art*, (NY: Dutton, 1972) and Lucy Lippard, *Six Years: The Dematerialization of Art* (NY: Praeger, 1972) as early anthologies; Roy Ascott, *The Telematic Embrace*, edited by Edward Shanken (Berkeley and Los Angeles: University of California Press, 2003); for John Cage, see David Nicholls, *The Cambridge Companion to John Cage* (NY and Cambridge: Cambridge University Press, 2002).

3. See Elias Canetti, *Crowds and Power*, translated by Carol Stewart (NY: Viking Press, 1966).

4. Kenneth Goldsmith, **http://epc.buffalo.edu/authors/goldsmith/**

7 BEYOND INFRASTRUCTURE: RE-HUMANIZING DIGITAL HUMANITIES IN INDIA

Nishant Shah

One of the ironies of the local-global divide is that certain practices within the local sphere often precede the global nomenclatures that are assigned to them. Digital humanities is a prime example of this phenomenon where a clutch of practices that emerged with the rise of digital technologies and their integration into the national policies on higher education and learning are now retrospectively understood. So even as the term "digital humanities" was gaining currency in the European and North American context, becoming one of the buzzwords through which new conditions of pedagogy and education were imagined within the universities in the North and West, it had almost no takers in the emerging knowledge industries of South Asia in general, and India in particular. There were a range of practices, largely inspired by the "Vision 2020" report by the Planning Commission of India (Gupta 2002) that the Indian state adopted for itself in its quest to harness the power of ICTs and global communication networks to solve the endemic problems of illiteracy, low education penetration, and the growing rates of educated but unemployed workers who are not easily accommodated in the shift from a manufacturing based economy to a service based economy (22), and introducing knowledge as one of the most important determinants of development (23).

The report identifies Information Technologies (IT) not as a sector but as a significant power that seeks to transform the fields of "employment, education, infrastructure and governance" (24). As the report stridently declares, "If only we could break free from the limitations of out-dated curriculum and out-moded deliver systems, we could utilize the opportunity to close the education gap that separates the world's most prosperous communities from their poorer cousins" (26). The report thus calls for a "computerization of education" in order to "dramatically improve the quality of instruction and the pace of learning" through "distance education ... for affordable higher education" (93).

These desires for reform of the education sector closely mirror the rhetoric, ambitions, and execution of digital humanities projects around the world, and it might be safe to recognize them as such. Largely because there are three direct implications that come out of this report, which also share the problems of digital humanities worldwide. The first is that there is a need to ensure that our education system is in alignment with employment opportunities being created for a globalizing India. There is a demand that professional and vocational courses that offer skills which find easy employment in the new knowledge industries need to be created out of existing humanities and social sciences curricula so that they can "leapfrog directly into a predominantly service economy" (93). The second is that "trade, technology, and investment" have to be the primary focus of the higher education systems to allow India to become a major player in the international information societies. The third is that in order for this to be achieved, a massive infrastructure has to be created in collaboration with different private and nonpublic actors to provide better access and education to those who must be built as new citizens for the country. The questions of professionalizing and mainstreaming humanities and social sciences education are almost universal right now, and indeed it is one of the ambitions of digital humanities projects to find validity for education that does not prepare a global information workforce. The realignment of the market with the education system has been critiqued by theorists of neoliberal globalization, who have pointed out how it enables state disinvestment from education and the privatization of learning resources. However, even in these existing critiques of digital humanities (whether or not they use that term), there seems to be a consensual agreement that infrastructure building is necessary and must happen.

Infrastructure as the Backbone of Digital Humanities

It is this call for infrastructure development, which the Vision 2020 report betrays, that I want to address in this chapter. It has now become natural for all talks about education to eventually veer toward infrastructure. There is enough reason for that, when we look at the pitiful lack of resources in a country like India where the larger problems endemic in higher education today are tied down to a massive infrastructure deficit that fails to serve the huge population. According to a World Bank Report on higher education enrollment in 1994, while industrialized developed countries had an average enrollment ratio of 51 percent in higher education, in India, the numbers in 2002 were as low as 6 percent (Katiyar 2002). Development Plans in the last two decades of the twentieth century have envisioned the emergence of India as a global player in the emerging information markets and hence encouraged the privatization of education, creation of new universities and teaching institutions, and the increase of government investment in building education infrastructure, without any significant change in this ratio.

Simultaneously there has always been a severe fragmentation and compartmentalization of knowledge systems within academe that is not restricted to the humanities, producing work-forces for a global finance-driven market.[1] As. P. M. Bhargava and C. Chakrabarti (2003), in their review of Indian science since Independence, note: "a distinguishing feature of development of knowledge in the last century has been that the borders that compartmentalized it have all but disappeared. Our educational system has been blind to this important phase-shift in the educational paradigm around the world.... Consequently, we are unable to produce people who would have a global comprehension—an important requirement for leadership in the present century."

There has been a severe re-hierarchization of education and its roles where the already limited infrastructure is diverted largely to the support of professional and technical education, which aligns itself uncritically with the economic goals set by the rhetoric of an "India Shining" campaign. This new educational infrastructure is geared largely toward a growing middle class, especially in the urban centers, positing education as nothing more than a means to making money. Any method that may be used to achieve what is viewed as success in education gains legitimacy and this can be seen in the wide range of ancillary education practices like private tuition and coaching classes that are tailored to crack exams rather than gain knowledge. In his review of Integrated Higher Teaching in India, K. Sridhar (2010) asserts that "the most talented young minds in this country get trained to emerge successful in examinations but do not get trained to develop an understanding of their subjects" (2010: 18).

This obsession with creating education that can immediately translate into socioeconomic indicators of employment consequently generates a crisis for education in humanities higher education in India. The crisis either paints humanities education as redundant in the imagination of a new economic powerhouse country, or forces it to restructure the disciplines as "soft-skill building" training labs, which equip students with the requisite skills to join the global labor force within technology support and outsourcing industries in the country. The crisis becomes the raison d'être for the Indian state to disinvest from these disciplines that are clubbed under humanities and encourage them, under the rhetoric of professionalization and modernization, to tailor their ambitions and rationale to the logic of the neoliberal market.

Digital humanities, then, as a part of the techno-utopia babble that we are witnessing in India right now, has emerged as a solution to these problems. It has been welcomed both by the state disinterested in questions of humanistic education as well as by the libertarian stakeholders who see the digital as a terrain by which the form, function, and role of humanities in education can be restructured to serve market expansion. Thus, at the ICSSR international conference on Indian Social Sciences in the Changing World, held in Delhi in February 2012, the HRD Minister Kapil Sibal argued:

> The ICSSR would... create a network of eminent academics to collaborate on creation of academic content including publications of texts, digests and manuscripts in specific areas in social sciences. These manuscripts would provide an inexpensive aid to teachers and students in social sciences and would be available in different languages. The manuscripts would be delivered electronically leveraging on the gains of the National Knowledge Network, which aims to interlink all institutions of higher learning with an information super highway.

In the short span of its discursive and small-pilot existence, this emphasis on digitizing education resources has been looked at as the responsibility of the humanities and social sciences. It led to a reallocation of resources where, on the one hand, the humanities and social science schools suddenly had more access to funds but, on the other, the funds were given to them to collaborate with technology and computing departments in order to create this digital repository and network. This reemphasis on what constitutes the infrastructure of contemporary humanities in India has taken predictable and alarming tones. There has been the inevitable talk of infrastructure development, fueled by projects like the One Laptop Per Child (OLPC) and the Akaash Tablets, that have led to massive digital infrastructure building premised on the idea that if we build the infrastructure for digital connectivity and access to resources, it will translate into increased literacy and education. Public-private partnerships have been built to digitize existing curricula and textbooks so that they can be delivered through the INFLIBNET like structures of content delivery. New university models have been proposed in collaboration with education technology companies like NIIT, that posit the teacher as a facilitator in the classroom, replacing him/her with video content produced by "experts" in remotely located studios. The Citizen Service Centres for public delivery of goods are imagined as outreach classrooms that reach the hinterlands of India, thus removing the last-mile problem that education seems to be perpetually haunted with.

As Tejaswini Niranjana points out in the report on Indian Languages in Higher Education, what we are witnessing is an increasing availability of "relevant" material in English, in a digital format, catering to the middle-class expansion of career oriented education. She argues that it now seems that the major goals of tertiary education are to be achieved primarily through the delivery of information using technologies, and with an emphasis on English as the lingua franca of education and that "the very idea of a knowledge society stands instrumentalized in this understanding" (2012: 3).

In the hyperconnected post-mobile world, digital humanities seems to be the answer to the question of what we should do with our education system. In this particular model, humanities research becomes important because it would help build digital infrastructure that would then provide professional education to the new educated workforces we are building for the nation. It is a double whammy for humanities education: on the one hand, the departments and schools are facing budget cuts because they have failed to modernize their outputs, and on the other, they are offered support only if they transform themselves into becoming think-tanks to further the growth of the digital nation, thus taking on questions of infrastructure, development, growth, and expansion over the larger sociohumanistic concerns that are at the core of their existence. It is thus a systemic and structural eradication of humanities concerns from the curricula, exploiting these disciplines by offering conditional support, promoting the state agenda without any space for critical negotiation or civil society interventions.

It is from this murky arrangement of hegemonic state power and flattening out of the critical voices the humanities have been producing historically that I seek to address the need to re-humanize digital humanities in India. Moving away from the debates of infrastructure building, which have found very strong critique in the ICT4D discourse in the country, I seek to explore how the current imagination of digital humanities flattens out earlier histories of science-technology-humanities-society studies that were initiated by the state. I further look at the technology driven aspirations and indexes of education, which are being built under the rubric of digital humanities in the country, and offer alternative and resistive frameworks that have produced a better integration of these various impulses toward a more holistic educational future.

Integrated Science, Technology, and Society Studies in India

While many education policy actors and practitioners buy into the shining promise of newness that accompanies the digital turn in the country, this is not the first attempt at bringing together the natural and human sciences; there is a much longer, albeit problematic, history of integrated education. In his broad review of Science, Technology and Development, A. Rahman (1995) charts out the various ventures that were initiated to bring about a synergy between the capabilities of scientists and the concrete plans of the political leaders. This resulted in an earnest attempt to bridge the separation of the scientific and the technological from the social and the cultural. There was a growing recognition among the scientific community that the practice of science and technology lacked insights into "linkages of scientific and technological problems with the social, economic, political and cultural dimensions of Indian Society."

In 1964, the Kothari Commission Report on Higher Education in India spurred the University Grants Commission—the regulatory body of Higher Education—to set up courses on science and society and also the social studies of science. Apart from introducing these courses in mainstream university curricula, there were special attempts made at building humanities and social sciences infrastructure in the premier technology institutes, like the Indian Institute of

Technology, across the country. This attempt at bridging the gap between limitations was inspired by a firm belief that scientific thought, discovery and practice are a part of our quotidian social, cultural and political lives and they should shape as well as be shaped by the specificity of this context. The Yash Pal report from 1990 also echoed this sentiment in its views on science: "Science should not be treated as an additional activity, not only as a tool not as something which necessarily must arise in some better developed countries and come to us second hand, instead it must be treated as a part of the culture of society, integral to our living, integral to our thinking, connected with our dreams, connected with concepts of ethics, beauty and spirituality" (Yashpal 1990).

However, these early integrationist experiments were very limited in their success. Rather than creating a space for the study of the natural and social sciences in an integrated way, they kept on reinforcing the separation of the two, with a clear hierarchy, both in institutional structure and curricula design where humanities were secondary to the sciences. The Integrated Science Education initiative at the Higher Education Innovation Research Application (HEIRA) at the Centre for the Study of Culture and Society commissioned a history of ideas of integration as has been practiced in Indian universities. The study, conducted by Dhruv Raiana of the Jawaharlal Nehru University, New Delhi, looks at the national body regulating universities and higher education in the country—the Union Grants Commission's (UGC) efforts at establishing integrated science—society programs and argues that (1) these efforts were not successful because of lack of well-defined course material and (2) these courses were not taken seriously by students since their performance in these courses did not count in the assessment of their overall performance.

In its study of the IIT experiment, the report acknowledges that with its vastly superior infrastructure, smaller student populations and a better faculty profile, the IITs have a relatively better track record in science studies as compared to the universities. However, the study notes that institutes like the IITs fail "to nurture a sense of democratic citizenship" or attention to the social context, and consequently these courses do not manage to achieve a holistic education program. The study reports that the students at IITs continue to take a rather instrumentalist view of these social and human science courses and remain unquestioning and uncritical of their own disciplinary worldviews and biases.

The current trend of digital humanities builds on and strengthens these biases where the nation's own technosocial aspirations of economic growth and power have superseded its human and social concerns. The responsibility of the state toward the citizen seems to be in building the infrastructure of growth, whereas the core concerns of life, justice, rights, and so forth, which have been the domain of the humanities, have been rendered secondary or obsolete. It leads to a new kind of technology mediated social sciences which is no longer critical of this changing role of the state and its implications on building just, open, and fair societies.

Instead, the humanities, in the name of integration again, are being asked to reformulate their core agendas, skills, and outputs in the service of the futures that the technological turn has demanded. Philosophically speaking, it is a regressive step, where the older integration dualism of the natural and social sciences has been dismantled to establish a straightforward validation of the goals of science and technology, forcing humanities to recalibrate to fit these new shapes of the future.

The Digital in Humanities: The Gross Enrollment Ratio

Much of the euphoria around digital humanities in India is still very new and concentrates, more than anything else, on the role that the digital would play in shaping and modernizing our concerns in the humanities and social sciences. It was only in February 2009 that the Ministry of Human Resource Development, Government of India, announced its first systemic initiative called the National Mission on Education through Information and Communication Technology (NMEICT), which marks the first foray of higher education reform using the digital technologies. The NMEICT begins with a strong mission statement claiming that it will convert "our demographic advantage into a knowledge powerhouse by nurturing and honing our working population into a knowledge or knowledge enabled population (and thus) . . . enable India to emerge as a Knowledge super power of the world in the shortest possible time." This mission's objective was that digital resources would arrive by 5 percent at the 11th Five-Year Plan's target of enhancing the Gross Enrollment Ratio (GER) in Higher Education, thus ensuring that by the end of this program, 16 percent of the total Indian population qualifying to be in university would indeed be in universities.

The Gross Enrollment Ratio has been a concern that reflects the failing numbers in higher education in India. It is a new indicator that evaluates the massive higher education infrastructure of more than 500 universities and 25,000 colleges to look at how they are able to progress from elite to mass and finally to universal education infrastructure in the country. Drawing from the American sociologist Martin Trow's idea that historically higher education is an elite occupation, the Gross Enrollment Ratio proposes a simple mathematical algorithm: that higher education systems that enrolled up to 15 percent of the relevant age group should be described as elite systems, those that enrolled between 15 and 50 percent of the age group were mass systems, and those that enrolled more than 50 percent were universal systems. The role that was envisioned for ICT based education systems—of which the digital humanities is a landmark experiment—was that it would negotiate the increase in GER within the higher education age group to reach what Ashish Rajadhyaksha calls "requirements of massification" (2011: 140).

Rajadhyaksha, in his exploration of the "ITification of the higher education" sector in India, points out that this concern about enrolment rates was not new. However, the dependence on ICTs to ensure this rise in enrollment rates was definitely new for India and it contained an inherent paradox:

> (1) Making higher education available to all people of society is ideally a good thing, but this is difficult to achieve under present conditions. (2) The problems as they present themselves to State structures are twofold: (i) As states have trouble making access to higher education (as against primary education) into a *right,* and (ii) since making higher education available to all is sure to make *quality* suffer, higher education should go down the tube, (3) and finally, higher education can only be delivered by private, or at least autonomous institutions, who will certainly accept no restrictions on either fee structures or on who they choose to admit. (Rajadhyaksha 2011)

The GER has emerged as a new index of human development alongside economic criteria like standard of living, development of the industrial base and per capita income. Jianxin Zhang, quoted by Rajadhyaksha in *The Last Cultural Mile: An Inquiry into Technology and Governance in India,* argues that the enhancement of the GER leads to the production of a "rights based" understanding of education where the "originally clear boundary between higher education institutes and society gradually disappears" (2011: 141). Higher education, in this particular model, becomes increasingly a *right,* linking up to the employability of students, thus producing a massive number of graduates who compete for limited employment opportunities. The universities approximating massification only produce students for a "massification employment era," whereas high research and quality education still remain in the realms of the elite, who will be able to afford that infrastructure from private sources.

The entire NMEICT project, Rajadhyaksha posits, is propelled by its imagination of technologies as neutral and hence intrinsically beneficial. But in the process the NMEICT is able to view students as primarily availing themselves of distance education, living in non-urban settings with major shortages of electricity and without access to technical assistance. Thus the mass education, or the delivery of education to the masses, clearly has a bias toward those who will be in the urban elite centers, able to afford expensive education, and against those who will be on the fringes, trained to be the workforce for the massification employment industry.

The paradox that Rajadhyaksha points out is also inherent in the digital humanities initiatives that are being structured. It is recognized that humanities and social sciences have positive contributions to make to the society, especially in conversations about stronger civil and political society building in the country. It is hence suggested that more people should have access to humanities studies, following the ideas of mass and universal education. However, there is a recognition of the fact that there isn't enough infrastructure right now to grant mass access to humanities education. ICTs and digital technologies become the route for the mass delivery of humanities studies, evaluated through the lenses of the Gross Enrollment Ratio. In order to achieve this massification, though, we will have to dilute the quality of our humanities education, which will suffer in the face of an infrastructure deficit and ill-prepared teachers.

To prepare for the massification, the humanities departments moreover will have to be made relevant to the career aspirations of the new demography it has to reach. It will also have to be reconfigured as a professional discipline to support the costs of infrastructure development that is being initiated for the betterment of the discipline. Eventually critical research and thinking will again be restricted to private and experimental spaces that only the elite can afford, whereas the masses, in the name of digital humanities, will be equipped with instrumental skills that could help them find employment and a role in the state's plans restructuring of governance and market configurations.

This is the essential paradox of the digital in humanities, where the perceived neutrality of technology, and the depoliticized mapping of humanities, leads to its self-eradication under the language of professionalization, efficiency, productivity, and enrollment, which the digital technologies build in. The focus instead remains on building increased infrastructure of access and interaction, without actually looking at the ways in which the digital can help rethink new approaches and frameworks to address the core concerns of these disciplines and how they can be addressed in emerging information societies. While this situation is specifically drawing from the Indian context, it also has resonances with a global trend that deploys technologies as neutral, and often ends up investing in the infrastructure of digitization and access, without looking at the core politics and values that the human and social sciences stand for.

Provocations for Digital Humanities in India

Things are not as dire as they sound. While the larger trends, dictated both by market trends as well as narrowly focused state policies are often alarming, they also build the possibilities for the future of higher education in its interaction with digital technologies. The digital turn, in small but substantial ways, has led to a dramatic questioning of older forms of knowledge infrastructures that suddenly have started appearing hegemonic, closed, and exclusionary in the ways they operate. Even in conservative state projects, which build the infrastructure of connectivity with the student in remote interiors of India, there is an acknowledgment of the fact that the student can no longer be seen merely as a recipient of education. Students are capable of being active participants in creation of the knowledge industries.

In the new information societies, as platforms and technologies evolve to change knowledge production, storage, access, and dissemination, forms of learning will be transformed. Already the transformation has been rapid in the quotidian practices of the Web. However, the new platforms, spaces of belonging, and modes of interaction that have produced transformative shifts, have not yet found their way to a successful integration with the old vanguards of knowledge and education within academe. Attempts to incorporate the digital as a way of thinking about our academic, pedagogic, research, and archival practices, rather than merely as an instrumental interface of interaction or a process of resource building, can help in reconfiguring the knowledge industries—the form, function, role, and currency of knowledge in our networked societies—and producing new structures of education outside a moribund university system more interested in sustaining itself than building alternatives.

Several grassroots movements that focus on indigenization of technologies, development of local language input and browsing devices, building of open source and open access communities, and producing collaborative structures of low-cost co-creation are rapidly changing the traditional ways of the stakeholders within the knowledge industries. Incipient small projects that question the naturalized process of knowledge production through print technology—in the gray zones of piracy, remixing, sharing, and the like—are taking a digital turn that promises be the core of digital humanities in the country.

Different research initiatives, as well as policy and implementation projects, have worked to encourage ICT driven regularization of education. There are four interventions that suggest what the future of digital humanities could look like when rescued (but not divorced) from the excessive conversation around infrastructure building:

1. *Integrating technology in core values* The Higher Education Innovations and Research Applications (HEIRA) think-tank at the Centre for the Study of Culture and Society (CSCS), along with the Centre for Internet and Society in Bangalore,[2] have been considering the ways in which digital technologies could help replay, recalibrate, and recuse the ideas of social justice in higher education. Working with students who are identified as "socioeconomically disadvantaged," their project on "Pathways to Higher Education" looks at the problems of exclusion and learning, which had been considered resolved in the education policies in the country.

Using the technology as a means of digital storytelling, helping the students to narrate, document, and build on their everyday experiences as politically motivated and shaped, the project uses multiple interfaces to bring back the politics of exclusion into the debates on access and learning within the classrooms. The project envisions digital humanities not merely as an interface between digital technologies and learning environments but as an opportunity to see how certain problems of the social and political fringes, which are made opaque in a language of rights and access within education, can be revisited in order to identify structural and systemic changes in the higher education system.

2. *Historicizing the digital turn* New media practitioners and researchers at SARAI at Delhi University in New Delhi, and at the Alternative Law Forum in Bangalore,[3] have helped identify the digital turn as having antecedents with other technology crises in the past. Looking at the history of pre-print cultures in the thirteenth and fourteenth centuries, they see how the questions of authorship, authority, knowledge production, intellectual property, and so forth, as are being challenged with the digital turn, were also under negotiation with the advent of the print cultures.

   In mapping the processes through which the current regime of knowledge production has matured, they unpack the forms and functions of knowledge stakeholders and intermediaries to see the possibilities of new knowledge commons and platforms of exchange. Instead of thinking of knowledge as residing in repositories like books or artifacts to be saved in museums, their work has reconceptualized the building of a new public sphere of interaction and engagement within the information societies. Digital humanities groups need to stop justifying themselves through practices of sharing, remixing, collaborative content creation, and iterative publishing to device platforms, but take up the intellectual challenge of understanding what new forms, formats, functions, and processes of knowledge production and distribution are going to look like in the hybrid online–offline worlds we live in.

3. *Conditions rather than infrastructure of knowledge* The knowledge program on "Digital Natives with a Cause?"[4] between Hivos Netherlands and the Centre for Internet and Society Bangalore has proposed that the question of infrastructure is not redundant, but needs to be thought anew. In their work with young people in the Global South who are growing up with technologies and strategically using them to make changes in their immediate environments, they propose that one of the ways to rethink the question of infrastructure is to reposition the researcher and the knowledge producer in the knowledge industries.

   The project shows that digital humanities can be a new framework of knowledge production that embraces the multiplicity, fragmentary nature of information, crowd-sourced and distributed across communities and collectives, and leads to new ways of identifying knowledge sources and formats. This fractal nature of research does not reduce its claims to truth but rather helps build platforms and sites that assist such processes of knowledge production. This is a systemic change in research practice that seeks not only to harvest the information but also to build affective, legal, educational, support systems through policy and practice that can support, sustain and facilitate collaborations and exchange.

4. *Openness and accessibility* It has been frequently argued that most the infrastructure building initiatives are guided by market-driven forces and end up producing new gated knowledge communities that are tied to expensive, inaccessible, or closed technology ecologies.

These initiatives might perform some of the sophisticated ambitions of digital humanities but end up reproducing older knowledge exclusions and hierarchies in the new technology environments. The older forms of centralization, which are being challenged by the new organizational collectives, often give an idea of democratization of knowledge. However, ignoring the technology infrastructure and the licensing regimes that operate on the digitized resources often means the production of new central and opaque power structures that restrict access to knowledge by putting it in formats and licenses that do not allow for knowledge exchange or collaboration.

The challenges for the future of digital humanities in India are twofold. On the one hand, they need the support of the state driven infrastructure, which builds the conditions of connectivity, access, and exchange within the formal and informal spaces of higher education teaching and research. Given the massive scale of the education landscape in India and the severe lack of infrastructure even for traditional educational practices, it has to remain integrated in the state's imagination of higher education in the country and align itself with the agendas of the neoliberal state. On the other hand, it needs to re-politicize the stakes of education by rescuing it from the market-driven language of careers, employability, access, and efficiency, which distances humanities from its core values of human and social interest. However, this cannot be a radical disavowal of the larger flows of global capital and ideology. Instead, digital humanities needs to be seen as an opportunity to question, contest, and remap the ways in which conditions of life, labor, and language are rapidly changing in the widely varied and uneven advent of digital modernity in the country. In its negotiation with the state infrastructure and its re-humanizing of the infrastructure centered debate, digital humanities can serve a dual function of building relevant and sensitive infrastructure of production and access to knowledge as well as build a critical voice that resists the dehumanizing principles of networked societies that reduce all human beings to actors and all human modes of engagement to actions and transactions.

Notes

1. The divisions between natural and social sciences are further stratified into various different hierarchies. For example, there is a growing schism between natural and applied sciences within the universities. Similarly humanities and social sciences cannot always be partnered. Liberal arts, philosophy, and professional courses in marketing and business have separate niches. There is a growing body of work that designates human sciences as a way of describing the new turn, and an equally strident voice that resists the "science" prefix for humanities and related disciplines. While this is interesting, for the sake of this chapter, I will place all science and technology based knowledge clusters into one group, called

science, and all human and social studies knowledge clusters into the other, called humanities, while acknowledging that this grouping is tense and needs more unpacking.

2. The Centre for the Study of Culture and Society is a research and academic think-tank based in Bangalore and can be accessed at http://cscs.res.in. The Centre for Internet and Society is an independent research and advocacy think-tank and can be accessed at http://cis-india.org.

3. SARAI is a new-media collective based in New Delhi and can be accessed at http://sarai.net. The Alternative Law Forum is a lawyer's collective that engages with research, policy and interventions, and can be accessed at http://altlawforum.org.

4. The "Digital Natives with a Cause?" is a Knowledge Programme initiated by the Dutch Humanist Development Corporation HIVOS, and details about it can be accessed at http://www.hivos.net/Knowledge-Programme2/Themes/Digital-Natives-with-a-Cause.

8 TOWARD A TRANSNATIONAL ASIAN/AMERICAN DIGITAL HUMANITIES: A #TRANSFORMDH INVITATION

Anne Cong-Huyen

We can agree that the digital humanities have arrived, and thus I will not start by talking about the emergence and triumph of digital humanities;[1] instead, this chapter begins with the understanding that as the digital humanities continue to evolve, we, as scholars and practitioners, must reflexively insert the heterogeneous and interdisciplinary critical lenses and approaches of ethnic studies (my own specific focus here is Asian American Studies) into the theories, methodologies, and practice of the digital humanities. At first glance, these two fields seem (at best) unrelated, or (at worst) mutually exclusive. As I will endeavor to demonstrate in the following pages, however, this need not be the case. Instead, I challenge my fellow digital humanists to ask: Can a field that encompasses a range of disciplines within the social sciences and humanities, that is rooted within a history of civil rights activism, devoted to the specificity and heterogeneity of its populations, and materially invested in global, national, regional, and local communities, critically inflect digital humanities practice in a productive manner?

The invitation that guides this chapter is one that I have struggled with as I try to bridge what seems like a wide gap between the digital humanities and Asian American Studies. Personally, I feel bound to actively participate in both, but I also feel it is imperative that I merge the digital humanities with the historicized and situated knowledge production and engagement offered by the gamut of Asian American Studies. In doing this, I must acknowledge that unlike the digital humanities, which celebrates greater institutionalization and investment, Asian American Studies, along with other ethnic studies programs, finds itself in increasing danger of defunding and even criminalization.[2] These studies are threatened in an era of resurgent neo-conservative "post-racial" ideologies that see danger in acknowledging difference and alternate histories. Despite this uncertainty, Asian American studies can be as beneficial to digital humanities, as digital humanities is to Asian American Studies, though the emphasis here will be on the former.

Some (often administrators or lawmakers) have hailed the digital humanities as having the potential to save the humanities, and even the university.[3] Such uncritical acceptance is problematic, and it is therefore crucial that we interrogate the assumptions that underlie scholarly practice in digital humanities, and the technologies that drive, motivate, and make this work possible. These digital and electronic technologies are of particular importance because they are often perceived as being neutral, without any intrinsic ethics of their own, when they are the result of material inequalities that play out along racial, gendered, national, and hemispheric lines. Not only are these technologies the result of such inequity, but they also reproduce and reinscribe that inequity through their very proliferation and use, which is dependent on the perpetuation of global networks of economic and social disparity and exploitation. Because these technologies are invisibly embedded within global structures of inequality, it is especially important that we use the social and cultural concerns of Asian American Studies—with its emphasis on situated knowledge, context, and politics—to critically inform our engagement with our material tools, the scholarship we produce with them, and the communities we teach and live in.

In general, it is important to integrate Ethnic Studies critique into the digital humanities, as the focus on historical legacy, power structures, economic, political, and social inequity, and their relation to race, gender, sexuality, and class, are necessary to complicate the positivist view of digital technologies with which digital humanities work often unproblematically engages. In particular, though, I want to argue for Asian American Studies, as an especially relevant Ethnic Studies framework to bring to the digital humanities. Asian American Studies, as a field, is shifting from its early domestic project of claiming a long history of "Asians in America," toward a more expansive decolonizing project that draws from Third World Feminism and Postcolonial Theory. Likewise the digital humanities should also be invested in the deeper histories of the digital technologies that are central to it as a field. These histories continue to have significant material consequences in our lives as students and academics, whether or not we identify with digital humanities. Moreover these conditions are of utmost importance for Asians, Asian Americans, and diasporic Asian subjects who manufacture, program, consume, appropriate, and use these technologies.

In the process, I want to continue the work of expanding Asian American Studies. As will become clear, my conception of "Asian" is expansive and calls into question the common understanding of the term, a holdover from European antiquity that highlights the power of naming. Furthermore my "America" is a broad one that takes into account not only all of the Americas but also the far-flung reach of the United States that has resulted from centuries of expansion along militaristic, imperialistic, economic, and cultural fronts. My transnational Asian American digital humanities is, I hope, an inclusive one that can encourage a diversity of historically situated, contextualized, and culturally critical dialogue and scholarship.

Awakening as an Asian American Digital Humanist

To convey the significance of infusing digital humanities with an Asian American perspective, I will share a recent personal history that took place at the annual Modern Language Association Convention in 2011. For me, and for the digital humanities, those few days at the MLA were formative ones. Matthew K. Gold (2012: x) claims as much when he writes, "In the aftermath of the 2011 Modern Language Association Convention, many members of the field engaged in a public debate about what it means to be a 'digital humanist.'" Gold relates Stephen Ramsay's now infamous talk, "Who's In and Who's Out," in which Ramsay provocatively states, "Do you have to know how to code [to be a digital humanist]? I'm a tenured professor of digital humanities and I say 'yes.'... Personally, I think Digital Humanities is about building things.... If you're not making anything, you are not... a digital humanist" (qtd. in Gold 2012: x).

As someone who has studied and taught about exclusion laws, discriminatory practices, and anti-miscegenation laws, I was particularly sensitive to the experience of hearing a tenured professor give a talk entitled, "Who's In and Who's Out," that energetically and aggressively defined the parameters for who can legitimately call themselves "digital humanists." Even if Ramsay's goal was to engender the heated discussion that followed, the experience of being in the room during his session was a rather jarring and marginalizing experience. Later William Pannapacker, in a blog for *The Chronicle of Higher Education*, extended this in/out discussion in relation to his own position by writing, "the field, as a whole, seems to be developing an in-group, out-group dynamic that threatens to replicate the culture of Big Theory back in the 80s and 90s, which was alienating to so many people. It's perceptible in the universe of Twitter: We read it, but we do not participate. It's the cool-kids table." But the comparison being made to a "cool-kids table" did not adequately describe my feelings of unwelcome. If anything, it was too benign.

The violence with which I experienced this particular panel, and many other digital humanities panels I attended at MLA that year, was a result of two main factors. The first, and most obvious, is the fact that the digital humanities seemed to be overwhelmingly white, and I was often one of a few people of color, let alone women of color, in those rooms. The second factor was that many of my larger research interests in the global, the transnational, in race, gender, sexuality, and inequality seemed to have little overlap with work currently being done in the digital humanities. I do not mean to accuse the attendees of these panels of being themselves in any way exclusionary or overly utopian, since digital humanists tend to be quite welcoming, even when it comes to criticism, but the feeling of belonging is not experienced equally. And even the analogy of an exclusive lunch table trivializes the experience of exclusion felt keenly by women, people of color, and others who have no choice but to embody difference.

In her article, “Why Are the Digital Humanities So White? or Thinking the Histories of Race and Computation,” Tara McPherson recounts a similar experience of discomfit, albeit from a very different subject position. She writes, “I found myself reflecting on how far my thoughts had ranged . . . from diaspora to database, from oppression to ontology, from visual studies to visualizations. And, once again, I found myself wondering why it seemed so hard to hold together my long-standing academic interests in race, gender, and certain modes of theoretical inquiry, with my more recent . . . immersion in the world of digital production and design.” In my own novice experience, I found little discussion of social justice in digital humanities panels at MLA 2011 except for a single panel called “Planet Wiki? Postcolonial Theory, Social Media, and Web 2.0.” The rest of my conference was spent attending panels on human rights writing, postcolonial diaspora, and Asian American literature. These latter panels all featured little talk of the digital. Aside from “Planet Wiki?” not one panel I attended at that MLA explicitly addressed the intersections of race (let alone Asian American Studies) with the digital. More than a space for humanists and linguists to share work and debate the future of our fields, MLA had become a constant negotiation of balancing seemingly disparate fields, which necessitated a fair bit of compromising and prioritizing, and led me to worry: *would I have to do this for the rest of my professional life?*

Hope for an Asian/American Digital Humanist or Digital Asian Americanists

After returning from MLA, I realized that these issues had surfaced before, when I reflected upon interactions I’d had with faculty members in the department where I did my graduate studies. The question, “Why do you want *me* on your committee?” from a prominent, tenured, Asian American literary scholar, in response to my proposed multimedia, multimodal, and multi-sited project, had baffled me. It became less baffling with that visit to the MLA. In attempting to reconcile these two interests together in my post-MLA blog, I came up with the following:

> The critical significance paid to code, to networks, to interaction and play, to labor and practice that are foundational to DH can be used to inform an Asian American or global project. But what about the reverse relationship? Does Asian American criticism lend itself to larger, possibly non-Asian, more global, more digitally focused projects? . . . Can we not re-appropriate the disciplinary concern of Asian Am (and other specific ethnic) criticism with its investment in social justice, equity, ethics, and representational politics to expand the scope of what is considered “Asian American literary criticism?”

After more than a year and a half, time spent reading the work of scholars like Wendy Chun, Anna Everett, Jennifer Gonzales, Kara Keeling, Lisa Nakamura, Alondra Nelson, Minh-Ha Pham, and Tara McPherson, I am convinced that an expansive integration of Asian American Studies and digital humanities is necessary *and* feasible. There is a growing community of digital humanists invested in ethnic, gender, queer, and disability studies, and they found each other through those very tools and spaces that digital humanities have been espousing: online networks, collaborative projects, nontraditional meetings, blogs, and discussions. But this community is small, and the participants are often invested because of their own identities, their own marked bodies. Although such practice is powerful, it would be more so if this work were not limited to those who participated because of their personal identities. Those of us in ethnic, feminist, and gender studies have long discussed the importance of allies in advancing social change, and such alliances are even more important now when university education, as a whole, is seeing significant change.

Such change is happening in an era of economic cutbacks and increased privatization. Technology seems to be a key factor in positively addressing these obstacles, as seen in discussions regarding the "future of the university." Technology in the minds of some will supposedly democratize and save the university through open access publishing, Massive Open Online Courses (MOOC), networked classrooms, even increased productivity. Such talk makes technology out to be simply a tool to be utilized carefully, but we must remember that our tools are the products of long, often troubling, material histories. McPherson, for example, has argued that the histories of computational technology have resulted in problematic modular forms of binary thinking, which is felt keenly by those who attempt to stitch together their interests in difference and the digital:

> [T]he difficulties we encounter in knitting together our discussions of race (or other modes of difference) with our technological productions within the digital humanities (or in our studies of code) are actually an *effect* of the very designs of our technological systems, designs that emerged in post–World War II computational culture. These origins of the digital continue to haunt our scholarly engagements with computers, underwriting the ease with which we partition off considerations of race in our work in the digital humanities and digital media studies. (McPherson 2012: 140)

For some reason, when we think of race, gender, sexuality, or nationality, it seems incongruous with our work on technologies so often seen as "neutral" (McPherson 2012: 142), "raceless, sexless, genderless,"[4] and unmoored from material, geopolitical considerations. This, however, is not the case, and we must recognize the irony with which our digital technologies have become instilled with the promise and hope for the future of the humanities and the academy. These technologies, for example, are often rooted in a history of militarism and reflect the economic and cultural imperialism that exploits economically depressed regions and peoples.[5]

And it is these same technologies that allow us to stay connected with one another, to dialogue, to share our work, and to build off each other.

In light of this, we must actively insert cultural criticism into digital humanities, as Alan Liu has prompted, and to "theorize" it, as Natalia Cecire (2011), Alexis Lothian, and the #transformDH collective (Phillips 2011) have challenged us to do. Liu, in relating the potential of digital humanities to its social and political context, states, "While digital humanities develop tools, data, and metadata critically . . . rarely do they extend their critique to the full register of society, economics, politics, or culture. How the digital humanities advances, channels, or resists today's great postindustrial, neoliberal, corporate, and global flows of information-cum-capital is a question rarely heard in the digital humanities association, conferences, journals, and projects with which I am familiar" (2012: 291). Similarly Lothian clarifies that for a small subset of digital humanists, "we were not using the term ["theorize"] intransitively. We were talking about queer, trans, butch, femme, critical race, women of color, Asian American, Puerto Rican theory. . . . We were talking about marked bodies, systemic social hierarchies, and transformations in a very specific and material sense" (2011). These appeals to inflect digital humanities production and to situate it historically and culturally, calls to mind Kuan-Hsing Chen's move toward an analytical framework he dubs "geocolonial historical materialism" (2010: xiii). Such a framework is ideal for developing digital humanities production in a historically informed and socially transformative manner.

Comparisons between Asian American Studies and digital humanities have previously been drawn by Patrik Svensson (2012: par. 12–14), who cites parallels in terms of the long-term debates occurring in each, including struggles to define the disciplinary goals; "inclusion and exclusion" of subjects, objects, and practitioners; institutional structures; and, perhaps most importantly, the tensions regarding "political and community grounding" as the fields become institutionalized. As the editors of *Asian American Studies Now* point out, however, Asian American Studies scholarship is, at its core, an activist scholarship:

> From the outset Asian American Studies included a commitment to working with and serving Asian American communities in their struggle for a more just and equitable present and future. Since its foundation, the field has undergone significant transformation, reconfiguration, expansion, and crisis. . . . We recognize that while Asian American Studies has produced far-reaching critical analyses and a powerful epistemological critique of American history and life, analysis and critique must be linked to broader efforts for social transformation. (Wu and Chen 2010: xv)

The "transformation, reconfiguration, expansion, and crisis" referred to by Wu and Chen alludes to the tensions within the field that struggles with negotiating the field's radical activist history with the pressures of academic institutionalization. Despite these tensions, Asian

American Studies is based on linking work done in the academy to work done in the community, a model that can only benefit digital humanities, especially as its practitioners find themselves imperiled by the rhetoric of utilitarianism and economic value in the resurgent neo-conservative language of reform and retrenchment.[6] One promising example of this work in action can be found in the 4Humanities platform, co-founded by Liu, which works to advocate for humanities research and education. As the mission statement for the site states, "They [the digital humanities] have the potential to use new technologies to help the humanities communicate with, and adapt to, contemporary society" (4Humanities). This model and the work that it engenders is one that resembles work that has been done in Asian American Studies over the past four decades: scholarship that is legible and dialogic across disciplinary lines, is relevant and accessible to communities outside the academy, and curricula that often include service based learning components that are co-designed with local community based organization.

For an example, I want to turn to the multimedia PDub Productions collection coming out of Los Angeles's Pilipino Worker's Center, a local community service organization that serves the Los Angeles Historic Filipinotown (Hi Fi).[7] This dynamic project documents the ongoing sociocultural history of the Filipino community in Los Angeles and provides spaces, education, tools, and platforms for local youth to create and distribute content about their community. In addition the project is embedded within the larger Los Angeles Research Collection, which utilizes the temporal-spatial mapping technologies of the HyperCities platform that comes out of digital humanities work done at UCLA and USC. PDub Productions ultimately allows Filipino American youth to create empowering, interactive, and contextually specific media that can then be shared with visitors and community members through various media: GPS mobile Hi Fi tours, navigable HyperCities maps, and a physical multimedia installation in the Worker's Center. This presents an exemplary practical model that shows both scholars and community members responding to Liu's suggestion that, "The digital humanities . . . can create, adapt, and disseminate new tools and methods for reestablishing communication between the humanities and the public" (Liu 2012b: 498).

Alternately, Adeline Koh's Digitizing "Chinese Englishmen" archive offers another mode of Asian/American digital scholarly production (**chineseenglishmen.adelinekoh.org**). Citing projects like "Africa Is a Country," which works to complicate representations of Africa and African diaspora, Koh's project takes as its objective the preservation of historical documents that diversify representations of "Englishmen" and "Chinese" during the Victorian era. At the same time it endeavors to decolonize the archive, an institution that continues to reproduce the problematic politics of documentation, preservation, and exclusion. Accessible to a wider public through an open website, Digitizing "Chinese Englishmen" also engages with a larger community of students and scholars through social media, thus expanding the audience that might otherwise be excluded from accessing such historical documents and the associated theoretical critiques.

The work carried out in these two projects, whether done through the appropriation and transformation of digital technologies, like that of Nokia tablets on which the Hi Fi Tours are run, or the production of a freely accessible digital archive for the preservation and sharing of unacknowledged histories, offers two distinct but complimentary examples of community-oriented Asian American digital humanities. By taking on this paradigm in our own practice, we will be less likely, as Nakamura puts it, to partake in the "preservation of the status quo" (qtd. in Svensson 2012: par. 40). And this is important because, according to Tanner Higgin (2010), "Without a robust critical apparatus, DH has and will continue to unwittingly remake the world in its old image," that is a result of what he calls "technofetishistic obsession." This privileging of the techno-digital that Higgin points to is one that produces detrimental binaries of its own—binaries that must be resisted. Our enterprise in digital humanities is to transform and revive scholarship and teaching, not to perpetuate it as it has been.

Making the Digital Humanities a Social Justice Humanities

One important similarity between Asian American Studies and digital humanities is the work within each that breaks down rigid binaries and the possibilities that lie in the ambiguous in-between spaces. Asian American and other ethnic or area studies, for example, are often viewed from the outside as fields limited by their specific racial or national politics. This is not necessarily the case, and my goal is to try to move beyond the dualisms associated with race, ethnicity, and nationality. I want us to think beyond notions of self and other, of inclusion and exclusion, of center and periphery, of majority and minority, of citizen and alien. Within the digital humanities, dualisms also emerge, but this time between the technological and the biological, the human and nonhuman, the digital and the analog, the humanities and the sciences and social sciences. Juridical, cultural, and social conceptions of these different binaries have tangible, often detrimental, effects on the lives of people and communities, but thinking through and around them also adds an element of unpredictability and possibility.

The unexpectedness that comes from access to digital tools, for example, opens up a wide realm of potential. In a discussion with David Theo Goldberg for the UCHRI, Nishant Shah (2012) points out, in the example of India, that the digital allows for alternative identities, spaces, and modes of connecting. He states, "The digital has been interesting in finding the points of commonality [in the postcolonial moment], simply because so much of the focus there has been on social networks, which is not a space where normative subjectivities are formed, this is not a space from which you can make entitlement and claims, it suddenly opened a new space for us to look at how people are trying to connect with each other, belong to each other." Earlier modes of communication, production, and consumption (though not obsolete) are being redefined with digital technologies, especially in Asia and connected regions. Our work in examining such phenomena must take into account a *spectrum* of experiences and locations that binaries do not adequately address. It requires sensitivity to the specificity of local and linked contexts, something that can be learned from scholars of Asian/Asian American Studies *and* digital humanities. And it also demands a certain amount of flexibility and innovation, something that the digital humanities are already known for. As Kuan-Hsing Chen argues, "Knowledge production is one of the major sites in which imperialism operates and exercises its power," and we must, as a result, "transform [the] problematic conditions, transcend the structural limitations, and uncover alternative possibilities" (2010: 213). Chen's call to "transform," "transcend," and "uncover alternative[s]" here resembles the work of some early digital humanists who worked within rigid institutional structures, received little institutional support, and encountered such obstacles with flexibility and innovation in order to build and define the field from the ground up. Of course, different digital humanities scholars and programs now come from a broad range of histories in relation to larger institutions and disciplines that fall outside this narrative, but the rather powerful presence of the digital humanities as we see it today is a relatively new phenomenon. It is also a label that is heavily debated.

Digital humanists, like Mark Sample, have also pushed us to think of digital humanities as a kind of radical scholarship. Sample writes, "The digital humanities should not be about the digital at all. It's all about innovation and disruption. The digital humanities is really an insurgent humanities" (qtd. in Svensson 2012: par. 40). This comment speaks to the digital humanities as a revolutionary and activist project. Figuring digital humanities as a kind of tactics presents interesting possibilities for the application of digital humanities tools and methods within Asian American studies as well. It is not just a *new* digital versus *old* traditional humanities, just as Asian American studies is not necessarily only about people of Asian descent living in the United States. The initial subjects of Asian American studies in the 1960s, 1970s, and 1980s might have been Chinese, Japanese, and Korean communities in the United States, but the label "Asian American" can no longer be exclusively tied to such identities. The economic and cultural imperialism of the United States extends its reach and influence far beyond lines marked on a map, while American studies has expanded "America" to include *all* of the Americas (though US exceptionalism cannot go without discussion). "Asia" also no longer refers to just China, Japan, and Korea but includes Southeast Asia, South Asia, the Middle East, and the Pacific. Taking legacies of colonialism, migration, and diaspora, as well as the developing effects of globalization and transnational neo-liberal capitalism into account also alters the scope of "Asian" peoples to include a more expansive body of historically linked postcolonial, diasporic, and transnational people. The rigid definitions of earlier decades can no longer hold, and we must view Asian America in a more capacious, inclusive, and increasingly networked and adaptable manner.

Perhaps it is through collaboration, collective action, and community building, (something digital humanities has fostered through online networks, un-conferences, meetings and workshops, and something Asian Americanists have done within local communities and their own networks) that we can establish a cohesive but heterogeneous and interdisciplinary scope. This scope should be informed by the potentials of the digital, and dedicated to investigating and alleviating material inequalities, especially as we observe these inequalities being reproduced in the digital realm. My hope is that the identities of "digital humanist" and "Asian Americanist" will take on manifold meanings, and our various engagements will result in a diversity of transformative digital humanities endeavors that come from and recognize a multiplicity of identities, practices, and locations.

Conclusion: Digital Asian Americanist Futures

I will end this chapter by reflecting on another visit to the Modern Language Association, this time in 2012. This year there were fifty-eight digital humanities related sessions at MLA (there were forty-four in 2011, twenty-seven in 2010), and four were related to ethnic studies, including my panel, "Old Labor, New Media."[8] Unlike my first MLA (2011), I was much more comfortable there. It was only a year later, but having found a small community of digital humanists, Asian Americanists, and #transformDHers who shared a vested interest in combining their ethical and political concerns with their academic and scholarly ones, made the experience more productive and inviting. Although my audience was small, it was an audience invested in the issues of race, labor, inequity, and the digital. Engaging with the audience and connecting with new and old colleagues about these shared topics positively transformed the MLA for me.

Similar experiences of community and productivity at recent academic conferences have demonstrated the importance of flexibility and openness to scholarship, which is becoming less strictly tied to disciplinary boundaries. Elsewhere, David Theo Goldberg has defined our current scholarly environment by saying, "The new modality about knowledge creation has to do with the fact that it is no longer disciplinarily bound, we're bound by problematics, we're bound by themes, we're bound by questions" (Davidson and Goldberg 2007). It is how we establish these problematics, themes, and questions that will unify us as a field and mark the contributions we make as we move forward. As utopian as it sounds, sharing and valuing the same ethical and political stakes have made the "big tent" of digital humanities that much smaller and navigable for me. It has also helped me (and others, I'm sure) in defining myself, and my work, and in connecting me with allies, comrades, and collaborators. Digital humanities may well be the savior of the academy, yet the digital humanities shouldn't just be about saving jobs and institutions; it should also be about social justice. In a climate where everything is increasingly temporary and precarious, those people on the margins should not be the only ones fighting. Those of us in the burgeoning field of digital humanities—the junior, the tenured, the overworked, the underemployed—need to recognize our counterparts outside of digital spaces and the university. And we need to apply the socially and politically engaged model of Asian American and ethnic studies to our own work, regardless of whether or not we are of Asian descent, because, ultimately, it is the responsible and informed way of doing publicly relevant scholarship and activism.

Notes

This chapter is informed by an earlier blog entry, "Toward an Asian American Digital Humanities," but has been dramatically revised and expanded. Many thanks to Adeline Koh, Konrad Ng, and additional #asiandias Twitter contacts, for helping me think through the relationship between Asian American and Asian studies. I am also grateful to Alexis Lothian, Amanda Phillips, and my fellow #transformDH cohort. And to Steven Pokornowski, Lindsay Thomas, Sharon Tang-Quan, Tassie Gniady, and Kim Knight for providing feedback on earlier drafts of this piece.

1. The state of the digital humanities, its relevance, its methodologies, its future, and the like, have been discussed by a large number of people and in numerous outlets and forms including a variety of MLA panels and roundtables, in personal and institutionally affiliated blogs, columns and pieces in the *Chronicle of Higher Education*, *New York Times*, and, most notably, in the recent collection *Debates in the Digital Humanities*, edited by Matthew K. Gold.

2. The most publicized example of this is Arizona's controversial law, House Bill 2281. This law banned ethnic studies from public schools. According to Gary Y. Okihiro, "That law bans schools from teaching classes that are designed for students of a particular ethnic group or that promote resentment, ethnic solidarity, or overthrow of the U.S. government." More subtle, less obvious, examples include Okihiro's account of transformations at Columbia University, which saw the collapsing of multiple distinct programs, majors, and research centers into generic disciplines defined around issues of "difference." "The Future of Ethnic Studies," *The Chronicle of Higher Education*, July 4, 2010 (website). http://chronicle.com/article/The-Future-of-Ethnic-Studies/66092/.

3. Sample entries (of which there are many) in this debate include: "Digital Humanities Manifesto 2.0" from the UCLA Digital Humanities Center or Steven Kolowich's "The Promise of the Digital Humanities" for the website, *Inside Higher Ed*. In my own experience at a small liberal arts college, an experience that

in many ways aligns with those at larger institutions elsewhere, we can observe administrators conflating digital humanities with online learning, MOOCs, and other digitally mediated profit-seeking ventures meant to increase institutional revenue.

4. This specific phrasing comes from a roundtable proposal for American Studies Association, "Transformative Mediations?" in Baltimore, Maryland, 2011. I collaborated on this roundtable with Alexis Lothian, Amanda Phillips, Tanner Higgin, Marta S. Rivera Monclova, and Melanie Kohnen.

5. The Apple iPhone is a perfect example of this, as seen in the recent work of Lisa Nakamura. "Economies of Digital Production in East Asia: iPhone Girls and the Transnational Circuits of Cool." *Media Fields Journal,* http://www.mediafieldsjournal.org/economies-of-digital/2011/2/28/economies-of-digital-production-in-east-asia-iphone-girls-an.html.

6. This type of language was especially visible during the period leading up to 2012 presidential elections.

7. PDub Productions can be found online through the HyperCities Los Angeles Research Collection or through their standalone site. (website) http://hypercities.com/LA/. (website) http://hypercities.com/pdub/.

8. This count was done through a perusal of Mark Sample's annual compilations of MLA digital humanities panels archived via blog, *Sample Reality*. "Digital Humanities Sessions at the 2012 MLA Conference in Seattle," October 4, 2011, (blog) http://www.samplereality.com/2011/10/04/digital-humanities-sessions-at-the-2012-mla-conference-in-seattle/.

9 BEYOND THE ELBOW-PATCHED PLAYGROUND

Ian Bogost

At the request of one of his colleagues, the Stanford comparative literature professor David Palumbo-Liu wrote a blog post making a case for why the humanities are indispensable. It's one in a long history of such justifications, a task that seems as necessary as ever. Yet, as with so many such justifications, Palumbo-Liu's speaks declaratively. Consider his closing charge—one I saw excerpted frequently and with enthusiasm on blogs and social media in the days after he wrote it:

> Lowering the bar for the humanities, or even dismissing the humanities as not having anything specific to teach us, is not only abrogating our responsibilities as teachers, but also ignoring the very patent evidence that the humanities are our solace and aid in life, and we need them now more than ever. (Palumbo-Liu 2011)

Among that evidence, Palumbo-Liu cites the continued presence of students in humanities courses and degrees, where complex topics get discussed in traditional literary form. In other words, the crisis in the humanities is not one of interest, but one of support. "People still care passionately about the humanities," argues Palumbo-Liu, but "today's students have been raised in an intensely competitive atmosphere," one focused on hireability over all else.

Palumbo-Liu goes on to tell several stories about students in Stanford continuing studies programs, those who take time out of their work lives to discuss "life in a way that it is able to be discussed through literature." Palumbo-Liu concludes that only the patient reading of the complex art known as the novel is capable of getting to the heart of significance.

In his final book, *The Conflict of the Faculties*, Immanuel Kant ([1798] 1992) discusses the relationship between the university and the state. Kant makes a distinction between "lower faculties," those oriented toward theoretical reason, and "higher faculties," those oriented toward practical reason. The higher faculties serve state and mercantile interests, and they are therefore bound to external ends. By contrast, the lower faculties are autonomous activities, separate from the interests of law or business. Kant's position on this matter influenced Wilhelm von Humboldt's design for the University of Berlin, an institution that in turn influenced the structure of the modern university, with its separation of professional schools and faculties of arts and sciences.

Here, between the higher and the lower faculties, lies a trap the humanities often falls into when self-justifying in the way Palumbo-Liu does in his justification of the humanities' indispensability.

On the one hand, humanists want to retain a place in the lower faculties, arguing that their work cannot be probed for predictable value. But then on the other hand, humanists constantly claim to have measurable value propositions. And worse yet, those value propositions are usually so vague as to become meaningless: "critical thinking," "lifelong learning," "communication," "cultural perspectives," and so forth. Palumbo-Liu's "solace and aid" is a reasonable candidate for this list as well.

As tempting and appealing as this logic might be, it signals a troubling move. For starters, it simultaneously embraces the high faculties' logic of predictable usefulness while also offering relatively weak examples of utility. But worse, when humanists comport themselves according to the tentatively useful values they espouse, the results tend mostly to service intellectualism anyway (e.g., "critical thinking" mostly takes the form of fashionable censure).[1] Communication" about "culture" tends toward cryptic self-reference and directs itself at insiders alone. Indeed the fact that ideas like "critical thinking" and "solace and aid" feel so vague even as they also ring true could be seen as a tiny object lesson in the problem. The terms we espouse mostly circulate within humanities disciplines, acting as comforting code words. Humanism has professionalized, and the interests it serves most often are its own.

For another part, there's nothing necessarily *humanistic* about skills like critical thinking or lifelong learning, or communication, or even culture and solace anyway. These are qualities to which almost any discipline could reasonably lay claim. Who is to say that linear algebra is any less of a candidate for critical thinking than is Latin? Or that the computer scientist can't develop an interest in lifelong learning as much as the art historian can? Or that civil engineering isn't cultural? Thus the humanities' stock self-justifications both embrace the high faculty's frame of utility, and in so doing, they offer responses that don't really answer the question.

One could simply refuse the challenge entirely. Famously, Stanley Fish (2008) did exactly that: "To the question 'of what use are the humanities?', the only honest answer is none whatsoever. And it is an answer that brings honor to its subject. . . . An activity that cannot be justified is an activity that refuses to regard itself as instrumental to some larger good. The humanities are their own good."

A lot of people didn't like Fish's answer (a lot of people just don't like Fish), but at least it was definitive. It held the line. Still, it's not the sort of argument that works anymore (if it ever did), unless you're someone like Stanley Fish.

But even if you are, would you want to make such an argument? That the only use your field serves is to serve itself, to reflect on itself, to return its spoils home, like hoarders or profiteers? Does being of use really threaten humanism so that it must insist on being "above" accounting? Is being on the books really the problem? Or is the problem rather that humanists have systematically removed themselves from the domain of human practice, mistaking participation for adulteration?

It's a situation created by a fundamental misunderstanding of what the "lower faculties" are meant to do. The humanities are not meant to run "off the books" as an elbow-patched playground. Instead, they are meant to represent and nurture a populace in the face of the governmental and organizational interests served by the higher faculties. The humanities are meant to be *populist* rather than statist. They shouldn't stand "against usefulness," but rather "toward the world."

And here, despite their name, the humanities have generally failed. I've argued that humanists bear active disdain for actual humans, whom they often perceive to be ignorant suckers, willing interpellees too far outside the "honorable" inner sanctum of Fishy humanism to be capable of the reflection the humanities claims to offer them (Bogost 2011: 241). Humanist intellectuals like to think of themselves as secular saviors working tirelessly in the shadows. But too often they're just vampires who can't remember the warmth of daylight.

Admittedly, there are not always obvious worldly correspondences for humanist interests like there are for engineers and lawyers. But the public has its own concerns, and often those would seem to intersect with matters of humanistic interest: "thinking and reflection on the human condition," to use Palumbo-Liu's words. But how can one think and reflect on the human condition while assuming either that it must be done apart from those conditions, since to do otherwise would be to be "useful" or "accounted for?" Alternately, how can the material for that reflection materialize in isolation from the world in which it exists? The value of the humanities is assumed to be intuitive, unchanging, and hermetic.

The result amounts to a puzzle: we have something to offer, but only to ourselves or to those who volunteer to join us. To offer something different would either transform the lower into the higher faculties (thus destroying them) or offer such a weak and disconnected account of utility so as to reveal its sequestration. It is a Neverland.

There *is* an inherent conflict among the lower faculties because the state sponsors their practice, whether through governmental or private support. But that conflict is part of the point of the lower faculties, not a structural calamity doomed to undermine them as so many humanists seem to believe. Kant called it a creative conflict, but the philosopher Stephen Palmquist takes things further, interpreting Kant's levels as circular rather than stacked:

> The "highness" of theology, law, and medicine connotes a royal calling, a direct link to the "high officials" of the government. The "lowness" of philosophy, by contrast, connotes a direct link with the general public. *There are no professional philosophers*.... That is, the academic philosophers is (or should be) like the general public's *spy*, strategically positioned at the heart of the university in order to collect information and serve as the public's most reliable informant. (2006: 235)

Humanists tend to task themselves with the production of others just like themselves ad infinitum. If we're being generous, we might admit that some humanists play the role of saboteurs, injecting skepticism into future professionals before sending them off to serve the higher interests. But only a minority *earnestly* begin their quest outside, among the public, relying on contemporary matters as compass-bearings for their intellectual work.

I understand "public" in much broader terms even than Palmquist. It's not just citizens, not just human beings, not just living creatures or even natural orders. The humanities should orient toward *the world at large*, toward things of all kinds and at all scales. The subject matter for the humanities is not just the letters and arts but every otherworldly practice as well. *Any* humanistic discipline can be oriented toward the world fruitfully, but its practitioners most choose to orient inward instead, toward themselves only.

Humanists can be private educators *and* public spies. But the latter role is far too rare, because humanist intellectuals do not see themselves as *practitioners of daily life*. Their disparagement comes largely from their own isolation within the institutions that reproduce them, a fate many humanists despise out of one side of their mouths while endorsing it with the other. The humanist corner of the university becomes, in Palmquist's words, "just a safe haven for half-witted thinkers to make a comfortable living" (2006: 235).

The humanities needs more courage and more contact with the world *in addition to* a continued commitment to removed reflection. It needs to extend the practice of humanism into the world, rather than to invite the world in for tea and talk of novels, only to pat itself on the collective back for having injected some small measure of abstract critical thinking into the otherwise empty puppets of industry. As far as indispensability goes, we are not meant to be superheroes nor wizards, but secret agents among the citizens, among the scrap metal, among the coriander, among the parking meters. We earn respect by calling in worldly secrets, by making them public. The worldly spy is the opposite of the elbow-patched humanist, the one never out of place no matter the place. The traveler is at home everywhere, with the luxury to look.

If we accept the premise that the humanities should be oriented toward the world and not toward a private, scholarly sanctuary, then what trends are already facilitating that process? One candidate is the "digital humanities," a topic about which I have remained silent for too long, despite having directed a digital media graduate program and teaching in a computational media undergraduate program, both housed solely or partly in a liberal arts college at a technical institute.

Digital humanities is a category that defies both definition and description. The Digital Humanities Manifesto 2.0 calls it "an array of convergent practices" surrounding both the transition from print to "multimedia" and the use of new digital tools in the arts and humanities (Presner, Schnapp, Lunenfeld, et al.). Such a moment implies novelty. But as both Patrik Svensson (2010) and Matthew G. Kirschenbaum (2010) have observed, digital humanities also has roots in humanities computing, a decades-old name for applying computational tools to humanities research and teaching.

I have no desire to offer a new definition of digital humanities—after all, the eclectic set of definitions proffered by those who participate in the annual Day of Digital Humanities shows that we already have too many ("How do you define Humanities Computing/Digital Humanities?" 2011). Instead, let's revisit Svensson's condensation of such perspectives into a few categories: tool, object of study, expressive medium, laboratory, and activist venue.

Some of these categories seem more central than others. Svensson's idea of an "exploratory lab," for example, is largely speculative, and his examples of digital humanities as an "activist venue" seem more like subspecies of the "objects of study" or "expressive medium" categories. The digital humanities have an encroachment problem. After all, using computational media as expressive tools is a common practice in art and design, whereas their digital humanistic applications mostly involve new methods of scholarly creation and dissemination. And as for objects of study, one can't help but note that scholars studying forms of digital media from a humanistic perspective—including the Web, social media, videogames, interaction design, and interactive narrative—don't tend to call themselves "digital humanists." Or if they do, they do so for reasons of personal survival. When all the job ads ask for "digital humanities," all the CVs update accordingly. For digital humanities, "the digital" is a kind of orientalist fantasy, whereas for the humanist scholar who works on digital media, it's just the office.

One of Svensson's categories thus stands out above the others: tools. Tools still occupy a wide berth in the digital humanities, including everything from open-access journals, scholarly collaboration tools, media digitization and archival efforts, and the creation of new digital approaches to studying traditional artifacts. While self-styled digital humanists seem eager to include anything that touches both the humanities and computers under their umbrella, I find such an approach aspirational and rhetorical. Instead, it is far more useful to identify digital humanities more modestly, as the spiritual successor to humanities computing, a practice intended to advance the existing practice of humanism through computational methods.

As far as things go, there's nothing terribly surprising or upsetting about the idea, save that it would have taken the humanities so long to embrace it thoroughly. After all, many other disciplines have experienced considerable change at the hands of the computer over the last half-century, and many have articulated that change in the same terms (including humanities computing). We have computational linguistics, computational biology, computational physics, computational neuroscience, computational chemistry, computational finance, and so many others.

Note that most of these fields use the prefix "computational" rather than "digital." That difference is more than just semantics. Computational methods tend to emphasize information processing and analysis over the creation and dissemination of information assets. But even so, other fields have also pursued new digital collaboration and distribution tools. In fact one of the earliest "open access" initiatives is arXiv (arxiv.org), begun in 1991 as a preprint repository for articles in physics, later expanded to include mathematics, computer science, and other fields. The National Science Digital Library (nsdl.org), a free education resource, was also created that year. Digital humanists eschew the label "computational" because it draws an uneasy connection to computer science, whereas scientists embrace it because, hey, who doesn't use computation?

This desire to separate things digital from things computational is both intellectual and political. But it's also tactical: by creating a new domain within the humanities, digital humanists reset expectations. Some will bristle at my characterization of digital humanities as mere tool use, arguing that their work involves a different or greater engagement with digital matter than I'm allowing. I agree, but with the addition of a very important codicil: *for them*. That is to say, if we zero the scales for the humanities, given full knowledge of its penchant for hermeticism and stasis, then digital tools for creation, collaboration, and dissemination do indeed represent a significant change. But it's embarrassing that they do.

When it comes to working as public spies among the lower faculties, there's certainly something to be gained from such adoption. After all, if the humanities are meant to be oriented toward the world, then they need adequate tools with which to do so, tools compatible with the present moment. Speaking directly to and with the public is an important way to overcome the isolationism and self-reflexivity of humanistic practice. In that respect, the digital humanists are making strides in rescuing the humanities from their vampiric roost. As Kirschenbaum puts it, "the digital humanities today . . . is publicly visible in ways to which we are generally unaccustomed" (Kirschenbaum 2010: 6).

Fair enough, but let's not be too satisfied just yet. The very fact that we are unaccustomed to it is tragic. *Making it customary* should not be mistaken for taming a new wilderness. Instead, it must be seen as just the tiniest baby step in reclaiming of a lost worldly responsibility.

Can't we just face it: it's mortifying how far behind the times the humanities really are, computationally speaking. Remember the 1980s, when everyone (save the humanists, apparently) got personal computers, modems, BBS's, and online services? Remember the 1990s, when everyone (save the humanists, apparently) got gopher and the Web? When humanities computing really took off, when mainstream digitization first became commercialized? Really, it's idiotic to pat oneself on the back for installing blog software or signing up for Twitter.

Let's imagine the best scenario. If the humanities are an agency of espionage, then the digital humanities would be its Q Division, the R&D arm that invents and deploys new methods in support of its mission. But we're not there. We're not close. How come?

For starters, the digital humanities more frequently adopt rather than invent their tools. This is a complicated issue, related to the lack product development and deployment experience in general among humanists, and their lack of computational and design abilities in particular (By contrast, most scholars of physics or biology learn to program computers, whether in FORTRAN or MATLAB or with even more advanced and flexible tools.) As a result digital humanities projects risk letting existing technologies dictate the terms of their work. In some cases, adopting existing technology is appropriate. But in other cases, the technologies make tacit, low-level assumptions that can't be seen in the light of day.[2] While humanists can collaborate or hire staff or otherwise accomplish technical novelty, it's often at a remove, not completely understood by its proponents. The results risk reversing the intended purpose of the humanities as public spies: taking whatever works from the outside world un- or under-questioned.

Furthermore this process of development creates a vicious circle of conflict and loathing. As lower faculties, humanists often see their work outside the logic of technological improvement or efficiency. Usefulness should not be anathema to the humanities. But since predictable usefulness is still commonly held in disregard, creating and deploying digital humanities tools explicitly involves servicing an instrumental end. This creates cognitive dissonance, as it causes the lower faculties to appear to be acting according to the logic of the higher faculties. And that dissonance results in anxiety and conflict.

These two factors combine in a surprising and perverse way. Kirschenbaum calls digital humanities simply "a term of tactical convenience" (Kirschenbaum 2011: 415). Kirschenbaum's point is that "digital humanities" is a concept that helps get things done. Things like getting a faculty line or funding a staff position, or revising a curriculum:

> At a moment when the academy in general and the humanities in particular are the object of massive and wrenching changes, digital humanities emerges as a rare vector for jujitsu, simultaneously serving to position the humanities at the very forefront of certain value-laden agendas—entrepreneurship, openness and public engagement, future-oriented thinking,

> collaboration, interdisciplinarity, big data, industry tie-ins, and distance or distributed education—while at the same time allowing for various forms of intra-institutional mobility as new courses are approved, new colleagues are hired, new resources are allotted, and old resources are reallocated. (Kirschenbaum 2011: 415)

Despite the apparent enthusiasm in this passage, Kirschenbaum is circumspect about such tactics. On the one hand, they allow humanities programs to pivot. But on the other hand, because of the nature of the academy, those pivots risk ossifying into long-term institutional structures that become difficult to change.

This is a bittersweet pill. It's encouraging that the digital humanities look to the outside for inspiration and influence—it's one example of a re-orientation of humanistic practice toward the world and its interests. But at the same time the rationale for that orientation is somewhat perverted; it is motivated primarily by an inward-looking reformational interest. This is why so much of the talk in digital humanities is about digital humanities. This is institution-building, not world-building.

If the internal Q Division is the best-case scenario for the digital humanities, its worst case is a kind of techno-liberalism, a weird inversion of Silicon Valley's techno-libertarianism. In this scenario the digital humanities becomes an organizational-political lever to advance arguments for the reformation of the humanities, but whose means of reformation is primarily self-reflexive, and whose manner of executing on that self-reflexive reformation relies largely on imported materials and methods to bulk up the ramparts that would protect humanism from the world it might otherwise enter. Recalling Stephen Palmquist's quip on half-witted thinkers, it seeks to make a sullied haven safe again for comfortable living: novels *and* computers.

In a *New York Times* editorial, USC Annenberg fellow Neal Gabler (2011) laments a world in which "big, thought-provoking ideas" without obvious purpose are in decline. Among the factors Gabler cites is "the retreat in universities from the real world, and an encouragement of and reward for the narrowest specialization rather than for daring—for tending potted plants rather than planting forests."

There's a place for potted plants. Every practice has to spend time reflecting on itself and reorienting. There's nothing wrong with importing solutions from the outside, from which there is always much to be learned. But the lower faculties must resist the temptation to partake of daily life only just enough to mine convenient resources into makeshift parapets. It's neither a cowardly move nor a treacherous one, but it's neither a courageous nor a righteous one either. The digital humanities must decide if they are potting their digital plants in order to prettify the office, or to nurture saplings for later transfer into the great outdoors. Out there, in the messy, humid world of people and machines, it's better to cast off elbow patches for shirtsleeves.

Notes

1. See Michael S. Roth, "Beyond Critical Thinking," *The Chronicle of Higher Education*, January 3, 2010. http://chronicle.com/article/Beyond-Critical-Thinking/63288/

2. For more on this topic, see the books published under the MIT Press series "platform studies," for which I serve as co-editor, http://platformstudies.com.

10 WHY YACK NEEDS HACK (AND VICE VERSA): FROM DIGITAL HUMANITIES TO DIGITAL LITERACY

Cathy N. Davidson

From Thinking to Making

If I were giving a final exam in my Humanities and the Digital seminar, I would ask my students to produce collaboratively an online, interactive humanities project accessible by scholars, students, professors, and the general public. They would turn their semester's research on a humanities topic into a tool or an app, a fully interactive website, a multimedia archive, a robot, an installation, a touch-screen apparatus, a game, a new interactive peer-driven badging system, leaderboard, or other customizable credentialing tool, or some combination of all of the above and more. They would be required to include an extensive "resources" component that would help others contribute to the project and they would make it open access so others could use and "modd" the source code for their own purposes. Different students would contribute different skills, content, knowledge, and insights, including computational abilities, Human Computer Interaction (HCI) skills, aesthetics, law, copyright, policy, and social networking, and all would contribute subject-area content. Students would also master collaborative management skills, using some form of peer-accreditation system to give feedback to one another and note contribution (or lack of it) in the group. Finally, they would create a business plan for the sustainability of their project.

I begin this discussion of the inherent and radically interdisciplinary project of the digital humanities with a hypothetical final exam to underscore what happens in the translation from thinking (what programmers call "yack") into online, digital Web-making ("hack"). That translational component should not only be embraced by those working in the digital humanities but also should be part of the field's evangelism on behalf of an educational revisioning for the twenty-first century.

In key ways digital humanities addresses a structural problem in the contemporary educational system, the division of knowledge into the "two cultures." As defined by C. P. Snow in his famous 1959 essay, one culture includes the arts, humanities, and social sciences while the other academic culture is made up of the natural, biological, and computational sciences and engineering (Snow [1959] 1993). It will be remembered that when Snow delineated these categories he noted that they were at least thirty years old, and he argued that they were laden with values, privilege, and hierarchies of prestige. While they left the humanities in a lowly and disrespected position, they also isolated the sciences from what now we might describe as culture and context. It is hard to imagine thinkers such as Galileo, Newton, or Leonardo da Vinci being able to understand such a shortsighted bifurcation of human knowledge.

As I have argued elsewhere, the two cultures binary is a product of a specific historical moment, namely the Industrial Age and the Taylorist model of productivity (Davidson 2011b). The research university and much of its accompanying institutional apparatus were specifically designed to train farmers to be factory workers, shopkeepers to be corporate bureaucrats and middle managers, and scientists to serve the needs of a specialized, technologically advancing workplace. Such features of the US curriculum as required courses, electives, distribution requirements, majors, and minors were all developed in the late nineteenth and early twentieth centuries. They are founded in the implicit assumption motivating all research universities (then and now), that the goal of an undergraduate education is to prepare the best or exemplary student for post-baccalaureate training, either in professional school or graduate school (both of which are also a product of the Industrial Age and the creation of the research university; see Cole 2009; Newfield 2003, 2008). The separation of the liberal arts or general education component of undergraduate education from the specialized major is one manifestation of this intellectual segregation and this pre-professional educational focus.

If ever it made sense to segregate the human and social sciences from the natural and computational sciences, it certainly does not anymore. In the famous words of Apple CEO and visionary, Steve Jobs, "technology alone is not enough—it's technology married with liberal arts, married with the humanities, that yields us the results that make our heart sing" (Jobs 2011). Jobs went on to insist that the liberal arts were part of the DNA of his company, part of what makes Apple a great innovator, beloved of customers and with a brand loyalty few other corporations can match. Surely part of Jobs' success is in realizing that the old binaries are over, that the DNA of the digital age is, quite precisely, its melding of the technological, artistic, and the human.

Beyond the glossy surfaces and the apps of Apple products, however, is the interactivity of the Web that blurs the difference between the consumer and producer, the scientist and the artist. The most definitive and revolutionary feature of the World Wide Web is its open architecture that allows, invites, and, indeed, necessitates participation by individuals who can contribute to its robust code and content via Hypertext Mark-up Language (HTML). If we are to fulfill and preserve the World Wide Web, then we need to be giving students, from kindergarten through professional school, the tools—computational, aesthetic, legal, and social—to contribute to it rather than simply download from it. To make that educational transformation possible requires not just teaching HTML in schools but also a paradigm shift about what constitutes "computation" and what counts as "human and social life." It means rethinking such binaries (implicit in the two cultures divide) as production and consumption. Indeed the translation from thinking to making requires competency on both sides of the two cultures divide.

Prescriptively, that paradigmatic final exam with which this chapter began is intended to encompass the ambitions of digital humanities. Digital humanities is not content or method exclusively, not solely theory or practice. Digital humanities is the embodiment and communication of ideas online, with the implicit goal of inviting community participation in the co-creation of knowledge. In the language of *histoire du livre*, digital humanities merges the differentiated roles played by individuals and corporations that contribute to the communication cycle—writer, printer, publisher, reader, reading community (Febvre and Martin 1979; Darnton 2010). HTML and the open architecture of the Web allow anyone who has an idea and access to the Internet to post that idea, without the intervention (or skilled expertise) of a publisher or an editor, thus making global, instantaneous self-publication a reality for those who have an Internet connection. That new opportunity also entails a challenge, one that educators have not yet fully absorbed as part of their civic educational mission, except perhaps in digital humanities. Digital humanities takes that blurring of producer and consumer as a serious opportunity and, I am arguing, an interdisciplinary mandate.

The potential of open source, multimedia, multi-lingual self-publication also entails a serious responsibility. Once consumers have the potential to be producers and publishers, consumers have to become informed about what it means to publish or to go public. Such issues as security, privacy, and identity are not simply matters of designing default switches but of deep cultural, legal, and social significance. So are matters of credibility (which is why credentialing and feedback suddenly take on a central role). If the ideal digital humanities final exam is about making things ("hacking"), I would suggest that digital humanities, to be truly realized, also requires heightened attention to theory, to exploring and understanding the intellectual, legal, philosophical, and personal issues of going public. What is subjectivity in a connected world? What is authorship? These are not trivial questions. Yacking is also necessary. In fact the catch phrase "more hack, less yack" does not fully comprehend the pedagogical and, sometimes, activist responsibilities of digital humanists. Humanists need to explain why we are committed to public knowledge, multimedia forms of representation, open access to the knowledge we create, and peer-forms of connected learning and assessment if we are going to contribute to the redesign of the university for the digital age.

Digital humanities is a relatively new field with a potential to disrupt traditional and siloed disciplinary divides. Beyond its historical situatedness, digital humanities, I would argue, must take responsibility for modeling new modes of thinking that transcend the two cultures. As John Unsworth has argued, digital humanities has long moved beyond the simple goal of digitizing archives or creating searchable databases (although that is one useful function) (Unsworth et al. 2006). Digital humanities is also about realigning traditional relationships between disciplines, between authors and readers, between scholars and a general public, and, in other ways, re-envisioning the borders and missions of twenty-first century education.

Making Requires Thinking

To understand fully why we need both "hack and yack" in the digital humanities, let's posit a different kind of final exam in an advanced undergraduate seminar on digital humanities. Let's say I worked in a traditional humanities department and wanted to ensure that my students measured up to the traditional text-based communication and rhetorical standards of my discipline. Let's say it's an English Department. Here's the kind of essay question I would ask on a final exam in digital humanities: "If the Stop Online Piracy Act (SOPA) and the Protect Intellectual Property Act (PIPA) had been in existence as US Law in 2002, would Wikipedia exist today? If either proposed law had passed through the US Congress in the winter of 2012, would Wikipedia still exist in 2022? Why or why not? *Discuss*" (Wikipedia 2012).

In thinking about such a hypothetical essay exam question, one confronts the inherent interdisciplinary nature of humanities in the digital age. Prompted by the translation from thinking to making, from theory to public practice, digital humanities requires the desegregation of knowledge that the Industrial Age imposed. In a truly re-imagined humanities of the digital, one could not make a project without understanding the implications of the proposed SOPA/PIPA legislation and other aspects of Internet intellectual policy, copyright law, regulation, and commercial law.

Digital knowledge, in other words, is polyglot. Since it also has to be in-depth, sophisticated, and of the highest standards to actually work, it inevitably also has to be collaborative. It is too much to expect that any one person can perform at the highest standards in all fields at once, either on the computational or the theoretical levels.

Part of the problem with the proposed SOPA/PIPA law was that Congress did not fully understand the relationship of domain names to noncommercial open Web architecture. The way the law was written, it would have granted law enforcement the right to block access to whole Internet domains if infringing content had been posted to even a single blog or webpage, effectively allowing for the shutting down and even prosecution of any site that did not control its contributors and any domain that contained such a site. Wikipedia could not exist in such a world. In fact digital humanities projects that were open to participation and connection beyond their originating designers and authors would also have come under the purview of the SOPA/PIPA legislation being proposed. Passage of such a law would have severely compromised the key open feature of the World Wide Web. Yet who teaches students today about such implications? In what department is this considered crucial subject matter—law, public policy, or economics?

The point is that experts from several disciplines—from programmers to legal theorists to experts in the regulation of online commerce and business—are needed to understand the profundity motivating the digital knowledge project. To fully address and preserve the potentialities of the participatory, connected digital world we live in now requires dismantling and then re-imagining the systemic, structural, material, financial, and intellectual divides of modern education, kindergarten to lifelong. Most especially, it requires challenging the siloed structures of the modern research university.

To return to our hypothetical SOPA/PIPA exam question: on some level, if you cannot answer it, you are not in control of your life in 2012. Because life in 2012 affords you constant opportunities to contribute online, the rulings of the Federal Communications Commission (FCC) no longer apply to us abstractly as *consumers* of radio, TV, and other media. Now, they apply quite personally to our potential as producers and connectors, participants in a virtual "town square" where, digitally, our intellectual, social, medical, religious, political, and other parts of life are in constant, continuous interaction. Broadband regulation as well as government surveillance pertains to every aspect of contemporary everyday life, as evidenced by the fact that the Department of Homeland Security has a Bureau in the FCC. Yet few of us are fully educated—Web literate—enough to understand all the dimensions of these factors that indeed govern much of our everyday lives.

To illustrate the extensive and invasive implications of legislative change, on January 17, 2012, Wikipedia took the extreme measure of leading a digital protest against the SOPA/PIPA proposed legislation. Followed by several other Internet giants, Wikipedia went dark to demonstrate to the general public how drastically those two poorly worded bills could have curtailed the World Wide Web as we know it. No polemic ("yack") could have been more effective than the blackout was in demonstrating to the public how a badly worded bill would have a direct impact on our individual, everyday lives, not in some general way but immediately. If you clicked on www.wikipedia.org on January 18, 2012, you were greeted with a black screen and the words "Imagine a World without Free Knowledge" (Wikipedia 2012).

There were 115,000 websites that participated in the SOPA/PIPA protest. The actions were successful. Around 8 million people looked up their representatives' names and around 4 million signed petitions to protest these two bills (Wikipedia 2012). By morning, support for SOPA/PIPA had dwindled so drastically in Congress that the proposals died. Will they next time? Everyone knows that some kinds of copyright and piracy regulation are needed but it is not easy, even for honest and scrupulous lawyers, to come up with protections that don't prohibit. What will the next legislation look like? Will we understand its consequences? Will Wikipedia exist a decade from now? Legal theorists such as Yochai Benkler (2012) have used the opportunity of the SOPA/PIPA protest to explain larger legal, intellectual, social, and regulatory issues to the general public, with the warning that unless we all learn enough to participate as activists, we will lose the affordances of the Web that are key to our lives.

The crucial, intertwined implications of the SOPA/PIPA protest for the digital humanities are that you need to understand a good deal of theory to be able to appreciate what you can and cannot make online. Digital humanities leaves the realm of "the academic" to perform a public function in the world and thus have a responsibility to educate students about that public role. Hackers squeamish about yacking need to realize that there is a polemical, pedagogical point to understanding and explaining what the Web is and does in order that an informed citizenry can fight for it. Yackers squeamish about hacking—theorists who aren't interested in learning the mechanics of making digital humanities and doing some of their own coding—will never have quite the same understanding of how theoretical ideas become matters of urgent policy and even philosophy. Translating ideas into a publicly accessible site on the World Wide Web forces a confrontation with these issues.

Collaboration by Difference

There is a long-standing debate in the field about whether or not you can be counted as a digital humanist if you do not write your own code. Recently the debate ignited again when, at the 2011 Modern Language Association Meeting, Stephen Ramsay noted: "Personally, I think Digital Humanities is about building things.... If you are not making anything, you are not... a digital humanist" (Ramsay 2011; Alvarado 2012: 54).

I had my own firsthand experience with this position on the way to the National Science Foundation for what would be our second major national meeting of HASTAC ("haystack": Humanities, Arts, Science, and Technology Advanced Collaboratory), a virtual network dedicated to designing new media, critical thinking about the implications of technology in our life, and interdisciplinary forms of pedagogy that think about, with, and through new media. In 2002, David Theo Goldberg and I came up with the idea for HASTAC, and we were soon joined by a dozen or so other scholars in various disciplines who were dedicated to new forms of learning for a digital age. In 2004, we were invited to convene a distinguished group of humanists, scientists, and technology innovators at the headquarters of the National Science Foundation. En route, some of our members sent an email to the listserv indicating that they wanted to leave the meeting with an agreement that a requirement for participation in HASTAC was that you had to write your own code. Within minutes, David and I were receiving panicked messages from a range of the famous scientists and engineers attending the meeting. It should be noted that at least two or three of those world-renowned scientists would appear in any history of the invention of the Internet and the World Wide Web. They threatened to turn around and go home. They insisted that they hadn't written their own code in decades and had no intention of wasting their time with a bunch of humanists who thought their great contribution to the world was that they knew some HTML. The scientists were indignant, even furious that we were wasting their time. They had wanted a high-level meeting of distinguished humanists and social scientists who could help address the most pressing challenges to the Web. They were seeking a new kind of virtual organization that brought together the finest minds across the two-culture divide to envision new projects beyond the imagination of any one field.

David and I immediately responded to the listserv that there would be no such edicts about "writing your own code" for HASTAC. Programmers were welcome but we believed in something much harder than knowing HTML: we believed in the importance of respecting and valuing others' expertise, in learning how (against all the odds of the academy) to listen well across disciplinary differences, and to ensure that, from the beginning, all those with a stake in the final product were engaged in co-creation of a research or project agenda. From that early disagreement was born HASTAC's method of "collaboration by difference" (Davidson and Goldberg 2011; Davidson 2011a "Collaboration").

To my mind, making everyone a coder can be a prescription for mediocrity (conceptually and as code) in all but a very few notable cases. To be equally excellent as a humanities scholar and a programmer is possible, but given the bifurcated educational system, the odds are against it and, unfortunately, the results of the one-person-does-it-all model are not necessarily of the highest standard.

Some whimsical analogies might be useful here: I remember the first time I visited Lethbridge, Alberta, Canada, which prided itself on being in the Guinness Book of World Records for having the "longest highest bridge": there were longer ones and higher ones all over the place, but at that time there was no higher bridge in the world that was also longer (and no longer bridge that was also higher). To my eyes, the bridge didn't look very impressive. I'd seen significantly longer ones and higher ones, and didn't really even have a category for one that was both. At another time I dined at the restaurant of an amateur opera singer still trying to make it into the big time. The motto on the menu was "As a chef, I'm a great tenor." Digital humanities, to succeed, must aim higher. We should not be staking our claims in these ways ("as a humanist, I'm a great programmer").

A better goal for digital humanities is to value collaboration by the best minds working across all fields in order to create something that no one person (no matter how talented) could have created alone. That, after all, is the implicit premise of the World Wide Web. In terms of impact on higher education, certainly learning how to make complex interdisciplinary collaboration work is more transformative than simply teaching English majors HTML and CSS.

The coding humanist is also, in many ways, a reification of auteurism and a glorification of individual achievement. That Romantic Era paradigm of the solitary genius has been productively deconstructed by the collaborative impulse of the Web and indeed within the majority of digital humanities centers, projects, and classrooms. Collaboration by difference, the HASTAC method, is by no means accomplished easily or naturally. It requires its own form of training, community practice, and reward structure to teach anyone how to collaborate with others who possess entirely different skill sets, worldviews, educational levels, backgrounds, and points of view. Yet learning how to do this, how to collaborate across difference, is part of the open architecture of the Web. In his foundational essay "The Cathedral and the Bazaar," Eric S. Raymond (2000) makes the point that the Web is much more like the disorderly, constantly fluctuating, participatory, haggling, energetic bazaar, with its constantly changing architecture and participants, than it is like the cathedral, whose architecture is as fixed as the hierarchy of church personnel charged with its efficient operation, from the Cardinal down to the choir boys. Raymond's famous mixed metaphor for the advantages of the bazaar over the cathedral in the world of the Internet is that "With enough eyeballs, all bugs are shallow." By that he means that if the same programmers with the same skills look at the same code over and over, they will not find the bugs in the code. You need an open, bazaar-like architecture that promotes use by the widest possible group in order to find the unexpected flaw and, conversely, to make the breakthrough that brings you to the next level. This is almost opposite of the idea of hyper-specialization that has characterized the university since the nineteenth century's creation of the corporatized, research university, or what Walter Mignolo (2003) has called the post–Kantian/Humboldtian model of the university or, more simply, the "corporate university" (Reading 1997).

Most research on collaboration focuses on close collaboration, participation by those whose fields overlap (Gutmann 2012; Rhoten and Caruso 2004; Wellman 2001; Wellman et al. 2003). In scientific research it has been shown by Cummins and Kiesler (2003), among others, that collaboration can sometimes result in radical delegation rather than a meeting of the minds, especially when distance is added to the multidisciplinary mix.[1] Unfortunately, in many cases, rather than real collaboration across the two cultures divide, a researcher (in the sciences as well as in the humanities) might "outsource" the coding, data mining, and visualization of results to a programmer or an expert in data curation and visualization. The result is that representation is treated as if it is a given; it is not scrutinized and deconstructed for such things as ideological bias, tautological thinking, or a mechanistic replication of the status quo as opposed to an imaginative, critical redeployment of new research data to a new intellectual end. These scenarios can cause premises in the epistemology of knowledge to go unquestioned, thus avoiding the deep analysis that contemporary humanists often consider to be their hallmark, especially in regards to assumptions of race, gender, or other factors embedded in the structure of programming itself. "Critical code studies" is a new field that addresses this issue (Marino 2006; HASTAC Forum 2011). As Tara McPherson (2011) argues, racial assumptions are deeply embedded in the ones and zeroes that encode technology, a subject that is not analyzed if the practices and theory of programming are simply delegated rather than incorporated as a crucial part of interdisciplinary digital humanities collaboration.

With collaboration by difference, the computational *insights* of a project are integrated with the computational *tasks* because all parties (including programmers as well as end-users) are peers, participating in the design and implementation of the research agenda throughout the project. This form of radically diverse interdisciplinarity is key to the procedural or operational design-based thinking required by digital humanities. The issue, to my mind, is not whether or not one can program one's own research project but whether a project team has fully and respectfully come to terms with the intellectual contribution that its programmers, designers, researchers, media artists, and all other participants make to the final design of a collaboration.

I began this chapter with two hypothetical exam questions—one that requires making (hacking) and one that requires theoretical writing (yacking)—to underscore the interdependence of content and form. Hacking and yacking operate in complex, diverse interaction in a fully realized humanities of the digital age. The hypothetical exam questions highlight the interdependence of areas of expertise in collaborative work. Because our universities and even our funding structures are rarely designed to encourage collaboration across the two cultures, thoughtfulness and mutual respect have to be built into the projects from the beginning or collaboration by difference fails. Not every digital humanist needs to do his or her own programming, but writing the code and building the website should be as much part of the intellectual conversations engaged in as the theories and specialized forms of knowledge that other researchers contribute.

Rather than a one-size-fits-all model of digital humanities—where everyone does a bit of everything—I'm advocating something more profoundly disruptive: that digital humanists lead the way in modeling for the academy the benefits of cross-disciplinary participation in all aspects of content, expertise, skills, problem-solving, assessment, and representation.

Collaboration by Difference in Action: An Example

To illustrate how collaboration by difference—a model of interdisciplinary collaboration across radically different specializations and interests—works for a collective good that is greater than any of its component parts, allow me to turn to a major HASTAC initiative currently being pursued with support, once again, from the National Science Foundation, a project entitled "Assessing the Impact of Technology-Aided Participation and Mentoring on Transformative Interdisciplinary Research: A Data-Based Study of the Incentives and Success of an Exemplar Academic Network" (Davidson 2012). This collaboration requires digital humanities experts to gather together with a team of data experts, social scientists, and management experts to mine HASTAC's own social networking data with a goal of understanding the benefits of collaboration to the development of future careers. We will be using computational analysis, data extraction, and social network analysis, combined with theoretical understanding of biases implicit in several existing data-base studies, to undertake one of the first large-scale organizational studies of the interplay of cyberinfrastructure and scholarly communication in an academic peer-produced network. The study requires highly technical expertise in data analysis and highly technical expertise in "humanities studies." By "humanities studies," I mean meta-humanities, a parallel field to "science studies" that situates the humanities in broad social, philosophical, cultural, and historical contexts. Like science studies, humanities studies interrogate the relationship between the humanities and society, as well as looking at the implicit purposes that under-gird humanistic claims and that drive national humanities and arts public policy.

Using HASTAC as our exemplar virtual institution, we will be studying what over 10,000 network members have contributed to innovative, technology-aided forms of collaborative research and teaching. Our study mobilizes six years of the extensive HASTAC social networking activity, including the 176MB MySQL database of individual and institutional member profiles. These profiles are made up of extensive, accessible, anonymized "clean" individual and institutional data in a Drupal open-source database. Acquia, developer of Drupal Commons, has designated the HASTAC network the most complex and trafficked Drupal Commons-driven open source network on the World Wide Web. It is an exemplary digital humanities project since it is the embodiment, communication, and open-source delivery system for a worldwide networked community dedicated to "learning the future together." It is a discussion platform, an information commons, and a site that holds groupware for online courses, forums, and other events.

We have reason to believe it is one of the most interdisciplinary academic networks on the Web. But it will require sophisticated data analysis, data modeling, and network visualization to verify if this is the case. It will also require careful theoretical and historical insight to inform the analysis of disciplinary structures and what impact a virtual network may have had on interdisciplinary formations. If any single field or person attempted a study of this magnitude, across so many domains, it would necessarily miss its mark. To my mind, this is an exemplary digital humanities project with intellectual and theoretical implications for all of higher education. It requires deep statistical and mathematical or quantitative skills for its realization—and the contribution of many specialized scholars in domains that do not often converse together.

Our goal is also to make informed policy recommendations as a result of these data investigations and explorations. We want to pioneer methods for using heterogeneous network data to help understand current and future capacities of institutions to support the interdisciplinary collaborations essential to meeting what the NSF designates as the world's Grand Challenges (e.g., energy, global warming, infectious diseases, water sustainability, cyber-security, human sciences, and cultural policy design). The challenges looming in the future are so enormous that they require collaboration among scientists working with humanists and social scientists. More to the point, as we have seen in the resistance to much contemporary science (from studies of global warming to the use of stem cells), "data" are not enough. To be adopted, research has to be presented in a way that does not violate societal and cultural factors. In other words, for scientific findings to be effective, insights and expertise about human context and conditions are essential. Scientists need humanists in order to be effective. It is an indispensable partnership, and we are hoping to learn more about how to make that partnership work. Collaboration by difference is the methodology at the heart of this endeavor.

An expansive humanities that fully comprehends the digital—in theory, practice, and method: hacking and yacking—can provide leadership in complex collaboration that is necessary in a globally connected world facing enormous challenges. There are not many places in the contemporary academy that have the reach, focus, span, and ambitions, especially in dealing with heterogeneous data and methods, of digital humanities. Or, to quote another HASTAC motto, this one coined by Fiona Barnett, Director of the student-run HASTAC Scholars program: "Difference isn't our deficit; it's our operating system." This should also be a motto embraced by digital humanities. To be an exemplar of interdisciplinary problem-solving in action—fully grounded in historical and theoretical ideas—is the goal to which digital humanities should aspire. It is a far more expansive ambition than teaching humanists how to code.

Digital Humanities and Digital Literacy

And yet . . . I have sympathy for those who insist that digital humanists write code. However, I believe that they make the case too narrowly: I would argue that everyone, kindergarten to graduate school, should be learning the basics of HTML and CSS.

I consider Web-making to be a fourth literacy, as essential to our era as reading, writing, and arithmetic. Like those skills, you can learn Web-making in a playful form early on and then build on it for a lifetime. For those not destined to be programmers, the basics of Web-making give us functional insights into the workings of a digital system that now is crucial to every aspect of our lives and our work. It is imperative that every digital humanist be enough aware of the basics of HTML and CSS to understand how Web-making works, what it requires, and how the architecture of the Web itself affords or does not afford certain modes of presentation and representation.

We would be better informed, prepared, and more creative as a society if everyone, beginning in kindergarten with delightful open source programs like Scratch, learned to code as a fourth literacy (Davidson and Surman 2012). As Douglas Rushkoff (2011) says, we can either program—or be programmed. In a world of apps and proprietary sites, we can either fight for our right to contribute to the Web, or we can lose those possibilities that were built into the architecture of the Web but that, unless we protect them, will be closed down (Zittrain 2008) The SOPA/PIPA protest worked in 2012. It is not clear that it will work the next time.

Digital humanities should be championing digital literacy. By that I mean not just being able to download apps or write a comment on someone's blog, but a true understanding of what the Web is as well as strategic, deliberate, creative incorporation of Web-making into one's personal and especially into one's intellectual and educational life. It may seem implausible that anyone below the age of 30 doesn't know the basics of social media practices, and that may well be true. But when we look at the *scholarly* use of digital media, we find some disheartening data. A 2012 study, "Researchers of Tomorrow," undertaken jointly by the British Library and the UK's Joint Information Systems Committee (JISC), indicates that most doctoral students in England born between 1982 and 1994, classic "Generation Y," tend to use digital applications "only if these can be easily absorbed into existing research work practices. Current institutional engagement with open Web and Web 2.0 technologies does not convince the majority of Generation Y doctoral students of the credibility of using such applications in a research setting, and reinforces their feeling that actively using, for example, social media and online forums in research lacks legitimacy. New Web-based and other tools and applications may also challenge their traditional and conservative research working practices." The study found that over 70 percent had never either maintained or contributed to a Wiki and 58 percent had never posted a comment online. It is hard to imagine that this *new* generation of intellectuals can be relevant to the students of today given such lack of engagement as users (Education for Change 2012).

We in the digital humanities have our work cut out for us. What I am advocating is not only that digital humanities are interdisciplinary and connected by their design but that there are at least two different public roles that the digital humanities can and must play. First, is what might be called the "public intellectual" or translational role of digital humanities. Because the scope of digital humanities projects can be wide and because collaboration requires cross-disciplinary respect, digital humanities are in an unusual position of being able to translate a range of scholarly research in many specialized fields to a general public. This in itself contributes to the general public good and helps in advocating to a general public the importance of the human, social, natural, and computational sciences. The second role pertains more to the importance of connection within the academy itself. Digital humanists can take a leadership role in championing new modes of technology invention, adaptation, and dissemination within and across disciplines, in research and in the classroom. Digital humanists have an exemplary institutional role to play in championing digital literacy for the general public and in advocating for new forms, content, and scholarly application of technology within higher education.

Not only do digital humanists have the skillset and the intellectual commitment to help us decide the wisest and most efficient ways to reform education for the twenty-first century, digital humanists, I would argue, have the *responsibility* to do so. We have had to deal with the key issues in our organizations, in our modes of publishing, in our systems of credentialing across boundaries (multimedia, theoretical, humanistic, aesthetic, computational, digital). The rest of the academy is just coming to terms with issues that we have been promoting for well over a decade.

That's the challenge. That's the opportunity. Digital humanists *can* help lead us to the digital literacy and educational reform our society needs now and demands. If we do not, we are not taking on our work as educators as responsibly as we should, nor are we seizing this important, yet vulnerable, moment in the history of higher education.

Note

1. Cummings and Weisler have found that geography is a more significant factor than disciplinarity in managing multidisiciplinary projects with multiple Principle Investigators. They note: "Projects with principal investigators in more disciplines did not appear to suffer more coordination losses and reported as many positive outcomes as did projects involving fewer disciplines. By contrast, geographic dispersion, rather than multidisciplinarity, was most problematic. Dispersed projects, with principal investigators from more universities, were significantly less well coordinated and reported fewer positive outcomes than collocated projects. Coordination mechanisms that brought researchers together physically somewhat reduced the negative impact of dispersion. We discuss several implications for theory, practice, and policy."

11 TOWARD PROBLEM-BASED MODELING IN THE DIGITAL HUMANITIES

Ray Siemens and Jentery Sayers

Most understandings of digital humanities include acknowledging that there has been, is, and will continue to be something important happening at the intersection of arts and humanities inquiry (into the nature of human experience, through its artifactual remains) and computational methods. Such understandings typically contend that this intersection is necessarily a point of interdisciplinarity, requiring both consideration beyond what is provided within most extant institutional or disciplinary structures and an embrace of all pertinent perspectives of this interdisciplinarity that includes an acknowledgment of what brings value to each group involved. Further such understandings typically recognize that, while the recent professional media attention that has surrounded digital humanities is valuable from a number of important perspectives, attention is potentially unhelpful if it detracts from the recognition that digital humanities has a rich, significant history that needs to be considered and contextualized. In the following pages, we account for both this media buzz and history as we gesture toward what we call "problem-based models" for digital humanities work. The motivation for these models is a need to build upon existing traditions in content and process modeling, and to ask how digital humanities—through a commitment to public humanities—might treat computational tractability as its very object of inquiry.

A View of Computing in the Humanities

As a research area, computing in the humanities is best defined loosely, as the intersection of computational methods and humanities scholarship. For some time now, it has been considered a quickly growing field, its importance being seen in its increased integration into undergraduate and graduate curricula, in its increased representation in the work of those who carry out research across the humanities today, and in the recognition by researchers and research supporters at local, national, and international levels. One foundation for current and past work in this area is our growing understanding of another loosely defined area, knowledge representation, which draws on the field of artificial intelligence and seeks to produce models of human understanding that are tractable to computation. While fundamentally based on digital algorithms, knowledge representation privileges traditionally held values associated with the liberal arts and humanities: general intelligence about human pursuits and the human social/societal environment; adaptable, creative, analytical thinking; critical reasoning, argument, and logic; and the employment and conveyance of these, in and through human communicative processes.[1]

More specifically, in the activities of computing humanists, knowledge representation manifests in issues related to archival representation and textual editing, high-level interpretive theory and criticism, and protocols of knowledge transfer—all as modeled through computational techniques. The results of modeling activities of humanists and the output of humanistic achievement with the assistance of computers are found in what are often considered the exemplary tasks associated with the core of computing in the humanities: the representation of archival materials, analysis or critical inquiry originating in those materials, and communicating the results of these tasks. Archival representation involves the use of computer-assisted means to describe and express print-, visual-, and audio-based material in tagged and searchable electronic form. Associated with critical methodologies that govern our representation of original artifacts, archival representation is chiefly bibliographical in nature, often involving the reproduction of primary materials, for example, in the preparation of an electronic edition or digital facsimile.

Among other things, critical inquiry involves the application of algorithmically facilitated search, retrieval, and critical processes that, originating in humanities-based work, have been demonstrated to have far-reaching applications. Related to critical theory, such inquiry is typified by interpretive studies that assist in our intellectual and aesthetic understanding of humanistic works, and it involves the application (and applicability) of critical and interpretive tools and the analytic algorithms they incorporate on those artifacts produced through processes associated with archival representation made available via resources associated with processes of publishing and the communication of results. The communication of results involves the electronic dissemination and electronically facilitated interaction about the products of such representation and analysis as outlined above, as well as the digitization of materials previously stored in other archival forms. It frequently takes place via codified professional interaction and is traditionally held to include all contributions to a discipline-centered body of knowledge; that is, all activities that are captured in the scholarly record associated with the shared pursuits of a particular field.

Understanding Digital Humanities Today, in a Time of Growth (and Considerable Attention)

The foundation we described that has typified humanities computing work just under a decade or so ago, in what we more readily call "digital humanities" now, is a key starting point for understanding movements in digital humanities today. Indeed many have moved to consider that most all computing across all humanities—from the point of code-level programming and algorithmically analytic computing to media-based social application use—takes place under what has been termed the "big tent" of digital humanities (Fitzpatrick 2010), even if they ultimately reject some of this work as nonacademic or noncomputational enough to meet their understanding of the point at which computational methods and humanities scholarship converge. One can see such tension, for example, when comparing the thematic billing of the "Big Tent Digital Humanities" at the 2011 international conference of the Alliance of Digital Humanities Organizations (ADHO) at Stanford (https://dh2011.stanford.edu/) with some of the debates and discussions that took place largely among a very similarly constituted group of computationally engaged humanists at the Modern Language Association in Los Angeles the same year.[2]

Abundant exemplary evidence suggests that, regardless of the specific point of debate, the movement is a positive one, and it is increasingly open, whether configured as a big tent or less controversially as a widely interdisciplinary field. The trajectories of digital humanities carry us well beyond humanities computing's word-counting, enumerative roots as documentarily parodied by David Lodge in his novel *Small World* (1995) and well into broader concerns including such diverse practices as data mining, word clouds, GIS, film studies, consideration of media cycles, applications pertinent to non-textual musicology, and beyond. Digital humanities bridges between past and present through research into reading technologies that span scroll, manuscript, print, and electronic media; and with remediating extant materials and no longer extant physical spaces, as well as creating new virtual worlds and networks for engagement of academic and non-academic kinds alike. Overall, the movement is toward embracing a larger scope of research while privileging difference, with a focus on issues of open public communications and engagement, not to mention founding collaborative networks and building methods-centered communities; and as part of all this, toward organizing professionally at various levels to achieve common goals, the greatest example of this being the international umbrella organization for digital humanities: ADHO.[3]

There is no succinct way to scan the current state of the digital humanities but, toward suggesting a trajectory of today's digital humanities and the vast array of those defining the field as it now stands, we offer the following sample of pertinent examples, notable instantiations, and heralded practices. What used to be termed *humanities computing* boasts deep enumerative roots and, while Rommel (2004: 90) might engage Lodge's implicit critique of these activities, considerable and profitable examples of such work and facilitating tools can be found in Ian Lancashire, John Bradley, Willard McCarty, Michael Stairs, and T. R. Wooldridge's *Using TACT with Electronic Texts: A Guide to Text-Analysis Computing Tools* (1996), a suite of powerful interrelated programs for literary textual analysis, distributed on open access/shareware principles since 1989, as well as other exemplary enterprises, including: R. J. C Watt's *Concordance*, a comprehensive application with a number of powerful features, such as multiple language support, user-definable alphabets and contexts, multiple-pane viewing, multi-format export options, and the ability to statistically analyze selected texts; *AntConc* (Anthony 2006), a free concordance program that performs a variety of linguistic analyses; and *MonoConc Pro* (Svedkauskaite 2004), a concordance program offering full-text and tag-based search and comparison capability. Beyond concordances, digital humanities practitioners have also developed or engaged in initiatives that are primarily based on data mining and visualization. Among many pertinent instances of textual analysis projects, one might consider *TAPoRWare* (tapor.ca; Rockwell et al. 2005), a collection of online and desktop tools designed to assist users in performing computational textual analysis on XML, HTML, and plain text files in order to develop concordances, tokenize texts, analyze collocates, and extract metadata. Further *Voyant* is a suite of Web-based textual analysis tools able to analyze and visualize text via document reader, term frequencies, collocation, word cloud, and scatterplot (voyeurtools.org).

Increases in digital humanities's growth has also brought an increased modality, including that reflected in a wide body of work revolving around notions of place and mapping like *ArcGIS* (Guldi 2010), a platform for building GIS systems for users to create, edit, publish, and analyze geographical information in the form of maps and models, and *HyperCities* (Presner 2010), a digital research and educational platform for exploring, studying, and interacting with the layered history of cities and global spaces using Google Maps and Google Earth. Multimedia-focused endeavors include *Mondrian vs. Rothko: Footprints of Evolution in Style Space* (Manovich 2012), a prototypical instance of image plot visualizations comparison that compares paintings of Piet Mondrian and Mark Rothko, and *The Rossetti Archive* (McGann 2013), which facilitates the study of nineteenth-century painter, designer, writer, and translator Dante Gabriel Rossetti via electronic access to all of Rossetti's pictorial and textual works, as well as a large variety of contextual materials encoded to high standards and (the majority) accompanied by high-quality digital images. Still other initiatives embrace sound, as in the *International Music Information Retrieval Systems Evaluation Laboratory* (music-ir.org; Downie 2008), which is invested in establishing resources for the development and evaluation of Music Information Retrieval (MIR), Music Digital

Libraries (MDL) techniques and technologies, and the creation of secure and accessible large-scale digital collections of music materials in a variety of forms; IMIRSEL includes projects such as the *Music Information Retrieval eXchange* (MIREX), the *Networked Environment for Music Analysis* (NEMA), and *Structural Analysis of Large Amounts of Music Information* (SALAMI). Moreover, the *Electronic Literature Collection* (Hayles et al. 2006)—as a periodical publication of current and older forms of electronic literature—expresses the creative past, present, and future of literary studies. Research projects predicated on the use of computing technologies in order to highlight previously unremarked on phenomena proliferate, such as *Implementing New Knowledge Environments* (INKE; Vandendorpe 2009; Siemens et al. 2009), an interdisciplinary initiative seeking to understand the future of reading through reading's past and to explore the future of the book through the history of the book.

Digital humanities practitioners are also at the forefront of the movement to open up the academy to public concerns, participation, and critical interventions. Various digital humanities initiatives straddle the real or perceived divide between academic and non-academic communities. Digital humanities actively works to blur this line, and these attempts are evinced in initiatives like the *Simulated Environment for Theater* (humviz.org; Roberts-Smith et al. 2012), a 3D environment for reading, exploring, and directing plays. Moreover this movement is visible in projects like *A Social Edition of the Devonshire MS, BL Add 17,492* (Siemens et al. 2013), an edition that uses the Devonshire MS—a courtly miscellany of ca. 1530s, notable because it presents the earliest sustained instance in the English tradition of a literary writing community comprising both men and women—to explore the use of existing social media tools such as Wikibooks, Twitter, blogs, and social media spaces toward its creation and sustenance. Urban exploration projects like Lisa Snyder's real-time visual simulation, *The World's Columbian Exposition of 1893*, further engage with public spaces and players to affect tangible social knowledge creation. Public communication extends through, among many others, initiatives like the GIS-based project *Mapping the Du Bois Philadelphia Negro* (Hillier et al. 2004), which uses archival information to recreate W. E. B. Du Bois' 1899 survey, *The Philadelphia Negro*, to tell the story of Philadelphia's Seventh Ward using interactive maps integrating extensive local, social outreach. Furthermore the *Orlando Project* (artsrn.ualberta.ca/orlando, Brown et al. 2009), a large, highly developed electronic textbase for research on and discovery of women's writing in the British Isles, seeks to produce a full scholarly history of women's writing in the British Isles by integrating biographical entries, bibliographic listings, contextual historical material, and more. The *Brown University Women Writers Project* (Flanders 2005) is a long-term project devoted to making early modern women's writing accessible to a wide audience of teachers, students, scholars, and the general public, and includes a variety of educational outreach efforts in the area of electronic textual encoding and preservation. The

Public Knowledge Project is dedicated to exploring how new technologies can be used to improve the professional and public value of scholarly research, investigating—across a partnership of multiple institutions and a variety of faculty members, librarians, and graduate students—the social, economic, and technical issues involved in online infrastructure and knowledge management strategies (see Lorimer et al. 2011). Finally, *MediaCommons: A Digital Scholarly Network* (Fitzpatrick 2006) is a community network for scholars, students, and practitioners in the field of media studies, promoting the exploration of new forms of publishing, and supports the production of and access to a range of intellectual writing and collaborative media production. *4Humanities: Advocating for the Humanities*, too, is a platform and resource for advocacy of the humanities that draws on the technologies, new-media expertise, and ideas of the international digital humanities community—at the same time highlighting the role digital methods have in effectively demonstrating to the public why the humanities need to be part of any vision of our societal future.

Networks and communities of support have grown concurrently with the relatively recent boom of interest and participation in digital humanities. Groups like the *Humanities, Arts, Science, and Technology Advanced Collaboratory* (HASTAC) have become established networks of individuals and institutions that promote and facilitate networked, collaborative research extending across traditional disciplinary, intellectual, social, geographical and institutional boundaries. HASTAC's Web space operates as an open forum where individuals can share, plan, and execute work, ideas, events, conversations, or collaborations. In a similar vein the online infrastructure project *Canadian Writing Research Collaboratory / Le Collaboratoire scientifique des écrits du Canada* (CWRC/CSÉC), is designed to enable the collaborative study of Canadian literature and literary history. CWRC/CSÉC incubates large-scale, cross-disciplinary collaborative engagements to connect scattered and siloed information; to investigate links between writers, texts, places, groups, policies, and events; to advance understanding of past and present cultural change; and to make new knowledge accessible to both Canadian and global audiences. Further initiatives reflect the potency of developing and sustaining professional networks, including centerNet's *DHCommons*, an online hub to facilitate the matching of digital humanities projects needs with scholars and individuals interested in project collaboration. The *Digital Humanities Summer Institute* (DHSI; Bialkowski et al. 2011), is a week-long digital humanities training and discussion institute consisting of intensive coursework, seminars, lectures, and colloquia that brings together faculty, staff, and students from the arts, humanities, library, and archives community, as well as independent scholars and those in public cultural institutions. The *Text Encoding Initiative* (TEI) is a consortium that collectively develops and maintains standards for the representation of texts in digital form (Clement 2011). *THATCamp* (The Humanities And Technology Camp) is a user-generated unconference for humanities scholars, technologists, librarians, museum staff, cultural professionals, and interested individuals, wherein a collaboratively produced agenda provides a blueprint for meetings and workshops.

An undercurrent of professional organization runs through the multiple initiatives and movements considered as digital humanities today. Among many positive instances, we wish to acknowledge the following examples of professional organizations to illuminate breadth: the *Museum Computer Network* (MCN), the recently founded (2011) *Japanese Association for Digital Humanities* (JADH), and the recently constituted *Red de Humanidades Digitales* (Digital Humanities Network; see also Galina and Priani 2011). ADHO represents the most visible and unified movement toward organization and across the field success, gathering as it does a number of national and international digital humanities related organizations whose goals are to promote and support digital research and teaching across arts and humanities disciplines. The association effectively draws together humanists engaged in digital and computer-assisted research, teaching, creation and dissemination, and this is reflected in a diverse membership that practices textual analysis, electronic publication, document encoding, textual studies and theory, new media studies and multimedia, digital librarianship, applied augmented reality, and interactive gaming. These varied activities are found in many academic departments, cultural institutions, public and private sectors, and countries in every hemisphere.

After the Digital Humanities Hype Subsides

The vastness and abundance of this field, even as suggested via exemplary evidence reflecting a very small percentage of its true size, does not in the least aid in the construction of a definition of digital humanities that properly captures its breadth and depth in full, accurate, and comprehensive ways that achieve universal agreement. One can find evidence of this, say, in the discussions at the 2011 MLA in Los Angeles mentioned earlier. Even so, universal agreement is not required to engage digital humanities in an actionable way, around which one might situate an individual or group program of research, a lab, a center, an institute, and/or an organization within an extant set of disciplinary and institutional structures. In that environment, for better or worse, one tends to arrive at discipline-specific definitions that have some gain with reference to extant disciplinarity but, at the same time, considerable loss in innumerable ways associated with decreased diversity and perspective. Other extra-disciplinary approaches can be viewed as unsubstantial or revolutionary—the latter of which has been seen across many attention-getting

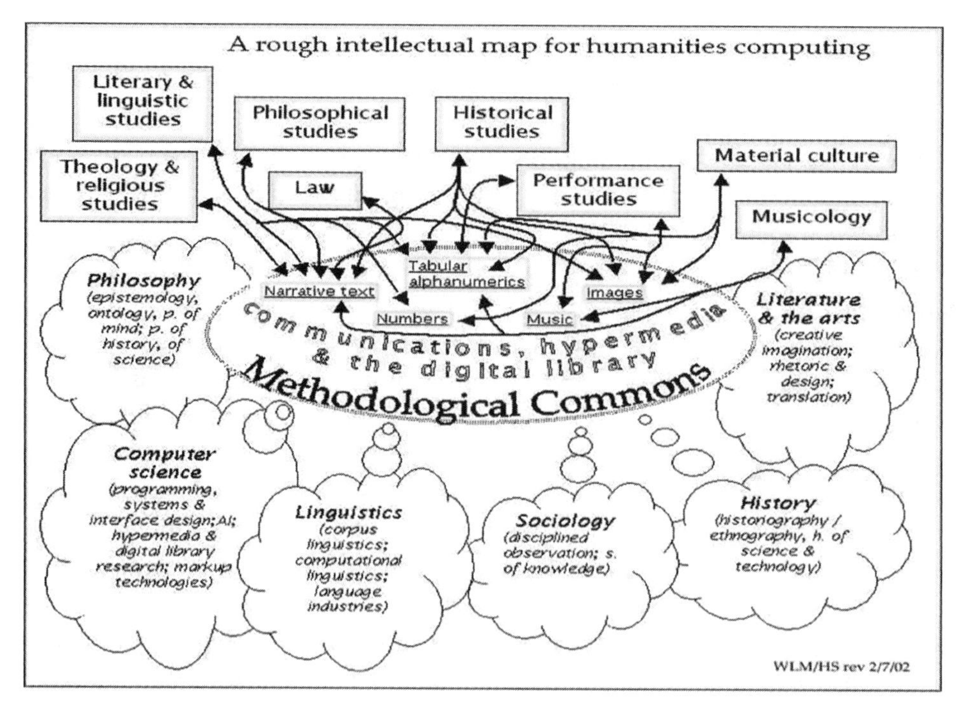

Figure 11.1
Mapping the field. From Willard McCarty and Harold Short (2002),Report of ALLC meeting held in Pisa, April 2002. http://www.allc.org/content/pubs/map.html.

and hype-oriented discussions of digital humanities in a number of prominent venues of late. Restated: we need enough of an articulation of digital humanities to engage in actionable ways and develop environments that foster this engagement, but if we embrace something with fixity, with the predictability of what has already happened, then we lock ourselves into a static vision. And if we embrace the rhetoric of current hype—one most often of revolution—at the same time as we may be freeing ourselves from the past we are limiting ourselves to manifesto-style thinking which, most often, is not easily implemented in a sustainable fashion even in the smallest and most manageable of personal or professional contexts. An optimal approach in this situation—one practiced by many of the leading digital humanities scholars, institutional bodies, and organizations at the moment—is to seek to understand digital humanities's past at the same time as engaging the nature of its current, sometimes-styled contemporary revolution, in this way drawing on disciplinary and interdisciplinary roots and structures that allow one to plan for a sustainable intellectual direction, after the rhetoric of revolution and hype subsides.

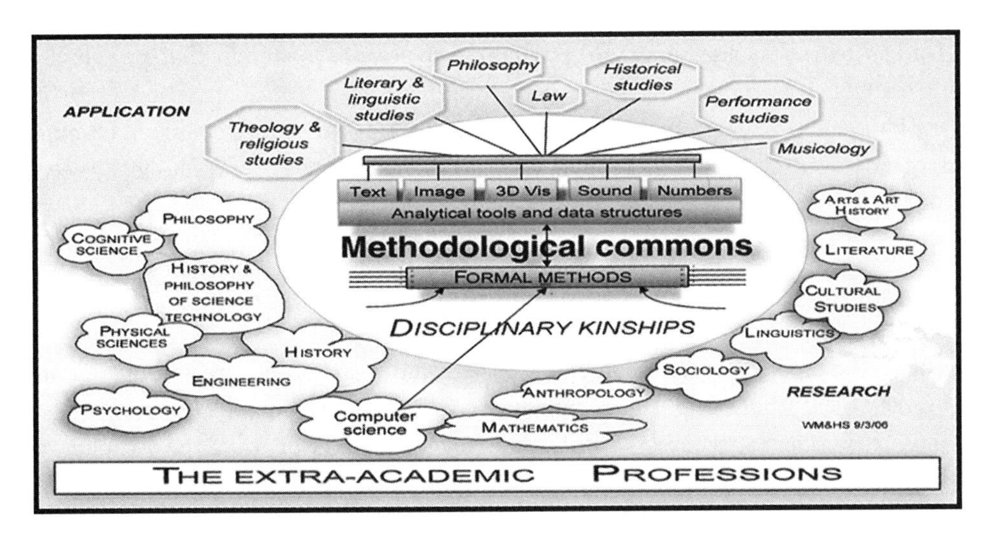

Figure 11.2
McCarty, rev. From *Humanist*.

From Content- and Process-Based Modeling, to Problem-Based Modeling

Without dropping the precepts of intersection and (inter)disciplinarity that are central to a broad-based understanding of digital humanities, we need to consider altering our institutional modes of thinking from a focus on *what and how it is we (and others next to us) do what we do* to understanding *where (the [inter]disciplinary and [extra]institutional location) it is that we do what/how we (and others next to us) do what we do*. The notion of the methodological commons, asserted by Willard McCarty and Harold Short some time ago, is very helpful in this context.[4] This commons is best imagined as a loosely modeled and iteratively evolving series of convergence points among disciplinary groups that support the ways in which they represent, analyze, and disseminate the knowledge that lies at their core.

Key to this model are emphases, initially, on *content modeling* (how we digitize and represent our data) and then, building on that foundation, *process modeling* (how we do what we do with that data; analytical process, with more inclusivity across media types and extra-academic partners than in initial attempts at content modeling). Current research in digital humanities, across the plethora of examples one can provide in support, suggests that within the methodological commons we might imagine a next step that builds on content modeling and process modeling, toward what might best be described as *problem-based modeling*. This approach to modeling is something: (1) reflecting greater scope in digital humanities's content- and process-modeling activities, (2) privileging application-oriented solutions that may span sectors (rather than solely disciplines), (3) extending into areas beyond those that are computationally tractable (e.g., inquiry involving the nature and influence of computational tractability); and ultimately (4) something moving the digital humanities centered commons into the environs of the larger societally based community that the humanities have traditionally been understood to serve.

Some Examples of Problem-Based Modeling

Enactments of the problem-based vision of digital humanities are indeed already apparent in existing projects and scholarship. The defining features include mapping the methodological commons onto pertinent structures, with attention to clouds of knowing and the disciplinary groups representing them; understanding what is important to the humanities and to a wide variety of audiences (academic, alt-academic, and nonacademic); reflecting that understanding in open, flexible structures, activities, and processes (including governance, assessment, and revisioning); and examining how problems emerge as well as how they are ignored or repressed (often through computation). These existing projects and scholarship may be best understood through five particular trajectories of current research: large-scale collaboration, linked data, nonempirical inquiry, computational culture studies, and physical computing and fabrication.

While these trajectories are by no means intended to exhaust possibilities for emerging work in digital humanities, they do highlight lines of inquiry that are no doubt gaining traction across institutions and sectors. More important, they build on, envelop, and complicate the content- and process-modeling legacies of the methodological commons rather than simply replacing them or responding to them in a reactionary fashion.

Large-Scale Collaboration

From *Implementing New Knowledge Environments* (INKE), the *Canadian Writing Research Collaboratory / Le Collaboratoire scientifique des écrits du Canada* (CWRC/CSÉC), and the *Humanities, Arts, Science, and Technology Advanced Collaboratory* (HASTAC) to projects such as *Hacking the Academy*, *Transcribe Bentham* (blogs.ucl.ac.uk/transcribe-bentham), and *MediaCommons*, large-scale collaborations typically engage a pressing, humanities-based issue (e.g., the future of the book) and mobilize methods like crowdsourcing, social networking, and interdisciplinary team research toward potential responses or possible interventions. Through projects like these (and to echo a now common gesture), digital humanities research troubles the "individual scholar" model of academic production.

Yet large-scale collaborations invested in problem-based modeling further critique the individual scholar framework in several ways. Perhaps most obviously, they experiment with the affordances of research and scholarship produced in multiple locations, across geographic distances, through networked technologies. As but one example, INKE teams consist of practitioners from institutions in different time zones and places. These teams rely heavily on innovative protocols and Web-based applications for collaborative research in order to gather up-to-the-minute research, share it, incorporate it into research workflows, and proceed accordingly. Since INKE practitioners (both academic and alt-academic) are affiliated with departments in the arts, humanities, and sciences, the content, style, structure, and delivery of this research varies tremendously, particularly when sources range across print journals, monographs, interviews, Twitter, blogs, archives, and online repositories.

In this sense "large scale" is about the scope and diversity of resources as well as the size of a team and its geographical and temporal range. How to address a pressing issue in the humanities (e.g., how should the networked scholarly edition behave?) becomes a matter of not only consulting a wide variety of practitioners from an array of disciplines but also negotiating the disparate conventions of their scholarly communications and—most crucially—articulating new approaches to scholarly communications that persuasively correspond with large-scale, problem-driven, collaborative research intended to be shared and updated often across numerous domains.

Likely for this reason, how scholarship looks, reads, sounds, and circulates is a core question for a problem-based digital humanities. Consequently venues such as MediaCommons are rethinking scholarly expression through networked, multimodal, and iterative means (McPherson 2008; Fitzpatrick 2011; Kraus 2012). How audiences attend to these venues—how scholarly communications are accessed, engaged, interpreted, repurposed, and archived—by necessity becomes a concomitant area of inquiry. Or put differently: in a problem-based digital humanities, networked media are not only the means through which practitioners articulate and apply their research. The media also become the subject of research (Kirschenbaum 2008; Manovich 2011). For instance, in 2012 HASTAC received US National Science Foundation (NSF) support to study its own social network and the interplay of cyberinfrastructure with scholarly communications (Duke 2012; Davidson, chapter 10 in this volume). This recent NSF support of a digital humanities project demonstrates how we are now only beginning to understand and analyze the effects and implications of large-scale collaboration in the humanities, especially when we understand digital scholarly communication as a process anchored in cultural and computational tractability. Ideally, such network analysis will give us a sense of how particular problems and pressure points emerge in the field, not to mention where those problems and points are taking us. Indeed, a collaborative, problem-driven digital humanities is also predictive in character. Frequently through large data sets, it works to anticipate the development of cultural and social networks like HASTAC, without assuming that their emergence is unicausal or that their future is determined.

Linked Data

Similar to large-scale collaborations, linked data projects are invested in the potential of a "big humanities." As scholarly Web-based content continues to proliferate, a key question—or problem—is how to structure the data related to it (e.g., through the Resource Description Framework). How should online repositories, scholarly editions, and exhibits speak to each other? Through what ontologies? How can data in them be slurped and interpreted in new contexts? What happens when scholars account for not only the organization of their own projects but also how that organization is interoperable with other initiatives and data? Projects like *Digging into Data, Linked Jazz*, the *Indiana Philosophy Ontology Project*, and the *Joint Information Systems Committee's Step Change* give digital humanities practitioners a sense of how these questions are shaping research (diggingintodata.org). All of the projects seek to define connections across subjects, concepts, and objects through specific disciplinary perspectives and controlled vocabularies. And they give practitioners a tangible sense of how network analysis (both historical and contemporary) may inform inquiry of all kinds. Additionally, Linked Jazz visualizes the networks structured through its ontologies (e.g., Friend of a Friend), asking audiences to discover and navigate the history of jazz in a fashion impossible without process-oriented computational modeling. As with the case of *Linked Jazz*, large-scale analysis and linked data frequently rely on crowdsourcing or surveying a significant number of domain experts in order to leverage collective knowledge toward problems that only appear at scale (Liu 2012a). Following Franco Moretti's work in distant reading (2005), we find these problems to be prosopographical, geographical, economical, or cultural in character, yet the processing and labor they require exceeds the time and capacity of an individual scholar's lifetime.

Nonetheless, as Cristina Pattuelli notes in "FOAF in the Archive: Linking Networks of Information with Networks of People: Final Report to OCLC" (2012), it is important to remember that linked data projects are simultaneously human and machine driven. They are also problem driven because they ask how gathering information, structuring it, expressing it, and opening it to reinterpretation and reuse can be understood across disparate instances. If this issue were only a matter of computation, then it would most certainly be resolved. But, in practice, it is largely an issue of how to motivate scholars to share their data in similar and agreed-upon ways. As such, platform and application development in digital humanities has focused on how to spark such motivation through authoring and composing environments, including environments intended for collaborative work, curation, gaming, and graphical expression. For example, how might the act of composing with metadata in mind gain traction in the humanities at large, beyond communities invested in markup and preservation (Sayers et al. 2012)?

In response to questions such as these, both the Omeka and Scalar platforms encourage authors to add Dublin Core metadata as well as item relations to a given project. In the case of Scalar, these relations are then visualized using the D3 JavaScript library. This integration of metadata considerations into scholarly workflows is ultimately an attempt at modeling the often invisible or overlooked procedures of data structuring: determining how to link data through the everyday routines of writing and composition rather than applying it to works post-production. To be sure, this modeling involves a tremendous amount of exciting guess-work, trial-and-error included. Determining how practitioners as well as algorithms turn this into that—and then facilitating such processes across sectors and large groups of people—is no small task.

Nonempirical Inquiry

There is most certainly an element of nonempirical conjecture in problem-based modeling. Digital humanities practitioners are increasingly turning toward work that anticipates how people, data, and intelligent objects behave. And during at least the last few years of digital humanities research, words such as "play," "tinkering," and "screwing around" have gained their own sort of popularity (Balsamo 2009; Ramsay 2010; Sayers 2011). These terms as well as the methods related to them can be traced through the legacies of speculative computing, gaming, and OuLiPo-like experimentation in the humanities (Samuels and McGann 1999; Drucker and Nowviskie 2004; Bogost 2006; Sample 2012). Despite the differences among these traditions, they share a common question, posed by Johanna Drucker and Bethany Nowviskie in "Speculative Computing: Aesthetic Provocations in Humanities Computing": "can the logic-based procedures of computational method be used to produce an aesthetic provocation?" (para. 44). Related to problem-based modeling, Drucker and Nowviskie continue with a compelling statement: "Speculative approaches make it possible for subjective interpretation to have a role in shaping the *processes*, not just the *structures*, of digital humanities" (para. 45).

Through an emphasis on processes, not to mention provocative engagements with pressing problems, such speculative approaches are now being integrated into digital humanities tools, platforms, and applications. Consider the playful character of *Voyant Bubbles*, which not only reads words in a document and displays those with the highest frequency (through proportionally large bubbles) but also adds a layer of sound to the text processing. As words are read by the computer, audiences hear computer vision at work: "Bubbles . . . emit[s] an eerie theremin-like sound as it calculates word frequency in the text" (Goddard 2012).

Elsewhere, the geotemporal exhibit-builder, *Neatline*, underscores the importance of "handcrafted, interactive stories" to digital humanities research (Neatline 2012). Rather than treating history homogeneously through objective and disembodied representations of time and space, Neatline allows practitioners to add subjective elements to the interpretative process. In so doing, the relevance of narrative, perspective, ambiguity, and speculation—as provocative problems—are foregrounded, destabilizing common assumptions about the truth claims typically made by abstract representations such as chronological timelines and geographic maps. However, the aim of such speculative ventures is not, in a reactionary fashion, to critique computational approaches for their ostensibly objective, empirical, or procedural tendencies. It is to conduct work with computers on a spectrum of objectivity and subjectivity, empirical and nonempirical, procedure and play. Problem-based modeling thrives on these ambivalences. It recognizes the contradictions inherent in digital humanities inquiry, embraces them, and works through (rather than outside) them.

Computational Culture Studies

In fact, ambivalence is foundational to ongoing work in computational culture studies, which—as the journal of the same name suggests—takes computational tractability as its focus, frequently with an emphasis on software and interface design. Recent publications in the field reconfigure traditions in critical theory and cultural criticism to ask, for example, how new media at once endure and appear ephemeral (Chun 2011), how instances of software may be studied as physical objects (Fuller 2008), and how a computer program's technical particulars shape the cultural imagination (Manovich 2011). This research informs problem-based modeling by complicating the historical divide between knowing and doing (or theory and practice). It also understands technologies and media as tangible processes beyond absolute understanding but still subject to repurposing, forking, and other such modifications. Comparable to the ambivalences of nonempirical conjecture, computational culture studies operates on a spectrum of transparency and opaqueness. Software—such as markup and text analysis tools, exhibit builders, and media suites—now familiar to digital humanities practitioners demands specific competencies and become more familiar with practice over time. Yet, resonating with arguments made by scholars of thing theory and object-oriented ontology (Brown 2001; Bogost 2012a), these instances of software can never be completely controlled or known. There is always a limit to the expertise of the practitioner, and the routine glitches, surprises, hacks, and bugs of software are reminders of that limit.

That said, if digital humanities is increasingly invested in problem-based modeling, then it must also account for how processes and politics change across sectors and conditions: how, in short, computation is tied to culture, ideology and social justice issues. Computational culture studies extends this critique into the hands-on study and production of digital artifacts (a practice historically outside the purview of cultural criticism), with multimodal journals such as *Vectors* acting as tutor cases for how fields like American Studies may intersect with digital humanities. See, for example, Sharon Daniel's digital collaborations with Erik Loyer (2007, 2012). They blend community-based research (e.g., interviews with women in California's state prison system) with a multimodal expression that is highly computational in character. Or consider David Theo Goldberg and Richard Marciano's work with Chien-Yi Hou (2012). Their "Testbed for the Redlining Archives of California's Exclusionary Spaces" project blends a commercial platform (i.e., Google maps) with a geospatial archive of residential security maps, which once ranked urban zones using ethnoracial criteria. Additionally groups such as #transformdh (2012) are enacting modes of cultural criticism through digital humanities methods in order to critique the "privileging of certain gendered, racialized, classed, able-bodied, Western-centric productions of knowledge" (Lothian and Phillips 2013). Many projects that fall under the #transformdh umbrella stress technoliteracy (Ross 1990), whereby groups historically marginalized, disenfranchised, or ignored by scientific and technological development are engaging those processes publicly and directly, often through making, coding, composing, play, and other forms of transformation. Here, examples include the *Fembot Collective*, *Women Who Rock*, the *Crunk Feminist Collective*, and HASTAC forums like "Queer and Feminist New Media Spaces" (2010). These examples correspond with a problems-driven digital humanities because they mobilize theories of social constructivism (Haraway 1985, Stone 1995) toward the material processes of computation and networks, blurring the boundaries between form and content, practitioner and artifact. Put this way, a problem in digital humanities is irreducible to either technology or culture.

Physical Computing and Fabrication

Contemporary practitioners of humanities physical computing and fabrication, such as Kari Kraus, Anne Balsamo, William J. Turkel, Matt Ratto, and Bethany Nowviskie, also understand how the very notions of technology and culture—of materiality and embodied expression, for example—are changing across digital humanities landscapes. Central to experimental work in circuit-building (e.g., Arduino), 3D modeling and printing (e.g., MakerBot Replicators), intelligent exhibit production (e.g., using Max/MSP), and the Internet of Things is the claim that tacit knowledge, or learning through doing, is crucial to engaging the trajectories of digital methods and computational tractability. In order to best understand and model real-world problems, digital humanities practitioners must constantly oscillate between the abstract and the concrete. As such, problem-based frameworks in humanities physical computing and manufacturing involve a high frequency of hands-on, trial-and-error approaches, with each iteration thoroughly documented for self-reflexivity as well as future reference by those who want to replicate the approaches in new contexts. By extension, problems are articulated as necessarily material, not only in the sense that they are tied to bodies and physical conditions but also in the sense that software and hardware are tangible artifacts capable of surprise and mutation. Steeped such as it is in communities of makers, this growing investment in material culture and tacit knowledge may be interpreted as a move away from the screen, toward a digital humanities curious about the historical and contemporary functions of everyday technologies. If this is indeed the case, then interaction design will surely play a significant role in the development of the field, its approaches to modeling, and how it defines problems.

Some Implications of Problem-Based Modeling

While the directions of problem-based modeling are still difficult to predict at this juncture, there are at least four specific implications for digital humanities. The first is a more direct engagement with the public humanities, beyond making texts and artifacts accessible online and toward more community-based research and outreach. In this case, initiatives like *4Humanities: Advocating for the Humanities* are persuasive examples of possible trajectories, which will continue to blur the line between academic, alt-academic and nonacademic knowledge-making. Second, digital humanities is increasingly overlapping with the lived spaces of research and everyday life, especially in museums and libraries. Of course, there are already innovative frameworks such as the Scholars' Lab at the University of Virginia, and public history work in interactive exhibit-building will continue to reconfigure how we perceive and engage sites of cultural production. Certainly much is to be learned from the arts, archeology, and performance studies as digital humanities moves in this direction. Thirdly, graduate training and pedagogy are—arguably for the first time in digital humanities history—being foregrounded by practitioners, namely because very few digitally-inflected methods courses exist in the books of humanities curricula (Nowviskie 2011a). As curricula are reconfigured, models such as the Digital Humanities Summer Institute and THATCamp offer examples for problem-based learning across the disciplines. And scholarly organisations, institutes, and collaboratories will need to continue investing in hybrid pedagogies that spark exciting correspondences between content and process, critical theory and technical competencies, online and offline spaces, and history and current issues. Finally, more and more projects appear to be focusing on real-time analysis, prediction, and contemporary issues. As Alan Liu (2012) suggests, such engagements are necessary to bring digital humanities into a present relevant to public audiences. What is more, they demand critical methods, approaches to data structuring, forms of expression (e.g., graphical expression), and literacies for which there is little precedence in the humanities. The key then is determining how to generate sustainable research practices and infrastructure in response to these exciting developments. To be sure, we are in a moment when we need to repeatedly ask ourselves what remains after the buzz subsides.

Notes

1. See, for example and further treatment, Schreibman, Siemens, and Unsworth, "The Digital Humanities and Humanities Computing," in *A Companion to Digital Humanities* (2004); also Siemens and Vandendorpe, "Canadian Humanities Computing and Emerging Mind Technologies," in *Mind Technologies: Humanities Computing and the Canadian Academic Community* (2006).

2. For the 2011 international conference of the Alliance of Digital Humanities Organizations at Stanford, see https://dh2011.stanford.edu/. At MLA see, for example, panels like "309. The History and Future of the Digital Humanities," "436. The Institution(alization) of Digital Humanities," and "474. Social Networking: Web 2.0 Applications for the Teaching of Languages and Literatures" via http://hastac.org/blogs/marksample/digital-humanities-sessions-2011-mla, as well as comments by digital humanities stalwarts like Steve Ramsay in debate and as documented in the Chronicle of Higher Education. http://chronicle.com/article/Hard-Times-Sharpen-the-MLAs/125905/

3. The authors would like to thank Daniel Powell and Alyssa Arbuckle for their assistance with this aspect of our work.

4. See Willard McCarty and Harold Short's "Mapping the Field" (from the report of ALLC meeting held in Pisa, April 2002; http://www.allc.org/content/pubs/map.html; retrieved 20 September 2010) and from the HUMANIST Discussion group (http://www.digitalhumanities.org/humanist/; retrieved 20 September 2010), including images.

12 DEPROVINCIALIZING DIGITAL HUMANITIES

David Theo Goldberg

Digital humanities has become a given. Perhaps much more a given than the humanities at large, which seem ever in crisis, ever under duress. The humanities, if you believe *The New Republic*, is even under "deathwatch" these days, as it seems to have been almost perennially (newrepublic.com/tags/humanities-deathwatch). There is now a "#dayofdh" where practitioners celebrate the practices in which they engage in the name of "digital humanities." Perhaps nothing so much signifies the supposed taken-for-grantedness of the "digital humanities" today as the fact that within the context of the humanities no one really needs to ask what the ready use of the moniker "dh" stands for. The tables are turned when digital humanities is moved to advocate for the humanities at large, for the humanities under duress. There was never a great deal of reciprocation; the humanities at large, at least early on, was more often disdainful of the digital, viewing it as a convenient technology, nothing more—not a principled commitment or intellectual practice for which to advocate. The brother from the same mother but perhaps a distant father was never a sibling considered quite one of us. Especially when, resentfully observed, the costs of digital humanities' production outstripped multifold that of the humanities conventionally understood save when grants more comparable to the applied sciences were produced with some regularity. The prodigal son made good gets to be a cherished family member after all. The model child to be emulated by all the siblings, to save the rest of us—from ourselves.

What has come to pass under the title of "digital humanities" seems to be of roughly four related if conceptually discrete kinds. The first concerns the development of technology aimed in some broad sense for use by humanists. Zotero is the best known and probably the most widely used; the text mining tool Voyeur another. But text mining and visualization tools, search engines, concordances, more recently a multimedia publishing platform like SCALAR, smaller apps and the like all exemplify the undertaking here. A couple of things bear pointing out. While perhaps designed with humanistic work in mind, these technologies need not be and the best mostly are not restricted to such use. This implies that technology built by humanists—in the sense of those trained formally in the humanities—does not automatically make for humanistic technology, for technology that best advances the humanities, that the humanities may be proud to call their own. On a strictly technological register, *digital* humanities is not per se humanities. Something else needs to be at work to make it so (see Natalia Cecire's critique—Cecire 2013—of the "more hack, less yack" sensibility widespread among digitalists; by contrast, for a more techno-driven statement, see Guiliano 2013).

Besides the purely tech development projects that have become the flag bearers of digital humanities production, there are, second, what we could call the "born-humanities" projects, to reverse engineer the phrase. These are the humanistic projects that digital technology helps realize, enable, that it supplements. They are projects that in conception don't require digital technology but their fulfillment may be fully or partially enabled by the technology, made easier if not possible. Along the way the project may well be reconceived, shifted, or transformed by the application of the technology. But the development of knowledge qua knowledge in this case is humanistically formulated, and would be so before and after the technological intervention. Here the technological is fully instrumental, in service of an independently conceived undertaking. It may help to refine the vision, after the fact, but this is taken always as enablement or refining rather than as constitutive or defining.

A third distinctive formulation of what commonly comes under the umbrella of digital humanities concerns interpretative and critical reflection on the digital. "Critical" here, of course, does not mean dismissive so much as seeking to comprehend, defining and refining possibilities and limitations, to interpret them for ourselves and others. Here the digital is the object of analysis and interpretation. Humanistic analysis of the digital is not restricted to the digital's contributions to the humanities. It takes as its object of interpretative and critical focus all of digital conception and application. The abiding humanistic contribution here—the *critical* meaning of "digital humanities"—lies in its interpretative and critical analysis rather than the delimitation of its focus only to digital contribution to or in the name of the humanities. Here work not conventionally taken by the techno-digital humanists to belong under the category reveals the stakes of the count: Bernard Stiegler's critique of the technological, for instance, or (closer to home) the pressing work of the postcolonial digital humanities group, concerned as it is with countergenealogies not readily recognized by those digitalists taking themselves to determine that there is *a* field, its scope and membership.

We can point to a fourth conception, a sort of hybrid set of practices that get identified as digital humanities. There are projects that are simultaneously born-humanities and born-digital. They are born-humanities in the sense that their humanistic content is characteristically no or little different than it would have been absent the application of digital technology. And it is born-digital in the sense that the digital is germane to the design of the content collection and presentation. There are many archives, for example, that fit this description. The technology might enable a larger archive than otherwise possible, quicker access, more revealing comparisons and robust networks of relation to be drawn with ease, and in those senses to produce new knowledge. But the *humanistic* nature or quality of the content is little if any different than had there been no or little involvement, or only instrumental application, of the digital technology.

This suggests too that there are any number of people who engage relatedly, sometimes even interactively, messily, in all four of these pursuits: a technology developer in the morning, a technological enabler of or contributor to humanistically conceived projects in the afternoon, and critical commentator on all things digital in the evening. To put it in these historically referencing, manifesto-like terms conjures both the revolutionary and utopian aspirations at play though the somewhat discrete temporalities at once belie the intersections—the interfacings—between these not quite discrete articulations. That there are any number of figures today who engage in all four, if we take the hybrid as a sometimes interactively productive combination of aspects of the others, places in question the privileging of the technological, the enabling, in relation to the conceptual and especially the critical. If there is any privilege at all regarding the humanistic, it is that of the interpretive and critical endeavors around all issues of meaning, value, and significance, of translating ourselves (in the broadest sense) to ourselves (inclusively) for which humanistic commitment has always stood. A compelling example of the latter is "JustPublics@365," the CUNY Graduate Center's critical public intellectual engagement with key social concerns of poverty, inequality, incarceration, public health, and racism constitutive of which are the uses of social media and digital technologies, including media training for academics to engage publics around such issues (justpublics365.commons.gc.cuny.edu).

This interfacing points in addition to the appeal to some scholars, frustrated with the delimitations placed upon them by their disciplines, of the openings enabled methodologically and conceptually by the digital to think about their work either generically or in a project at hand in innovative ways. Here the creative possibilities to think anew about subject matters operate very much like the sorts of affordances of the interdisciplinary (see the Midwest Faculty Seminar 2013). They refuse established limits in favor of advancing novel insight and knowledge about the object(s) of composition, curation, or analysis at hand.

The driving distinction at work here, then, is that between working *with/in* technology, on one hand, and working *on* technology, on the other. Working with technology does not have to be taken narrowly as the humanist her- or himself doing the tech development so much as working with a technologist—a coder, engineer, computer scientist—to develop what the humanist has in mind. The conception might change as the humanist deepens comprehension of the technological frames and possibilities, and as the technologist recognizes the humanistic provenance, aims, and purpose. In this sense humanists can engage in digital work without knowing how to code, just as an effective philosopher of science need have no experimental experience in the lab, so long as he or she has sufficient understanding of what coding—the logic of coding in general as well as the specific language in which the programming partner is working—enables and cannot do, is beyond its ken. Working *on* the digital suggests an interpretative and critical disposition, analytically understood, in relation to assessing the broader social contributions and impacts of the digital on working, learning, teaching, and recreational lives. If there is a defining condition uniting all of the humanities it is in fact the interpretive and critical.

In the early new millennium "techno-humanists" seriously interested in the interface between the digital and the humanistic insisted that knowing how to code—demonstrating capability of programming a language—is crucial to computational humanities. Some still do. As the "humanities and computing" movement of the 1990s gave way to the "digital humanities" in the new millennium, fueled in part by some very vocal debates around the founding of HASTAC and "humanities computing" resistance to what I along with Cathy Davidson, Kevin Franklin, and others were insisting on in its name, the tension around techno-facility and coding in particular remained a touchstone. It was symptomatic of the broader and longer-standing tension concerning the level of technological sophistication to be a card-carrying member of this emergent field. The tension exactly, that is, between the digital and the humanistic (see here also Sean Michael Morris's provocation, Morris 2013).

The more dogmatic among those counting themselves, and reserving the right to determine of others whether they are "digital humanists," insist that a narrow version of the first is a necessary and perhaps even sufficient, if not exhaustive account, of being so. The insistence on knowing how to code is the strictest and most exclusionary of such considerations, almost to the point of discounting all humanistic commitment, at least rhetorically. One would never say that being able to read and write fluently in French is a sufficient condition of one nationally defined form of literary scholarship. The insistence on knowing code is made by those who are largely silent about the driving debates in the humanities, or indeed in any significant field of the humanities, though some also have been known to be vocally dismissive precisely of interpretive and especially critical drives so central to humanistic commitment.

It is interesting that in this formulation of "digital humanities" the humanities have been largely taken as given, long established, whereas the digital is presumed to change and to produce change. The humanities are static, the digital innovative; the humanities backward-looking, the digital prospective; the humanities become object to be worked on, the digital the producer, maker, bearer of agency. The humanities provide the archive—the otherwise dusty collection of the past, of the already produced—that the digital gets to work *on*, *with*. This sense of the relation, if it could be called that—less an interface, less co-defining, less mutually transformative—reinforces the presumption of the humanities in need, if not in crisis, of the (latest) way of saving the humanities from itself (in the presumptive singular). But it also reinforces the presumptions that, as active agent, coding is a necessary if not sufficient condition of digital humanities *and* that critical reflection on the digital and by extension on digital humanities itself doesn't belong to the self-identity. It is something alien, the discardable margins of the humanities, their past that has passed that may get to be incorporated into some archive or other. But even the archivability of the critical is taken, where taken at all, as marginal, as uninteresting, merely words on a page that present the least—which may also be to say the most—of challenges to a coding humanities. "No more Foucault," as a dismissive contributor declared to a group committed to using the digital to advocate for the humanities. You could hear the snickers of delight reverberating across the listserv, nary an expression of objection.

The digital humanities, then, is sometimes taken, whether explicitly or more readily implicitly, as *the*—or even *a*—compelling solution to the current institutional challenges of the humanities. A cash cow comparable to the sciences, more readily understandable or approachable by the uninitiated in the abstraction of much contemporary humanistic production, high in dazzle value, the instrumentalities of the digital have steadily cohered under an assumed discipline apart. It has developed its own specialist language, technologically inspired, its driving concerns, its associations and networks, practices and privileging. It has, in short, taken on all the bad habits of self-containment and, by extension, constraint, even the arrogant self-assurance of a reifying practice in denial.

Digital humanities, it turns out, has become its own discipline: it speaks its own language, normativizes the sorts of questions and projects to be pursued, has its own journals and associations and networks and book series, and often its own divisions or offices at major humanities funding agencies and foundations. The question, of course, is what kind of humanities is being pursued, practiced, advocated for in the name of the discipline. What kind of knowledge formation and learning does the disciplinization of digital humanities stand for, and what sorts should its dedisciplinization be committed to promoting? In short, what sense of the humanities is it committed to advancing?

I am suggesting that what takes itself as digital humanities more or less silently exercises a series of judgments about the nature and scope of the humanities, normative assessments if not presumptions about the application, scope, reach, and practice of what counts as humanistic endeavor. Digital humanities shapes humanistic pursuit: what objects to be analyzed, what questions to be asked, what counts as appreciable if not also viable responses, in short, what knowledge is rightfully (to be) produced in its name. "Digital humanities," perhaps perversely, at the limit delimits by homogenizing the humanities, by excluding or at the very least marginalizing not just from "digital humanities" but from the very conception of what properly belongs to the humanities and should be practiced in their name.

The "digital humanities," it could be said accordingly, reinforces a now soundly debunked notion of technologically determined progress. Of course, the humanities have to recognize that the digital has deeply, even constitutively in some ways, impacted how we engage in the practices of the humanities, and perhaps too what it is the humanities take up in the name of the humanistic. In the wake of the digital the humanities can no longer be engaged and exercised as though the digital revolution never transpired. Post-digital, the state of the humanities is analogous to the transformation of the commonplace of painting after photography. After the digital, how we communicate, read and write, construct and deconstruct, compose and remake have altered, our practices of searching and archive formation, our ways of relating, of curating, of critical commentary, our temporalities and modes of relation, the contrast between the "real" and the "virtual" have all been profoundly affected. This is not just a matter of content, or of volume, but a deeply qualitative shift in modes of socio-human engagement and relation. If the human has been so dramatically impacted by the digital in all its constitution, relation, and expression, how can the humanities not be deeply challenged by these changes to their principal object of scrutiny as well as in their own interrogatory and analytic practices?

Digital technologies have indeed made possible new modes of cultural production, new forms of aesthetic expression, multi-modal compositions, novel articulations of meaning, value and significance. The technology, it could be said, is the great enabler, and in the enablement becomes transformative. Expressive remaking is not new with digital technology. Picking up on Dadaist practices in the 1920s as well as collage in art, William Burroughs cut up typewritten text printed on (news)paper in the late 1950s and 1960s, reshuffling the words, producing new relations among them, and in the "cut-up" process both newly expressed meanings and rhythms of articulation. Digital technology allows for greater immediacy, enhanced degrees of randomness, of the reshuffling and remaking, shifting rather than erasing the mediation and its sites of enactment. The technology allows for remakings, drawing new relations between the elements both manually and programmatically, the manual perhaps itself the outcome of the programmed, thus at once shuffling the processual relations as well.

If one generalizes this point to all digital production or contribution in relation to cultural expressiveness, to meaning-making, value and significance, it still leaves in question what exactly is humanistic about the digital here. Cultural expression and articulation give rise to meaning, value, and significance, however produced or enabled. This is not to say the materiality—material media—is unimportant. But the digital contribution is not itself the meanings, values, or significance in play, though as the means or media of expressibility and articulation they help to shape and "color" the modes of signification at work. And so the humanistic is not simply reducible to the digital, nor the humanities to their media, as important as they may be to the shaping, to the possibility of that particular form of the articulation, that specific meaning production. The medium, alas, is *not* the message, at least not in this *sense*, the sense as such, its meaning, its value, and yes, too, its significance.

Obviously. I am not. Rejecting. The digital. It is just that, qua digital, digitalists are not humanists, the digital as such is not the humanities, though may be both a tool and an object of the humanities, of humanistic interpretation and critique even as the latter may be produced by, with the help of, the digital. The digital sometimes may even be part of the meaning, value, and significance produced. But where the digital—the technology, the code—is all there is in its raw state, the humanistic drops out. We see this as soon as code is taken to be a language, taken to generate meaning, value, and significance through the uses and relations which it enables, to which it gives rise. Random, non-rule bound formulation of code (if one could even elevate this by referring to it as *form*ulation) would produce meaning, would be of value, would be found significant merely as anomaly, as calling attention to its abandonment of the rules. Meaningless in itself, so to speak, it becomes meaningful only by pointing to that which it is not. Its being, one could say in Heidegerrian terms, is a function only of its nonbeing.

The humanities too are changing. Humanists never spoke of medieval or classical or renaissance *humanities*, of philosophical or cultural or literary humanities. Mainly because to do so would be somewhat oxymoronic; they were—are—ipso facto aspects of humanistic analysis, ways and objects and styles of analysis taken *as* the humanities. They are just what the humanities (or its earlier formations of and formulations as liberal arts or even earlier as philosophical, literary, religious, or historical studies) have always done. Perhaps we can call this definition by use, by practice. Today we hear of environmental humanities and urban humanities, but what are they other than taking the central humanistic concerns with interpretive and critical analysis and applying them to those sets of objects in all their complexity, like film studies. The object may shift but does the object transform the driving analytics of the humanities per se?

In this regard what has come to be called "public humanities" is revealing. In the altogether conventional sense it has come to mean translating the abstract and abstruse technical language of much contemporary humanities for nonacademic publics. Here the digital may serve as just one set of media among others for enabling this. But this conception leaves the humanities conventionally expressed largely untouched while giving in to the presumption that the likes of Arendt, Derrida, Butler, Foucault, and even Said are too complicated in their own voices for general publics to engage. The reductive caricatures aside, the trouble with this view is that none of these critical thinkers ever had any trouble making themselves understood in public appearances before broad publics, and have done so without dumbing down their views. But this merely instrumentalizing sense of public humanities likewise forecloses a far more compelling sense for which contemporary digital technology is constitutive, as the "JustPublics@365" example earlier attests. Here the digital makes for modes of interpretive and critical expression and connection around key public issues. Which is to say that the digital both makes for the possibility of public expression, it actually makes—constitutes—publics in ways otherwise not existing (for instance, through crowdsourcing), and it makes for genuinely new and collaborative humanistic production. Consider here, among various possibilities, the UC Berkeley War Crimes Studies Center site, *Virtual Tribunal,* which serves at once as archive informing international justice tribunals and criminal trials such as those in Cambodia and Sierra Leone, as modes of public historical memory, as contemporary public learning and scholarly tools, as objects and the exercise of historical critique, and as platforms in service of civic reconstruction (War Crimes Studies Center 2013).

It is revealing too in this regard that there is no talk of digital physics, or of digital astronomy, and so on. What would be the point? Physics and astronomy and so much else in knowledge making today require the digital for their exercise. The digital is second nature, just what we do when we engage in these objects of analysis, more or less complexly. The humanities, I am saying, is and should be no different. It is not that the humanities had no technology prior to the digital: quills, pens, typewriters, yes, the voice too. No "digital humanities," then, so much as humanities for which the digital is—or should be—second nature, which digital technologies help to enable. And the digital here may range from the everyday and banal—what pretty much most any humanist would take up to do their work effectively today—to the technologically specialized and sophisticated, the complex of technologies necessary to carry off a specific project requiring devoted technologies because of the size of the data sets, visualizations, modes of analysis, speed of computation, bringing things at a distance into relation otherwise not possible, tracking the knowledge flows, and so on.

If this is right, all that the *digital humanities* would amount to exactly is taking the digital as the object of critical analysis. If digital humanities has taken itself to stand humanistic practice on its head, what my argument entails is turning the humanistic back, right side up. Software studies or screen studies are humanistic undertakings not because they involve telling us how we should use software or screens instrumentally—the how to, painting-by-numbers, follow-the-rules-of-use manual—but because they address the significance of software or screens to political economy and cultural expression at a given historical moment, the meanings and values represented by screens or software in specified contexts. Similarly with cultural analytics, which doesn't tell us how to produce culture in a technical sense but what this or that cultural set—its data-set, to be precise—means, what its historical or contemporary significance and value amount to, how cultural expressions relate both to each other and to political economic considerations prevailing at their time of expression, or contemporarily. There is no such thing as digital humanities as a stand-alone, *humanistic* discipline because the digital is now more or less—but pretty much *not* not at all—germane to humanistic work. At the more than less end of the spectrum perhaps the focus of concern begins more and more to look like its own undertaking, with all the trappings of disciplinarity. But what I am insisting is that at that point it begins to lose its humanistic interest for its instrumentalization.

So we have here the digital in and for the humanities and a critical and interpretive humanistic analysis of the digital. The most compelling projects, those likely to have transforming and lasting impact, are those that bring the instrumental and interpretive-critical, the shaping and deeply meaning generating, the resourceful and radically signifying possibilities into productive engagement with each other. Many of the contributions to this book represent exactly those sorts of projects. Where digital practices for the humanities and humanities for and of the digital interface to produce meaningful knowledge for the times we inhabit. And that serve to produce a collaborative, connected, and relational knowledge production, of making and learning and learning through making concerning matters of meaning, value, and significance that are at once meaningful, valuable, and significant for the times we now inhabit.

What this thinking at the interface, the intersectional co-making, reveals is a range of creative and insightful practices reducible neither to the instrumentalization of the technological nor to a pre- or altogether nondigital production of the humanities, whatever the latter might mean today. It will pressure humanistic knowledge making as much as it undermines the merely instrumentalizing contribution of the digital by placing in question the national framing of historical culture and its disciplinizations so central to humanities as we have inherited them. And it undercuts the radically individualizing auteurist presumptions underpinning much conventional humanistic self-conception.

It is not that some humanistic production today is digital humanities, and some is not. Nor again that we are all digital humanists now. The driving question, as it has always been, is what kind of humanists we choose to be in and for our times. And the new(er) question is how, in exercising this renewed and renewing vision of the humanities, in realizing their potentialities, we engage the digital to common, collaborative, creative, productive, critical-interpretive purposes and practice.

II INFLECTING FIELDS AND DISCIPLINES

To be or operate between humanities and the digital presupposes a relation between the digital, broadly speaking, and the humanities as an intellectual platform and institutional site. Several of the contributions to part I of this volume note that digital humanities often has a fairly marginal impact on the humanities. This implies less that there is no good work carried out at this intersection than that there is substantial untapped potential. This is particularly the case in relation to deep-going, interpretative-technological scholarship with impact on the disciplines. This lack is not one-sided, and there has arguably been a paucity of far-reaching engagement from both the digital and the humanistic sides.

This section is concerned with what is or takes place between humanities and the digital, focusing on challenges and questions that arise from the humanities disciplines and fields rather than from or through the digital humanities. This is admittedly a blurry line and many digital humanities projects unsurprisingly have a relation to the institutionalized humanities. However, the humanities manifested by the digital humanities can often seem static and traditional. In a provocative blog entry, Andrew Prescott maintains that digital humanities centers in the United Kingdom "are busily engaged in turning back the intellectual clock and reinstating a view of the humanities appropriate to the 1950s" (Prescott 2012). Prescott proposes in response that digital humanities develop more of its own scholarship and scholarly identity. While this is an understandable sentiment, it would seem at least as important to turn to the disciplines and disciplinary perspectives as drivers of digital humanities. The digital humanities alone arguably can never be an intellectual powerhouse, but needs to work closely with the disciplines, while also having intellectual direction and maintaining integrity.

A significant factor underlying Prescott's analysis is that the centers central to his account are steeped in a particular conception of digital humanities, namely one that is project driven and invested in creating archival sites and tools in collaboration with the humanities and different cultural heritage institutions. This work has traditionally focused on texts, on cultural heritage, and on making materials available to researchers and others. There is largely, if not exclusively, an instrumental relation to the digital, and technology tends to be seen as a tool rather than an object of inquiry or an expressive medium. This primary mode of engagement is embedded in the established genealogy of the digital humanities, which means that some contact points between the disciplines and the digital humanities are more likely than others. For instance, disciplines such as English and history are much more commonly integrated than media studies or philosophy. Furthermore the focus on tools, structured data, and platforms tends to lead to foregrounding research materials and access rather than specific research questions and interpretative frameworks.

In a discussion of the limited connection between philosophy and the digital humanities, Lisa Spiro (2013) traces resources such as encyclopedias, textbook platforms, and journals, and the subsequent collaborative discussion is mostly concerned with such methodological and instrumental perspectives. There is very little in the discussion about the core issues in philosophy that may benefit from a digital engagement or what central thematics—such as subjectivity, epistemology, the human-technology interface, autonomous warfare, and medical ethics—may have a technological inflection. While this lack is particularly evident for a discipline such as philosophy, with its relatively limited reliance on large sets of materials and big data, the general lack of disciplinary core perspectives in the digital humanities adversely affects both the humanities disciplines and the digital humanities.

If the humanities disciplines engage with the digital in different ways, it is imperative to find ways to accommodate different modes of engagement between the humanities and the digital. A broad engagement between the humanities and the digital simply allows for a larger range of disciplines to engage with information technology in a productive, technologically informed, and intellectually driven way. However, the balance between different interests, epistemic traditions, and modes of engagement is not clear-cut or necessarily easy. For instance, validation of academic work most often takes place within the disciplines, which means that almost any mode of knowledge production developed within the digital humanities must relate to the disciplinary context. There is substantive intellectual work associated with disciplines, inter- and transdisciplines that cannot be outsourced to the digital humanities or any other instrumentalizing institutional formation.

The disciplines and interdisciplines, largely critical in relation to their own traditional work, must be so also in their own engagement with digital platforms and initiatives. Such work, however, requires infrastructure, specific competencies, as well as an open disposition to interdisciplinary collaboration. It is not remotely sustainable for each discipline to build its own platforms. Here the humanities can definitely take a leaf from the sciences, while also insisting on other infrastructural needs and rationales. The digital humanities can help knowledge formations realize a common infrastructure and core set of competencies and thus facilitate a broad engagement with the digital. The field of digital humanities can also do things that the departments and disciplines cannot or will not do, and make use of its liminal position to foster dialogue, support development, and inspire hope. For this to happen, both the disciplines and the digital humanities need to be open to change and to changing. Most current configurations do not support this level of integration and reach. The chapters of part II point to the range of disciplinary perspectives and questions that are important to the continued buildup of rich humanistic engagement with technology.

An example will illustrate the challenges and process often involved in finding and developing technological, methodological, and intellectual points of connection. In the fall of 2013 and spring of 2014, one of the editors of this volume put together a large research proposal with an environmental humanities consortium. The discussion had started earlier, largely as a coming together of science and technology studies, environmental history, and digital humanities. The digital humanities component was fairly marginal and there was an instrumental focus, also in the way that environmental researchers with digitally inflected topics (e.g., the mediation of nature through location-aware technology) made a case for incorporating a digital humanities perspective into the proposal. They pointed to matters such as research presentation, building community websites, and supporting dissemination. The tendency to demonstrate the usefulness of digital humanities in instrumental terms is common and reminds us, by contrast, of how the humanities as a whole are defended and demonstrated (e.g., see During 2014).

In the case of the environmental studies proposal, it moved away from an instrumental model as it was felt there was more to be gained from a fuller collaboration. Such an approach does not exclude instrumental use of technology and expertise but entails looking at the collaboration in a deeper way. A core idea of the proposal (called "re-humanizing Nature") was to offer alternative, humanistic narratives of the environment and nature as a contrast to the predominant science and technology-driven narrative. The dialogue led to a realization of how challenging such narratives are: they also raise the question of infrastructure, both in the sense of critically engaging with the infrastructural level of these narratives, and employing infrastructure and expressive modalities in order to enact humanities and arts based narratives. This is not just a matter of presentation or representation, but of ontological, interpretative, and creative processes that are critical to the understanding, creation, and sociopolitical enactment of natural knowledge. This was seen as fertile ground for environmental and digital humanities to come together, and resulted in a common experimental platform driven by the intellectual questions at the center of the proposal.

In chapter 13, Whitney Trettien explores the materiality of texts from a media archeological and tactical media perspective through linking a book and a computer present in the same room as the author. The book is a Biblical concordance from around 1630, whose place as a technological artifact is accentuated through its appearance at the World's Fair in New York City in 1939, and the computer in the room shows a digital library entry of the same volume.

Trettien draws on this context and a focus on the mediation of time to make the argument that the greatest power of the digital is not changing how we study, but transforming what we see in the past. Her case study suggests that a media archeological approach, where the material artifacts play a central role, can be deployed to challenge Enlightenment era perspectives on authorship and textuality and thus provide a different way of understanding our own moment.

In chapter 14, Cecilia Lindhé does not primarily deal with textual objects, but rather with aesthetic objects such as sculptures and baptismal fonts. Lindhé argues that our understanding of these objects as represented by digital archives and platforms is typically conditioned by a post-romantic view of aesthetic objects. She focuses her attention on the representation of the Virgin Mary as a role model for lay people in Sweden during the Middle Ages as enacted in church spaces. Lindhé describes how a series of digital installations are used to enact the Swedish medieval church as a multimodal and experiential space that facilitates and encourages multisensuous involvement. The rhetorical concepts of memoria and ductus function as a structuring device for the installations. The spatial, architectural, and sensory components of the installations contribute to challenging the ocular-centricism in aesthetic and literary practices. Like Trettien, Lindhé suggests that the real power of digitally enabled work in the humanities lies in a deep and conceptual engagement with entangled material technologies and central research questions.

Nick Montfort (chapter 15) is also interested in aesthetic objects, but with a stronger focus on the inner structures and creative processes associated with such objects. His chapter addresses computational literature and one of his initial claims is that just as voice can be said to be critical to song, computation is essential to some literary projects. He describes computational literature as a radial category, where the prototypical member includes complex and extensive computation while also having a deep engagement with literary questions and techniques. Such literature is not restricted to the digital computer, and Montfort brings up writing organized around computational techniques, such as constraints (e.g., not using certain letters) and textual transformation. Apart from discussing different types of computational literature and associated processes, Montfort also looks at underlying systems used to create computational literature, including Inform, a tool used for writing interactive fiction. He demonstrates how methodologies taken from digital humanities can be used to analyze the fiction created by such systems. His analysis comparing works produced on different platforms, raises questions regarding how such different platforms structure and condition these narratives. Montfort argues that there is considerable analytical possibility at the intersection of digital humanities methodologies (such as distant reading) and computational literature.

Jenna Ng (chapter 16) takes us to a world of filmic expressions and the idea of the importance of spaces in between, especially the cut, as potent and creative negations that can open up new meanings. She grounds her analysis in a discussion of Sergei Eisenstein's work on montage as a collision of images and of the material ontologies of film-based cinema as dependent on the strips of black, not just as dividers but as a recontexualization of photography into cinema. Ng argues that the digital cut transforms not only the convenience of the cut but makes the violent undertones of the cut stronger and implies another sense of finality. The cut does not only expose what is in-between but also "the constitution of being in that betweenness, spaces that form being in their complex and fluid connections, dialogue, confluence, and relations." Ng focuses on digital remix to illustrate her argument, in particular, the so-called reflective remix. One of the key works analyzed is Kevin Macdonald's documentary film *Life in a Day*. The chapter demonstrates how the cut can help us understand digital materiality and support engagement with digital making as a productive practice. Furthermore Ng makes a convincing argument for thinking about and enacting the relationship between digital humanities and the humanities not in terms of ruptures but rather through collisions (and, one might venture, through cuts). Filmic cuts offer both metaphor and example for work between the digital and the humanistic.

In chapter 17 on locative media in everyday life, Larissa Hjorth also traces visual continuities, ruptures, and contexts. Her analysis is partly based on a series of studies carried out in the Asia-Pacific region, and a key argument is that second generation, location-based services and mobile use go beyond sharing, storing, and saving images. According to the author, this new type of visuality is no longer only about the visual (if it ever was). Hjorth points to the accelerated rate at which photos are taken, edited and shared. They are often produced and consumed in motion and overlaid onto places. A rich context is created though several means. First, GPS information places images in a geographic space and invites viewers to recall the place as well as overlaying the context of the image. Second, the co-presence on a site of a number of images organized by place or time creates a narrative. Furthermore the images taken by a member of a network is contextualized relative to that person. Drawing on Sarah Pink's work, Hjorth argues that such use creates a sense of place as a multisensory entanglement, enacting an emplaced form of visuality. As Hjorth's work demonstrates, in order to study such entanglements and approach the messy nature of mobile media more generally, there is a need to work across disciplinary boundaries, revise humanistic methodologies, and engage with rich empirical studies.

From a position in science and technology studies and gender research, Jennie Olofsson (chapter 18) addresses the politics of search engines in terms of inscription and de-inscription (borrowing terminology from Madeline Akrich). Investigating the structures and meaning making of search algorithms and databases would seem to be an important area for digital humanities not only in terms of specific research questions but also in the sense of investigating some of the conceptual and technological underpinnings for systems created in the field (and elsewhere). Olofsson's chapter analyses the structure and politics of Google with particular focus on the gendered implications of the algorithmic systems at play. She argues that power relations between men and women are both exhibited and reinforced through predicative suggestions made by Google through the auto complete suggestion and instant predication of search queries. While such systems inscribe values and meanings, users also de-inscribe search results and suggestions, and such de-inscribing activities can lead to a realignment of algorithms and rules that may appear static and opaque.

Jo Guldi (chapter 19) takes up the question of scale in terms of the scope and methodology of history research. She maintains that short-durée work and thinking is often prioritized at the expense of long-durée approaches to history. This question is deeply embedded in the fabric of the discipline and, against the backdrop of Irish agrarian and property history between the years 1870 and 1980, Guldi presents a range of tensions and perspectives at play. In doing so, she explores how digital methodologies and tools challenge and integrate with disciplinary research questions and epistemologies. The material encompasses thousands of articles on agrarian reform and property rights. Guldi argues that this is a kind of problem well adapted to digital history and that much digital humanities work operates on too small a scale and has little chance of discovering something that is not already known. What Guldi suggests, by contrast, is an intellectually and methodologically risk-taking digital humanities that engages with large problems and appropriate digital tools. And while she stresses the long durée, she also foresees that the greatest advances are likely to emerge on the borderline between the long and the short durée.

Geraldine Heng and Michael Widner's chapter 20 starts out from a teaching experiment (later relabeled as a learning experiment) called "Global Interconnections: Imagining the World 500–1500 ce" initiated in 2004 and well aligned with a long-durée approach to history. This seminar developed into the idea of a "global Middle Ages" humanities platform invested in experimental work across disciplines and specializations inside and outside the humanities proper. The authors emphasize the transactional nature of the analytical process in this trans-humanities laboratory and the importance of working with a range of competencies including computer scientists and social scientists. They see the value of methodologies such as computational analysis of global climate conditions and optical pattern recognition of fabrics and weaves, and argue that quantitative analysis can help us understand matters such as distributed industrial revolutions. Essentially Heng and Widner present both an idea in the process of development and infrastructure that has partly been implemented. The underlying conceptual foundation seems to have remained fairly stable over tim, and there is a set of basic theoretical and conceptual parameters discussed in the chapter. These include three overlapping trajectories: the mobility and routes of culture (how people, artifacts, and ideas circulate through the networks that interconnected the world), points of anchoring (e.g., cities and trading blocs), and time (including fictions that produce premodernity as the antithesis to modernity).

Tim Hutchings (chapter 21) notes that religion is currently marginal in most digital humanities contexts, although computational analysis of religious texts is often seen as the origin of humanities computing. Careful and structured study of religious texts has a very long history and the first Bible concordances appeared in the thirteenth century (a somewhat later example appeared in Trettien's chapter 13). Contemporary projects continue this tradition, but the most noticeable growth in terms of access to Christian texts seems to be in remediations of the Bible such as Bible verses on Facebook and platforms such as YouVersion. The app YouVersion has been downloaded 60 million times since it was launched in 2008, and allows commenting, highlighting and sharing of verses. Hutchings argues that such remediations transform the structure, materiality, and dissemination of the text. Data rich materials such as YouVersion call for digital methods, but Hutchings notes there is little engagement with such methods. More generally, there is a need for tools that allow critical investigation of how religious media shape knowledge, subjectivities, and networks. There is a strong connection and challenge to the digital humanities here. Conversely, the digital humanities could probably benefit from engaging more with work done in religious studies on the theology of (digital) technologies.

Maurizio Forte (chapter 22) represents another discipline with substantial technological engagement which is often not seen as a central representative in the digital humanities landscape. Forte argues that the shift in archeology from what he calls "virtual archaeology" to cyber archaeology is highly relevant to the digital humanities. Early 3D reconstructions are examples of virtual archaeology, and as Forte points out, these were often created to demonstrate the potential of computer renderings and graphics, so they did not really focus on people and their lives. The reconstruction process was often separated from the archeological interpretation, and highly photorealistic models tend to be difficult to validate and leave little room for alternative interpretations. In contrast, cyber archaeology is more focused on interaction and environment, and the digital space can be seen as a simulation space. Data are born digitally through technologies such as digital photogrammetry and laser scanners, and can be used on site—in the excavation context—instead of necessarily waiting until the laboratory post processing. The enacted nature of this kind of work resonates well with Larissa Hjorth's chapter 17 on emplaced visualities. Forte illustrates such work and processes through the archeological site of Çatalhöyük, where his group has been collecting data which have been processed and visualized in the lab with 3D projectors as part of a methodology that supports real time interaction between observation and data simulation.

Chapter 23 by Natalie Phillips and Stephen Rachman explores the emerging intersection of literary studies, neuroscience, and digital humanities. Humanistic engagement with neuroscience is a contentious area and, according to the authors, objections to literary neuroscience can be divided into three categories: it is just a fashionable trend and a source of funding; science is likely to impose empiricism on and a reductionist framing of literary studies; and science is a pseudo-apolitical discourse that is not inflected with race, gender, history, and aesthetics. The authors argue that important humanistic work cannot be done unless we engage across the humanities and sciences, and that this has to be a reciprocal relationship. Phillips and Rachman question what happens in the brain when we read novels and how different types of reading may be cognitively distinct. They deploy functional magnetic resonance imaging (fMRI) and eye tracking to explore the cognitive dynamics of the different types of focus that we can bring to reading novels. The readers in the study also wrote final essays, which were used to support qualitative analysis. One of the central hypotheses is that close reading would generate more neural activity than pleasure reading. While preliminary results show that this hypothesis holds true, there was also heightened neural activity associated with distinct areas for both reading styles, which would seem to point to these two styles of reading having their own cognitive demands and distinct neural patterns. Such findings relate to recent research on reading practices, attention, and digital reading. Phillips and Rachman present an intriguing example of the complexity and possible gains of deeply integrative work across multiple knowledge domains where the digital can be a connector, enabler, and carrier of humanistic sensibility.

13 CIRCUIT-BENDING HISTORY: SKETCHES TOWARD A DIGITAL SCHEMATIC

Whitney Anne Trettien[1]

To its millions of visitors the Fair says: "Here are the materials, ideas, and forces at work in our world. Here are the best tools that are available to you; they are the tools with which you and your fellow men can build the World of Tomorrow. You are the builders; we have done our best to persuade you that these tools will result in a better World of Tomorrow; yours is the choice."

—*Official Guide Book of the New York World's Fair 1939*

In 1939 New York City staged the first explicitly future-themed World's Fair. Built around the iconic Trylon and Perisphere—an enormous sphere-and-spike structure, featuring the world's longest escalator and a utopian diorama of "Democracity," a future metropolis—the Fair transported visitors to the "World of Tomorrow," where corporate innovation in transportation and communication technologies transformed every aspect of modern life. Fairgoers lined up to drive colorful Ford Zephyrs along the "Road of Tomorrow," tour General Motors' famous Futurama, make their first long-distance telephone call at AT&T's pavilion, or catch a glimpse of the new, mirrored projection cabinets called "televisions." As the official guidebook described the theme: "the eyes of the Fair are on the future—not

I'm in the Houghton Library. It's July. Outside, the temperature is a blistering 103ºF; inside, though, the reading room is cool, dark. Quiet.

Two things are open before me. To my left, foam wedges cradle a book—a big volume bound in gold-tooled red Morocco leather. To my right, my laptop is open. On the screen, I've called up the HOLLIS (Harvard Online Library Information System) entry for the book I'm reading, titled "Bible. N.T. Gospels. English. Authorized. 1630." A second title in HOLLIS qualifies this vague concatenation, though with square-bracketed caution: "[Little Gidding concordance]." In the gap between these two fields lies a more accurate description of the book itself, a biblical Harmony of the New Testament produced at the religious community in the sense of peering into the unknown and predicting the shape of things a century hence—but in the sense of presenting a new and clearer view of today in preparation for tomorrow" (*Official Guide Book* 1939: 36).

Debuting alongside TRK-12 television sets and Vocoders was, oddly enough, a newly discovered 310-year-old Biblical concordance. Produced at the Anglican community of Little Gidding around 1630, this "Harmony" (as such books are called) had been discovered several years earlier when—in a story almost too charming to be true—a man noticed it advertised as "an excellent blotting pad" in an old bookseller's catalog used to wrap a package sent to him. After purchasing the book, the man, Mr. Bernard George Hall, realized it was not only a product of Little Gidding but was in fact the early Harmony borrowed and annotated by King Charles I. Unfortunately, Mr. Hall died shortly after his discovery, and this "excellent blotting pad" soon found itself displayed in the British Pavilion's Royal Room, alongside replica Crown Jewels and the Magna Carta.

of Little Gidding around 1630. The women of the community made the volume by cutting apart and pasting together different printed editions of the Bible, alongside engraved illustrations, to form a single narrative of Christ's life.

An old book, a new computer. The temporal gap we perceive between them obscures the simple fact that here, right *now*, in this climate-controlled time capsule of a reading room, they share the same present moment; and that for me, occupying the same *now*, neither object can be approached independent of the ideological frameworks of the other. Notebook computer qua universal reading machine; cut-and-paste concordance qua pre-digital remix, a kind of early modern hypertext. Although both book and computer appear closed, singular, and self-contained, each is a collaboratively engineered assemblage where disparate moments converge in aggregated bits of matter. In fact, more than merely amassing different materials, both objects actively operate as machines of inscription, writing, rewriting, and shedding palimpsistic layers of pressed paper and digital data.

This chapter examines the temporal hinges—both actual and virtual—linking an old book and a new computer, a historic relic to the "World of Tomorrow"; "digital" to "humanities." More generally, it pulls recent ideas in media archaeology into conversation with digital humanities in an attempt to rewire how the latter imagines its relationship to historical materials.[2]

I'm examining a gap in the concordance open before me: a discolored square indicating where a small bit of paper, about ¼ by ¼ inch, was once glued. Below, a woodcut initial "A" indicates the start of a new chapter, but the space immediately after it is blank. Perhaps the cutouts fell out—not surprisingly, given this book was assembled almost 400 years ago. Or perhaps they were never pasted in. After a small space, the text resumes with "and take the young Child and his," then a blank line, then "gypt, and be thou there until"—conspicuous openings in an otherwise tight block of neatly pasted prints; except these gaps are not blank. At some point in time, an anonymous reader wrote in the expected verse, Matthew 2:13. Taking notes on my laptop, I piece the passage back together, using italics to indicate handwritten annotations:

> and take the young Child and his
> *mother and flee into* Egypt,
> and be thou there until . . .

Even with the italics, my electronic notes give a false sense of unity to the passage, masking the heterogeneity of the material text. I improvise an arrow next to "*E*gypt," and write: "<--- inserted, c17 hand, lrg; KCI? bt not noted in Craig."

What did World's Fair visitors see in this book? It's an unusual artifact, to be sure. While other *printed* gospel harmonies were published in the seventeenth century, the women of Little Gidding made their volumes using a unique process of cutting apart printed Bibles and engravings, then pasting them back together to create a single multimedia narrative of Christ's life. Later Little Gidding concordances exploit the form and function of differing typographies and colored inks to construct multiple reading paths through the story, acknowledging textual variance in their material design while nonetheless synthesizing these differences. One member of the community describes "this new-found-out way" of harmonizing the Gospels with scissors and paste as a technical novelty, "a new kind of printing" they have invented (Muir and White 1996: para. 99). Thus even as the World's Fair visitors of 1939 and 1940 came to gawk at the handwriting of the ill-fated King Charles I, they were also examining a relic of seventeenth-century innovation in communications technology—one that, like Ford's mechanical roads or Elektro the smoking robot, would prove short-lived in the history of media machines.

I have chosen to focus on the mediation of *time* for the simple reason that it is one of the most radically transformative and least examined concepts in current discussions of digital humanities. Through large-scale digitization projects, print-on-demand publishing, and electronic editions, recent initiatives in the humanities are recalibrating our experience of history, altering how its material weight bears on the present moment (Straw 2007). More immediately, blogging, microblogging, and social networking platforms are reconfiguring the temporal coordinates of our professional relationships, constellating new forms of interaction that are both exciting and exhausting. Yet we still lack a vocabulary for discussing these shifts that does not dichotomize the "old" and the "new," the humanities' past and its digital futures. Within digital humanities in particular, a gap is opening between the desire to "move the field forward," away from an older, more conservative model of the humanities and the conviction that its goals and methods signal a temporal *return* to core humanistic values.[3] The image emerging, then, is of a movement eternally pushing forward to the restoration of an always/already constituted Ur humanities. Thus digital humanities sets about (re)building the Ship of Theseus, one (new?) digital tool at a time.

Against this visionary rhetoric—borne of the field's desire to conceive of itself *as* a field—is Matthew Kirschenbaum's (2012, 417) descriptive claim that "digital humanities" is, in practice, a tactical term, "possessed of enough currency and escape velocity to penetrate layers of administrative strata to get funds allocated, initiatives under way, and plans set in motion." Embracing "digital humanities" as a tactical term ends the tedious custody battle of disciplinary ownership and, more to the point here, acknowledges the material novelty of electronic tools and methods without assigning this "new"-ness the burden of a revolution. In this model, "digital humanities" is not the future perfect tense of the "humanities" but a kind of disciplinary adverb qualifying modes and actions. Kirschenbaum's chapter implicitly participates in this repositioning of the historical relationship between the "humanities" and the "digital humanities" by tracing early uses of the term through a series of subtle tactical transformations rather than ruptures and returns.

Nudging Kirschenbaum's claim beyond its institutional utility, a media archaeology-inflected approach deploys not only the term "digital humanities" but digitality itself as a tactic within the broader strategy of humanistic inquiry. We might describe the humanities as a set of inherited disciplinary powers, or "strategies" in the de Certeauian sense of the word (de Certeau 1984), built presently upon the structures of mass print. Digitizing collections or producing electronic editions updates the humanities for the emergent knowledge economy but does little to actively rewire the politics of digital devices. By contrast, deliberately engaging the charged differences of electronic software and hardware—their *strangeness* in relation to other media—has the power to short-circuit scholarly conventions within the humanities, forcing current methods of reading, writing and communicating to run along previously unintended paths.

While this approach borrows much from the work of tactical media artists/archaeologists (Raley 2009, Wark 2006, Lovink 2005; see also Losh 2012b)—who, notably, are often cited by digital humanists but rarely describe themselves as such—its positioning within digital humanities helps to expand tactical media's focus on twentieth-century consumer technologies to deeper histories of objects and protocols, indeed to a Benjaminian engagement with history itself as a political imperative, "charged with the time of the now [*Jetztkeit*]" (Benjamin

(1968)2007: 261). Hinging the political tactics of media art to the institutional tactics of the term "digital humanities" also helps direct the brunt of new media art's cultural critique against very specific policies and practices that structure higher education. Most important, this tactical hybrid method demands that practitioners attend to the materiality of digital mechanisms and, following that, to the *alienness* of electronic objects at the level of both code and device (Bogost 2012a). Doing so short-circuits not only the conventional strategies of the humanities but also our sense of what history is, and what it can do for us.

tactics, n. Arrangement or disposition (*obs.*, rare). A term for the study of the relation of linguistic units. Any mode of procedure for gaining advantage or success.

Microsoft Word remediates the gap in this Little Gidding Harmony in strange ways. For others to understand what I describe will require high-quality images; so I have decided to have the book digitally photographed. What will be the material existence of this electronic representation of a seventeenth-century collage?

First, these digital images—frozen moments, mediated by both camera (then) and screen (now)—will be stored on a server and posted online for others to access. Though I have never seen or touched these servers, there is a noisy weight to them: they suck up electricity, blow out hot air, spreading noise and heat into their surrounding environment. The more Web users access the images, the more the servers work and thus the more energy they use—a fact that, in aggregate, has a global impact on the planet's flora, fauna, and atmosphere but no impact on the digital artifact itself. Thus, while my physical handling of the Little Gidding Harmony at the Houghton Library inevitably left minute traces of that moment on the book—of my

Tinkering with the digital photos already shot of Harvard's Harmony, I begin sketching a plan for a remixable Web-based project that honors the cut-and-paste methods of the women of Little Gidding. This temporal dislocation—a blurring of scissors, wheat paste, and CTRL-X—is tactical, designed to dredge up lost protocols through intentional anachronism. What do book historians learn (I want to ask) by dragging the mechanisms of remediation across material history like a raking light, pulling its texture into sharp relief?

The immediate lesson pertains to our own moment in time. Though less prominent in recent discussions, remix culture has been central to how many in the digital humanities have defined themselves against traditional humanities methods, with UCLA's "Digital Humanities Manifesto 2.0" claiming that "anything that stands in the way of the perpetual mash-up and remix stands in the way of the digital revolution" (Presner et al. 2011: 5). Many scholars have (as well-trained humanists are wont to do) sought precedents for body, the bodies surrounding me and the room we inhabited together for a brief afternoon—my online examination will leave the digital artifact unaltered in its appearance; yet it *will* still leave its trace on the globe by contributing, in a small way, to climate change and the depletion of the planet's natural resources, both fossil fuels for electricity and the metals and minerals used to manufacture the servers (Gabrys 2011; Koomey 2011; Sterne 2007).

The key point here is not that one form of interaction is less environmentally damaging than another (for the climate control required to preserve rare artifacts arguably has a greater carbon footprint than viewing a digital facsimile online). Rather, I want to underscore that this deferral of a physical impact away from the object or the moment of interaction indicates a major shift in how we experience and perceive history. Viewing the Little Gidding Harmony in the Houghton Library, I am witness to its existence across multiple centuries—to its physical aging and preservation—as well as to its co-presence in a *now* in which it continues to accrete the material stuff of its environment. Viewing the digital images, though, I observe an eternally returning present moment, the moment when the book was photographed. The representation is not subject to the same *now* as me, then, but drags its own frozen moment in time across a multiplicity of *nows*.

cut-and-paste criticism in the experimental artistic practices of Bryon Gysin and William Burroughs, in Jonathan Swift's combinatory literary engine or, more recently, in Alfred Jarry's pataphysics (Drucker 2009; Ramsay 2011; McGann 2001). As one might tell from this list, a desire for disruption, for what Mark Sample (2010) has now (in)famously described as an "insurgent humanities," lies implicit in the field's attraction to experimental remix methods; indeed, my own interest in tactical remediation and intentional anachronism is similarly inspired.

Yet "remix" is a capacious concept. Though composed as a multimedia mashup of printed materials, the Little Gidding Harmony did not use cut-and-paste methods to disorder but to *harmonize* meaning, weaving the many narrative threads of the gospel into a single tapestry. Situated within—in fact remediated by—the context of the Web, the seemingly conservative aesthetic of this seventeenth-century book unravels the intellectual ethos of radical experimentation woven around slicing, dicing and mashing up media objects in our own time. In doing so, this juxtaposition encourages those interested in historicizing digital humanities to hone the meaning of "remix" in ways that invite alternative histories of creative, cut-and-paste criticism—histories, that is, that do not begin and end with 1960s counterculture or the perpetual avant-garde.

Earlier I used the metaphor of short-circuiting; a more apt image might be *circuit-bending*. In an article on trends in media art, Garnet Hertz and Jussi Parikka (2012: 428) envision the practice of media history not as a narrative form of time travel, as happens in archives—those timeless vacuums where, shedding personal effects, the researcher "pages" the past—but as a short-circuiting of the technological residue of history. Through this temporal circuit-bending, "the black boxes of the historical archive and consumer electronics are cracked open, bent, and modified" in order to disturb, renew, or otherwise intervene in their operations (Hertz and Parikka 2012: 428). Importantly, Hertz and Parikka mean this statement literally. Rather than metaphorical actions, verbs like "cracked open, bent, and modified" describe the circuit-bending acts of the hacker (an)archaeologist, rewiring the past to hear or see or read or perceive something new. The radical and intellectually jarring juxtapositions that occur in such acts of "t(h)inkering," as Erkki Huhtamo (2000) calls them, spark a deeper appreciation of the presence, the material weight, of history and its significance to our own moment.

Circuit-bending works as an approach to media history precisely because, as Wolfgang Ernst (2011: 241) points out, a technical object like a radio or a computer "discloses its essence only when operating." In Ernst's example, merely observing a low-cost 1930s German radio in the basement archives of a museum, the way one might observe a painting, offers little insight into the culturally and politically significant role it played by disseminating Hitler's voice to the people. Switching it on, though—listening to contemporary pop music on this aged system—brings the relationship between cultural and technological functionality into focus, showing that

> there is no "historical" difference in the functioning of the apparatus now and then (and there will not be, until analogue radio is, finally, completely replaced by the digitized transmission of signals); rather, there is a media-archaeological short circuit between otherwise historically clearly separated times. (Ernst 2011: 240)

Although Ernst performed no physical "t(h)inkering" on this radio, the result was similar to the circuit-bending art described by Hertz and Parikka: namely, through the creative (re)activation of a device, our sense of history, as delineated by the physical archive, changed.

A focus on late-nineteenth- and twentieth-century consumer technologies, like radios, has occluded media archaeology's connection to trends in book history and the digital humanities. Yet an inverted form of Ernst's radio experiment occurs every time, for instance, a medieval manuscript is digitized and posted online in the framework of Web-based software. That is (to state an obvious but still underappreciated fact), an online e-book platform is, like a radio, an operational apparatus, such that certain aspects of it cannot be understood unless it is switched on; and though Ernst may not agree, so, too, is a medieval manuscript: it must be opened, *used*, for it to "disclos[e] its essence." When the former mediates the latter, it alters the operations, the "codes" of the medieval codex, physically and temporally rewiring its circuitry according to the software's designed logic. What represents the "past" and the "present" in this modified, mashed-up digital artifact? Scholarship abounds on the social transformations engendered by such tools; much remains to be said on the *material* transformations they enact, and even more on how their tactical deployment might spark new historical devices.

device, n. An invention or contrivance. Purpose or intention (*obs.*). Will, pleasure, fancy; now only in the phrase *left to one's own devices*.

Bringing the Little Gidding Harmony into a playful, digital space elucidates otherwise invisible moments in the book's history, including its appearance at the 1939 New York World's Fair. To my knowledge, there are only two bits of evidence that the volume was ever displayed at the Fair: an article in the *Harvard Library Bulletin*, published shortly after Harvard acquired the book, and the source for that article, a typewritten provenance record shelved with the book at the Houghton Library. Its appearance (on the fairly safe assumption that it appeared) seems to have factored little if at all in future scholarship on Little Gidding, and indeed may have left little impact on fairgoers themselves, overwhelmed as they were by the more glamorous sites of the "World of Tomorrow." Nonetheless, as technological innovations today invigorate an interest in these early modern examples of remix, the book's presence at the first future-themed World's Fair—an exhibition flanked on the one side by the Great Depression, and World War II on the other—suddenly seems significant. Remediating history can thus powerfully change its topography.

To state that digital facsimiles present an eternally returning frozen "now" is more than ethereal philosophizing. If I download a page image from the digitized Harmony to my local hard drive, or even just access it online, the date and time of the download may become the "date created" in the photo's metadata or the "date modified" in my cache, giving the appearance of an image updated with every viewing. Thus, while the image as a *representation* remains frozen in time, the image as *information* is temporally on the move, copied from my server today to the local disk drive of a website visitor tomorrow, to another set of storage hardware next month in a constantly fluctuating rematerialization of data. (These fluctuations occur even without my intervention as my hard disk drive shuffles data, cutting and pasting new over the old—a palimpsest of frozen moments in the history of my storage device.) Similarly, regardless of what kind of digital repository is used to store the electronic photographs, that system of hardware will require replacing anywhere from every three to every twenty years to avoid bit rot and technical obsolescence—yet another

Of course, this holds for the study of Little Gidding itself. While Little Gidding and its founder Nicholas Ferrar have long been lauded within Anglican histories, and even saw a brief spat of public fame in the late nineteenth century with the publication of Joseph Henry Shorthouse's historical romance *John Inglesant* in which the eponymous hero visits Little Gidding, scholars have neglected the community's remarkable Harmonies, those (in the words of one twentieth-century critic) "dreadful monuments of misdirected labour" (quoted in Muir et al. 1996: 18). Even recent work on the Harmonies tends to overemphasize Nicholas Ferrar's intellectual authority in composing the books, thereby minimizing the contributions of the women to their production. Approaching the Little Gidding Harmonies from a digital humanities perspective, though—a perspective that argues for the value of process over product, collaboration over individual authorship, and "maker" culture in general—the women's manual labor becomes significant in its own right, in fact becomes a form of *intellectual* labor as significant as that of Nicholas Ferrar. In this way, a shift in disciplinary values cuts new pathways into the past.

This form of recursive historiography—I am borrowing the term from Markus Krajewski, through Geoffrey Winthrop-Young (2012: 6–7)—journeys from the present to a related moment in the past, then back to the present, allowing each stop, each iteration of a particular *topos* like "remix," to transform our relationship to all others in way digital storage temporally and spatially defers its material impact.

Almost a century ago Walter Benjamin identified "reproducibility" (*Reproduzierbarkeit*) as a key feature of modern media, a concept updated by Lev Manovich (2001: 36) as "variability," the idea that new media objects "can exist in different, potentially infinite versions"; such terms, though, do not capture the *temporal* dislocations that occur when a digital representation of a historical artifact becomes the primary access point into the past. By hoisting entire archival collections online, digitization projects transform history into a series of recursively updated, selfsame snapshots. Though small and seemingly insignificant, these micro-moments of temporal and physical remediation add up to a large-scale revolution in what it means to do history. As critical theorists have argued over the past few decades, the contours of archives direct the flow of historical research, such that the artifacts an institution chooses to preserve and how scholars interact with them in their research influences—determines, even—the kinds of questions that the humanities can ask of the past. Digital collections do not obviate these lessons simply by "opening up" library treasures to the "public" (I put these words in scare quotes because "openness" and "publicness" are implicitly relative in such discussions) but extend their significance into a new and very different institutional arena, no more so than for unique, delicate, and access-restricted the series. Winthrop-Young describes this approach by quoting from T. S. Eliot's "Burnt Norton"; however, another poem in Eliot's *Four Quartets* may be more appropriate here: "Little Gidding," which he penned shortly after visiting its titular place. "We shall not cease from exploration," read Eliot's (1971, 59) now-famous lines; "And the end of all our exploring / Will be to arrive where we started / And know the place for the first time."

objects like Houghton's Little Gidding concordance. I, like others with similar projects, am interested in digitizing the Harmony in part to preserve it, in part so that other scholars can research it. This altered perspective thus changes the kinds of questions we can ask of the past. Yet *which* of this Harmony's "pasts," which frozen moment in its existence, will my set of digital photographs preserve and disseminate?

The ability to transform not only how we study but *what we see in* the past, deposing Enlightenment-era attitudes toward authorship and textuality, is perhaps the digital humanities' greatest potential. Whether or not practitioners in the field can mobilize this strength, though, hinges on their insistent awareness of mediation, its effects and affects. As Johanna Drucker has forcefully argued, too many projects employ digital mapping tools and information visualizations as if their interfaces were transparent, self-evident reifications of data, rather than contemporary graphical conventions. Not only does this blind adoption of "tools"—the very word implies a mere extension of human capabilities—countermand humanistic forms of creative, critical inquiry, it also tares history against our own media ecology. In this way the present becomes the inevitable outcome of a past in which Western culture has always traversed the routes constructed by Google maps, or Western thinkers like Newton always "intended" their writing to be read through highly mediated digital transcriptions.

Rather than projecting the past onto the present, a circuit-bending approach to digital humanities uses our present media ecology as a map for discovering the neglected corners of history, then plugs them into our own moment. It is powered by a recognition of "the madeness and constructedness that inhere in *any* representation of knowledge" (Drucker 2012: 91) and, as such, exploits the rich potential of electronic media without compromising the transformative power of the humanities to think and perform new ways of perceiving, experiencing, and being in the world. Pursuing this approach also helps bridge the perceived divide between building and thinking, making and theorizing; for here, a material engagement with objects enacts a theoretical relationship between past and present, reconfiguring our historical coordinates in the process. It is critical *and* creative, interpretive *and* interventional. While in this sense a circuit-bending approach shares much with the algorithmic criticism of Stephen Ramsay (2011) or the conjectural criticism of Kari Kraus (2009) or the notion of creative criticism expanded upon by Jamie "Skye" Bianco (2012), its potential is fully realized not as an interpretive *method* for "deforming" the dematerialized text but as an electronic schematic that diagrams the historical junctions where our sense of what is "old" both meets and diverges from our perception of the "new."

The Communications Building at the 1939 World's Fair featured Symbolizing Man, a twenty-foot plastic head linked by a gleaming light to a thirty-foot globe. Between them, a multi-paneled mural showed "the acceleration of inventions in communications from primitive beginnings to the 'World of Tomorrow,'" including "postal service, printed word, telegraph, telephone, motion picture, radio and television" (*Official Guide Book* 1939: 76). Of course, an object like Harvard's Little Gidding Harmony fills no gaps in this list, forms no missing links in the evolutionary process connecting sphere to Symbolizing Man. Its inability to fit into this technological history speaks volumes on the ideologies that shaped the exhibit, and that still shape how we conceptualize newer technologies in relation to older ones. The current hinging of "digital" and "humanities" has the power to short-circuit this narrative, if we deploy the former as a tactical re-wiring within the strategic devices of the latter. In doing so, we, with Eliot, "arrive where we started / And know the place for the first time."

Notes

1. Thank you to Mary Caton Lingold and Sarah Werner for reading, commenting on, and talking through earlier drafts of this chapter.

2. Practitioners in both fields have increasingly recognized the need for dialogue between them. In "The State of the Digital Humanities: A Report and a Critique," Alan Liu acknowledges the potential contributions of media archaeology to digital humanities (Liu 2012: 16), while Jussi Parikki, in *What is Media Archaeology?* asks (but leaves unanswered) the question: "How could media archaeology contribute to some of the debates in digital humanities?" (Parikka 2012: 134).

3. I hesitate to target specific texts or thinkers, as this conviction is manifest less as an explicit formulation, worthy of pointed critique, than as an implicit undercurrent within the discursive formation of "Digital Humanities." Nonetheless, two examples may elucidate the tension I describe. The first is the recent collaboratively written volume *Digital_Humanities*—a book that presents digital humanities as a *provocation* to the solitary, print-based scholarship of early humanists even as it self-consciously invokes the long history of humanistic inquiry as a way of grounding the field. In the historical logic of the book, new media forms and modalities are crucially different "harbingers of renewal" that force a "*re*interpretation of the humanities"—yet this "reinterpretation" is in fact a *return* to an earlier, pre-print culture of vision and voice for which "the art of oratory must be *re*discovered" (10–11, emphasis added). Thus the text's discourse deploys history to legitimate a field that is both always outside that history and

embedded at its center. This tension is also evident in Stephen Ramsay's *Reading Machines: Toward an Algorithmic Criticism.* On the one hand, Ramsay wants to insist, following Jerome McGann and Lisa Samuels, that all critical acts are, and always have been, deformative, "rel[ying] on a heuristic of radical transformation" (Ramsay 2001: 16, 38, 48); on the other, he argues that the deformative power of digital tools "represents a revolutionary provocation against the methodologies that have guided criticism and philosophy for centuries" (63). In this way the book ends up claiming that algorithmic criticism simultaneously transforms and fulfills the goal of the humanities as it "has always been" (68), a confusion Ramsay consciously dances around in paragraphs like that concluding chapter 1 (17). Similar tensions may be pinpointed in *Debates in Digital Humanities* or many recent blogposts; and while all of these works make clear and thought-provoking contributions to the field, the way each situates its own intervention historically is broadly problematic.

14 MEDIEVAL MATERIALITY THROUGH THE DIGITAL LENS

Cecilia Lindhé

The art of the Middle Ages does not hold up a perfect "globed fruit" but leads one in a walk along converging and diverging paths. (Carruthers 2010: 190)[1]

—Mary Carruthers

One of the mantras of SpecLab has been that nothing is self-identical—no text, no image, no object. All aesthetic objects are fields of potential. It is only through interventions and aesthetic provocations that a work is constituted as an act. (Drucker 2009b: 132)[2]

—Johanna Drucker

In David Small's (2002) interactive installation piece *Illuminated Manuscript*, text emerges when the viewers move their hands over one of the twenty-six blank pages. Sonar sensors track the movements and an overhead video projector responds by virtually printing text on the white surface. As fingers run over the oversized book, pages are turned, typography is disrupted, and textual elements are continually reconfigured. David Small writes about his work that "[t]he piece explored new types of reading in tune with human perceptual abilities."[3]

However, *Illuminated Manuscript* also refers back to older types of reading, as it orchestrates what might be called a pre-Romantic aesthetics (Carruthers 2006: 17n.26). This is an aesthetic practice that is *not* to be equaled with generalized ideals of the beautiful, or with a philosophy of idealized taste, but rather, with the meaning it had until the late eighteenth century of "pertaining to things perceptible by the senses," especially by experiencing and feeling, but also by seeing and hearing.[4] In this older vocabulary *aesthesis* had to do with perception, feeling, and embodied experience. It was physiological in nature and emphasized immediacy, interactivity and tactility. For example, in ancient times, cult statues and images were not mere representational objects, physical interaction was expected, and, according to Caroline Walker Bynum (2011: 37, 38), during the late Middle Ages, art was "material" in the sense that it probed "to be touched more than seen" and even "meditating was as much tactical as visual."[5] When viewers are encouraged to touch art works, "the tactiloclasm" that permeates post-romantic aesthetic theory, where the work of art is viewed as untouchable and concerned with ocular scrutiny only, is evaluated (Huhtamo 2008: 71–101).[6]

The post-Romantic view on aesthetic objects also informs websites and digital archives of medieval artifacts such as sculptures, panel paintings, and baptismal fonts. They are displayed as art objects in a museum—not to be touched or interacted with—as a "perfect 'globed fruit.'"[7] Further, in a similar manner to how the Gutenberg Bible remediated manuscript culture, when the printed book displayed the activity of manuscript copying, digital websites of medieval texts and images often replicate the printed page.[8] It is not just that, in the words of Jerome McGann (2001), we "think in textual codes," but these environments seem to be designed according to a hierarchical relationship between technological tools and humanities tools, where the former have been granted greater authority (ix). Johanna Drucker (2009b) describes this accordingly: "Computational methods rooted in formal logic tend to be granted more authority in this dialogue [between computational technology and the traditional humanities] than methods grounded in aesthetic and subjective judgment" (xi).

Consequently in this chapter I attempt to invert this power relation by using humanities tools, such as *memoria* and *ductus*, in order to challenge the conceptual foundations of existing websites and digital archives of medieval artifacts. Inspired by Drucker's book *SpecLab* (2009b), I will regard this challenge as an aesthetic provocation "grounded in a serious critique of the mechanistic, entity-driven approach to knowledge that is based on a distinction between subject and object." (Drucker 2009b: 21). This chapter thus has as its wider scope the deconstruction of the filter of print technology with which we look at cultural history, and instead employs digital interactive installations to renegotiate a view on medieval aesthetic practice that emanates from print technology and a post-Romantic notion of aesthetics. In other words, it moves from treating the medieval church as text to treating the medieval church as experience, and explores the relationship between word, image, and performance during the Middle Ages.[9] In this context, "the digital" is not only a phenomenon that could be tied to certain digital objects or used as a tool but as an approach to history, with strong critical potential.

The discussion to follow is derived from a current project about the representation of the Virgin Mary in Swedish medieval image and text.[10] The aim of the project, which will be described in greater detail below, is to raise innovative research questions about digital representation of cultural heritage collections. By developing new methods and tools, the project investigates how digital technology could function as a critical perspective on medieval materiality and, more generally, on the humanities as such. Consequently it is vital to understand "the digital" not only as the latest and currently predominant technology but also as a critical perspective that could revitalize media historical approaches in the traditional disciplines. This view is in accordance with what John Guillory (2010) maintains in his article on the genesis of the media concept: "If a new instauration of the cultural disciplines is to be attempted, it is all the more necessary that scholars of culture strongly resist relegating the traditional arts to the sphere of antiquated technologies, the tacit assumption in the losing competition between literature and the new media" (321–62). Accordingly, I argue that the humanities (and the digital humanities) would benefit from engaging much more strongly with the digital as a critical perspective on aesthetic concepts and cultural history.[11]

Memoria: Architecture as a Trope for Invention

The trope of architecture has constituted the conceptual basis for the four interactive installations that have been developed in order to explore the digital as lens on medieval materiality.[12] These installations are part of the ongoing cross-disciplinary research project *Imitatio Mariae—Virgin Mary as virtuous model in medieval Sweden*, that aims to study the representation of the Virgin Mary as a role model for lay people in Sweden during the Middle Ages (c. 1100–1500). The researchers focus on a selection of scenes and stories from the life of the Virgin Mary and how they are depicted in the interplay between medieval Marian motifs in Old Swedish texts and visual representations painted, carved, or sculpted on the church walls, as well as on three-dimensional objects such as Madonnas and baptismal fonts.

An essential part of the project concerns aesthetic theory and the reconsideration of medieval materiality through the digital lens. Instead of thinking in terms of collecting, preserving, and exhibiting collections of medieval artifacts, this project works out a model of humanistic research infrastructure that goes beyond databases and 3D modeling. The Swedish medieval church is modeled as a multimodal space that encourages multi-sensuous involvement. This aspect of medieval materiality thus proves to be best interpreted through a digital lens that does not strive toward authenticity. The embodied movement of spectators or interactors triggers representations of the Virgin Mary as developed by digital interactive installations. The rhetorical concepts *memoria* and *ductus* constituted the basis for the interactive installations. I will return to these concepts below.

Interactive installations and virtual reality art have evolved from and in relation to architecture, sculpture, and performance.[13] These kinds of artworks are realized in virtual spheres that generate tangible spatial experiences. Naturally this is not an entirely novel practice. Hellenistic visual art and poetry also created modes of viewing in order to involve and integrate viewers and readers visually as well as spatially into compositions and in medieval monastic rhetoric the concept of architecture was used as a crucial metaphor for reading and experiencing art works.[14] For example, the twelfth-century Benedictine Peter of Celle writes about how the reader should think of reading Genesis as a kind of map and "journey through the greater part of your reading [of *Genesis*]," then "enter the Ark at the time of the flood," and "with a deliberate but light step [go] through the contents of this book" (Carruthers 1998: 100). He contends further:

> Whenever you enter a pleasant meadow of prophetic blessings, loosen the folds of your garment, stretch your belly, open your mouth and extend your hand.... Then come to Exodus [grieving for] the entry into Egypt.... Then admire the foreshadowing of our redemption in the blood of the sacrificial lamb. Observe how the law was given on Mt. Sinai and how it is open to spiritual understanding. By progressions of virtues run through the forty-two stopping places [in the Sinai] with what they signify. With an angelic mind construct within yourself the Tabernacle and its ceremonies. (qtd. in Carruthers 1998: 109)

Peter of Celle emphasizes variations of tempo and mood—"the *colores* of the journey determined in each site"—when he encourages the reader to "walk," "enter," "travel," as well as to look, touch, taste, and hear (Carruthers 1998: 109). Architecture was also considered the best source for remembering in the ancient art of memory—*memoria*. The mnemonic system developed in the ancient period was based on *loci*, the placement of allegorical images within constructed mental architectures such as a building or a landscape. *Memoria* and the practice of *loci communes* thus stressed the importance of linking spatiality to memory images.[15] In the context of medieval monastic practice, architectural mnemonic is, according to Carruthers, to a great extent founded on a text from St. Paul (Corinthians 3:10–17) who describes himself as a master builder and uses architectural metaphors not for the storage of memories but rather as a trope for invention (Carruthers 1998: 17). The goal of *memoria*, then, is not about repeating previously stored material—much like the way websites reproduce medieval artifacts. Rather, it is about "creative thought" and, according to Carruthers, it is "crucial to understand ... that what is important to [for example] early pilgrims is not the site as an authentic, validated historical object. ... What is authentic and *real* about the sites is the *memory-work*, the thinking to which they gave clues" (Carruthers 1998: 42). This means that the sites such as a temple or a city, and in this case the digital installations, trigger imagination and creativity through embodied and sensorial interaction. Monastic *memoria* was an art of image-making—to the mind's eye—that were performed in close connection to existing images and actual words that, "someone had seen or read or heard—or smelled or tasted or touched, for all the senses, as we will observe, were cultivated in the monastic craft of remembering" (Carruthers 1998: 10).[16] Central to this way of thinking is not how truthfully the images are represented, but rather, their "cognitive utility," that is, if they can be used as sites upon which it is possible to build or invent new content, such as thought or prayers (Carruthers 1998: 72).

Figure 14.1
The Digiti

Ductus: Journeying through a Digital Interactive Installation

In the installations that have been developed so far, the concept of *memoria* and the medieval monastic emphasis on architecture have been essential to the working process. The material that is used in the installations consists of approximately four thousand photographs and videos from almost one hundred Swedish medieval churches. Also sound, such as the silence in the baptismal fonts, the movements of bodies and footsteps, have been recorded. The prototypes have been developed in the digital humanities laboratory, HUMlab at Umeå University. HUMlab has a unique screenscape that consists of eleven dispersed screens, a sound system, as well as a LED-based modular light ceiling that can create special effects, shapes and images, on the lighting plane and project different colors.

The first installation piece—*The Digiti*—allows the user to juxtapose image and text on a 70" touchscreen. Several thumbnails are available on top of the screen and it is possible to compare two images at a time by using touch.

The Magnifier is developed for a 170" high-resolution, back-projected screen. The user can choose between several different thumbnails and with the aid of a mouse move the image, as well as zoom in and out. The high-resolution screen makes it possible to zoom in on ceilings paintings and other areas that could be difficult to reach on site. These installations have mainly been developed and used as tools for the project members, since

Figure 14.2
The Magnifier

Figure 14.3
The Calendarium

Figure 14.4
The Sensorium

they make possible the juxtaposition of images as well as the magnifying of details in paintings.

Two installations—*The Sensorium* and *The Calendarium*—are more experimental and emphasize to a greater extent the necessity of bodily movement and active participation of the user for the realization of the piece. In *The Calendarium,* a Madonna with child functions as the basis for a spatial exploration of different Madonnas. With the aid of a mouse, the user can find different *loci* in the digital image that contain other Madonnas in smaller variants. Naturally this piece could be a useful research tool as it makes possible the juxtaposition of Madonnas in different styles, from different locations and times, and also, perhaps more important, shows *how* the images are found. When a new, previously hidden Madonna becomes visible, the search patterns in the image are connected to the lighting in the ceiling that changes color depending on which image the user locates. This is a journey that is connected to different *colores* or ornaments that function as keys or obstacles for finding one's way through the composition (Carruthers 2010: 198). What is experienced here, then, is the activity of moving the mouse over the screen to find the links. This suggests that it is not the artwork as an authentic historical object that is important but rather the aesthetic object as a field of potential, as a process, or *loci* through which the reader, user, or viewer passes while inventing new thoughts and memories. This installation thus actualizes the notable statement of Thomas Aquinas, that "art imitates nature in its operation" (qtd. in Barbetti 2009:13). By this, Aquinas implied that art does not imitate nature in the sense of copying, but rather imitates the way nature works—its *processes.*

The installations' emphasis on movement and process is derived from another concept crucial in the context of medieval monastic practice and thinking about art: *ductus*. The core principle of *ductus* is movement, not *stasis*. It concerns process rather than the finished product and could be translated as "directed motion" (Flynn 2010).[17] It emphasizes finding one's way through a composition by arranging it as a journey through linked places where each such site or loci has its own mode or color. Carruthers summarizes: "For a person following the *ductus*, the 'colors' act as stages of the way or ways through to the *skopos* or destination. Every composition, visual or aural, needs to be experienced as a journey, in and through whose paths one must constantly *move*" (Carruthers 1998: 81). In *The Sensorium* we have *ductus* used to analyze and represent the experience of artistic form as an ongoing, dynamic process rather than as the examination of a static or completed object (Carruthers 2010: 190). This installation registers the movement of a body that approaches the screen and triggers new images accompanied by a church choir singing *Ave Maria*. The lighting continually changes in relation to a body's movement toward the screen. In this installation the senses of sight, sound, and touch are aroused—as were central to the multimodal and tactile medieval experience of the mass.

The representation of medieval text, sculptures, paintings, and baptismal fonts, not as static objects but as an orchestration of artistic form as an ongoing process, foregrounds the audience rather than the author. As the body interacts with the technology in the form of touchscreens and sensors, embodied movement becomes a prerequisite for the realization of the images. By active participation, viewers not only can behold but can interact with the installations. This way the Mary digital interactive installations appear to exploit the connection between the physical and the virtual and recreate a multisensory experience close to that which was critical for medieval aesthetics.

From Sight to Touch

The Mary installations have also presented us with some theoretical and critical questions and prompted us to reflect on traditional scholarship. For example, rereading medieval materiality with an emphasis on the multisensory and by exploring tactical and auditive aspects in relation to medieval materiality, we can begin to inquire into, as Mark Paterson notes, "another critique of the discourse of visualism, analysing certain practices historically assumed to be visual and abstracted; and ... thereby reveal the underlying haptic ... aspects of spatial experience and reinscribe them into that cultural history" (Paterson 2007: 59). This could open up discussion of medieval mythic and visionary literature in relation to the Mary examples of interactive installation art. Hitherto these genres have been interpreted in literary criticism with regard to the *mimesis* concept and print based theories (Carruthers 1998: 24). Accordingly, to stimulate cultural disciplines that are founded on the older scheme of the arts, it is vital not only to discuss, analyze, and understand nondigital artifacts in relation to "older technologies" (e.g., how print technology influenced the novel and the concept of genre) but also to investigate how the digital perspective could challenge theory, method, and our view of history.

In this regard I want to emphasize that classicists and medievalists were early adopters of digital technology and that scholars within these fields have been at the forefront of digital humanities research. Early and ambitious projects—both from 1972—within the fields are the *Thesaurus Lingua Graecae* and *The Lexicon of Greek Personal Names* (Cane 2004: 47–55; Brodard and Mahony 2011). Other projects are creations of digital 3D-models and digital archives such as *Rome Reborn: A Digital Model of Ancient Rome* at the University of Virginia, the *Digital Roman Forum* at UCLA, the *Stanford Digital Forma Urbis Romae Project* at Stanford University.[18] However, these efforts seem to be driven by discourse and practices that either strive toward authenticity or to provide digital methodologies for dealing with large data sets in analyzing and archiving non-digital cultural material. Still other efforts, more in line with the Mary projects described here, have shown how digital environments—in the visualizing and analyzing of geospatial information—can help us in re-thinking concepts such as those of "space" (Dunn 2010: 53–69).

Digital interactive installations that attend to the physical interaction and the materiality of a work encourage a rediscovery of multisensory experiences. Our engaging with digital interfaces creates a *vervremdungseffekt* that provides insights into traditional art forms and practices. Digital interactive installations can align with the medieval experience of art forms as process and particularly in the way that its meaning is about effect, immediacy, aurality, and tactility. Installations of the type discussed above, stage through a space–body–word–image nexus the medieval multimodal patterns of performativity. What we today call art objects, such as the Madonna or sculptures of Christ, were used in liturgical processions where music, spoken words, bodily movements, and audience interaction were crucial for the experience (Bynum 2011). But digital interactive installations also foreground the importance of space and architecture in a way similar to how ancient rhetoric was performed in public spaces such as the agora.. As Mary Carruthers puts it: "[T]he heart of rhetoric, as of all art, lies in its performance; it proffers both visual spectacle and verbal dance to an audience which is not passive but an actor in the whole experience, like the chorus in a drama" (Carruthers 2010: 2).

In the last few years substantial criticism has surfaced against the hegemony of vision in our Western cultural tradition.[19] Indeed, while sight was considered the noblest of senses during antiquity, as we find in the writings of Heraclitus, Plato, and Aristotle, for example, the emphasis on aurality and physical engagement was as important in oral, rhetorical presentations. In *The Senses of Touch* (2007), Mark Paterson shows how Greek histories of measurement were multisensory and dependent on the body. Paterson writes:

> Before it becomes an abstracted, visual set of symbols on a surface, at one stage geometry involved the actual bodily process of measuring space. In the measuring process, the hands, feet, eyes and body are all involved in spatial apprehension and perception. Spatial relations mediated through the body become represented in abstract form through a set of visual symbols. As we know, such visual symbols become part of a whole system of representation, geometry, which is subtracted from the original, embodied measuring process. (Paterson 2007: 60)

The development of geometry into an abstract form moreover meant an active forgetting of the senses, which implies a move from "the variability of the senses and sensory experience to the static invariability of a desensualized, abstract space" (Paterson 2007: 65). This is, according to Paterson, symptomatic for how the body has been written out of the cultural history of the West and instead emphasis has been put on the visual sense. The development of this geometry is evident in print technology, and in databases and digital websites medieval materiality is still being channeled into a concern with ocular scrutiny and so the dimensions of the pre-print culture are excluded; that is, the sense of touch, movement within these environments, and daily use of objects. The digital interactive installations of medieval spaces developed within the *Imitatio Mariae* project emphasize the processes through which the user passes through and experiences the historical environment, and indeed the relations between space and its embodiment of memory and performance.

With the aid of the digital lens, it is therefore not only possible to analyze what is new in digital interactive art but also to experiment with new ways of reading and viewing. It is possible to defamiliarize—affirm, correct, deepen, or overturn—our understanding of and approach to more traditional art forms and practices. But perhaps most important, the digital lens makes possible the exploration of the limits print culture has imposed upon pre-print culture and so to destabilize those critical limits.

Notes

1. Mary Carruthers, 2010, "The Concept of Ductus, or, Journeying through a Work of Art," In *Rhetoric beyond Words: Delight and Persuasion in the Arts of the Middle Ages,* Mary Carruthers, ed., Cambridge: Cambridge University Press, pp. 190–213.

2. Johanna Drucker, 2009, *SpecLab. Digital Aesthetics and Projects in Speculative Computing*, Chicago: University of Chicago Press.

3. Small Design Firm, Inc., "Documenta 11: Illuminated Manuscript," accessed June 27, 2013, http://www.davidsmall.com/portfolio/illuminated-manuscript/.

4. "Perception by the senses" is its root meaning in Greek: *aisthesis* (αἰσθησις), see Carruthers (2006: 17n.26).

5. See also Sarah P. Morris, 1992, *Daidalos and the Origins of Greek Art*, Princeton: Princeton University Press; Polly Weddle, 2010, *Touching the Gods: Physical Interaction with Cult Statues in the Roman World*, Durham: Durham University Press; S. Alcock, and R. Osborne, eds., 1994, *Placing the Gods: Sanctuaries and Sacred Space in Ancient Greece*, Oxford: Clarendon Press; Nigel Spivey, 1995. "Bionic Statues," In *The Greek World*, ed. Anton Powell, London: Routledge, 442–59.

6. Huhtamo writes, "We could speak of "tactiloclasms"—cases where physical touching is not only absent, but expressly prohibited and suppressed (75). The emphasis on sight in Western cultural practices is well documented. See also Mark Paterson, 2007, *The Sense of Touch: Haptics, Affects and Technologies*, New York: Berg; Constance Classen, 1993, *Worlds of Senses: Exploring the Senses in History and across Cultures.* London: Routledge; Jonathan Crary, 1990, *Techniques of the Observer: On Vision and Modernity in the Nineteenth Century*, Cambridge: MIT Press; Martin Jay, 1993, *Downcast Eyes: The Denigration of Vision in Twentieth Century French Thought*, Berkeley: University of California Press; Erkki Huhtamo, 2006, "'Shaken hands with statues ...': On Art, Interactivity and Tactility," *Second Natures*, Faculty Exhibitions of the UCLA Design-Media Arts Department, Los Angeles: University of California, 17–21.

7. See the epigraph of this chapter and Carruthers (2010: 190).

8. See Elizabeth L. Eisenstein, 2005, *The Printing Revolution in Early Modern Europe*, Cambridge: Cambridge University Press; Martin K. Foys, 2007, *Virtually Anglo-Saxon. Old Media, New Media, and Early Medieval Studies in the Late Age of Print*, Gainesville: University Press of Florida, p. 7.

9. See Flynn (2010: 250–80).

10. *Imitatio Mariae—Virgin Mary as virtuous model in medieval Sweden* is a project funded by the Swedish Research Council and based at Umeå University, Sweden.

11. The digital as a lens or critical perspective have been the focus of a recently completed postdoc-project about *ekphrasis*. For a re-interpretation of the concept of *ekphrasis* with the digital as lens, see Cecilia Lindhé, 2013, "A Visual Sense Is Born in the Fingertips: Towards a Digital Ekphrasis," *Digital Humanities Quarterly* 7: 1.

12. The prototypes were developed in 2011 at HUMlab, Umeå University, Sweden by Jim Robertsson, Emma Ewadotter, and Cecilia Lindhé.

13. See Bruce Wands, 2006, *Art of the Digital Age*, London: Thames and Hudson.

14. See Graham Zanker, 2003, *Modes of Viewing in Hellenistic Poetry and Art*, Wisconsin: University of Wisconsin Press, p. 27; Carruthers (1998: 7–22).

15. See Mary Carruthers, 1992, *The Book of Memory: A Study of Memory in Medieval Culture*, Cambridge: Cambridge University Press, p. 9; Carruthers (1998); Frances A. Yates, 2006, *The Art of Memory*, London: Pimlico; Quintilian, *Institutio Oratoria*, transl. H. E. Butler, 1953, Cambridge: Harvard University Press, xi, li *sqq*.

16. For the difference between monastic memoria and pagan rhetorical practice, see Carruthers (1998: 7).

17. See also Carruthers (2010: 196).

18. “Rome Reborn,” Virtual World Heritage Library, accessed June 27, 2013, http://www.romereborn.virginia.edu.; “Digital Roman Forum,” University of California, Los Angeles, accessed June 27, 2013; http://dlib.etc.ucla.edu/projects/Forum; “Stanford Digital Forma Urbis Romae Project,” The Stanford Digital Forma Urbis Romae Project, 2002–3, http://formaurbis.stanford.edu.

19. See note 7.

15 COMPUTATIONAL LITERATURE

Nick Montfort

Certain literary projects have computation as an essential aspect, as the voice is essential to a song. The development of works of this sort—of computational literature—involves turning the versatility and power of computation from the business, scientific, and military uses that typified the early decades of digital computing. Computational literature brings the power of symbol manipulation to bear on poetics, the imagination, and the conceptual and aesthetic capabilities of language.

Other useful and inclusive terms indicate similar categories: "digital literature" and "electronic literature," for instance. The latter category, as officially defined by the Electronic Literature Organization, includes "works with important literary aspects that take advantage of the capabilities and contexts provided by the stand-alone or networked computer." Grouping literary work in this way is very sensible for authors seeking to show how print literary production, reception, and culture relate to new developments in digital media. Authors of all sorts of electronic literature share many concerns, including concerns about the preservation of their work and the ways in which they can develop their practices as digital writers.

In this discussion, I focus on a closely related but slightly different category, "computational literature," because it is of particular relevance to the digital humanities and to methods for literary analysis that are currently being developed and used. Work in this category uses explicitly defined computational processes, either at the time of presentation or beforehand. When the computation is done before the time or presentation, the work can be presented on the page, screen, speaker, or in many other ways. The term I am using is analogous to the visual art term "computational art."

How does "computational literature" compare or connect to existing definitions used to discuss digital media work? An influential framework for the study of digital media is Espen Aarseth's (1997: 62–65) typology in *Cybertext* of 576 different media positions. Aarseth defines seven dimensions (dynamics, determinability, transiency, perspective, access, linking, user function) by which texts (or text-producing systems) can be assessed. This typology considers the work as it is ultimately presented rather than the process used in creating it. Whether we consider the whole set of computational literature or only those works that do computation at runtime, computational literature can be of all 576 types. The typology captures some important aspects, such as whether or not a system uses randomness (via the variable "determinability"), but computational works are not restricted to particular parts of it. It is possible, for instance, for a computational literary work, even one that computes at runtime, to be effectively static and deterministic, as seen in my "ppg 256-7," which is essentially just a computationally compressed version of a single Beckett-inspired text (Montfort 2012).

Computational literature presents challenges and opportunities for the humanities. Can humanistic study in the digital age, including large-scale projects and work within departments based on traditional disciplines, focus on contemporary work of cultural significance? Can it do so even when the practices in question have technical aspects that are more commonly encountered in computer science, computational linguistics, and statistics? When an understanding of computation and some facility in programming may be necessary? When there is only a very slim "cultural heritage" of computational literary work before the second half of the twentieth century, and this earlier work is of a very different sort? When the often-touted techniques for the analysis of "big data" may not be enough because the data in question are not text, images, video, or the like, but include source and executable code?

The questions would be very similar if we were to inquire about another of the many sorts of cultural production to which computation is essential: Computational art, hobbyist programming, computer and video games, demoscene productions, mobile apps, and programmed Web systems from free software ones such as Wikipedia and WordPress to corporate social media sites. Humanists (along with scholars from the social sciences and researchers of other sorts) have certainly done work in these areas and have produced insights, but many of the methods associated with the digital humanities have not yet been used on computational media. We have an ideal opportunity to bring these new methods together with new types of media that are based on computation. The unappealing alternative is to maintain the current hundred-year waiting period for the study of media forms.

Computational literature may be multimedia, may be networked, may work on large stores of data, may react to specific interface conventions and contexts, may be interactive, or may not have any of these attributes. Furthermore computational literature can be created with or without a general purpose, electronic, digital computer. The use of such a device is certainly important in this type of work, as it allows computations of tremendous complexity to be carried out with great alacrity, but there are other ways to compute, and certain relevant pre-electronic examples of computational literature do exist. The important quality of literature in this category is simply the use of computation, and that computation is made extremely significant, if not central, to a literary project. The question of whether the creator of a particular work is mainly seen as an author or poet is not directly relevant to whether that work is an example of computational literature, which can be created by artists, video game developers, and many others who engage in projects with significant computational and literary qualities.

As described, then, computational literature is not sharply defined, but is a radial category. The prototypical member of the category would involve extensive, complex computation while also engaging deeply with literary questions and techniques. As the computation involved becomes less significant, or the literary nature of the work becomes less clear, what is being considered is less likely to be identified as computational literature, but may still be considered in these terms in some ways. In other words, instead of declaring that there is a bright line separating computational literature from the rest of the world, I assume that things can be more or less central examples. A video game or art project that deeply involves computation and that to some extent explores language and literary questions can be considered computational literature, even if the literary aspects of the project are not the main ones, and whatever the original context in which the work was presented.

My discussion ranges through historical and contemporary work. Since I am a creator of digital poetry, interactive fiction, and other computational literary works, I discuss the work I have done alongside other work that I approach critically.

This is not the only mode of critical discussion, but it is a type of discussion that has been productively undertaken in digital media by (for instance) Ian Bogost, John Cayley, Mary Flanagan, D. Fox Harrell, Michael Mateas, Stuart Moulthrop, Janet Murray, Scott Rettberg, Warren Sack, Phoebe Sengers, Stephanie Strickland, and Noah Wardrip-Fruin. Such writing has been a major contribution to our understanding of electronic literature and related practices, and is particularly appropriate in a field of emerging practices.

Qualities of Computation

In an introduction to the most important issues in computation, Ian Horswill (2008: 1) writes, "Computation isn't tied to numbers, acronyms, punctuation, or syntax. But…in all honesty, it's not entirely clear what computation is." Of the several perspectives on computation, it will suffice in this discussion to consider the imperative one, which defines a computation as the work that one or more commands do in manipulating symbols. Addition and other arithmetic operations are computations, then; so is sorting a set of strings alphabetically, and so is determining which rectangle in an image is most likely to represent a human face.

Of course, computation can be done without a digital computer, as when people compute a tip at a restaurant or place a new book on a shelf alphabetically. As Lev Manovich (2003a) notes, however, the implementation of preexisting computation on our modern, electronic platforms does more than simply speed up the work that was done before, since "a substantial change in quality (i.e., in the speed of execution in this case) leads to the emergence of qualitatively new phenomena" (21) The rendering in *Quake* may be based on perspective drawing techniques known since the Renaissance, but the game (along with other first-person shooters) can hardly be seen as just a rapid series of Renaissance paintings.

Consumer-grade computers can now manipulate symbols billions of times faster than humans can. Even leaving the interactive, cybernetic aspects of the computer aside, their computational ability is tremendous when compared with that of people or early electronic computers. In 1945, Vannevar Bush (2003: 41) predicted, quaintly, that computers "will have enormous appetites. One of them will take instructions and data from a roomful of girls armed with simple keyboard punches." In 2012 Google's networked system and other computers on the Internet take instructions and data from a worldful of people—more than 1.5 billion—providing search queries, webpages, scans of books, video productions, emails, tweets, updates, and collaborative writing on Wikis and other systems. While computers operate interactively, they also do a huge amount to process and index stores of data, to filter and digest, and to generate and transform data in various ways. It is this type of work that comes into play in computational literature.

Understanding computation as it relates to literary practice, then, means acknowledging the connection to the pre-computer past of manual, rule-governed, symbol manipulation while also recognizing that our recent digital, electronic speedup has enabled completely unprecedented types of computing.

Computational Literature before the Computer

Before "the computer"—the general purpose, electronic, digital computer—there were computers of other sorts, including analog electronic computers and, before that, human beings who worked to calculate with pencil and paper. Some textual works with literary aspects were assembled, at the time of reading, by paper machines such as the rotating volvelle (Trettien 2009). Card catalogs have been cited as capable of computation, thanks to their ability to have elements removed, added, and rearranged (Krajewski 2011), and literary work on index cards has been undertaken by Judy Malloy (e.g., in her 1980 *The TV Blew Up*) and Robert Grenier (1978). The cut-up method as described by William S. Burroughs (2003: 90) (cutting a sheet in quarters and placing the quarters in a new configuration) is also a well-defined symbol-manipulating process that can be implemented by hand or by computer.

Beginning in 1960, a French group of mathematicians and writers, the Oulipo (Ouvroir de la Littérature Potentielle or Workshop for Potential Literature) have explored procedural and constrained writing and have had a remarkable influence on the avant-garde and on writing in many languages (Mathews and Brotchie 1998). In 1981, members of the Oulipo began working on computer-generated literature, founding Alamo (Atelier de Littérature Assistée par la Mathématique et les Ordinateurs or Workshop for Literature Assisted by Mathematics and Computers). But from the beginning, the group concentrated on ways of writing that included computational ones. Some of the most famous techniques are constraints (write a novel without the letter e, write a palindrome) that are underspecified if considered as computational procedures. Other techniques for textual transformation (Harry Mathews's algorithm to rotate elements of texts, Jean Lescure's N+7 technique of replacing each noun with the one found seven places ahead in a dictionary) are definitely computational, as they are functions mapping one literary work to another.

Combinatory Literature and Exhaustion versus Sampling

A common, unhelpful reaction to the idea of computational text generators is to dismiss the processes of such systems as nothing more than randomness. First, not all text generators are random. Second, there are many successful systems that incorporate randomness in insightful ways, bringing it into frameworks of form, probability, regularity, and convention. While one can certainly find systems that use randomness and are unimpressive, the flaw in these systems cannot be randomness itself.

A poetry generator—consider a noninteractive one, for simplicity's sake—can be seen as defining a certain set of possible poems or outputs. For a given haiku generator, there is a possibly very large but finite number of haiku that can be produced. For a system that permutes a particular text, say one that is five words long, there are 5! = 120 possible texts.

In fact Byron Gysin and Ian Sommerville's permutation poems do define a set of texts. Their poem "I AM THAT I AM" is a list of *all* the possible permutations of that phrase, including those that are lexically the same, and so is 120 lines long. This poem is a prime example of a computationally generated text—created with an electronic computer, although it could have been created with more difficulty by hand as earlier permutation poems were. The poem is not random at all. It is *exhaustive*, providing every possibility in the order generated.

This type of exhaustive provision of every possibility is also described, for a much larger text, in Jorge Luis Borges's story "Library of Babel." Another example of a fairly short text that is combinatorial and exhaustive, and has actually been realized, is Jean-Michael Espitallier's "De la guerre civile," which plays on the maxim that "The enemies of my enemies are my friends," going through every combination of enemies and friends and extending the length of the chain three times.

In visual art, John Simon's *Every Icon* is a clear and intriguing example of an exhaustive system (numeral.com/eicon.html). It is programmed to generate every possible 32 × 32 black-and-white icon and can be seen at work on the Web. An instance was started in 1997 and has already gone through every configuration of the 32 pixels in the first line. Completing all of the possible configurations of just the first two lines, however, will take six billion years.

When creators of systems do choose to provide a single element of the set rather than enumerating the entire set, this choice is called sampling. The probability of picking an element does not need to be the same for each element—if it is, the distribution so defined is called uniform. But the probability for each element can vary, making certain choices more likely than others. Because of this, what was described earlier as a set is best understood, more generally, as a distribution, one in which every element has a particular weight of probability. Since one can either print out every element in a distribution or sample an element one at a time from a distribution, "random" or sampling systems can be easily converted back and forth from exhaustive ones. Listing everything has its own rhetoric and aesthetics, as does sampling one element at a time. Both are suitable techniques in different situations. Additional ways in which randomness has been incorporated in computational literature will be discussed in the "Poetry, Story, and Conversation Generation" section.

Simulating Story Worlds

A different approach, very related to computer gaming, has been to use computational power to simulate an environment or microworld, often simply called a world. Literary works that do this include interactive fiction, which originated in the text adventures of the 1970s and in which the physical nature of the world, portrayed in text, is particularly significant. There is also interactive drama, which shares many features with interactive fiction but emphasizes character relationships, behaviors, and affinities.

Interactive Fiction

Will Crowther created the Fortran program and cave simulation *Adventure;* Don Woods expanded it into the canonical text game of the same name in 1976. Since then, the form of computer "interactive fiction" has thrived in various ways and has simulated environments and often special physical or magical laws as well. Works in this form, which are usually text-based, often exist within the adventure game genre, which takes its name from *Adventure*. One of the earliest follow-up games was *Zork,* created at MIT, which was the basis for the well-known game company Infocom. There were other interactive fiction games created at universities, and then many commercial games of this sort, and finally, beginning in the 1990s, a large number of games created by individuals and released for free online.

Interactive fiction simulates locations, the containment of objects in other objects, and the rules by which a particular environment operates. Hence it is often recognized as modeling a microworld or world, rather than simply a map or a layout of different locations. Within this world, the interactor commands a character who can explore and manipulate the environment and (in limited ways) interact with other characters. Computation is usually deployed in two main ways. First, to understand, given the context of the world and situation, simple, natural-language commands that include, in different games, "pick up ax," "give the password to Gus," "hitchhike," and "draw the Feverleaf on the bark with the bone." Second, computation is used to simulate the player's action, the automated actions of other characters, and the effects of these (along with other phenomena) on the world.

While interactive fiction has an obvious connection to fiction and the novel, the interactor is also invited to solve puzzles. In the best cases these puzzles systematically relate to the simulated, fictional world, and it is necessary to read carefully and understand the world appropriately in order to solve them. Since the interactive fiction world is often constructed figuratively, well-crafted puzzles often function in the ways that literary riddles do (Montfort 2003, 2007).

Interactive Drama

A related category of computational literature is interactive drama, which focuses more on the rich simulation of characters and on generating dramatic situations via their interactions. While the stereotypical interactive fiction is text-only, interactive dramas are usually presented graphically, often with cartoon-style animation. Complex interactive dramas were first developed at Joseph Bates's Oz Project at Carnegie Mellon University, which built on Brenda Laurel's work and other ideas about drama, film, and narrative. Early Oz Project creations included "text worlds" that looked a great deal like interactive fiction and visually expressive but mute characters called Woggles.

The last system associated with the Oz Project was *Facade* by Michael Mateas and Andrew Stern (2003), an award-winning and highly influential work which was touted as the first "fully realized" interactive drama. *Façade* allows the interactor to take on the role of an old friend visiting a couple for drinks in their apartment. As the couple banters, it becomes apparent that their relationship is not in good shape and the interactor is called on to react—to try to mediate or make light of the situation, for instance, or to take sides. A more recent project is *Prom Week*. This interactive drama does not allow free-form typed input, as *Facade* does, but has a larger cast of characters and a variety of social interactions that play out over a longer period of time.

Poetry, Story, and Conversation Generation

Sampling texts independently at random from a distribution has already been discussed. Poetry, stories, and conversation can be generated in this way or using a world model of the sort described in the previous section. However, many computational literature developers have based their systems on techniques that are neither combinatorial nor based on a simulated world, but instead pertain to models of language, metaphor, and writing.

Markov Chains

A Markov chain is a random process in which the outcomes are based on conditional probabilities. Imagine trying to generate words by repeatedly choosing among 27 options (the letters and a space symbol). It would clearly be better not to select "q," "z," and "j" as often as "e" is selected. So, instead of assigning 1/27 probability to each outcome, consider weighting the system so that "e" is most frequent and so that all the probabilities correspond to occurrences of letters (and the space) in English text.

Even with this adjustment made, this scheme has obvious drawbacks. After selecting a "q," it is clearly not appropriate for "e" to be the most probable next letter. A quick consideration of English reveals that the probability of a letter is not independent of the previous letter. After selecting "q," almost all the weight should be on "u." (Just as it's possible to count letters to determine the original weighting, a reasonable probability can be determined by counting how many times each letter is seen to follow "q" in a corpus of text. If "u" follows "q" 98 times and there are 100 occurrences of "q" overall, the probability of "u" given "q" can be set at 98 percent.) A first-order Markov chain takes into account the letter that has just been selected and determines probabilities according to it. A second-order model considers the last two tokens, and so on.

Markov chains were developed by Andrei Andreyevich Markov early in the last century; they are well known today because of Claude Shannon's use of them in his 1948 *A Mathematical Theory of Communication*. There, Shannon introduces Markov chains and uses the generation of English as an example. This treatment, no doubt, contributed to the widespread use of Markov chains for recreational and literary text generation.

One Markov chain system that attained some fame by contributing to USENET newsgroups was Mark V. Shaney, which began posting on net.singles in September 1984. The system, which was a hoax or in-joke, depending on one's perspective, was discussed in a *Scientific American* article (Dewdney 1985) and has been re-implemented in different languages. Since the early 1980s there have been at least many dozens of Markov chain chatterbots that interact with people on IRC, on MUDs and MOOs, through Web interfaces, and even through SMS and mobile phone apps. The precedent for such conversational systems was the early non-Markov program Eliza, running the famous *Doctor* script, which was created by Joseph Weizenbaum in the mid-1960s and simulated (or rather parodied) a psychoanalyst. An early real-time chatterbot that used a Markov process for language generation was MegaHAL (1998). This system has also been used to generate poems.

Markov chain text generation has been used for all sorts of purposes recently, including political ones. During the US presidential election year of 2008, at least two systems, a speech generator and a chatterbot, parodied the language of vice presidential candidate Sarah Palin (Bogost 2008). Since the mid-2000s, John Trowbridge and Eric Elshtain have been continuing to develop the Markov chain system Gnoetry (gnoetrydaily.wordpress.com). This system operates in specified ways on source texts selected by human collaborators, and has been used by several authors to generate fourteen chapbooks of poems. Edde Addad's charNG is a simpler Markov chain system that provides a Web-based interface. The system is particularly easy to use and is ideal for understanding how Markov processes work on text.

Computational Creativity

Other systems participate in the field of computational creativity—they are attempts to computationally model the creative process. Computational creativity is related to artificial intelligence in some ways, but is also distinct from it. We usually know what the outcome of an intelligent action is supposed to be, while the outcome of a creative action is, by definition, novel. Instead of trying to simulate a world, as interactive fiction does, a computational creativity system attempts to model a cognitive process, such as one involved in figuration or some process that people use to write stories.

D. Fox Harrell's (2008) GRIOT system is based on conceptual blending, a process that generates new metaphorical ideas. The system has been the basis for interactive text-only and graphical systems, including *Loss, Undersea* (in which everyday experiences of life and work are blended with an underwater domain) and *The Girl with Skin of Haints and Seraphs,* in which concepts associated with Africa and Europe are blended. Another system based on the process of literary composition is Rafael Pérez y Pérez (2001) MEXICA. This is a plot generator that tells stories about the pre-hispanic inhabitants of Mexico by going through an engagement-reflection cycle of the sort that writers are hypothesized to use. It too has been reworked in a variety of ways, to produce daydreams, for instance, and even to do interior decorating.

Systems, Code Reuse, and Remixing

Computational literature systems are sometimes coded from scratch in general purpose programming languages—and this is particularly the case with simpler and more straightforward systems—but many of them are developed using platforms and development systems that were created with particular literary goals in mind.

There are now numerous free systems for creating state-of-the-art parser-based interactive fiction. Today the most popular of these is Graham Nelson's Inform 7 (inform7.com). This system is unusual in providing a syntax like that of natural language, so that one specifies a room by writing what looks like a paragraph of English. Other interactive fiction systems, including all versions of TADS (The Adventure Development System) and earlier versions of Inform, have a syntax that looks more like that of C or other contemporary programming languages.

In poetry generation, no platform or system has emerged that has been used to develop many different interesting generators. But Gnoetry and charNG are both free software systems that are available for anyone to use and modify. A resource listing many systems and platforms used to create electronic literature, including computational work, is July Malloy's "Authoring Software" site, which also provides comments by authors on the software they use.

Looking beyond the use of development systems, there is the site *R3/\/\1X\/\/0RX* (remixworx), organized by Chris Joseph, Christine Wilks, and Randy Adams. This site is a blog-based community for the remixing of many forms of digital media, including computational literary works. Since it began in November 2006, more than five hundred pieces have been created and posted. Although this intentional online community of remixers has thrived, remixing of code and media elements does not always happen as planned. I collaborated on the 2005 *Mystery House Taken Over* project, for instance, which featured an initial collection of eight remixes and a painstakingly crafted kit to allow anyone to make their own version with relative ease. The kit was used in some classes, but the site received only two submissions of new remixes after its launch. On the other hand, my short and simple poetry generator, "Taroko Gorge" (2009), which was not specifically intended for remixing, has been the basis for at least twenty published works by others.

The Digital Humanities and Computational Literature

The new, special techniques for large-scale analysis that have come to fruition in the digital humanities (e.g., distant reading, culturomics, and cultural analytics) have been applied only in very limited ways, as yet, to computational literature.

In a study at MIT and the National University of Singapore, a collaborator and I applied some of these techniques to 1,333 of the games stored on the Interactive Fiction Archive, those that were written in either Inform or TADS 2. The results we reported involved the reading, play, and analysis of a few individual games; an analysis of these two IF platforms; a discussion of how each favored class structures of different complexity; an analysis of source code; and a consideration of how a larger category of games, wordplay games, related to platform (Mitchell and Montfort 2009). In working on this project, we also developed techniques for automatically reverse-engineering large numbers of games, and we considered the outputs of this process in light of measures from software engineering and other disciplines. However, although we did reverse-engineer and statistically analyze the 1,333 games in the data set, we did not develop the results fully. Future work in this area has the potential to help us understand how the functional qualities of code relate to the text and other media that is embedded in or used by games and other creative programs.

Work in the areas of digital media and electronic literature has revealed the importance of how computational literary systems compute. There are already large-scale systems for the utilitarian analysis of large numbers of computational systems—anti-virus programs and malware detectors are some of the most widespread and straightforward examples, but there are well-developed techniques in software engineering that are particularly relevant when source code is available. The time is right to bring our existing ability to understand programs together with large corpora of creative literary programs, which we already recognize as culturally significant.

16 THE CUT BETWEEN US: DIGITAL REMIX AND THE EXPRESSION OF SELF

Jenna Ng

> *"WHAT AM I," then? Since childhood, I've passed through a flow of milk, smells, stories, sounds, emotions, nursery rhymes, substances, gestures, ideas, impressions, gazes, songs, and foods. What am I? Tied in every way to places, sufferings, ancestors, friends, loves, events, languages, memories, to all kinds of things that obviously are not me. Everything that attaches me to the world, all the links that constitute me, all the forces that compose me don't form an identity, a thing displayable on cue, but a singular, shared, living existence, from which emerges—at certain times and places—that being which says "I."*
>
> —Invisible Committee (2011, emphasis in original)

Provided that we take a text's existence in an objective sense as an independent and culturally produced artifact (Blanchot 1995), we may acquire its meaning in any of three ways. The first is through a conventional reading of the text, whereby the reader uses various methods, interpretations, histories, politics, and social contexts as lenses through which the text is understood to contain meaning. The second way is to transpose the text into a different contextual matrix (e.g., space, time, media, and culture) so that it can be read anew and thereby take on new meaning. A dominant illustration of this is found in art: Duchamp (1951), as one example out of many, transforms a bicycle wheel from an ordinary object out of ordinary life into an art object by upending it on a stool and exhibiting it in a gallery. (Dis)placed thus, the wheel acquires new meaning not because its fundamental objective existence—as a wheel—has been changed in any material way, but because it has been shifted from one cultural understanding to another: "it [is] transferred from the sphere of tools or use objects to that of aesthetic contemplation" (Sonesson 1988: 85). Another example is found in cinema, or what Eli Horwatt (2009) defines as found footage filmmaking: "the practice of appropriating pre-existing film footage in order to denature, detour or recontextualize images by inscribing new meanings onto materials through creative montage" (76). Used in different spaces and times, found images become an alternative media text that is conceptually, if not visually and aurally, different as previously understood in the images' prior space and time. A text may also acquire new meaning by being turned from one media form to another, such as cinema becoming performance when "film jockeys" show, loop, edit, re-work, perform live soundtracks to, and otherwise manipulate images in real time, with their modulations of media creating new meanings, as suggested by one group's tagline: "expect cinema but don't think you will see a film" (Mindpirates 2012). Machinima, or real-time films made in virtual worlds, likewise recontextualize the video game into other media forms (e.g., cinema, performance, and television), which Henry Lowood (2008) identifies as "found technology." (169) Using video games or virtual worlds in new communicative ways subverts their original purpose and content, thereby creating new meanings and ways of understanding the text.

The third way—and the focus of this chapter—is through the text's *interstices*. Here I want to think about how a text changes when put into a state of *between-ness*—between lines, between words, between letters, between edges, between pages, between frames, between sounds, between marks, between binaries (light/darkness; sound/silence; line/space). How may between-ness be a useful concept to think about authoring in the digital humanities? In particular, I want to think about *the cut* as the operative tool for understanding between-ness, as the act which crystallizes the fissure represented by that space of the in-between (the cleaving of two positive spaces literally delivers a space betwixt—a fissure, a breach, a cleft). As a central issue, I want to think about the cut, not in terms of production, but destruction—or, more specifically, creation in destruction—and ultimately to link that, if paradoxically, to content creation in the digital humanities. A cut is ordinarily destructive: it carves up rather than constructs; it dissects rather than constitutes. It is the converse of more conventional production tools which put things together—the pen which assembles words and paper to create a text, the brush paint and canvas to create a painting, the camera light and film a photograph, and so on. Cutting, instead, renders asunder.

In this chapter I want to argue the cut as a productive method, one that can create an active space of between-ness, enabling new meanings to be formed, new juxtapositions to be made, new ways of seeing and new content to be created. I will use as my primary example the digital remix, a media form created precisely by shredding up its original contents and re-organizing them to bring about new meaning. As I explore the digital remix, I am also mindful of my discussion as a part of what David Berry (2011) proposes as the third wave of digital humanities, "that is . . . to look at the digital component of the digital humanities in the light of its medium specificity, as a way of thinking about how medial changes produce epistemic changes" (4). To that extent, I also want to consider remix specifically in its digitality as an agent of such changes in the way we understand authorship, content creation and identity transformations in the digital humanities.

The Cut: Between Violence and Creativity

In thinking about the cut, I look to two ideas in particular as a backdrop: the first is Sergei Eisenstein's essays on montage, and the power of collision between images which he advocates as montage's main characteristic: "What then characterizes montage and, consequently, its embryo, the shot? Collision. Conflict between two neighboring fragments" (Eisenstein [1922] 2006: 144). For Eisenstein, the key to montage is the interaction—more specifically "collision"—in that space of between-ness amid "two neighboring fragments." This perspective effectively shifts the locus of meaning from the images to their interstices, whereby our understanding is derived not from what is shown but from what is *not* shown. To Eisenstein, that negation is a potent one as it gives rise to montage. Between-ness here is thus more than just a defined absence—what is not in the boundaries—but an active and significant place of encounter from which meaning arises. The second idea is the material ontology of film-based cinema, namely, a reel of individual still frames segmented by strips of black that, when run through a projector along with persistence of vision, creates the illusion of moving images we are now familiar with as cinema. Again, the interstices of black here are not just divisions but creative loci that re-contextualize photography into cinema, concocting new meanings through the playing of between-ness in stasis and movement, light and darkness, images of color and narrow slivers of black.

In both cases the key lies in conceptualizing between-ness as spaces that actively open up new meanings, rather than being mere conduits or boundary markers. In a sense this positivity of between-ness runs counter to its nature, which is to exist not as a discrete space but one that relies on other entities for its existence; it must be defined *relationally*. As Elizabeth Grosz (2001) observes, between-ness, or what she terms "the in-between," is a space

> ... whose form is the outside of the identity, not just of an other (for that would reduce the in-between to the role of object, not of space) but of others, whose relations of positivity define, by default, the space that is constituted as in-between. (90)

Grosz's observation is significant as it renders the relational feature of between-ness a specific place for action and potential: it "is the only place—the place around identities, between identities—where becoming, openness to futurity, outstrips the conservational impetus to retain cohesion and unity" (91). To that end, the in-between is bound up with movement and realignment precisely because it is defined only in the flux of boundaries, intentions, and identities. It is in this potentiality that I want to explore the making of meaning not just in the space of between-ness, but also in the cut which, as explained, formalizes that space. Moreover, to Grosz's characterization of the in-between (as the place for becoming), I wish to add the idea of violence. Like the Higgs-Boson particle,[1] between-ness can only be conceived as a flash point;

it is extant only by a collision against others. Because it is such a flash point, it is also a violent space; indeed, Eisenstein identifies it primarily with collision and conflict. The destructive force of the cut thus also converges with this violence, for the cut as both act and metaphor ("I feel cut up"; "your words are cutting"; "to cut one out"; "to cut one off") connotes hurt, pain, and injury. Violence can also be conceived in terms of friction and resistance, insofar as between-ness is a space for movement ("one could say that the in-between is the locus of futurity, movement, speed" [Grosz 2001: 93]). Without qualifying Grosz's characterization, I argue that movement can evoke not just speed but also resistance in terms of confronting friction, chafing and traction. Movement (and, for that matter, transformation) is not just blitzing down a smooth highway, but also an overcoming of innumerable points of contact, each one requiring effort and energy.

The in-between is thus a space of two possibly conflicting elements—the potential transformations in its openness and futurity, and the violence it entails in achieving that potentiality. In that sense between-ness is also a delicate space: it can only exist relationally, so it cannot be taken too far from the objects between which it exists. The violence of the space thus also threatens its existence—taken too far, the in-between becomes instead an unbridgeable chasm. To that extent, too, these ideas of the cut and of between-ness can also be applied to the digital humanities in general in terms of developing alternative critical and creative bases for understanding. We can understand and create texts not just in themselves, but also in the between-ness of spaces they occupy with other texts and other data. This gives us further avenues and spaces for critical reading and making, whereby a work may be more than just its positive and in-between spaces, but also the fundamental creative frictions it engages in collision with others. Thinking about the cut and its complex destruction/construction binary can also change our perspectives of authorship in the digital humanities, paving the way for more merged creative roles in digital media between "making" and "cutting," to occupy the paradoxical binaries of constructive destruction/destructive construction. The primary task in the rest of this chapter then is to think about ways in which that destructive mode can, without implosion, be made into a creative and constructive force. This I will aim to do here using illustrative cases of digital remixes and, in particular, video remixes.

The Digital Remix

The remix can be better understood today as a discourse rather than any particular set of cultural practices or techniques. The remix has its roots as a musical practice starting from the late 1960s in Jamaican music (Brewster and Broughton 2000; Poschardt 1998) and developed in the 1970s with the growing practices of sampling in disco and hip hop culture. Since then, however, and particularly so in the twenty-first century, it has been appropriated for various contexts and applications, ranging from literature to architecture to pedagogy to Web 2.0 technologies (Sonvilla-Weiss 2010; Navas 2010) so that it is now almost a language, connecting and pointing with its own semiotics at its original texts (or allegorizing them, as Navas puts it).

Due to the wide range of applications, the exact parameters of practices constituting remix have become a little muddy. In its most basic form, "remix means to take cultural artifacts and combine and manipulate them into new kinds of creative blends" (Knobel and Colin Lankshear 2008). Navas divides the remix into four categories. The first two categories are the extended remix (where long instrumental sections are simply added to the original composition "to make it more mixable for the club DJ" [159]); and the selective remix (a more complex version of the extended remix, where new material, sections or sounds are not only added to but also subtracted from the original composition, resulting in a more elaborate derivative work but whose original composition is still recognizable). Significantly, these two types of remixes, despite the changes they make, retain the essence and identity of the original composition, or what Navas calls its "spectacular aura." The third category of remix—the reflexive remix—does not. The reflexive remix "allegorizes and extends the aesthetic of sampling," typically from numerous sources. Its root in material practice here is the collage, where different materials are used to invoke a uniting groove or idea or emotion, rather than to point to any one original composition. Navas's main example of the reflexive remix is the music megamix—a medley of sampled brief sections of preexisting songs ("often just a few bars, enough for the song to be recognized") to basically form a musical collage that does "not allegorize one particular song but many," recalling not necessarily a single artist or composition but "a whole time period." The fourth category is the regenerative remix, something Navas identifies specifically to new media and networked culture, whereby "remix as discourse becomes embedded materially in culture in nonlinear and ahistorical fashion":

> Like the other remixes [the regenerative remix] makes evident the originating sources of material, but unlike them it does not necessarily use references or samplings to validate itself as a cultural form. Instead, the cultural recognition of the material source is subverted in the name of practicality—the validation of the Regenerative Remix lies in its functionality. (162)

Here the allegorical ethos of remix is subverted "not to recognize but to be of practical use," and its operative principle is of periodic change and updates, an issue that is particularly in keeping with the fluidity of digital media.

What I am interested in here are the critical implications of between-ness in the third category of the reflexive remix, where new texts are made with their own creative charge without allegorizing the original composition. I see the reflexive remix co-opting the space of between-ness by operating across boundaries in two ways. The first is in the between-ness of texts, or their sampled materials. As explained, the general principle of allegory in remix specifically points to its original text in various ways—by extending it, by adding to or subtracting from it, by sampling and so on. Film scholars tend to emphasize the connections or commonalities presented by remix, particularly in terms of intertextuality and community: for example, Chuck Tryon (2009) discusses how remixes can be (to take Matt Hills's term) of "semiotic solidarity" with other fans, and used to express their affection, fandom, or even geekdom for the original texts; Henry Jenkins (2003) writes of remix as a practice that forms part of a participatory culture, tying the remix to a larger culture or community.

Without qualifying their arguments, I argue that the semiotic of remix also, as with Eisensteinian montage, relies on a certain friction in the space between texts for their creative energy. Remix is a unique combination of simultaneously stitching together commonalities and ripping them apart with collisions, contrasts and conflicts, so that meaning arises precisely from that complex negotiation between conflict and alignment. For instance, in "Buffy vs Edward: Twilight Remixed" (2009), Jonathan McIntosh remixes the *Buffy the Vampire Slayer* TV series with the first *Twilight* movie, specifically editing together scenes featuring Buffy with Edward Cullen from *Twilight* so that they look like they are talking to and interacting with each other. The editing is, not surprisingly, highly flawed—the eyelines do not match, let alone the backgrounds—but these discontinuities are smoothened out by their commonalities: both deal with vampires and the remix makes good use of voice cuts and shared situations such as telephone calls, school caféteria lunches, and to-the-death fights. However, the success of the video stems primarily from the friction created by pitting the respective texts against each other, specifically their gendered politics: *Buffy*'s strong, independent, take-no-nonsense female protagonist against the brooding romantic hero represented by Edward and the general sexist gender roles in *Twilight* (male = strong protective hero; female = helpless and in constant need of being rescued). As the write-up on its YouTube video site judicially states: "Ultimately this remix is about more than a decisive showdown between the slayer and the sparkly vampire. It also doubles as a metaphor for the ongoing battle between two opposing visions of gender roles in the 21st century." The two texts as spliced together are more than an alignment of two popular texts about vampires, but about a clashing dialogue (above the literal one created in illusion between Buffy and Edward) of their conflicting politics. Meaning thus arises not out of the connections or allegories to the original texts, but out of the violence of how they conflict, contrast and collide with each other.

The second mode of between-ness in the remix lies in the information surrounding the materials. This can include anything from genre knowledge to additional music and sound. Examples include what Chuck Tryon (2009: 161) calls the "genre remix," where scenes from a film are remixed and recombined to convert it from one genre to a significantly different genre. The most common illustration, also cited by Tryon, is the remix in 2005 by Robert Ryang of Stanley Kubrick's horror film, *The Shining*, to create a fake trailer portraying it as a family drama-comedy. In this case the remix allegorizes only one text, so it does not operate in the between-ness with another text. Instead, it engages with additional data about the film, creating meaning in the between-ness of the information we have about it (a classic horror) and the genre-establishing data we receive from its "trailer," and deliberately creating friction in that space. Tryon points out how scenes have been edited "to suggest bonding moments between father and son, while Jack Nicholson's feverish dancing is recast as cheerful exuberance, in large part through careful visual and sound editing" (161). Music is also a part of the data: the song in the fake trailer, Peter Gabriel's "Solsbury Hill," evokes romantic drama, used "most memorably in ads for the Topher Grace–Scarlett Johansson romantic drama *In Good Company* and the Tom Cruise–Penelope Cruz romantic thriller *Vanilla Sky*" (161). Other elements include the deep and chipper voice-over of the unseen narrator ("Meet Jack Torrance... he's a writer looking for inspiration!") and the use of "a quiet moment" (shots fading into each other, accompanied by soft keyboard notes played slowly in the background) to present an emotional hook, here presented as a father-united-with-son drama. As the narrator intones, "but now... sometimes... what we need the most... is just around the corner," his statement finishes with an eyeline shot from Danny to Jack, suggesting that what Danny needs most, earlier shown complaining that he has no one to play with, is his father, Jack. It is precisely in the collision, rather than in the solidarity, with the original text (in which a crazed Jack ends up trying to murder Danny) that the shot is now imbued with new meaning. The creative force of the remix's irony, humor and subversion come through in contrasts and juxtapositions such as shown in that eyeline shot: Danny really needs Jack as he needs the proverbial hole in the head, or, more accurately in this case, an axe in the skull. This conflict deliberately invoked in the between-ness of meta-data is not limited to Ryang's remix; virtually all genre remixes use this mode of conflict and collision (*Mary Poppins*, *Sleepless in Seattle*, and *Ferris Bueller's Day Off* as horror films, *Back to the Future* as a Brokeback-style gay romance film, etc.).

Remix thus operates effectively in the friction created in this space of the in-between. Yet between-ness is also a paradoxical space, for one of the main operative tools (on top of the effective use of music, well-written narration, etc.) for bringing about this creative space in the digital remix is the somewhat destructive operation of cut and paste. This would be so in its literal sense if the remix had been spliced from an analogue reel, as the frames would have to be first snipped from the reel and then subsequently glued or placed together to form a separate continuous reel. In the context of the analogue, the cut is also of particular violence and finitude. For instance, these descriptions of Godard's jump cuts, and probably one of the most dramatic uses of the cut, as featured in his classic representation of the French *nouvelle vague*,

A Bout de soufflé, feature strong language: "Godard *chopped* it up any which way . . ."; a "disconnected cutting" as "pictorial *cacophony*"; "[Godard] chopped it about as a manifestation of *filmic anarchy, technical iconoclasm*." (Raskin 1998, emphasis added) While these comments are obviously directed toward the anarchistic aesthetic of the jump shots so in contradiction to the smoothness of continuity editing then in conventional film language, the violence and finality underpinning those descriptions and the fundamental nature of the cut—as the chop, no less—is clear.

However, in the digital sense, cut and paste takes on new meanings. It obviously ramps up the ease and convenience of the edit, underscoring the fluidity of the digital text and the unprecedented ways in which we may navigate its new spaces. With digital technology, we can now find, manipulate, and put together films more easily than ever. However, scale here is a minor issue. Rather, I argue that the digital cut, in enabling frames to be effortlessly found, re-worked, and pasted together, transforms the violent undertones of the cut and the finality of the act it implies: in its digital form, the cut is no longer about finitude but about endless possibilities. In that sense the continued icon of the scissors for the "cut" function in common digital video editing interfaces such as Windows Movie Maker and Final Cut Pro is anachronistic: the cut is no longer annihilative but the conduit for continuous transformation. More than that, the cut in relation to the immateriality of the digital text signifies a fundamental change to the nature of text and authorship. A film strip can only be literally cut once, but a digital text can be "cut" endlessly. No longer destructive or terminal, the digital cut is instead an act signifying innumerable second chances, creative acts that can take into account all kinds of mutability and change, resulting in a final product that need never be final but be instead a constantly resurrecting entity. With digital editing, the text is stripped of both finality and finitude. Making is now in the flow of these changes, struck up by a destructive force that is paradoxically also its very agent of creativity and transformation.

The example that I think takes this paradoxical creative energy to its highest limits so far is Kevin Macdonald's documentary film, *Life in a Day* (2011a). On July 1, 2010, Macdonald launched an open call on YouTube for contributors—"thousands of people, everywhere in the world"—to upload onto the video sharing site a recording that they were to make of one day in their lives, the day of July 24, 2010. Macdonald and his team would then use the material "to make a film that is a record of what it's like to be alive on that one day"; it is to be "kind of like a time capsule . . . a portrait of the world in a day" (Macdonald 2011b). The result presents some astonishing numbers: over 80,000 videos were submitted from 190 countries, amounting to over 4,500 hours of footage, out of which the 95-minute film was ultimately "made," apparently featuring more than 1,000 clips from YouTube contributors (Moner 2011). Its premiere was eventually screened at Sundance in January 2011, before being generally released in June 2011.

Life in a Day is an innovative project in a number of ways. For one, it is unprecedented in the scale it calls for and implements user-generated content.[2] William Moner, distinguishing *Life* from other amateur and small-scale productions proliferating online, calls it "the first major United States motion picture to engage crowdsourced labor for content." In the same vein, it is also one of the most collaborative creative film projects to date, technically involving more than 80,000 parties who submitted footage for the film (albeit a collaboration of which Moner is also highly critical, arguing that the contributors were effectively under-compensated labor, having signed away all rights and receiving no monetary payment for their work).

For the purposes of this chapter, I note that the film is also innovative as an example *par excellence* of the cut as its primary operating tool. Needless to say, *Life* must have involved extensive editing in order to cut and put together a coherent 95 minutes out of 1,000 clips. Yet the average Hollywood blockbuster, with its hyperkinetic action, probably contains more or at least as many cuts: a standard movie today, after all, has about 5,000 cuts (Murch 2004). In any case, what I am interested in is not the scale, but the role of the cut. Here the overwhelming operative mode of making the film, as created by Macdonald, is by cutting and pasting together submitted clips. It does not matter so much *what* was being filmed as opposed to *how* it was being put together.[3] The job of the director effectively converges with editor as he "directs" a film with footage he had *not* directed, nor indeed ever seen before. Yet for *Life* this is all a matter of semantics now as it is all in the editing. *Making is now in the cutting.*

Self-reflexively or otherwise, this mode of cutting resonates with the themes of the film. Taking Robert Plant Armstrong's (1975) ideas of synthesis and syndesis (the former an accumulative linear progression; the latter an accretion of repeated units), the digital cut and paste of *Life in a Day* seems to be simultaneously both. All films are edited to be synthetic: one frame is pasted to the next so that they all follow each other linearly. Yet for *Life*, there is also a sense of inevitability that marks the repetitive nature of syndesis: the film progresses linearly, starting from the break of dawn to midnight. While we only see a small fraction of all the videos submitted, we cannot help but realize the existential repetitions with which this day is composed: fleeting (and fleeing) seconds to minutes piling upon each other to hours turning on slow axles to dawn lightening into day before darkening into night, only for it all to be repeated the next day, and the next, and the next. The endlessness of cutting and pasting in this case becomes Sisyphean, mirroring the temporal labor that is the making not only of the film but indeed life itself: frame upon frame, clip upon clip, time upon time, age upon age. It is not only the central creative act of the film's making but also a self-reflexive statement of its truth: a slow march to a relentless rhythm of sunrises and sunsets, toward an end that we only faintly comprehend in the stillness of others' passings and the chilling glimpses of our mortality.

Conclusion: Remix and the Expression of Self

I have argued here for how between-ness, in particular as exposed by the cut, may be creative and meaningful spaces, and uses the digital remix to illustrate this. In that sense the cut echoes Pier Paolo Pasolini's (1967: 6) poignant account of the edit as *precisely* that which renders meaning, because the cut passes the present into the past. It completes action—the practical equivalence of death, which makes our lives meaningful by converting a chaotic, infinite, and unstable present into a clear, certain, and stable past: "It is thanks to death that our lives become expressive." The cut exposes not only the space of the in-between but, more important, the constitution of being in that between-ness, spaces that form being in their complex and fluid combinations of connections, dialogue, confluence, and relations.

This understanding of cutting and being may thus relate to the digital humanities in three ways, with the first two already discussed at length: first, in understanding digital materiality itself, of both its fluidities and fluencies in changing our thoughts on meaning, time, space, and realities; and second, in opening up spaces for digital "making," so that creativity lies not only in the discrete production of things but also in the in-between spaces of cutting, collision and violence. The third way, then, may be in thinking through (re-)alignment of the digital humanities to humanities study in general, a relationship in which a schism threatens to emerge: as one example, William Pannapacker (2011) describes the digital humanities as "the cool-kids' table," seeming "more exclusive, more cliquish, than they did even one year ago." Yet, even in allowing conceptual leeway for differences (in output, content, method, etc.), the space of separation need not be divisive—the limbo of between-ness can also be fun, creative, and constructive. Might not more (in ideas, understanding, and knowledge production) be achieved via collisions rather than ruptures? Might there not be greater potential in more positively identifying the digital humanities by making use of the destructive force of the cut, drawing on violence and acting on conflict in precisely those spaces of the in-between? Recent research revealed a new scientific understanding of what the human being is in biological terms: the conventional view is that a human body is a collection of cells—ten trillion strong, to be precise, produced from 23,000 genes. However, scientists have now discovered the microbiome, which is bacteria of "several hundred species bearing 3 million non-human genes," in the numbers of hundreds of trillions. In other words, by counting the microbiomes too, "humans are not single organisms, but superorganisms made up of lots of smaller organisms working together" ("Microbes maketh man" 2012; "The human microbiome" 2012). On this account, the space of between-ness intrudes on our fundamental ideas of identity and thoughts about who we are: the expression of the self being not in who I am but in all the ties from myself to everything else, "to places, sufferings, ancestors, friends, loves, events, languages, memories, to all kinds of things that obviously *are not me*." Like the remix, what is significantly meaningful is not the entity in itself but the friction, the collisions, the contact with all others. I am not in the entity of the human being; I am in the microbes that make up literally the biological ecology in which I exist. I am not in "I"; I am in everything that flows from, into and out of me.

Notes

1. Discovered as a breakthrough in July 2012, the discovery of the Higgs-Boson particle could only be achieved by smashing subatomic particles together at high energy levels in the Large Hadron Collider, based at CERN in Geneva, Switzerland, and observing their crash to see if a Higgs particle "would momentarily pop out in view of the detectors" (Connor 2012).

2. Notably in the sense of requesting for and using moving images. Note that there have been previous photographic projects in the same style, albeit via the use of still images rather than videos (see Schiller 2012).

3. Note earlier examples, such as those by William Burroughs, Brion Gysin, and Antony Blanch (who worked in advertising): see The Cut Ups at https://www.youtube.com/watch?v=MMQSDwQUwWM&feature=youtu.be (1966).

17 LOCATING THE MOBILE AND SOCIAL: A PRELIMINARY DISCUSSION OF CAMERA PHONES AND LOCATIVE MEDIA

Larissa Hjorth

As the young, Korean girl Mi-Hyun sat waiting for a friend in a busy café in Hongdae she toyed with her iPhone's photo apps. Mi-Hyun took a "retro," analog-looking picture of her coffee with Instagram and uploaded the picture to the social media application Kakao. On viewing Mi-Hyun's image, her late friend took a picture of her feet walking quickly, which she then uploaded to Kakao so Mi-Hyun knew she would be there in a few minutes. They shared a laugh in the Kakao space as their two co-present physical locations quickly converged.

A young female named Hyunjin was walking to her university in Shinchon (South Korea) when she saw her close friend Soohyun having coffee at a café and looking at her phone. At first, Hyunjin thought she would run up and surprise her. But on second thought, Hyunjin decided to take a camera phone picture of Soohyun, geo-tagged the location through GPS (global positioning system), and uploaded to Kakao. She watched as Soohyun quickly did a double take of her phone and her Kakao app and then looked around. Within a minute Soohyun spotted Hyunjin standing outside and the two girls laughed. Soohyun gestured for Hyunjin to join and then the girls shared lunch.

Grandmother Su-Bin had never been much into technology. She grew up in a generation where technology was deemed men's business. However, her husband recently bought her an iPad and uploaded her favorite novels. Before long, Su-Bin was hooked. She started to use the various media applications including Facebook and email. She especially loved looking up places on Street Views. She would often Google places she was curious about. She would look up places she had visited years ago. Each time she felt she was transported. She would often run her fingers across the roads under the digital screen out of habit from using paper maps.

These opening vignettes present just three stories of the many locative media practices in everyday life. Once upon a time, we navigated our ways to a place via paper maps. Now, the mobile and digital portal of the smartphone provides a frame for travel. We electronically travel across social media spaces. The physical and geographic is increasingly overlaid by the electronic and social. Various forms of intimacy and mobility are amplified and transported. This mobile frame means the ways in which people experience and conceptualize place, presence, and intimacy has changed.

While mobile media has always been intimately tied to a sense of place (Ito 2003; Hjorth 2005), now we are seeing different tapestries involving a variety of modalities of presence (co-presence, tele-presence, net-locality) that transform notions of place into a dynamic series of "placing" (Richardson and Wilken 2012), entanglement (Ingold 2008), and emplacement (Pink 2011). As we move further into the twenty-first century in which increasingly everyday spaces are overlaid with the digital, we are required to revise our methods and definitions within the humanities. The digital humanities sees a redefining of the boundaries of fields and disciplines as they converge, intersect, and inflect with digital cultures and practices in different ways. This is particularly apparent in the case of locative media which harnesses the emergent, interdisciplinary field of mobile communication and the various issues around changing methodologies (Gordon and de Souza e Silva 2011; de Souza e Silva and Frith 2012; Farman 2011). Indeed, in the case of mobile media, it is not only the *content* and *object of study,* it also provides a *context* that is shifting. The unruly and messy nature of mobile media—which is increasingly the main or only context for online media for much of the world—means that context moves across various platforms, media, and spaces (Dourish and Bell 2011).

With the rise of high quality camera phones, accompanied by the growth in in-phone editing applications and distribution services via social and locative media, new types of co-present visuality are possible. In the first series of studies on camera phone usage by the likes of Mizuko Ito and Daisuke Okabe (2003, 2005, 2006) in Japan, they noted the pivotal role played by the three "S's"—sharing, storing and saving—in informing the context of what was predominantly "banal" everyday content (Koskinen 2007). For Ilpo Koskinen (2007), camera phone images were branded by their participation in a new type of banality. While this banality can be seen as extending the conventions and genres of earlier photographic tropes (i.e., Kodak; Gye 2007; Hjorth 2007a), they also significantly depart by being networked and recontextualized. As camera phones become more commonplace, as smartphones become more popular, and as new contexts for image distribution such as microblogging and LBS become more commonly used, emergent types of visual overlays on geography become apparent. In this cartography, images are given ambient, networked contexts in which the geographic place is overlaid with social and emotional affect.

While camera phone genres such as self-portraiture have blossomed globally, vernacular visualities that reflect a localized notion of place, sociality and identity making practices are also flourishing (Hjorth 2007; Lee 2009). Smartphone camera apps like Hipstamatic and Instagram have made taking and sharing photographs easier and more interesting.

Instagram evokes nostalgia as part of its aesthetics by borrowing from pre-digital (i.e., analog) photography like Polaroid. The ease with which people can take and share images has seen a growth in the volume of images circulating globally. Many of our interviewed respondents noted an increased interest in camera phone image capture and sharing. With gamified location-based services (LBS) like *Foursquare* and *Jiepang* we see a further extension of overlaying place with the social and personal whereby the electronic is superimposed onto the geographic in new ways. Specifically, by sharing an image and comment about a place through LBS, users can create different ways to experience and record journeys and, in turn, impact how place is recorded, experienced and thus remembered. This is especially the case with the overlaying of ambient images within moving narratives of place as afforded by LBS.

The rapid uptake of smartphones has enabled new forms of distribution and has provided an overabundance of apps, filters and lenses to help users create "unique" and artistic camera phone images. Although the iPhone has been quick to capitalize on this phenomenon through applications such as Instagram, other operating systems like Android have also had their share of success in this expansive market. So too, social media, like microblogs and LBS, have acknowledged the growing power of camera phone photography by not only affording easy uploading and sharing of the vernacular (Burgess 2007) but also providing filters and lenses to further enhance the "professional" and "artistic" dimensions of the photographic experience (Mørk Petersen 2009).

The Asia-Pacific region has been noted as one of the early adopters of camera phone practices, especially in Japan (Ito and Okabe 2006) and South Korea (Hjorth 2007; Lee 2005). With these locations boasting some of the key global mobile camera phone manufactures such as Samsung, Sony, and LG, the uptake was far from surprising and camera phone practices in these locations were quickly adopted as an integral part of everyday life. However, we must acknowledge that behind the so-called media revolutions of smartphones are uneven forms of participation in a broadband society across technological, social, economic, and cultural differences and distances (Goggin and Hjorth 2009).

For Bo Gai (2009) in her study of camera phone practices in Beijing, mobile media practices are part of expansive media tactics that have seen Chinese pedestrian mobile users granted a public voice. They are, as she observes, both reinforcing and departing from Chinese notions of media participation by providing public, collective commentary, but also individual voices. In China, participation can take many forms, including what in the west is seen as non-participation: lurking (Goggin and Hjorth 2009). Mobile media evokes a particular kind of ambient participation that is configured through and by place in specific ways (Hjorth 2012). With the convergence of social, locative, and mobile media, how we conceptualize camera phone visuality and affect is changing. So how do we frame these new visualities in motion?

One way, as this chapter suggests, is through the movement from "networked" to what Pink (2011) calls "emplaced" visuality.

Beyond the Snapshot: New Visualities in Motion

In studies of first-generation camera phone use in South Korea, Hjorth (2007) noted tensions around the camera phone's relationship to place and mobility in what she called "snapshots of almost contact." The increasingly quick edit and deletion of images *in situ* has created a different relationship between recording and the mediated experience. This tension around mediation, reflection and engagement is amplified in second-generation camera phone studies, with the growth of locative and social media that converge the aesthetics of the captured image, with the sociality of the sharing of the image, the communicative power of accompanying text, and the specificity of place provided by GPS location. As previously noted, these social-visual performances are about mobile intimacy and how these practices get mapped across a variety of intimate and social publics, where the performativity of place becomes a process of perpetually "placing" various forms of presence (Richardson and Wilken 2012).

For Daniel Palmer (2012), iPhone photography is distinctive across three areas. First, it creates an experience between touch and the image in what Palmer calls an "embodied visual intimacy" (2012: 88). While "touch has long been an important, but neglected, dimension in the history of photography . . . the iPhone, held in the palm of the hand, reintroduces a visual intimacy to screen culture that is missing from the larger monitor screen" (Palmer 2012: 88). Second, the proliferation of photo apps for the iPhone has meant that there are countless ways of taking, editing, and sharing photos. No longer do camera phone images have to look like the poorer cousin to the professional camera—increasingly camera phones have higher technical capabilities. Third, the role of GPS capability with the iPhone automatically "tagging photographs with their location, allow[s] images to be browsed and arranged geographically" (Palmer 2012: 88).

As Goggin (2011) has noted, the increasing prominence of the citizen journalist has been epitomized by the camera phone revolution, to the point where even some professional photojournalists have opted for camera phones instead of professional cameras. As Daniel Rubinstein and Katrina Sluis (2008) note, the rise of the camera phone has been part of a shift from print-based to screen-based photography. As Rubinstein and Sluis further observe, this "rise of photo sharing sites has created a context," something "vernacular photographers have always lacked: a broad audience" (2008: 18). For example, moblogging, that is, mobile phone blogging sites like Weibo and Twitter, have enabled the mass distribution of camera phone images, many of which evoke much more than 140 characters, or a text blog. In China, an example of this can be found in Weibo, a rich media Chinese version of Twitter. Here millions of camera phone images are shared each day. The participatory elements of sites like Flickr with what Burgess (2007) has called their "vernacular" and "situated" creativity have created new forms of what Søren Mørk Petersen (2009) calls "common banalities."

Rubinstein and Sluis also emphasize the multilayered, multimedia composition of images that are bundled up with text, GPS, and a means of social distribution. They point out that the increasing "reliance on tagging for organization and retrieval of images is an indication of the importance of textuality for online photographic procedures.... Tagging provides a substantially different way of viewing and interacting with personal photography" (2008: 19). According to Rubinstein and Sluis (2008), these networked images get transformed as part of metadata in which the original context is lost.

For Chesher (2012), the rise of smartphones like those made by Samsung and the iPhone—with their attendant software applications like Instagram, Google Goggles, and Hipstamatic—have created new ways in which to think about camera phone practices and their engagement with both image and information. For Chesher, the iPhone universe of reference disrupts the genealogy of mass amateur photography. Applications like Instagram, which allow users to take, edit, and share photos, partake in what could be called a second generation of camera phone and photo sharing social media. With "vernacular creativity" (Burgess 2007) sites such as Flickr being the precursor, Instagram heralds a new generation of visuality in which the cult of the amateur is further commercialized. Launched in October 2010, Instagram quickly grew to boasting over 160 million uploaded images (Nitrogram 2014). The virtual and viral nature of Instagram was illustrated by a graphic design firm in Italy who recently built a physical digital camera prototype that looks like the Instagram icon, called the Socialmatic. With these new applications, often working in collaboration with social and locative media, camera phone images have been given new contexts.

> The iPhone camera mobilises longstanding subjective impulses for making images, common not only to Kodak, but also to the motivations for cave drawings, oil paintings and daguerreotypes (a long-term, constantly changing abstract machine). However, iPhone users take up these affective forces and aesthetic values in different ways, with different materials, forming different connections. (Chesher 2012: 100)

As Chesher (2012) notes, Kodak's "long moment" that spanned most of the twentieth century was disrupted by the iPhone camera. Kodak had been "synonymous with the amateur image-making industry for over a century" (Chesher 2012: 102) as represented by the invention of the Brownie camera. Through advertising that linked photography with memory, Kodak colonialized the domestic by implying that a holiday or an event without a Kodak, wasn't one (Gye 2007). The Kodak was on hand to record the cultural changes in the twentieth century, creating particular, "official" discourses of the unofficial place of the domestic. In the 1990s the birth and rise of digital photography created momentary ruptures in the analog universe of references.

However, although the digital has displaced the analog, theorists such as Lev Manovich (2003b) observe that the digital remains "haunted" by specters of the analog. The ghosts of analog genres, content, and conventions were all transferred into the digital world. This "simulation" of nostalgic prints is continued by camera phone apps like Hipstamatic, in which "auto-nostalgia" comes in an arty and romantic visual package (Chesher 2012: 107). Ruptures are also evident though. As Chesher notes, while the iPhone's camera "opens onto similar universes of reference as the Kodak . . . with apps it reconfigures, or even gets rid of images altogether" by mediating a significantly different set of "practices in making and consuming images" (2012, 107). For Chesher, "many apps take the camera beyond its photographic heritage to use it as a data input device, collecting information instead of making conventional photos" (2012: 107).

With elements such as augmented reality becoming part of camera phone culture, Chesher argues that this "regime of vision has less connection with Kodak culture and more resemblance to a gun sight or a head-up display in a modern fighter jet. It visually interprets the surrounding space with a view to action" (2012: 112). As Chesher concludes, "it is not yet clear what the implications of this emerging Universe of imaging will have for individual and collective senses of identity, place and memory" (2012: 144). For Pink (2011), the rise of the smartphone's networked visualities requires us to rethink the relationship between the image and its context. Rather than images being viewed as snapshots that seize a moment, these emergent databases of millions of images are about a diversity of moments electronically framed within particular locations in time and space. Taken as a whole, these temporal-spatial visual configurations are no longer isolated and frozen as suggested by a snapshot but are part of a vast, moving set of image–text–metadata media objects, orchestrated by social and locative services.

The Place of Visual Intimacy: Social, Locative, and Mobile Media Visualities

Rubinstein and Sluis's (2008) study of the networked image operates as a precursor to the second generation of camera phone practices in which GPS creates new levels for recording the "original" context with a geographic coordinate. As Pink (2011) observes in "Sensory digital photography: re-thinking 'moving' and the image," the recent "sensory turn" in visual cultures scholarship requires an examination of meanings and materialities associated with the image. Calling for a need to destabilize the authority of the image, Pink argues that visual production and consumption need to be conceptualized through movement and place. For Pink, locative media and the photographic image require a new paradigm that engages with the multisensoriality of images. Drawing on Tim Ingold's (2008) critique of the anthropology of the senses and of network theory as well as *Doing Sensory Ethnography*, Pink (2011) argues that by exploring the visual in terms of the multisensorial, the importance of movement and place are re-prioritized.

Rather than viewing networked images as part of a contextualized metadata as suggested by Rubinstein and Sluis (2008), Pink argues that locative media provide new ways in which to frame images with the "continuities of everyday movement, perceiving and meaning making" (2011: 4). By contrasting "photographs as mapped points in a network" with "photographs being outcomes of and inspirations within continuous lines that interweave their way through an environment—that is, in movement and as part of a configuration of place" (2011: 4–5), Pink argues that we must start to conceive of images as produced and consumed in movement. Here we can think about how images are being transformed in light of various turns—emotional, mobility, and sensory. Indeed, of all the areas to be impacted and affected, camera phones—especially with their haptic (touch) screen interface and engagement, along with their locative media possibilities—can be seen as indicative of Pink's (2011) call for a multisensorial conceptualization of images. As Pink notes, the particular way in which text, image, and GPS overlay creates a multisensorial depiction of a locality.

This shift can be viewed as the movement from camera phone visuality that is networked to camera phone images that are "emplaced" (Pink 2011). An image that is socially networked, tagged, and GPS located is "emplaced" in a number of ways. It is emplaced as one of many images captured by a particular member of our intimate or social public and is contextualized by our relation to that person. The co-presence of many images arranged by time or place on the one site places each image in the context of others to constitute a narrative, and thus another context. The GPS coordinates place the image in geographic space and invite the viewer to *recall* the place in question as well as *view* the image captured in the place in question, thus overlaying another context. The social distribution of the images creates a social public for those images, thus overlaying another context, and the image tags entered by the public overlay yet another context.

In fieldwork conducted in Shanghai with Kay Gu and Michael Arnold, the uptake of gamified LBS, *Jiepang*—where users can "check-in" to online spaces and win prizes when they visit offline places—is creating out-of-game visualities (Hjorth and Arnold 2013; Hjorth and Gu 2012). For many of the generation Y (*ba ling hou*) respondents, the playing of the official *Jiepang* game was secondary to social motivation. In this social dimension, sharing camera phone pictures—with accelerated frequency—had become a key practice. The popular use of rich-media moblogging sites such as Sina, Weibo, and SNS Renren (China's equivalent of Facebook) is evidence that the compulsion to photograph, edit, and share is growing, and in each of the different social media spaces, different types of photographic genres could be found. For example, many used Sina for self-portraiture, Weibo was for more political or news-worthy images, while Renren was a space for reflecting inner feelings. In the case of *Jiepang*, visuality was more about new types of place-making. These place-making exercises are like diaries (Ito and Okabe 2005), but with their active use of filters and perspectives, *Jiepang* is demonstrating emergent forms of creative practice for the *ba ling hou*.

In the fieldwork in our six Asia-Pacific locations—Melbourne, Shanghai, Seoul, Singapore, Tokyo, and Manila—as part of an Australian Research Council Discovery grant project *Online@ Asia-Pacific*, Michael Arnold and I noted the significance of camera phone photography in the place-making exercises of LBS (Hjorth and Arnold 2013). We interviewed sixty respondents in each location every year for three years. In LBS photo albums we noted that while respondents used traditional genres such as food and places, they did so by using filters and lenses to create highly aesthetic images, often using filters that not only made the pictures look analog but also gave them a sense of nostalgia. Far from banal and boring, these images were often unique and creative.

These images are not only about the vernacular qualities of user-created content (UCC), they signal new types of emplaced visualities. While the first generation of camera phone studies emphasized the three "S's"—sharing, storing, and saving—of the "networked" images, the second generation sees an embrace of Pink's (2011) aforementioned idea of "emplaced images." This "fastforwarding present/presence" (Hjorth 2007)—that is, the accelerated way in which images are taken, edited, and shared, often during the experience of an event—of camera phones as images in a movement and event, took on a specific "emplaced" ambivalence with LBS. Through the overlaying of highly edited camera phone images and comments, respondents can narrate place in new ways. For example, in Shanghai, respondents tried to take more poetic pictures of locations in order to provide a unique comment on that location. Often the visuals were deployed to present a unique image of the place, whether through the image genre or, more often, using filters and lenses to create a mood.

In our Seoul case study, LBS camera phone images are indicative of Massey's (2005) notion of place as a series of "stories-so-far" that reflect disjuncture as much as presence. In LBS camera phone images, we see that place is a process of perpetual oscillation between "placing" (that is, actively situating or contextualizing phenomena) and "presencing" (i.e., "being there" through telepresence, co-presence, located presence, and net-local presences; Richardson and Wilken 2012). The rise of "emplaced" rather than "networked" visuality apparent with LBS camera phone practices is a manifestation of the changing performance of co-presence, whereby binaries between online and offline experiences don't hold.

Moreover these images evoke the multisensory experience of place as a process of perpetually placing and reflects an intersection of senses that are not just visual. By overlaying the social with networked GPS, images provide a multisensorial depiction of visual cultures. For example, our respondents noted that they often took pictures of shared food. In one interpretation, this is just another example of a virulent consumer ethos, making the private consumption of food into conspicuous consumption through a public statement. In another interpretation, the common practice of food imaging is a response to the need to create multidimensional and ambient experiences of place that are overlaid by intimacy, or, the need to convey, in the Korean context, a sense of *bang* (space/room) with the *ilchon* (intimate relation). In a location like Seoul famous for its high use of *sel-ca* (self-portraiture; see Hjorth 2007; Lee 2005), it was interesting to note that this genre was on the decline in LBS contexts. This shift was due to the impact of governmental and corporate examples of privacy breaches (Lee 2011; Wallace 2012).

Another example of LBS camera phone practices evoking a particular, localized form of intimacy and place was in Shanghai (Hjorth and Gu 2012). Through thirty interviews with male and female *Jiepang* users, we found gender influenced the ways in which users recorded and reflected upon places and intimacy. We asked respondents to take us through their shared pictures and discuss their motivations and how their relationship to the image and experience changed once it was shared. According to Ai (25-year-old woman), *Jiepang* use was in order "to record where I go to everyday. It's like a diary with location." The other respondents shared Ai's sentiment—many viewed recording locational information via *Jiepang* as both for their own and other's benefit. For Bai (27-year-old man), *Jiepang* was primarily used to "record where I had been" and to have this information "synchronized to social networks such as Weibo to share with my followers." For others, they didn't record everywhere they went but, rather, used *Jiepang* only for those occasions when they went to new places.

For Bai, *Jiepang* was important in recording and archiving his activities and journeys. Unlike Ai, Bai didn't record each place he went to everyday, just a few highlights. Here we see the way in which gender inflects the ways in which *Jiepang* relates to ongoing endeavors to narrate the everyday as part of movements through spaces. Bai viewed *Jiepang* as a tool for showing where he was when he wanted people to know. Sometimes checking in on *Jiepang* was accompanied by taking pictures of the place, an activity Bai definitely viewed as gendered, stating, "usually females would spend more time on it than men, and take more photos." Both these informants noted that recently there had been a growth in different LBSs on smartphone and PC platforms, and that groups of friends would use similar ones—in short, the deployment of an LBS reflected Chinese social capital/network, *guanxi* (Wallis 2011). With the additional dimension of camera phones interwoven with locative media, *guanxi* and place can take on more complex cartographies that place, emplace, and embody visualities.

With camera phones now an essential part of smartphone culture, we have witnessed not only an expansion in the "professionalization" of the amateur through a plethora of photo filters and lenses, we are also seeing an explosion in the number of images taken and shared as part of a "poetic" intervention that emplaces images within a place. In particular, through LBS like the Chinese *Jiepang*, users are taking, editing and designing camera phone images that, as part of representing a place in a unique way, are then shared to reinforce social networks (*guanxi*).

Just as *Jiepang* is "more about the journey," *Jiepang* images are not only akin to what Ito and Okabe (2005) identified as visual diaries but indicative of an "individual" experience that is then emplaced. Through images of food and scenery, and through accompanying text, users communicate the experience and subjectivities of a place to their publics, while through metadata, the technology communicates about time and precise GPS place. While respondents noted that they used Renren (the Chinese equivalent of Facebook) for albums and archiving, *Jiepang* visual narratives were more about "*where* I am and doing/feeling what." All respondents noted that *Jiepang* inspired them to take more camera phone pictures; and thanks to smartphones it was easier to take, edit and share images.

Moreover, with *Jiepang*, respondents progressively felt the need to make visual and textual comments about places, especially emphasized through the idea that images were part of an event or movement. Thus *Jiepang* images are part of what Pink (2011: 4) identifies as the emplacement of images whereby they are located in "the production and consumption of images as happening in movement, and consider them as components of configurations of place." Interestingly, it was through its majority of users predominantly sharing camera phone images that saw *Jiepang* rebrand itself into be a visual shared diary in 2013.

In the case of gamified LBS, camera phone images are contextualized through both multisensoriality and movement. Not only is the genre and content about narrating place as part of a journey, but also the frequency and its link to reinforcing social capital suggest a complex representation of place (through image, GPS, and text tags), publics (through sharing and distribution), self (through choices made in all of the above), sociality (through participating in all of the above), and visuality (through the reading of the image and its aesthetics). In the case studies, the role of camera phone content constructing types of gender performativity—akin to Judith Butler's (1991) notion whereby cultural norms about gender are enacted through a series of iterations and regulations—and normativity is evident. Moreover many of the gamified LBS not only became a site for gender performativity but also a way in which users could be active in conveying the multisensorial movement of the context of the image so that it was not just located, networked, or placed but also emplaced.

In case studies in Melbourne and Singapore, the impact of smartphone apps, especially those for the iPhone, dramatically increased the number of images taken and shared. Respondents noted that the rise of photo apps like Instagram made camera phone picture taking and sharing more compelling. Some respondents with no interest or experience in photography found themselves becoming compelled to send and share pictures. Often pictures were an easier and convenient way to comment on an experience; they provided ambience and emotion. With the apps providing those lacking photographic skill with quasi-talents, suddenly the "bad" amateur photographer could become like an artist. With the apps equalizing a lack of skill and also affording easy ways to share, these new visualities impact not only how place is encoded with particular social and emotional vignettes but also how these records affect collective and individual memories.

In Singapore, Shanghai, and Manila, camera phone LBS practices often were used to negotiate diaspora and place. Often respondents used camera phone images, overlaid onto place via Facebook Place, to give distant intimates back home a sense of ambient co-presence. In Seoul and Tokyo—the two places with the longest history in camera phone practices—the relationship between image and place was further complicated by LBS. In Seoul, there was a decline of self-portraiture in the face of growing anxieties around trust online. Instead, there was a rise in inanimate objects like coffee cups in what our informants said was an attempt to embrace friends and evoke non-visual senses like smell. So too, in Tokyo, there was a growth in sharing pictures that is often obscured or abstracted to create ambient intimacy.

In many of the locative camera phone images, the social dimension of the image is brought to the forefront, first, in the sharing or distribution and, second, in the overlaying of geographical or locational knowledge among social and intimate publics. Echoing Ilkka Arminen's (2006) point that, when it comes to SNS and social media more generally, social context rather than "pure geographical location" is of greatest user interest, what seems to be most at stake in LBS games is new knowledge about particular sites and what these are likely to signify within social network settings. LBS, like *Jiepang*, highlight that the various dimensions around ideas of place—as imagined and lived, geographic and psychological—are contextualized by social capital. Thus many of those surveyed revealed that the higher the perceived level of novelty or uniqueness that is seen to be associated with a place (e.g., in the words of one respondent, a "supreme" restaurant or hotel), the higher the likelihood that their presence in and knowledge of this place will be recorded via *Jiepang*, as "routine places like home or [their] company are not worth checking in." In this way, it is geographical (or "environmental") "knowing" and an appreciation of the "capital" that is invested in and carried with this knowledge that is paramount and which forms a vital resource within the participants' wider peer network.

Conclusion: Emplaced Visuality and Geospatial Sociality

The use of smartphone apps has transformed the texture of visuality by rendering it no longer only about the visual. Instead, LBS camera phone images and genres are about evoking a sense of localized place as a multisensorial entanglement, or what Pink (2011) has called "emplaced" visuality. This transformation of visuality is also about enhancing existing intimate rituals and sociocultural notions of place in new ways. As we discussed in the case studies of Shanghai and Seoul, each location fused the visual with their own forms of place (*bang*) or intimate relation (*guanxi/ilchon*).

While initial studies into camera phone visuality discussed it as part of networked media (Rubinstein and Sluis 2008; Villi 2010), this second generation of visuality is about new types of place-making exercises. These exercises are emotional and electronic, geographic and social—highlighting the complex entanglement that is the ever-evolving of place. In each location, camera phone images are overlaid onto specific places in a way that reflects existing social and cultural intimate relations as well as being demonstrative of new types of emplaced visuality.

Many of the examples of LBS in the different locations were predominantly the preoccupation of the young: Generation Y or X. The exception to this was Seoul where LBS are deployed by parents, governments, and corporations for different forms of friendly and not-so-friendly surveillance (Lee 2011; Wallace 2012). In gamified LBS, like *Foursquare*, *Jiepang*, *Line*, and *flags*, new types of emplaced visuality and geospatial sociality are implemented almost exclusively by the *ba ling hou*. Through the locative and microblogging experience of *Jiepang*, users are creating new forms of intimate publics whereby the importance of network pales in comparison to the significance of the *guanxi* in providing ambient contexts. As part of the smartphone phenomenon, *Jiepang* is accompanied by an accelerated rate of camera phone image capture, editing, and sharing. Far from banal a-contextualized images, these pictures deploy the newest of filters and photographic tricks to give a sense of the poetic and unique and are then overlaid onto places via *Jiepang*. In *Jiepang* camera phone images, images are edited and contextualized so that the *guanxi* is clearly overlaid onto place. This is not a mere practice of networked visuality as noted by the first studies into camera phones, rather, we see emplaced and multisensorial visuality that creates and reflects unique forms of geospatial sociality.

Gamified LBS like *Jiepang*, *flags*, and *IN* enable users to be part of intimate publics in which the personal practices of the everyday become, on one hand, commodified and, on the other hand, further tethered to a sense of place that is as much emotional and social as geographic and physical. Social relations are thus played out in public domains in ways that foreground both networked social and place-based settings as they are negotiated in combination. In these practices, the unofficial play of camera phone images has become a compelling motivation for users.

With new photo apps marketed all the time, respondents found that beyond the "official" play of the LBS games, "unofficial" camera phone image sharing afforded richer experiences of place and co-present sharing. These cartographies are overlaying onto the LBS games in ways that both rehearse older sociocultural notions of place and intimacy as they rewrite them. Through new types of visuality that are more than just networked but "emplaced," respondents use the images to reflect the multisensorial experience of place with their culturally specific notion of intimacy. The saliency of localized intimacies and practices of place also highlights that while LBS is creating new forms of ambient, emplaced, social visualities, it is also rehearsing and tightening older social ties.

18 "DID YOU MEAN 'WHY ARE *WOMEN* CRANKY?'" GOOGLE—A MEANS OF INSCRIPTION, A MEANS OF DE-INSCRIPTION?

Jennie Olofsson

The Politics of Search Engines Revisited

With the rise of the Internet, the use of computational means by nonspecialists has become habitual and integrated into everyday life. The Internet provides social platforms, online communities, and Web-based services that clearly affect how we communicate, learn, shop, and play (Jackson et al. 2001). Search engines are crucial to online information retrieval, and as a growing part of the data that traffics the Web flows through search engines, they are also decisive in the creation of new information flows (Grimmelmann 2007). The primary function of search engines is to identify relevant content and to organize the most pertinent results (Granka 2010). As the basis for indexing, and in order to determine the importance of webpages, search engines use their own preparatory algorithms. "Spiders" and "crawlers" are used to construct an indexing database. A crawler or a spider is a program that browses the Web, collects data, and finds related webpages by following links on the pages visited. The more links a site attracts, the more visible the site is to the search engine, and the more likely it is to be indexed and prioritized in search results (Vaughan and Zhang 2007).

However, far from being an objective means through which information is retrieved, search engines have profound effects on and even shape the Internet. Because online search engines match users with content providers, they can be seen as accumulators of knowledge. No user has time to browse through millions of webpages for a given query; thus users depend on search engines to facilitate and customize information retrieval by providing users with links to "relevant" webpages. To further assist in users' searches, some search engines even apply additional features such as instant search, auto complete suggestions, and automatic spelling correction; features that are based on user input but later affect the results users receive. Search engines then both influence what we see on the Web and are themselves influenced by the collective preferences of seekers (Introna and Nissenbaum 2000b). In their study *Shaping the Web: Why the politics of search engines matter* (2000b) Lucas D. Introna and Helen Nissenbaum note that because search engines systematically exclude certain sites in favor of others while relying on user input, search engine results tend to cater to majority interests, making the design of search engines not just a technical matter but also a political one. George A. Barnett and Eunjung Sung (2006) reach similar conclusions in their research on the irregular results and uneven coverage of webpages provided by search engines. They assert that culture, rather than economic relations within nations, is an organizing mechanism of the Internet.

As recently as 2007, Liwen Vaughan and Yanjun Zhang note that while the significance of the Web in various aspects of our lives is well known, the impact of online search engines on what information we find and consume has not been critically examined. Since then, however, there has been an explosion of interest in search engine impact by scholars from fields as diverse as Informatics, HCI (Human–Computer Interaction), Law, S&TS (Science and Technology Studies), Sociology, and Media Studies. Common for all of them is that they highlight the various implications of online search engines. Researchers have focused on many issues, including the politics of search engines (Introna and Nissenbaum 2000a, b), users' eye movements as they browse through search engine results (Pan et al. 2007), the inequitable representation of websites in the results of various search engines (Vaughan and Zhang 2007), as well as the structure of search engine law (Grimmelmann 2007). Despite this uptick in interest in the politics of search engines, it is important to remember that the images that circulate in cyberspace are often passed off as "information" and "knowledge"—abstracted from uneven social relations (Gonzalez and Rodriguez 2003).

While search engines are one of the most popular means through which information on the Internet is retrieved, most users are unaware of the conditions that determine what information is retrieved through search engines (Introna and Nissenbaum 2000b: 176; Pan et al. 2007). At the same time, the selective coverage of the Web by search engines has many social, political, cultural, and economic implications (Vaughan and Zhang 2007). Search engine optimization, for example, allows webpage designers to modify the ranking process of search engines, thereby improving the position of a particular webpage (Bar-Ilan 2007). In a similar way, search engine editors seeking additional references within a certain category tend to lower the entry barriers so that more webpages are listed (Introna and Nissenbaum 2000b). As the battle for rankings is fought between search engines, Web designers, and organizations wishing for webpage prominence[1] (Introna and Nissenbaum 2000b), the information obtained through search engines must be seen as a product of these conflicting and converging interests. As Vernadette V. Gonzalez and Robyn Magalit Rodriguez (2003: 222) note, "[t]he process of creating 'objective' information is deeply subjective." The search engine's results then are necessarily the outcome of diverse interests and subject to continuous change as the interests of search engine companies, Web designers, advertisers, and users, constantly shift. In what follows, I explore some of the gendered implications of the information retrieval processes of Google's search engine services using Madeleine Akrich's (1992) actor-network theory concepts "inscription" and "de-inscription." Specifically, I focus on the ways in which Google's additional features, such as instant search, auto complete suggestions, and automatic spelling corrections services, contribute to gender stereotyping at the same time as users also actively respond to, and challenge these stereotypes. In investigating the gendered implications of online information retrieval processes, this article also finds inspiration in the digital humanities as it focuses on the incorporation, and mutual transformation of digitalized and born-digital materials and the notion of asymmetric gendered structures. As digital tools, techniques and media have altered production and dissemination of knowledge, online information retrieval processes are equally part of creating and reproducing, but also breaking with, gender stereotypes.

Google—A Means of Inscription, a Means of De-Inscription

The structure of Google's search engine service as well as its expansive growth during recent years has already received a great deal of critical attention (see for comparison Bar-Ilan 2007; Grimmelmann 2007; Pan et al. 2007; Tene 2008; Vaidhyanathan 2011). While Internet users do employ a variety of online resources to fulfill their information-seeking needs (Hargittai 2007), Google dominates the field, at least in Western parts of the world, and this dominance makes Google an excellent case-study for examining the gender politics of search engine services. In addition, unlike many other search engines, Google's search function is accompanied by a suite of additional online services, such as YouTube, G-mail, Google groups, Google maps, Google. docs, Google +, the Android mobile platform, and Google ads. Google also owns its own Web browser and an operating system that is based on its online services. Further the increasing number of Web applications founded on Google's platform reinforce its dominant position. In the book *The Googlization of Everything (and Why We Should Worry)*, Siva Vaidhyanathan (2011) investigates the effects of Google's supremacy. Although it is possible for the individual user to change the system's default settings—Google even provides videos explaining how to do this—Vaidhyanathan points out that the very act of changing these settings requires the active choice to do so. These default settings structure the user's experience of the Internet and the Internet itself and are an expression Google's particular ideology.

Of course, this ideology is informed by the economic interests that backcloth the structure of Google's search engine services. I explained above that Google offers a suite of online services. In return, the company collects information about user habits. Using this information allows Google to provide better services, but it also enhances Google's ability to efficiently target advertisements to users. In 2008, 97 percent ($21 billion) of Google's revenue came from online advertisements, indicating that Google's primary business is consumer profiling (Vaidhyanathan 2011).

To better understand the ways in which Google imposes meaning on its users and the ways in which users supply, adopt, resist, and deflect that meaning, I have borrowed Madeline Akrich's (1992) terms "inscription" and "de-inscription." According to Akrich, "inscription" is the process by which technologists, designers, and innovators define the characteristics of their technical objects and, as a result, also inscribe certain "tastes, competences, motives, aspirations, and political prejudices" onto the presumed users of these objects (208). "De-inscription" is the response users enact to the provided "script," or user manual, of inscription. These responses can vary greatly, and can both undermine or reinforce the imposed script of the technical object. This means that technical objects, far from being objective tools, define and are defined by human actors, the spaces in which they move, and the ways in which they interact. Akrich engages with the ways in which the scripts of physical technology, such as photoelectric lighting kits and electric generators, inscribe meaning on their imagined users and are, in turn, "de-inscribed" by users who refuse to follow the script. While Akrich focuses on physical, technical objects, I argue that cyber technology is also governed by inscription and de-inscription. Because the results provided by the search engine are a consequence of the interactions between search engines, Web designers, organizations, and, of course, users it is important to explore the negotiations between each of these parties in order to study how these interactions are translated into technological form and user practice (Akrich 1992). As the relation between the different stakeholders changes, the search results transform accordingly.

One of the primary mechanisms by which Google's inscription and de-scription occurs is its information retrieval algorithms. While Google does not disclose any details of its ranking algorithms (Bar-Ilan 2007), the search results give the impression of having been sorted solely by "relevance," namely by an estimate of the probability that it will fulfill the user's information need (Pan et al. 2007). However, as "Google's ranking algorithm takes into account the quality and quantity of links pointing to a given page" (Bar-Ilan 2007: 912), public opportunities to manipulate the search engine emerge. This capacity for manipulation is a classic example of de-inscription. Thus, while Google's search engine inscribes, that is "controls the moral behavior of [its] users" (Akrich 1992: 216), it is also subject to continuous processes of de-inscription in that "users may define quite different roles of their own" (Akrich 1992: 208). One example of this de-inscriptional role is "Google bombing," whereby the creation of links for particular webpages or keywords manipulates the ranking of search results. Google bombing does not just affect the targeted page, but the entire structure of the search engine (Bar-Ilan 2007). Importantly though, the algorithmic ranking system of Google does not hinge on unauthorized actions, but on Google's preparatory algorithms, which indicate "how access to the Web is preconfigured in subtle but politically important ways" (Introna and Nissenbaum 2000b: 170). Google's additional features, such as instant search service, auto complete suggestions, and automatic spelling corrections, show that Google itself stipulates the conditions by which user input contributes to search suggestions by adapting search results according to its policies. And while Google's policies are available for public perusal online (*Google Inside Search* 2012), much like the standard functions that a user must turn off to avoid, individual users are often not aware of, or cannot be bothered by, their implications.

Google's policies do not just condition processes of information retrieval through suggestion and correction services, but also, and more notably, through *the lack* of predicted searches for a particular word or topic. As stated on Google's information webpage, the lack of certain predicted search terms is a response prompted by "search term [that] violates our [Google's] autocomplete policies" (*Google Inside Search* 2012). One example of Google's exclusion of suggestion for certain terms is the way it filters piracy-related key words. In June 2011, the online forum *Torrentfreak* (Ernesto 2011) reported that Google, in an attempt to curb online copyright infringement, had begun censoring key words such as "torrent," "BitTorrent," "uTorrent," "MegaUpload," and "Mediafire," earlier that year. While the actual search results remained unaffected, the piracy-related key words were excluded from Google's auto complete and instant prediction services. Hence, as the user types in "BitTorrent," she will not receive suggestions on how to conclude the search, a factor that has proven to affect the search volume negatively. After the anti-piracy filter was implemented in January 2011, the searches for "BitTorrent" dropped by half. The online forum *Torrentfreak* notes that these search trends are the same for practically all the censored key words, which shows the magnitude of influence of Google's auto complete and instant prediction services. The *lack of* predicted searches clearly discourages users from continuing the information retrieval process thus inscribing behaviors on users.

Filtering piracy-related key word was not the first instance in which Google modified search results according to their policies. In 2007 Google attempted to defuse Google bombing, because "over time, we've seen more people assume that they [the results of Google Bombing] are Google's opinion" (Bar-Ilan 2007: 932). However, search results that are not produced by Google Bombs are not objective measures of mere relevance. As we have seen, Google is decisive in the creation of new information paths and is a far from neutral medium through which information flows. Instead, Google's search results and suggestions represent particular policies, opinions, and ideologies. The questions then emerge: what are the opinions and policies that steer the information provided to millions of Internet users? And how do Google's additional features contribute to the creation and maintenance of gender stereotypes? Here Akrich's concepts of inscription and de-inscription facilitate an understanding both of the gendered scripts that the users of Google's search engine service are expected to enact and of the extent to which these users are able to reshape and reformulate (de-inscribe) the inscription practices that pervade the search results.

To Filter or Not to Filter . . .

As recently as 2010 scholars have argued that the aggregate traffic through search engines and the results it produces merely reflect mass tastes (Granka 2010). According to this view, the information retrieved is assumed to indicate the interests of search engine users. As such, it is also presumably one-way communication. The argument, however, fails to take into account the processes by which search engines retrieve information and the many competing interests that affect those processes. While personalized and customized information retrieval is realized through instant search, auto complete suggestions, and automatic spelling corrections services, these search results are determined by societal norms and expectations of, for example, men and women. In order to outline the political implications of search engines, it is therefore necessary to understand that the act of locating information also informs what the user is looking for (Introna and Nissenbaum 2000b). James Grimmelmann (2007) asserts that "to the extent that a search engine is viewed as a speaker . . . its recommendations of content potentially become endorsements of that content's message" (31f). Hence, as technology and society mutually feed off and reinforce each other (Wajcman 2004), the recommendations provided by Google do not just reflect gender stereotypes input by users but also tailor them, an effect that has been the subject of much debate in online forums. In what follows, I will demonstrate some examples of these debates.

"Did You Mean: He Invented?"

Google's auto complete suggestion service is triggered whenever a user inputs search terms, such as keywords or sentences, into the search box. Google then offers potential ways of concluding the search. These suggestions are shown in a drop-down list and are displayed in bold type. Google's online policy document explains this feature by asserting that the search engine's auto complete suggestion service provides an opportunity for the seeker to rest his/her fingers (*Google Inside Search* 2012). As the seeker initiates a search by typing key words, possible suggestions of how to complete the search facilitate quicker and easier information retrieval. But Google's auto complete suggestion service also reinforces gender stereotypes. For example, the article "Google offers disheartening suggestion on 'Women bosses'" (2012) shows that the initial series of keywords "women bosses are" is completed by Google, in descending order, with the suggestions "women bosses are the worst," "women bosses are," "women bosses are better," and "why women bosses are bullies." The auto complete suggestion service then is not just a way to facilitate faster and more efficient searching; the search completion also funnels searches toward the most common continuation, meaning women are slotted into prefabricated categories of representation. And while women bosses are the worst, better (than whom?), or bullies, their male counterparts are rendered invisible.

In addition to the auto complete suggestion service, other users' search activities are componential to the instant prediction of search queries. Each prediction that is shown in the drop-down list has been typed before by users of Google's search engine. I mentioned above that Google's suggestions reflect the popularity of a particular search term, a feature made possible by the fact that Google keeps track of previous searches. As Google describes it: "[a]s you type, Google's algorithm predicts and displays search queries based on other users' search activities" (*Google Inside Search* 2012).

Another example of the ways predictive search terms affect and reinforce gender stereotypes is the article "Can Google Tell Us What Men and Women Are REALLY Thinking?" (Fox 2010, emphasis in original), which shows how Google's auto complete responds differently to the search terms "how can I get my boyfriend to" and "how can I get my girlfriend to." The top three suggestions activated by the formulation "how can I get my boyfriend to" radically differ from "how can I get my girlfriend to." Whereas the former prompts "propose," "spend more time with me," and "love me again," the latter prompts "give head," "sleep with me," and "lose weight." Recalling Judy Wajcman's assertion that technology is both a source and a consequence of gender relations, I argue that the marks of power relations between men and women are at once exhibited *and* reinforced as Google's auto complete suggestion service makes predictions, thus implicitly inscribing the preferred continuation. Women are subsequently confined to roles as passive recipients of men's love and affection or objects of men's sexual dissatisfaction.

In its online policy document, Google explains that "[q]ueries are algorithmically determined based on a number of purely objective factors (including popularity of search terms) *without human intervention*" (*Google Inside Search* 2012, my emphasis). Following Akrich, I note that Google's naturalization of the search results conceals the inscription processes through which information is filtered as well as the social norms that this inscription creates. However, while it is generally accepted that queries in auto complete suggestion functions are algorithmically determined, Google simultaneously asserts that certain search terms do violate their auto complete policies. This being the case, no predicted searches are provided for a particular word or topic. Google's filtering of piracy-related key words and the decision to defuse Google bombing are just two examples of how Google has modified the algorithmic conditions by adapting the search results according to their policies. In analogy with Akrich's example of photoelectric lighting kits, the filtering of piracy-related key words and the decision to defuse

Google bombing work by a process of elimination: Google tolerates only docile users and excludes others (Akrich 1992). Similarly Google has also taken measures to terminate certain spell-checking behaviors, specifically with regard to their gendered implications. In the spring of 2007, Google was reported to correct the search query "she invented," two key words that due to the automatic spelling correction service previously had resulted in the suggestion "Did you mean: he invented?" In a similar vein, the search queries "she scored," "she instructed," "she saved," "she discovered," and "she golfed" were all assumed to denote male practitioners, and were thus regarded as misspellings or less likely alternatives. Google's automatic spelling correction service then presents more common spellings, in this case "he invented," "he scored," "he instructed," "he saved," "he discovered," and "he golfed." While it is clear that Google's suggestions reflect the popularity of a topic—more people obviously search for "he invented"—however, the suggestion cuts a far wider swath than the aspiration to correct a misspelling. Importantly, the automatic spelling correction service grooms the user: "it offers a set of rewards and punishments that is intended to teach proper rules of conduct" (Akrich 1992: 218f).

The online forum *Google Blogoscoped* (2007) notes that the spell-checking behavior described above has been stopped, and concludes: "it seems clear that a search query like 'she invented' is not likely to be a misspelling [which is why] a 'did you mean' box is bad usability" (Lenssen 2007). Google's search engine algorithm then is not static or merely subject to algorithmic organization, but subject to political, cultural, and economic interests as well. In this case the decision to correct the search query "she invented" made female inventors, instructors, athletes, adventurers and golfers visible, and was the result of de-inscriptional user critique of Google's gendered spell-check script.

Concluding Remarks

This chapter has explored some of the disagreements and negotiations between search engines, Web designers, organizations, and users. Investigating the gendered implications of Google's search engine, and drawing on Akrich's actor-network theory concepts of inscription and de-inscription, I argue that Google's additional features, specifically auto complete suggestions and automatic spelling correction services, work in tandem with gender stereotyping. As "Google refracts, more than reflects, what we think is true and important" (Vaidhyanathan 2011: 7), it is noteworthy that "[t]he ways in which search engines participate in creating and reinforcing racial and sexual logics has serious ramifications for meaning-making for particular bodies on the Internet" (Gonzalez and Rodriguez 2003: 217). Google's additional features play an important role in constructing the gendered identities of users and gendered issues consistently emerge through their usage. The provision of these services then is part of continuous processes of inscription but also, through user critique, of de-inscription.

Note

1. Fifteen years after Introna and Nissenbaum's publication (2000b), I note that search engine users too are actively involved in the battle for ranking. As users produce large amount of the data that is being sought for they function as content providers rather than mere content consumers, something that affects, not only the input of data, but also the scanning procedure of the search engines.

19 TIME WARS OF THE TWENTIETH CENTURY AND THE TWENTY-FIRST CENTURY TOOLKIT: THE HISTORY AND POLITICS OF *LONGUE-DUREE* THINKING AS A PRELUDE TO THE DIGITAL ANALYSIS OF THE PAST

Jo Guldi

The history of property ownership as we know it is one of the best-documented subjects in the university, the subject of concentration by economists, historians, law school professors, philosophers, and anthropologists. It is also a subject with an impressive century-long hole in it, from approximately 1870 to 1980, during which time period a global debate about property ownership happened that no one has written about. It is exactly the sort of subject that digital tools for reading extensive archives were built to handle: global, massive, almost entirely uncharted, deeply relevant, and open to debate.

In 1870, a historian named James Godkin inspired a new phase of political resistance geared toward the distribution of land by publishing a new account of Ireland's struggles against England, *The Land-War in Ireland*. In it Godkin framed contemporary struggles in terms of ancient rights to land ownership abrogated by colonizing invaders, who, since Spenser's time, had used rent and eviction as the major tools for terrorizing the colonized. In the generation after Godkin's book, Irish resistance changed from guerilla struggles to organized rent strikes that placed land ownership at the center of their work. The era of the Irish "Land War" had begun.

In the 1880s and 1890s, rent strikes and other rebellions influenced by the Irish spread to Scotland, England, and New York, where organized campaigns against paying rent became the major tool of urban immigrants in protecting their incomes, and were successful in many areas where unionized labor strikes were still illegal and their gains meager. More important, Godkin had thrown into question the ultimate justice of colonizing powers' right to land and made possible the abrogation of claims of property ownership by previously excluded and oppressed citizens who, in the tradition of the French Revolution, restated their claim to economic and political participation on the basis of universal rights. In this case they were the right to own property and the right to be free from eviction, typically phrased as "the right of the tiller to the soil." Through the popular journalism of the San Francisco born international political pundit Henry George, previous legal traditions of public property were reworked through a broad challenge to the legacy of empire and class privilege. In the writings of Fabian socialists like Annie Besant, land reform became canonized into a new agenda for socialism. In the writings of legal reformers like Frederick Maitland and Paul Vinogradoff, the folk ownership of the land in the Middle Ages provided a precedent for state collectivization of land in the service of infrastructure and housing for the people.

This robust intellectual foundation provided the basis for national politics around the reform. By 1914, every political party in Britain supported some version of a land reform agenda in Britain's Parliament. In postcolonial Mexico, arguments for land reform echoing the Haitian revolution urged on land reforms capable of reversing imperial concentration of the land into haciendas. For similar reasons of reversing aristocratic control, the League of Nations urged land reform upon imperial China in the 1930s. By 1945, national programs of land management, mortgages, and public housing were on the agenda in every developed nation in the world. By 1946, the United Nations' Food and Agriculture Organization dedicated large portions of its administration to the collection of land tenure policies and statistics from the rest of the world, with the agenda of supplying legal and economic advisors to developing nations seeking to emulate land reform in Europe and America.

The result of so much policy shift was the manufacture, between 1880 and 1980, of masses of documents, historical, legal, and economic, examining the past and future of land reform in every country in the world. Hundreds of historians, sociologists, anthropologists and economists in Europe and North America took the global history of land as their subject for their dissertation research, assured that their work would find committed readerships in international policy.

Then, precipitously, around 1982, political opinion turned against the land reformers. The Right, influenced by Milton Friedman, argued that land was just like any other commodity, best left to the market rather than the state. The American Left repeated Richard Hofstadter's critique of the "Agrarian Myth" in nineteenth-century American; at the heart of their dismissal echoed something like Lenin's critique of the peasantry—against decades of Latin American and Asian experience—that the revolution would come from city dwellers, not from peasants, and therefore by union organization, not by land reform or rent strikes.

As a result of this about-face, a devastating silence followed, where barely a historian touched the global century or more of land reform for some thirty years. A few aging agrarian economists who had spent their lives working in development wrote memorials to land reform, pondering why it went away when it was doing such a good job of distributing incomes. But most historians pretended to bury the subject, until now. A few years ago, a few brazen Britons, Paul Readman and Matthew Cragoe, began to unpack the English example and to show that land reform really had mattered. But no one dared to touch the global question—what was land reform, and what happened to it? It was a formidable question, for it aims straight at a broad consensus in political science and economic policy that has a stranglehold over world events at the moment, not least the global political stagnation around environmental regulation.

The more we know about land reform, the more we realize that there is abundant political and legal precedence for broadcast regulation of land and water in the service of the people. Foundations of modern economic consensus begin to fall apart, namely that private property is an unchanging, easily formulated category; that it has ever been so and has always been apparent; that a golden age of pre–welfare-state capitalism existed in the glorious years 1880 to 1920 that we should turn back to; and that private property and land and water in particular cannot be made into public utilities without destroying the entire market economy.

Pursuing questions such as these is methodologically as well as intellectually difficult. It is hard for a few historians to take on a subject whose archives stretch over decades, let alone transnational centuries. There is far too much paper to read. And that makes it ideally suited to a methodological innovation at the heart of the American university right now. At the root of the questions of digital methodology—of which tools and data we collect and how—are questions about how we address silences such as these. If we use digital tools to address long-term questions, we raise the possibility that a mere historian (a *humanist* mind you, schooled in reading and writing and the digital, not in STEM) can beg to tangle with economists, indeed with a consensus all the way across political policy today.

In the era of big data, we face methodological opportunities for placing the long-term analysts and the short-term analysts in conversation with each other. This is particularly appealing to those of us embarking on a digital turn in which questions of big data and their analysis are at the forefront of our activity. In an era in which the manipulation of large-scale aggregate data over space and time has become easier than ever, scholars in disciplines with adverse proclivities to short- and long-term storytelling face the option of allowing their data to speak to each other. The question of long-term or short-term history has a methodological aspect in the digital humanities, and it is this tension that governs the rest of this chapter.

Under the domain of the short in time come the intensive digitization and analysis of a perfect corpus—the poetry of Gertrude Stein coded for different types of speech acts (Clement 2008), the plays of William Shakespeare coded as to the gender of the speaker (Hota et al. 2006), maps showing speech acts next to demographics in the pre–Civil War South, county-by-county (Thomas 2004). Into this category fall the digital editions favored by the Society for Textual Scholarship and a great deal of beautiful digital cartography. Deep studies informed by many documents, they are short merely in the number of lifetimes under study.

Of the *longue-durée* type there are far fewer. There's the theorizing by the occasional maverick like Franco Moretti (2012), whose *Maps, Graphs, and Trees* forced a hundred studies of the novel over five- and ten-year periods into a synthesis of a hundred years. There are the trade maps of the long eighteenth century made by Ben Schmidt (2012). There are collaborative projects like the Enlightenment of Letters. We might add N-grams, allowing, as they do, the imagination to analyze the use of the word "fancy" over three hundred years (Michel et al. 2011). These projects, like those in the first category, are also intensive in terms of corpus, but their analysis tends to be whittled down to one question—the social network, changes in word frequency, or the shape of geography.

Demographically, the two kinds of methods tend to break down by the place of the scholar in the institution. Dissertation-writing and junior scholars are pressured to pursue short-durée projects. The same could be said of many traditional historians working on their own with the help of a large grant, for example, the Digital Harlem project which covers two decades of that city's life. *Longue-durée* projects like the Old Bailey Online and "Connected Histories" tend to be the product of large-scale collaborations where an analyst has turned administrator, or toolbuilders, often working outside the humanities or even the academy. Here Google Books might serve as an example, the sources for which span the centuries (whether or not the f and s search works equally well across them). In short, they tend to be outliers: collaborative straddlers of disciplines, using occasionally messy and imperfect methods (oldbaileyonline.org; connectedhistories.org). To judge from the mass of presentations at the MLA or AHA or Chicago Colloquium on the Digital Humanities and Computer Science or THATCamps, these *longue-durée* inquiries hold second-rate status in the disciplines. There are fewer dissertations being written that exploit these methods, and they are less frequently held up as a model worthy of emulation.

The purpose of this chapter is to question the prioritization of short-durée over *longue durée*, to hold up to question the advisor's mandate that narrowing to an appropriate question requires narrowing over time. Digital methods, with their powers of mass aggregation, raise important questions about how these tools are best put to use—whether in reading the entire corpus of a single author or of a single nation. The point of this essay is to take quite seriously those questions of scale.

Reasons Why Short-durée History Prevails

Short-durée history dissertations have not always prevailed. One has only to think of Anthony Grafton's account of maps of centuries and millenia illustrated by medieval and early modern historians for perspective. Notoriously, Fredrick Jackson Turner's 92-page dissertation performed a sweeping history of the making of the American west and Kissinger's undergraduate thesis described the rise of Western civilization. Even methodological interventions could be quite long term in scale. In the institutional history of the 1950s, first books regularly handled the whole of the sugar trade or the institution of the customs over a two- or three-hundred-year cycle. But in the era of retreat from international policies of reform, academic history began to distance itself from questions of long duration. Other events that might have had an effect included the professionalization of the university disciplines in the era of the GI bill, the increasing dominance of a publish-or-perish mentality where short contributions came to be favored over deep theses long in the making (De Rond et al. 2005; Baneyx 2008), and the synchronized arrival into consciousness of social problems ranging from gender prejudice (Canning 1994; Hall 1991) to environmental catastrophe for which no adequate previous critical tradition existed. By the early 1980s, short-durée history became the model of good historical scholarship, the crucible in which to refine ideas, actors, and methods.

In the era of publish or perish, social historians carved out their niches by excavating unusual archives whose sheer arcaneness meant that an entire career could be dedicated to making sense of that set of records. To take three of the most distinguished careers of a generation, Carlo Ginzburg (1980; 1994), Bill Sewell (1980; 2005), and Natalie Zemon Davis (1983) built their reputations upon the exploration of arcane archives—peasant records in the provinces and early working-class organization in the south of France, all of which filled in major historical gaps in our understanding of subaltern life, to date only interpreted secondhand through the observations of officials and travelers. So dense and disparate were the records they called upon—court depositions, rumors, dreams, fairytales, and petitions—that each document had to be triangulated with the larger context of official legislation, agrarian struggle, gender identity, and power dynamics. The biography of each unknown individual had to be painfully reconstructed (Brewer 2010). And the whole of this archeological labor of reconstructing, piece by piece, the shattered vase of the past was a long-term, labor-intensive dedication to sitting alone in an unexploited archive, for the purpose of producing a vessel capable of holding the stories of peasants preserved nowhere else in the historical record.

For the generation of the cultural turn, exploiting an arcane archive became a coming-of-age ritual for a historian, one of the primary signs by which one identified disciplined methodological commitment, theoretical sophistication, a saturation in the historiographical context, and a familiarity with archives. Gaining access to a hitherto unaddressed archive signaled that one knew the literature well enough to identify the gaps within it, and that one had at hand all of the tools of historical analysis to make sense of any historiographical record, no matter how obscure or how complex the identity of the authors. In the generation that followed, almost every social historian experimented in some sense with short-durée-form historical writing in the genre of remarkable arcana—from Ian Hacking's (1998) discourse on fugue states in *Mad Travelers*, which departed from a twenty-year psychiatric fad of diagnosis, to James Vernon's (2000) work on early transgender individuals in the British military. In some sense, the more obscure or difficult to understand, the better: the more that strange archive proved the writer's sophistication within a wreath of competing theories of identity, sexuality, professionalism, and agency, the more the use of the archive proved the scholar's fluency with archives, commitment to long-distance train journeys, and saturation in the field.

The Return of *Longue-durée* History

There are limitations associated with the format of the short-durée dissertation that are becoming essential to consider in the age when digital methods pull scholars in different directions. Many short-durée histories fail in that their impact on the surrounding discipline is limited. They may contribute another brick to the wall of knowledge without formulating a turning point of consequence to the rest of the discipline. This is frequently the case with dissertations limited to a case study that turns out to not be that different from another study of working-class life or gender or Foucaultian relationships between vision and the state studied in another part of England another decade before or after in another dissertation written thirty years earlier. Short-durée, single-archive studies are limited in space as well as time. It is unclear how representative they are of the nation, let alone of modernity as a whole.

It is largely by consequence of these limitations that historians have begun to look elsewhere for dissertation topics. Short-durée history is gradually being displaced in history as a discipline by two other forces—transnational and environmental history—which are expanding Hegelian questions about the nature of agency and the balance of power (Smail 2008; Mazlish 1999). Environmental history has long positioned itself in light of *longue-durée* questions of industrialization and empire. Now it looks even more widely at the anthropocene, positioning human agency in terms of pollution and environmental devastation in light of our interdependency as a species in a constantly changing environmental matrix (Chakrabarty 2009; Koeberl 2009; Christian 2004). Transnational history, for example the historiographies of squatters on the edge of empire now being written for the sixteenth and seventeenth century, similarly destabilizes questions about agency in a long-term context. A new generation of scholars such as Jamie Belich (2009) suggest that it was squatters moving in advance of the state, rather than imperial bureaucrat and surveyor, who caused the great exploitation and fighting with indigenous people in north America, Australia, Africa, and South America. Belich argues that only squatter petitioning of the metropole led to the extension of imperial peace resulting from military protection of property claims as a way of ending cycles of violence already instigated by individuals moving in advance of the state (Banner 2007; Russell

2001; McLaren et al. 2005; Veracini 2007). In both of these emerging subgenres, *longue-durée* history allows us to step outside of the template of national history to ask about the rise of long-term complexes.

Totally aside from the question of digital methods then, we are seeing a slow conversion of the short-term historical inquiries typical of the 1970s and 1980s into long-term questions that destabilize the idea of agency in both the species and national contexts. These accounts break open room for new inquiries, for dissertations to be written exploring this borderline of where history changed—of what new transnational spaces and cultures of environmental exploitation really mattered, in aggregate, over the course of the centuries.

The questions that have appeared in environmental and transnational history are becoming all the more pressing in an era of digitized resource. Mass-digitization projects such as Google Books have dramatically changed the character of data in the hands of an average researcher. For the individual historian approaching questions of time and space, there is no lack of important historiographical issues testing our notion of what constitutes important history. Nor is there a lack of texts. Mass-digitized databases have already made reams of paper available. Tools, likewise, are not exactly the problem. Digital tools have a natural fit with such questions, for the tools scale: a tool for counting word frequency is just as effective with the corpus of Shakespeare as it is for the corpus of the English language over all of time. They scale over time—which is the case with N-grams, and they scale over space—which is the case with geoparsers.

The lack of work in the *longue-durée* humanities, rather, must be chalked up to questions of habit within the disciplines. We are dealing with a lack of comfort, a shortage of theory, and a need for basic skills. We need to inquire into the conjunction between algorithm, paper, and relevance: between what has already been scanned, digitized, meta-data'd by forward-thinking archivists, what tools the world of computer science has already delivered to our disposal, and which historiographical questions of deep meaning to the rest of the discipline can be answered in this way. There is a thin space in the Venn diagram between these three intersecting spheres where lies the realm of the meaty question answered over scales of time and space.

I embarked on the Long Land War project in search of data fit for the new tools of digital history, problems of a scale such that no individual scholar, working by traditional methods, would be willing to take on the sheer mass of paper required to form an analysis. Working with digital tools to create generalizations about large historical periods, my research took me to questions about the expanding social science matrix after the Second World War that framed land reform as a peaceful path between communism and capitalism suitable for foreign policy. In collecting bibliographies, I was struck by the sheer mass of paper produced by the generations writing about land reform from 1945 to 1980. It wasn't until I began physically visiting the departments that had contributed the most research—Ostrom's Workshop on Political Economy at the University of Indiana and the Land Tenure Center at the University of Wisconsin—that I began to conceive of the paper as itself a problem. Arrays of filing cabinets, a hundred long, lined dark corridors. Each filing cabinet was filled with hundreds of articles on land tenure, land titling, and peasant history in South America, Africa, and Southeast Asia. These cabinets, and the paper within them, constitute machines designed to instruct peasant leaders, agrarian ministers, and foreign policy experts alike, but their sheer mass made the possibility of coming to a synthetic understanding of the field nearly impossible. I began to wonder about how contemporaries made sense of this much paper, and about what digital tools could do for the paper as well.

In the 1950s and 1960s, as the rise of NGOs came to ring conversations about urban planning and international development with coordinated institutions—all of them creating more paper—even the process of finding one's way to the beginning of an argument became a labor reserved for the few and privileged. Introductory texts to these problems increasingly included an organizational flowchart, a diagram borrowed from business texts, which served as a map to tell would-be activists where, in the vast continent of NGOs, banks, and government organizations, they entered the conversation. In short, the power struggles of the twentieth century have produced a mass of paper, too much to be read by a single scholar or even group of scholars. The secrets of the paper archive are the record of the power struggles that determined the rise and fall of utopian movements in the twentieth century.

But, in modern history, revolutions in bureaucracy and the limitation of political participation have frequently been a reflection of the number of pages of paper that experts produce, a social and political clout with which to disarm and outrank their political opponents. Who produced the most paper, and how did their ideas travel? That is a historical question, and it can be answered by solving a quantitative question in the stacks: How much paper was produced and by whom?

The subject governs too much paper to read—thousands upon thousands of articles from researchers in sociology, anthropology, history, and economics, working between 1930 and 1990 on sites around the globe. Those questions are constrained around a single set of issues—agrarian reform and property rights. But virtually nothing has been written about the subject in terms of secondary historical studies. Almost any pattern we find in the papers will be a new pattern.

This then is the shape of the kind of problem suited to digital history—a problem of scale, where the subject of the question inherently requires turning to a body of texts so large that it would stagger traditional kinds of scholarship.

It rests on the strengths of work that's already been digitized—the officials I'm trying to understand at the FAO and its affiliated organizations were publishing in academic journals that are cataloged in Web of Knowledge, JSTOR, and other online repositories. The metadata are already clear. The question covers decades, within a project that spans a century and a half, and both projects span the globe in the reach of the bureaucracy and where they raised questions about government and property rights.

The Paper Machines project was designed from the beginning to be enticing to students in computer science. It represents a mass of text, too deep to read, about which there's very little historiography. Any pattern a computer can find is potentially an interesting pattern to historians. I can't guess what the social network diagrams will look like in advance, nor can I forecast the most important keywords of each generation—too little history has been done. The field is ripe for hacking—and not my hacking. Against digital historians like Bill Turkel, I'm no fan of teaching history graduate students to code. Graduate students in Computer Science are already incentivized to find new patterns in masses of text, create new visualizations, and improve user interfaces, working with the best advisors and mentors possible, motivated by the constraints of their own job market and conference circuit. I predict that much of the innovation coming in the digital humanities will rest atop collaborations like this one, exploiting unprecedented bodies of text, approached with the help of individuals in CS motivated to find new connections and work at scales of unprecedented size.

Problems in the social sciences are likely to benefit from digital tools in categories where the scale of papers to be read and examples to be consulted dwarf the tools of the solitary reader. Such questions may, like the Long Land War, concern the emergence of twentieth-century bureaucracies, corporations, and movements. The problem of scale applies all the more to the noisy institutions and movements of the twenty-first century, whose electronic archives will dwarf traditional paper archives in the sheer volume of text produced. Digital scholars will turn to tools of synthesis that freely generalize about these movements and their concerns. They may also exploit the technologies used by lawyers to find the "needle in the haystack" of outlier testimony, different from all other testimonies. They are likely to be the more relevant wherever scholars are concerned with particularly modern questions—the documentation of large-scale bureaucracies, formulaic legal languages, and swarming social movements—where the problem to be understood is one of the dynamic emergence of a fundamentally new set of ideas or patterns. Digital technologies may be the less useful in diachronic questions of individual biography or character, of the close reading or multiple meanings inherent in an individual text. But particularly for these reasons of generalizablility of scale and applicability to collective language, digital methods are likely to become crucial to historians seeking to define the tendencies of modern political, economic and social form. It is easy to imagine a frontier where doing history without them would be impossible.

That transition to large-scale questions of modernity and radical discontinuities in the historical record requires ingenuity in working with evidence and its analysis, forcing the historian to consider the skills of collaboration and interdisciplinary flexibility as an essential part of her task. In August 2012, ethnomusicologist Christopher Johnson-Roberson and I released the code for Paper Machines, a toolkit for visualizing large masses of text, designed with the vast output of social science over the long twentieth century in mind. The code capitalized on current work in computer science, drawing upon state of the art topic modeling and visualization to generate iterative, time-dependent visualizations of what a hand-curated body of texts talks about and how it changes over time. Adapting these tools to the hand-tailored corpora of an individual scholar curating inside Zotero, Paper Machines reveals large-scale tendencies over time via a probabilistic topic model fitted to the data, allowing unexpected correlations between topics to emerge from the data itself, rather than requiring the researcher to predefine her terms of inquiry (chrisjr.github.com/papermachines/). By defining questions around small corpora—for instance, the work of a single author—and larger corpora—for instance, the entire discipline of economics in the twentieth century—the toolkit allows the researcher to tack between focused and broad-based examinations of a corpus, drawing her own conclusions about the relationships between diverse, intermediate actors and institutions over a vast array of time.

The plugin was designed to allow a scholar, upon first glance, to wrap her eyes around a corpus too large to read, understand something about its themes and significance, and draw some informed conclusions about what questions to pursue further. Any researcher with full-text data stored in Zotero can make use of Paper Machines. Word clouds and phrase nets provide a first look at the language used in a group of texts, showing what words, bigrams, or phrases are most prevalent in the corpus. Its geoparsing function links a text's place of publication to the places it mentions, generating an interactive map that reveals shifting sites of interest and institutional authority over time. DBpedia annotation displays a network of named entities in the text, and will eventually enable the automated acquisition of additional quantitative data to aid in analysis. Finally, a topic model visualized over time or by category allows one to see trends and correlations in the corpus: for example, one might observe the rise of environmentalist discourse, or a persistent reference to income redistribution in texts that also discuss Latin America. All of these tools may be used for exploratory data analysis, and some can also be used for statistical hypothesis testing to a degree of significance defined by the user.

Toward a New Theory of Modernity

Technologies such as these are apt to make traditional historians uncomfortable most likely regarding stories like the one with which this chapter begins: the problem of the short and the *longue durée* and the seeming war between them, in light of the tangled political battles between different narrators of history, within the discipline and outside of it, over the long twentieth century. We are familiar with the tools and theoretical questions necessary to perform an analysis of identity and agency in a short archive, but questions of the kind I've just illustrated tend to tickle the nerves. Yet the problem may not be so much lack of theory as the wrong kind of theory.

When we read Davis and Ginzberg in history, we neglect other schools of historical writing. Problems of shifting resources, historical epochs, the rivalry of continents, and anthropological questions of gender, family, and identity were typical of the *longue-durée* generations of historians such as Turner, Toynbee, and Tawney.

But when we start to look to other precedents—the Annales School or French schools of cultural geography, for example—we start to be able to imagine what the cultural history enterprise would look like on an altogether different scale of collaboration. Laboratory-type work—Annales School questions—require the collaboration of hundreds of students, garnering material suitable for analysis, analyzing it, and comparing it over the course of a ten-year project.

If we are to truly ally ourselves with the potential of the digital humanities to tackle questions of scale at time and space, we will have to summon the courage to look into the theory behind *longue-durée* questions. Historical methods syllabi must start to shift, as the search for *longue-durée* questions analyzable by *longue-durée* methods becomes important.

Longue-durée diagnosis of the world of paper will not be the task for everyone, nor will it (or should it) consume all members of every history department. My reflections are delivered from the privileged position of a British historian working on the seventeenth to twentieth centuries. This is a world in which mass digitization projects have already delivered heaps of documents: ECCO, Making of the Modern World, Parliamentary Papers online, Google Books, the Times Newspapers, and Proletarian Lives, to name a few, cover a huge bulk of the circulating books, periodicals, newspapers, pamphlets, and official literature of the era. The OCR is good. It can be automatically downloaded, collected over time, then rendered in any of a number of analysis packages. There is no more favored position for using digital tools. It's a literature in which even proletariat voices have been, to some degree, consciously resuscitated by projects like the Old Bailey. But many fewer resources exist for listening to subalterns around the edge of empire. And few international literatures have been mass digitized to the extent of Britain's.

A profound opportunity is available to scholars of American and British history, the privileged few who by little fault of their own have been granted an extraordinary blessing, the room to freely experiment with new technologies with little or no training. If they are wise, they will see their place in the disciplines and seize the opportunity to become scholars of the *longue-durée* and so pioneer the types of questions that can only be asked at scale. It is imperative that they do so in dialogue with the secondary literature from scholars of China, Africa, and Native America, whose digitized projects will necessarily be more constrained in time and space.

There's also an opportunity here for scholars of the global periphery to experiment, in their framing of the question, in their introduction and in their intellectual context, with the largest possible questions that they can ask of the *longue durée*—positioning their study of police action in nineteenth-century Singapore, for example, within a *longue-durée* study of the idea of policing in the west, supported and expanded by N-grams or similar methods of historical context.

For our generation of scholars, new methods will be pioneered on the borderline between *longue-* and short durée. This is where we can expect the greatest advances in historical knowledge and the most sophisticated collaborations, and this is where introspection as to the limits and possibilities of each archive, each national tradition, and each methodology deserves the greatest attention.

These insights run counter to much work in the digital humanities. When we teach computers to read Shakespeare, there's a very low probability of the computer discovering anything that has not already been glimpsed or understood in some sense by the three hundred years of literary analysis already applied to that corpus. What we're looking at, in the digital Shakespeare, is a vanity project. It's tempting to the vanity of CS researchers to vie with Harold Bloom for superiority. It tempts the vanity of the humanists and archivists affiliated with the project to pretend to contribute something new to so rich a body of analysis. But ultimately the aims of such a project are decadent, in the sense that they enter into a realm where much has been done by traditional methods and the potential margin of profit is minimal.

It is far healthier, for our self-understanding as modern people as well as our understanding of the tools of analysis, that we should apply the new digital methods not to merely old canons of thought but all the more enthusiastically to large-scale corpora of a kind where we have few landmarks, where reading is daunting, where a historian working with traditional methods would be foolish to venture. We waste time and resources on projects of small importance and scale. Far more important to use these tools to analyze the unsynthesizable bulk of texts produced in the nineteenth and twentieth century, to make sense of the riot of paper that keeps us from understanding what officialdom does or how it works. Far more important that we as digital humanists should become the aid of ordinary people trying to make sense of the mess of legal and medical bureaucracy, to use our tools to forge a clear and simple guide through the masses of paper by which economists and other bureaucratic elites have shrouded themselves from public scrutiny.

What such an application of digital humanities would produce is far more than the toppling over of Harold Bloom. It would produce the use of digital history to take on present-day economics, public health, and social science as a whole, using the tools of distant reading to synthesize and to critique the unreadable morass of the paper machine.

This is a critical question in our time, attempted by the occasional project like IBM's Many-Bills, which tried to help citizens detect patterns in the mass of legislative paper produced by Washington with the end of creating a more engaged citizenry. But this is also a place where the digital humanities could earn its keep—proving itself as more than Harold Bloom's cheekier, flashier young rival, proving itself more than the face of young academia battling against the old for tenure. It would be a moral purpose in the sense of proposing an engaged academia trying to come to terms with the knowledge production that characterizes our own time.

20 AN EXPERIMENT IN COLLABORATIVE HUMANITIES: ENVISIONING GLOBALITIES 500–1500 CE

Geraldine Heng and Michael Widner

Our story begins with a teaching experiment in spring 2004. Seven faculty members came together at the University of Texas to introduce to graduate students a decentered world across a millennium, encompassing points of viewing in Europe and Dar al-Islam, sub-Saharan/Sudanic Africa and Maghrebi Africa, India and South Asia, the Eurasian continent, China and East Asia, in a transhumanities seminar called "Global Interconnections: Imagining the World 500–1500 CE."

This transhumanities work was a kind of laboratory: we met for six hours a week around a core syllabus of twenty-one pages, which we reshaped in response to seminar discussion, in an open-ended process of trial, correction, and experimentation. Despite the anxiety some of the instructors felt in attempting this, the experience for students and faculty alike turned out to be exhilarating and extraordinary, unlike anything we'd known before. Articles describing the experiment appeared in print, and as word spread, people outside Texas began to ask why this kind of collaboration did not occur more commonly, and on more campuses.[1]

In 2007 Geraldine Heng at the University of Texas and Susan Noakes at the University of Minnesota accordingly began three new initiatives: the Global Middle Ages Project (G-MAP), with a focus on research and pedagogy, the Mappamundi digital initiatives, to deliver a multimedia, interactive platform visualizing the planetary world, 500–1500 CE, and bring together online learning communities, and the Scholarly Community for the Globalization of the Middle Ages (SCGMA), a consortium that embraces anyone working in early globalities, and that is now international.

A "Global Middle Ages"— experimental work that asks scholars to step outside their discipline and specialization and engage with other humanities scholars, social scientists, computer technologists, musicologists, archeologists, designers, and others to make sense of an interconnected past—seems an idea whose time has now arrived. Undergraduate or graduate concentrations on a Global Middle Ages are gaining ground in the United States; in the United

Kingdom, Edinburgh offers an advanced art history degree on the Global Middle Ages. Oxford historians in 2011 convened a year-long series of workshops on the Global Middle Ages, held a conference to identify research objectives in 2012, and followed with more workshops in 2013 and 2014. Oxford's Centre for Global History now has a webpage announcing the creation of "a UK-based network of medievalists with interest in the global which has recently gained an AHRC network grant." In 2012 the University of Illinois convened a symposium on the "Medieval Globe," and announced their intent to inaugurate a journal of the same name, which they intend (to quote a conference organizer) as "the journal of record for this emerging field."

The study of a Global Middle Ages, it would seem, is now an emerging field. Many projects are planned, or underway—special issues of journals in various disciplines, a journal exclusively dedicated to the field, a year-long graduate/faculty seminar at the University of Minnesota in 2012 to 2013, a volume planned on global premodern literatures, and, of course, digital initiatives. All this work, of necessity, is improvised and experimental.

Even our acts of naming and description are provisional. *Teaching* experiments rapidly become *learning* experiments, in which faculty admit they have as much to consider and accomplish as the students in the classroom. There is continuing dissatisfaction with the term, "the Middle Ages," a Eurocentric construct fabricated by Renaissance historiographers to name an interval between two eras of glorified empire identified by their putative cultural supremacy and authenticity—Greco-Roman antiquity and its so-called Renaissance. Dissatisfaction *ab origo* with naming means that some of us—Susan Noakes comes to mind—*never* speak of "the Middle Ages" nor the "*European* Middle Ages" (a term that, even worse, extends the Eurocentric fictitious naming of time to the rest of the world) without placing quote marks around "the Middle Ages," to problematize the term.

How to name this emergent field of study was one of the first questions addressed in a 2007 organizational symposium at the University of Minnesota. Other names were critically considered. "Global *pre*modernity" retained attention too fixedly on modernity as the focal point and pivot of an implicitly linear temporality, and had the disadvantage of temporal vagueness. "Premodernity" could indicate the Bronze and Iron Ages, Biblical time, and Greco-Roman antiquity all the way through the Renaissance, which is still considered by some in the academy as a premodern era. Attempts at naming the past thus pointed to the inescapability of the conceptually and politically freighted nature of the language we have to use in order to participate in academic discourse.

In the end, in order to be able to speak at all among ourselves and with others, we agreed to continue the use of conventional terms, but also continue to critique and problematize them. That troublesome fable, "the Middle Ages," was thus embraced *under erasure* as a Eurocentric historical construct with little bearing for the not-Europe cultures and chronologies of the world, and perhaps with little bearing for Europe itself.

For Euromedievalists reflexively to export "the Middle Ages" to territories beyond Europe in naming cognate chronologies in other zones of the planet would inadvertently be a colonizing gesture of Euromedieval studies: the centrality of European time giving its name to asynchronous temporalities elsewhere, so that there is an Indian "Middle Ages," an African "Middle Ages," inter alia. Zones and cultures are asynchronous, following different timelines of change, development, recurrence, and transformation; differential temporalities characterize the many zones of the world.

However, some scholars of India, the Islamic world, Japan, and other areas, have *themselves* begun to conjure with terms like "the medieval" and "the Middle Ages" in identifying periods in the historiography of their region. Their references to "the Islamic Middle Ages," "medieval Japan," or "medieval India" can perhaps be seen both as the hegemonic ineluctability of European studies' influence in the academy and as efforts of goodwill in positing the utility of structuring overarching heuristic paradigms across geographic zones through attention to features and characteristics that suggest resemblance or analogy. Needless to say, situated terminology of this kind, issuing from *within* non-European studies and attached by their scholarly proponents to *their own* zones and periods of study, carries a different valence from its attachment by Euromedievalists willy-nilly to chronologies and zones around the world.

Leaving aside the naming of time, however, hardly anyone doubted that a consideration of *global temporalities* could revise that monolithic model of linear time that is the staple clone of academic and public discourse in the West. That model of temporality, which underpins many kinds of academic study and academic theory, sees modernity as a unique and singular arrival that ends the long eras of premodernity and instantiates the origin of new, never-before phenomena: the Scientific Revolution, the Industrial Revolution, the beginnings of colonization, empire, race, inter alia.

Global temporalities, by contrast, afford a recognition that *modernity itself* is a repeating transhistorical phenomenon, with a footprint in different vectors of the world moving at different rates of speed. One example is China: 700 years before Western Europe's "Industrial Revolution," the tonnage of coal burnt annually in eleventh-century Song China for iron and steel production was already "roughly equivalent to 70 percent of the total amount of coal used by all metal workers in Great Britain at the beginning of the eighteenth century" (Hartwell 1967: 122; see also Hartwell 1962). A "global Middle Ages" thus affords recognition of more than a single scientific or industrial "revolution" and more than a single geographic locale as the instantiating matrix. Alternate views of historical realities and temporalities can thus emerge.

From the vantage point of different locations on the planet, we're offered a view of history as overlapping repetitions-with-change, or history as oscillating between ruptures and re-inscriptions, as phenomena tagged "modern" or "premodern" recur, each time with difference, each time not identically as before, across the vectors of the world. China's modernities-within-premodernity also guide an understanding of the plurality of time—of temporalities that are enfolded and co-extant within a single historical moment. The example of China helps to make intelligible not only premodern worlds but also societies today around the globe, which can seem modern, postmodern, and premodern all at once.

Premodern China's past is a history that also attests to the difficulty of building on technological and scientific innovations in the context of repeated territorial invasion and political and social disruption. China's example thus restores an acknowledgment of the role of *historical contingency*—randomness and chance—as operative factors in shaping civilizational history, decentering tenacious narratives of a unique European genius, essence, climate, mathematical aptitude, scientific bent, or other environmental, societal, or cognitive matrices guiding destiny in "the rise of the West" in the so-called Scientific and Industrial Revolutions.

More adventitiously, our announced time parameters of 500–1500 CE—gathering stories from a thousand and one years, as one inspired graduate student in 2007 dubbed our efforts—points to its own self-factitiousness, and the unstable logic of arbitrary and neat temporal edges. Since a number of our initiatives necessarily begin earlier and end later than the millennium announced as a convenient rubric, the very implausibility of the announced time-parameters conduces to foregrounding our auto-critique of a Europe-based periodization hedged at one end by the collapse of the Roman empire and at the other by the incipience of the so-called European Renaissance in the West.

If global time usefully troubles and reconfigures the conventional temporalities of the West, the geographic span of the global imparts other advantages. Engaged with points of viewing gathered multilocationally across the planet, projects on early globalities privilege no academic discipline, no geographic region, or culture, but conjure an uncentered planet encompassing a multitude of formations seen as both interdependent and discrete, dynamically transforming themselves, and offering multiple kinds of worlding in deep time.[2] The study of *early* globalities moreover means that Europe is always already "provincialized" in Dipesh Chakrabarty's sense: a provincializing that is activated *ab origo* by virtue of our focus on multifarious locales on the planet—centuries before even the consolidation of "Europe" as a corporate entity.

It needs to be stressed that the study of globality by no means overlooks the national. For those of us trained in particular national literatures, languages, and histories, the investigation of globalities yields rewarding glimpses into how the national and the global interlock, in literature and in history. For instance: Hispanists can point to Spain's persecution and expulsion of Jews and Moriscos—a moment of self-purification constitutive of the early Spanish nation—as that moment in which Spain's global-colonial ambitions arose and began to spread their umbra across the world. As it forcibly emptied itself of people it saw as belonging elsewhere in the world, not in Spain, Spain under the Catholic monarchs also made its governance bloom elsewhere in the world. The spread of Spain's national boundaries outward, in the form of Hispanized colonies around the globe, thus affirmed Spanish national and global identities as mutually constitutive and interlocking.

Acknowledging the historical overlap of local and global undercuts the fantasy that an earlier Europe was the opposite of Europe today, a continent that contains global populations from everywhere. Early globalities bear witness to a medieval Europe that was already contaminated by the presence of people from everywhere—Jews, Arabs, Turks, Africans, Mongols, "Gypsies," steppe peoples, and others—and refuses the fiction that a singular, homogeneous, communally unified Caucasian ethnoracial population once existed in Europe. The notion that a "white" Europe existed as a historical inheritance—not as a concept manufactured by centuries of assiduous identity construction—is thus exposed as the fantasy of contemporary European politics and political factions. The study of the global past in deep time can thus productively speak to the racial politics of the contemporary European now.

If the study of the global past suggests to us that modernity is a repeating, transhistorical phenomenon whose signifiers recur, with difference, across deep time, that useful surmise does not in the same instant require a supposition that *all* periodizations then have no heuristic value. This is because the *phenomenality* of modernity is never co-identical in its *expression*, as varied modernities recur around the globe: expressions of modernity are never repetitions-without-difference. Historiographies of the global urge a rethinking of existing periodizations, but they do not work to render *periods*, as such, *in*-different, nor require that we overlook the specificities of difference in each *expression* of modernity, with its imprint upon its particular time. For us, the salient question to ask is not "is the Middle Ages really different from other periods?" but "what is at stake in saying that it is, and characterizing it thus-and-thus?" Or, conversely, not "is the Middle Ages the same as later periods?" but "what do we lose when we say it is?"

One example may suffice to point to why it's important to witness distinctions among time periods. In premodernity, *slavery* is an equal opportunity condition for all races, and dispositions of slavery vary significantly in time and place. The abject slavery experienced by Roma ("gypsies"), who became enslaved as they moved out of India and over continental Europe in the Middle Ages—so well-documented by scholars of the Romani—differs considerably from Egypt's Mamluk military slavery of the mid-thirteenth through fifteenth centuries, when the sultan of Egypt and Syria could only be drawn from the ranks of purchased slaves. Depending on time, place, ethnoracial groups, and political and economic conditions, slavery in the Middle Ages could mean abject, subaltern misery for an exploited population, or be a *sine qua non* condition for the highest offices in the land.

Premodern slavery is thus distinct from early modern and modern slavery, including the plantation slavery of North America, and the mutating forms of slavery (including child sex trafficking) that dog the twenty-first century. Caucasians, and Eastern and Western "Europeans," were sold at slave markets alongside other races all through the Middle Ages, marking off medieval slavery from later, Africa-sourced plantation slave labor. Household slaves were common and typical; plantation and field slaves were statistically less attested. In Dar al-Islam, extraordinary social mobility meant that being a slave could be an important first step to power, wealth, status, and authority—an avenue of upward mobility importantly open to women. This is not the case for slaves in the later American South.

The sheer variety of medieval slavery's conditions and opportunities thus attest to very specific differences *within* the medieval period, as well as *between* the medieval and other periods in the phenomena that characterized the institution we call slavery. Distinctions of this kind must be honored with acknowledgment that periods can be marked off differentially by institutions and phenomena that recur, but with different expressions, over *la longue durée*.

It is also important to disaggregate our study of early globalities from ongoing academic conceptualizations of globality, such as the world-systems analyses associated with Immanuel Wallerstein. Wallerstein's historiography of the global—devised, in part, to displace state-centrism and Eurocentrism, and to indict capitalism—schematizes a mechanics of economic power that grips the world in a network of periphery–center relations locking territories and polities in unequal dependencies of clientage and exchange from 1500, when new modes of production, it is argued, arrive. Wallerstein's view of the world as an economic machine specifies modern, not premodern globality, but has been retroactively extended by premodernists like Janet Abu-Lughod and others to earlier eras. Though widely influential, Wallerstein's world-systems model is not universally embraced.

For those averse to privileging the economic as an all-determinative, rational, and totalizing macro-structure, the explanatory functionalism of a view of the world as an economic machine captures less than satisfactorily the untidy fluidity and unpredictable micro-dynamism of lived realities (for economic historians or those who, like Abu-Lughod, track the flows and organization of trade routes from sources to markets, however, world-systems analysis appeals by virtue of its ability to organize commercial data). To be clear: Wallerstein's materialist critique is unquestionably valuable for its attestation of the qualitative difference that ensues when the economic attains supremacy above religion and ideology—when the economic *becomes*, in effect, the new religion and ideology in the era of "ceaseless accumulation" that characterizes modern to late capitalism, if not the trade capitalism of earlier periods (Wallerstein, "World System": 295).

However, materialist critique of this kind, of necessity, and perhaps inevitably, assigns little attention to *culture* or *the cultural*. While world-systems analysis clearly has its place and value in scholarship, its retrievals of globality also ineluctably carve up the early world through an optic of center–periphery relations that some find exclusionary, consigning certain zones of the world, like sub-Saharan Africa, to a status that over-recognizes the extraction of their natural resources and raw materials (identifying such zones as producers merely of "peripheral products") and under-recognizes their complex economic agency, thus reinforcing the impression of their under-performance relative to the all-important civilizational metropoles elsewhere (identified as the producers of "core-like products"), instead of recognizing the fine-grained complexity of their economic interactions.[3] The deck thus seems stacked against the zones some scholars study, not only by the great metropoles of the global past but also by the terms adduced and the definitions set in place by world-systems models themselves.

Processes of reading and interpreting globalities early and late are thus inhabited by a politics of reading that needs to be unpacked and foregrounded along with the political dynamics these processes of reading discover. An important case in point involves academic discussions of *globalization*—the specific name attached to the complex processes that characterize our contemporary moment, but a name that is nonetheless also sometimes claimed for the many forms of globality that have existed over the centuries, and even across millennia, in order for some to suggest that contemporary globalization is not new but exists on a continuum with the past, sans meaningful break.

In making such claims, those who say, for instance, that the *world systems* or large trading networks of the early past are tantamount to *globalization* (e.g., Ali Behdad), and thus indicate to us that globalization itself is nothing new—except by virtue of the speed, the scale, the intensity, the technological innovation, the planetary interconnectivity, the digital immediacy, and the sheer density that characterizes our current global moment—must willfully ignore the differences that such speed, scale, immediacy, interconnectivity, technology, and density instantiate.[4] This is not the kind of exercise in retrieving recurrence across temporality that premodernists should applaud.

When premodernists point to the recurrence-with-difference of phenomena identified as indexes of modernity, they carefully point to phenomena that in fact *share a basis* in technology, institutions, methods, or statistical reckoning across time. Thus early China's precocious modernity is witnessed by print culture in the form of blocks and movable type, paper money, gunpowder, scientific tracking of supernovas, linear algebra, massive use of coal in iron and steel industries, "modern" taxation, welfare and census systems, and demographic data, inter alia—all markers of a modernity typically located, in common understanding, only in the West, and only several centuries later.

To indicate how industrial modernity can recur across seven centuries, Robert Hartwell's use of statistics in meticulous, data-driven research compares the tonnage of coal burned in eleventh-century Song China's iron and steel industries with the tonnage of coal burned in early-eighteenth-century Britain's iron and steel industries: this is a comparison of iron and steel industries similarly driven by *coal*, the fossil fuel utilized in both eras. Hartwell did not claim the recurrence of an industrial modernity by comparing, say, a coal-based industry in China some 800 years ago, and a nuclear-fission-based industry today, to assert that we have always been modern.

Responsible premodernists recognize that technology makes a difference. A planet alive with social media, the Internet, cell phones, global positioning satellites, and supercomputers produces new outcomes materially distinguishable from a time when geographic distance forcibly delayed communication across long time periods of months and years. The compression of time and space experienced today, as David Harvey points out, is something new even for the undisputed centuries of modernity: the result of technological and material transformations of recent duration. Even as we rejoice in Yo Yo Ma's declaration that the Silk Road (really a braided network of several "silk roads") is the "Internet of antiquity," we thus retain the metaphoricity of his declaration without insisting on any literality in the analogy (The Silk Road Project). The "Internet" of the Silk Road/s, we understand, is not the same Internet that speeds social media, Wikileaks, Indiegogo, Reddit, Anonymous, Facebook, Twitter, or the Arab Spring.

Wallerstein's insistence on attention to modes of production and the temporally distinct phases of capitalism is helpful here. Karimi merchants who plied the Red Sea and Indian Ocean trade in the mercantile capitalism of the Middle Ages are not the same as a gargantuan corporate Apple or Microsoft outsourcing production and their supply chain to Foxconn at the other end of the world in the People's Republic of China, in today's post-Fordist global economy. Post-Fordist capitalism means that the production process itself is now fully global, supply chains are globalized, and the relations of production are globally dispersed.

Transmigrant labor's character moreover has changed with the advent of new technologies: transnational workers can now forge and maintain personal, sociocultural relations across time and distance, in their homelands and host countries in which they labor, connecting local and global in new manifestations of diasporic identity, attachments, and forms of locatedness. To insist therefore that *globalization* can be an always-already, catch-all category across deep geopolitical time is to empty the term of the significance it bears as a political, economic, social, and cultural analysis of the contemporary present: a moment that manifests not only the distribution of late capitalism and its cultural logic across the world but also the global relations that characterize the flexible, reversible colonizations Alan Liu accurately labels "post–neo-colonialism."

World history scholars have nonetheless sometimes suggested that the world has seen an "archaic globalization," a "proto-globalization," and a "modern globalization" in previous eras.[5] For a taxonomy like this, the term "globalization" functions largely to name an *interconnectivity* in earlier centuries and millennia that links places and peoples around the world in networks of communication and mobility. Such networks were emplaced through trade, war, pilgrimage, missionary activity, diplomacy, and so forth, and occurred at the speed of camel or boat, horse or wagon: at a pace that is quite different from the instantaneity of the Internet, fiber optic telecommunications, or the electronic speed of global financial circuits—the definitive meta-conditions undergirding the processes we call globalization today.

Used in a more general sense to describe the past, the term "globalization" is really a synonym for the interconnectivity of the world and its networks. What is usually intended, then, by "globalization" when historians apply the term to mean *a global interconnectedness*, might be better referred to as the *globalisms* of earlier eras, preceding the *globalization* of today, and preceding also Europe's *colonial globality* in the era of the European maritime empires of the sixteenth through nineteenth centuries, before the decolonization movements of the twentieth. To attest to the world's interconnectivity as forms of *globality*, or as the *globalisms* of different eras, perhaps better retains a sense of the variety and character of the global connectivities of those earlier eras, without yoking all to a single relationship with contemporary globalization and forcing a resemblance. In this fashion, varied globalisms are not recruited for a chronology monolithically aimed at the endpoint of the contemporary present.

The more careful of contemporary cultural theorists do not usually deploy the term *globalization* mainly to indicate the planet's interconnectivity and networks. *Globalization* points to the complex, often ironic, uneven and contradictory political-social-cultural outcomes produced by new technologies for which no premodern or early modern antecedents exist; to a post-Fordist outsourcing, subcontracting, corporatization of the world where production regimes in speed-sensitive economies of flexible accumulation have no precedents in premodernity or early modernity; and to a compression of space–time that ends distance and shrinks the planet, condensing life and interaction to the point where a single event—in the entertainment industry, in business negotiations, or in the new MOOC pedagogy—can be experienced simultaneously by people in Bangkok, Rio de Janeiro, and Vladivostok: this too has no premodern or early modern precedent. Under globalization, the very experience of time itself is transformed. Distance no longer exercises the constraints of old on human activity: when a family can video-Skype instantly around the planet, rather than wait for the next monsoon to bring news by ship across the ocean, the world has shrunk in a way that has no precedent in earlier globalisms.[6]

Just as we would want academic and public epistemology to afford recognition to the existence of more than a single scientific or industrial revolution in planetary history, we should accordingly afford recognition to our contemporary moment's own technological revolutions, with their particular imprint and signature upon our time.[7] We see thus that *periodization of a kind* can be useful for historiographies of the global, to enable the varied modes and patterns of globality issuing in different eras to be differentiated, and disaggregated, so that all need not bear the one name of *globalization* (whether of an antique, archaic, proto-, premodern, or modern kind), in discussing the world's interconnectivity. Critical scholarship in global studies leads the way in this, stitching together historiographies that are attentive to the specificities of the political, economic, and sociocultural outcomes of specific configurations of globalism, including today's.

The value of studying early globalities for those committed to a critical global studies lies partly in the capacity of such studies for crucial analytic revisionism: so that it is possible to attest, for instance, that globalism did not begin only when the West began to exercise maritime power, at the start of what became Europe's centuries of colonial globality. Honoring the differential character of varied globalisms with attentiveness and recognition, rather than lining up the globalisms of the past as precursors for the globalization of the present, ensures that variety and difference are not collapsed into an invariant chronology issued by a calendar embedded in Western perspectival interests.

At our contemporary moment, when even top US universities function as corporate businesses in globalization—exporting their academic brand around the world, and franchising educational enterprises in the Middle East and Asia, Cornell in Qatar and Yale in Singapore—such attentiveness to historiography is requisite for a critical problematics of the present. Neoconservative advocates of globalization, for example, who may seek to legitimate contemporary globalization by recourse to precursors in the global past are denied those precursors when it is seen that early globalities and globalisms cannot be lined up to explain and legitimate globalization today. With the corporate university implicated in globalization, the forms of autocritique that must, and do, issue from within the academy require the study of early globalities as part of a critical global studies. Preventing the co-optation of the past, even as we engage past and present in productive conversation, is a part of the institutional stakes involved today.

Finally, one question that loomed large in the 2007 Minnesota symposium revolved around methods of investigation, and the sheer variety of our objects of study, which could range from fabrics, poems, climate, topography, ornaments, disease, and people's lives, to walrus tusks, music, grave goods from the Eurasian steppe, or grains of rice in Africa. Because the *interactional demand* posed by the properties of each object varied considerably—you cannot read a grain of rice the way you would read a poem—the persistently *transactional* character of the process of study was highlighted in a way impossible to avoid. It became quickly obvious that we would need methods and strategies of all kinds, offered at all speeds and distances, plumbing many ranges of depth, and scaled differentially as needed. In transacting with a global Middle Ages no methods can be forgone. Computational algorithms can impart a view of global climate conditions and climate change across deep time, through the rapid parsing of data. Optical pattern recognition, and pattern matching, involving pixel counting and machine learning, can identify individual manuscript hands and illuminators, and can also pick out motifs and styles in fabrics and weaves scattered around the planet.[8] This kind of computational micro-analysis can transform our understanding of global commerce, artisanal dispersion, and the mobility of patterns and weaves as summaries of socioeconomic and cultural relationships.

Computer analysis, simulation, and modeling can help us study musical notations and instruments across a wide swathe of time and space. Quantitative analysis, a staple of sociological method, can disclose, via the mining of data, distributed industrial revolutions—as we saw with Robert Hartwell's statistics on the iron and steel industries of China—and economic, scientific, and demographic modernities that have erupted within global premodernity, to recalibrate our understanding of time itself.

In our digital projects, SCGMA's *Mappamundi* is intended as a digital and Web entity that, in its fully mature form, would enable an online user to walk around the world. Global premodernity is multifarious, surprising, and immense in its diversity: with cityscapes ranging from the bridges, canals, and markets of fourteenth-century Hangzhou in China to the gilded pediments of Constantinople, to the streets of fabled, storied Damascus. A user might want to view temple dance and intricate carvings in India, watch shadow puppets and listen to bronze drum tympani in the Malay peninsula, and have the ability to see how, in a twenty-first century classroom, that past is transacted, in original documents and in critical discussion, by students and teachers today. She might wish to listen to Chapurukha Kusimba, Chair of Anthropology at American University and a distinguished scholar of the Swahili coast, narrate the importance of women in premodern East Africa, or ponder what Roger Martinez, a historian from the University of Colorado, has to say about Jews, Christians, and Muslims living cheek-by-jowl in premodern Placensia in Spain.

We'd like our global citizen to be able to access critical pedagogy and scholarship on the Jewish diaspora, silk and spice routes, the migration of coinage or bacteria, browse Timbuktu manuscripts in African languages and Arabic, follow the Mongol postal system, or thrill to the epic of Antar, the black knight of African and Arab descent. She might scrutinize illuminated manuscripts and frescoes, follow the development of technologies like the compass, printing, water-driven automata, and clocks, and consider Chinese paper money, mathematics, and steel manufactures. She could read letters by lovers, popes, and far-flung missionaries, track caravan routes in Africa, reflect on the multiple histories of the Dome of the Rock or the Hagia Sophia, the economic effects of the Black Death, or the politics of space in Kublai Khan's summer palace. She could hear readings of Vinland sagas describing settlement encounters with Native Americans 500 years before Columbus, or thrill to the Decameron or the epic of the Cid. She might listen to the soundscape of a medieval global city, trace the ideological world-view of maps, study landscapes, topography, and built environments.

Massive as this digital undertaking might seem, numerous sites and digital projects already currently exist to offer data and resources, in scattered contexts. The Timbuktu Manuscripts Project (tombouctoumanuscripts.org) is a gateway to nearly a million African manuscripts from premodern Mali, and the result of African and international efforts to preserve, digitize, and grant access to documents from an otherwise vanished world. The International Dunhuang Project (idp.bl.uk), represents a transnational partnership of universities, libraries, and institutes that seeks to present digital images and resources, in several languages, on the paintings, carvings, sculpture, textiles, manuscripts, and artifacts of the magnificent cave complex of Dunhuang on the old silk roads. Rome Reborn (romereborn.frischerconsulting.com) renders the city of Rome itself digitally across a millennium and a half. The Harvard Yenching Institute, which owns the rights to Robert Hartwell's GIS database on premodern China up to the fourteenth century, has announced plans for wider user access online. Nezar AlSayyad's Virtual Cairo (jstor.org/stable/1576690) uses computer simulation and modeling to depict streets, buildings, and urban space in medieval Cairo from the early Fatimid through Mamluk eras. The late-Byzantine monastic church of Kariye Camii can be accessed through images, text, 3D animations, and VR panoramas (columbia.edu/cu/wallach/exhibitions/Byzantium/html/building_history.html). NubiaNet endeavors to engage all—even children—in the history of a significant swathe of premodern Africa across a generous timeline.

Mappamundi aims to curate, coordinate, and interlink available resources to see how they might narrate civilizational relations in a global web across time. Such work can be usefully supplemented by Mappamundi's own digital projects. At the time of our writing, there are eight Mappamundi projects in the pipeline: on Native American Cahokia and Mesoamerica; East Africa's relations with the world; Virtual Placensia; the Black Death as global pandemic; the global travels of Prester John; a database of Christian, Muslim, and Jewish poetry in medieval Seville; the "Discoveries" of the Americas; and a digital application, Augmented Palimpsest, based on Chaucer's *Canterbury Tales.* Waiting in the wings are projects on West Africa, the Mongols, the story of ivory, global Mamluks, and an extraordinary Kipchak grave on the Eurasian steppe with grave goods from everywhere in the Global Middle Ages.

The Mappamundi projects stand on the shoulders of digital humanities work that has been accomplished over long years by a variety of imaginative medievalists: medievalists who have digitized manuscripts and maps, digitally modeled the inside of a cathedral, closely analyzed art images, curated online collections, advanced codicological studies with densitometry, improved and recalibrated search engines, and centralized data across websites.[9]

One new tool currently being created will link research and pedagogy on a global Middle Ages seamlessly. The brainchild of Michael Widner, a medievalist who is a former HASTAC scholar, and Jason Yandell, an industry developer, *Bibliopedia* is a state-of-the-art research tool-in-the-making focused on joining primary source materials (images, documents, etc.) with aggregated scholarly citation data from articles and books published in the humanities, and found in databases such as JSTOR, Project MUSE, and the Arts and Humanities Citation Index, transforming these data into a format usable in the semantic Web, and allowing for community correction and elaboration via a Wiki format. *Bibliopedia* thus joins some of the most proven and most promising technological methods for developing a citation index, and provides space and infrastructure for a scholarly community to form to curate and iteratively improve the data. Because automatic data extraction is often inaccurate, the crowdsourced verification of data is essential to ensure data integrity as well as a way to generate community participation and involvement.

The value of aggregating citation data that currently lives in multiple closed silos like Google Scholar and JSTOR, and transforming the results into a linked open data format cannot be overstated. *Bibliopedia* will make available information that is currently closed off, not cross-referenced, and often hard to find. How many times must scholars at all levels—from undergraduates to senior faculty—search multiple databases and still miss works relevant to their inquiries? *Bibliopedia* will not only put all this information in a single, easily browsed location but also make it possible to find what library scientists call "forward citations"—later citations of earlier works. Right now, if we read an article written in 1976, finding every work that cites it in the intervening decades to the present poses a significant challenge, if not an outright nightmare. By automatically indexing these results from multiple locations *Bibliopedia* will make the task of discovery as simple as browsing a networked list of links. Most important, the creation of a desktop data-mining and visualization tool usable with any browser, compatible with Zotero, and capable of delivering rapid cross-referencing, linked data, annotation, full-text citations, and a fully automated bibliography-generation process, is extensible to non-European languages, so that *Bibliopedia* can eventually serve scholars working in a globality of languages.

Access to a global Middle Ages thus occurs through *both* disciplinary attention *and* transdisciplinary collaborative endeavor, in a kind of *humanities ecology* made up of many humanities disciplines in conversation with the social sciences and the sciences, especially computational science. In the end, the variety of methods, questions, tools, and projects are simply pathways to the education of our individual and collective desire as educators and researchers. Acknowledging the transactional demands issued by each particular cultural artifact, historical event, literary text, or material shard from around the world—responding to each *invitation* to transact—is also part of the education of our desire. And for feminists, queer studies scholars, race studies scholars, materialist critics, medievalists, and many others, the education of our desire is a continuing responsibility in all our transactions with global cultures past, present, and future.

In this spirit we'll end as we began: with a pedagogical experiment. "Global Interconnections," the 2004 University of Texas seminar, was mapped across three broadly overlapping trajectories. One trajectory involved the mobility and routes of culture: how people, material artifacts, and technology, commerce, armies, religion, money, and ideas circulated through the networks that interconnected the world, and the impact sustained at various locations of culture. Another trajectory centered on points of anchoring: cities and states, the imagined unities of countries, geographic regions, trading blocs, and empires. We compared social organization—the reproduction of administrative cultures and military castes, systems of gender relations, laws and institutions, technological developments, literary and cultural motifs, and ethnic diasporas—geospatially, macrotemporally, and at individual sites, to examine continuities and discontinuities. Our third trajectory turned on time—how the dynamics and relations of so-called modernity and so-called premodernity are identified across the world, to help us scrutinize the fictions through which premodernity is produced as modernity's antithesis.

This large and untidy blueprint continued as the context for subsequent learning. In 2012 to 2013, at the University of Minnesota, Heng, Noakes, Michael Lower, Maguerite Ragnow, Gabriela Currie, and other colleagues collaboratively convened a new, year-long, graduate seminar: "Early Globalities I: Eurasia and the Asia Pacific" took place in fall 2012, and "Early Globalities II: Africa, the Mediterranean, and the Atlantic" in spring 2013. In addition to the questions posed in 2004, "Early Globalities" experimented with the concept of *a human life cycle experienced globally*, as another axis of time positioning our investigations. Tracing the many kinds of worlding that existed in dynamic tension in the *longue durée,* we attempted some refinements in questions posed earlier to produce situated knowledges. We asked questions like the following:

- How does the world's climate—periods of cold or drought, earthquakes, floods, the monsoons, tsunamis—open the prospect of seeing the world's peoples as only one kind of agent who act within a larger network of forces, animate and inanimate, to produce environments?
- How does each particular human culture or society answer the most fundamental of questions: Who are we? Why are we here? What do we know?
- Climate and zonal conditions in the steppe lands of Eurasia produced human diets where cultures of horse and camel determined the basis of life, and where nomadism was an efficient environmental response: how do such responses compare with nomadism and transhumance that developed, say, on the North American continent in a similar swathe of time?
- What are the *consequences* of routes: of movements of slaves from eastern and northeastern Europe to Dar al-Islam; of fruit and vegetables from the Americas to continental Europe and Asia; of bacteria from China to the Mediterranean and Europe; of art motifs and ritual objects from Mesoamerica and central Mexico to the tribal societies of the North American southwest?
- What are the characteristics of global cities like Aksum, Baghdad, Cairo, Constantinople, Cordoba, Damascus, Delhi, Hangzhou, Jerusalem, Kilwa Kisiwani, Novgorod, Paris, Tenochtitlan, Timbuktu, Venice, and X'ian?

These and other embedded questions still need to be unpacked, turned over, and meditated on, reflectively and critically, through situated knowledges, and they no doubt have afterlives that we are yet unable to envisage. Not knowing the future, in recovering the past, and nurturing the dialectical conversations, the education of our desire in experimental transhumanities work continues.

Notes

A version of this chapter appeared in *A Handbook of Middle English Studies* edited by Marion Turner.

1. See, for example, Heng (2004, 2007, 2009) and Davidson (2011).

2. Dimock adapts the expressive term "deep time" from the physical sciences.

3. Wallerstein thoughtfully reminds us that "core" and "periphery" are intended to designate production processes, not particular zones, states, countries, regions, or polities: "one could use a shorthand language by talking of core and peripheral zones (or even core and peripheral states), as long as one remembered that it was the *production processes and not the states* that were core-like and peripheral" (*World-Systems Analysis* 17, emphasis added). In actual discussion, however, the distinction tends to be elided.

4. Behdad advances the notion that globalization largely denotes the world's interconnectivity and "global flow," and is thus synonymous with earlier world-systems, trade networks, and European colonial globality: "the condition we call globalization is not new if viewed historically in the context of colonial relations of power and other earlier world-systems preceding European hegemony since 1492. The scale of economic, cultural, and technological connectivity is certainly larger, and the speed of the current global flow of capital, commodities, peoples, and ideas is certainly faster as the result of technological advances, but the flow and interconnection themselves are rather old" (63). For Behdad, it would seem, twenty-first century technology and economics do not amount to a difference in kind of some import.

5. "Globalization" used as a general term is finely exemplified by Wilkinson's essay in Gills and Thompson's anthology—a volume in which the editors themselves waver over how to define "global history," "globalization," and even something called "world consciousness." In another anthology on globalization and world history, "archaic" and "modern" "globalizations" of the eighteenth and nineteenth centuries are discussed by Bayly; Ballantyne conjures with "proto-globalization" and "modern globalization," and Bennison considers "Western globalization" in the context of "Muslim universalism."

6. Brennan calls this "the meaninglessness of distance in a world of instantaneous communication and 'virtuality'" (44). Globalization, of course, does not reach or affect all and sundry on the planet equally; some parts of the world do not participate in globalization, or to the same degree as others: the notorious divide between the global north and the global south, for instance, testifies to differential impact and outcomes. We have argued for the co-existence of multiple temporalities in any historical moment, of which

early China's precocious modernity-within-premodernity is one example. In like fashion, premodern, modern, and postmodern temporalities, cultures, affects, institutions, and practices can coexist today, even within a single society and even under globalization.

7. For instance: the abacus and the supercomputer are both technologies for counting, and each is revolutionary in its time. But between the abacus and the supercomputer are such degrees of separation in speed, volume, scale, and complexity as to make for a qualitative difference between the two machines. Globalization today attests a difference of this kind in its separation from the globalisms of the past.

8. *Graphem* (Grapheme based Retrieval and Analysis for PalaeograpHic Expertise of medieval Manuscripts; http://www.digitalmedievalist.org/journal/7/cloppet/) is one such endevour, where medievalists, in collaboration with computer technologists, analyze digitized images of manuscripts pixel by pixel to group the features of each work, including color, direction, and curvature of a line, and the contour of writing, using advanced imaging techniques.

9. UCLA's Catalogue of Digitized Medieval Manuscripts (manuscripts.cmrs.ucla.edu) lists over 3,000 digitized manuscripts. Medieval Manuscripts on the Web, maintained by Siân Echard at the University of British Columbia, likewise links a wealth of collections around the world. Other projects—to name just a few—include the Digital Scriptorium (scriptorium.columbia.edu), the Early Manuscripts Electronic Library (emel-library.org), Manuscriptorium (manuscriptorium.com), and the Monastic Manuscript Project (earlymedievalmonasticism.org). Work on manuscripts includes the building of tools for crowdsourced transcription, online transcript and annotation, and the automatic recognition of folio layout, and even scribal styles. The Transcription for Paleographical and Editorial Notation (T-PEN) Project, for example, "is a web-based tool for working with images of manuscripts. Users attach transcription data (new or uploaded) to the actual lines of the original manuscript in a simple, flexible interface" (t-pen.org/TPEN/).

The Medieval Electronic Scholarly Alliance (mesa-medieval.org) is described by Matt Shipman as intended to be "a website bringing together scores of electronic resources on medieval subjects, including literature, history, theology, architecture, art history and philosophy." Shipman notes that the "creation of a centralized search engine for medieval materials would be a big step forward. At present…those interested in studying the medieval era may have to visit dozens of different sites to search for documents related to their research topics, from King Arthur to church history to the Hundred Years War. And that's assuming they know how to find those sites in the first place" (Shipman 2012). One similar centralized portal, "The Labyrinth: Resources for Medieval Studies," in place since 1994, has been of great value. Created and maintained by Martine Irvine and Deborah Everhart of Georgetown University, The Labyrinth links to a wealth of long-standing scholarly sites, is well organized and easily searchable and browsable, and has a straightforward interface (blogs.commons.georgetown.edu/labyrinth/).

At York University's Centre for the Study of Christianity and Culture Anthony Masinton is "creating a cathedral that evolves over time and is also an architectural history primer that will be executed and interacted entirely in a game engine in real time" (A. Masinton, personal communication to Widner, June 20, 2012).

21 DIGITAL HUMANITIES AND THE STUDY OF RELIGION

Tim Hutchings

Introduction: Religion and the Digital Humanities

The study of religion remains a minor participant in the digital humanities. Religious texts were analyzed in the earliest days of humanities computing (Busa 2004), but today religion is at best a marginal theme in digital humanities conferences and debates. Digitization projects continue to be undertaken, and the study of digital religion is a vibrant academic field, but methods of large-scale data analysis, visualization, and publication remain largely traditional. Religious practitioners have taken to digital media with greater enthusiasm than academics, undertaking their own digitization, data visualization, publishing, networking, and conceptual projects outside the academy with a considerable degree of success. This chapter will survey some of the work currently being done at the intersection of religion and digital technology, drawing attention to nonacademic work and calling for greater engagement with the methods and ideas of the digital humanities.

I will start by mapping current activity against models of the digital humanities, and then discuss three specific areas of particular interest from a digital humanities perspective: sacred text, digital religion, and digital theology. My own research has focused on Christian media, including studies of online churches (Hutchings 2011), proselytism (Hutchings 2012a), and digital Bibles (Hutchings 2014), and this chapter will emphasize those examples. I will not attempt in this chapter to provide a thorough overview of all Christian online activity (Hutchings 2015a), or a thorough literature review of the academic study of digital religion (Campbell 2012). For a comparative study of the digital engagement of different religious traditions, I direct readers to Heidi Campbell's *When Religion Meets New Media* (2010), the case studies in her edited volume *Digital Religion* (2012), and the 14-chapter section on "Virtual Worlds and the Visual Media" in Stan Brunn's forthcoming multi-volume edited work *The Changing World Religion Map* (Brunn 2015).

Patrik Svensson identifies five "modes of engagement" with digital technology (2010: para 101) within the "landscape" of digital humanities: "as a tool, as an object of study, as an expressive medium, as an exploratory laboratory and as an activist venue" (para 102). The boundaries between these "co-existing and co-dependent layers" are blurred, but each can be analyzed independently, and different academic disciplines engage more closely with some layers than others (para 103). In the study of religion, digital technologies have been engaged primarily through the first, second, and third of Svensson's modes. Researchers are exploring new digital methodologies, new topics of study and new ways to share their work.

Working in the "tool" mode, scholars of historical and contemporary religion have used computers to construct archives of their primary sources and to analyze those archives in new ways. Cecilia Lindhé's chapter 14 in this book is based on a Swedish research project that provides an excellent example of this approach.

Digital media have also become "study objects," key data sources for the study of contemporary spirituality. Religious practitioners were quick to embrace computer-mediated communication, and have established a presence in almost every area of online activity (Campbell 2012). Religious groups have sought to create new online communities, practices and rituals and to construct new digital technologies. The continual flow of conversation through Twitter, Facebook, and the blogosphere can help researchers track trends in contemporary spirituality and access real-time reactions to events, while the emergence of new faith-based design demonstrates the complex relationship between ideology and technology.

In some cases—still too few—new digital research tools and methods have been deployed by scholars to facilitate analysis of these different religious approaches to the digital. Sariya Cheruvallil-Contractor and Suha Shakkour organized a one-day conference on digital methodologies in the study of religion in Derby, UK, in 2012 (Cheruvalil-Contractor and Shakkour 2012), leading to a forthcoming collection of chapters (Cheruvalil-Contractor and Shakkour 2015). Contributions include considerations of methods for the study of mobile and social media, distance learning, AI approaches to the study of textual corpora, and debates over the ethical challenges raised by academic study.

The digital has also attracted attention as an "expressive medium" from religious studies researchers interested in using video or visualization techniques to share new kinds of data and analysis with audiences beyond the academy (e.g., see www.religiousstudiesproject.com). Religious communities have developed new kinds of digital pedagogy to inform, educate, and inspire their followers, investing heavily in production and distribution. Universities looking to explore the potential of online learning would do well to attend to the work of evangelical mega-churches, which are turning increasingly to multi-sited and online models for distribution of teachings (e.g., see LifeChurch.tv in Oklahoma).

Svensson's final two modes—laboratory and activist venue—have so far received less consideration from scholars of religion. Many academics are reluctant to engage alongside religious believers in activities intended to reimagine or promote religious interests. Theology should not be overlooked, however, both as an academic discipline and as a set of wider discourses. Religious believers have long considered the possibility that the adoption of new technologies might have consequences for their spirituality. There is a sociological dimension to this, focused on identifying and explaining trends in popular thinking and practice, including analysis of the spiritual and secular ideologies encoded within the technologies we use. Technologies are also evaluated to distinguish beneficial uses from those that must be forbidden, and—less frequently—the digital can become a provocation to new spiritual thinking.

The priorities of the digital humanities have also been modeled as a series of "waves," offering an alternative approach to Svensson's five "modes of engagement." According to Schnapp and Presner (2009), the digital humanities have entered a second wave. The first was devoted to efficient analysis of print archives. The second is "qualitative, interpretive, experiential, emotive, generative in character," using digital tools to pursue the core strengths of the humanities disciplines: "attention to complexity, medium specificity, historical context, analytical depth, critique and interpretation." Dave Berry (2012: 4) has invited digital humanists to push on again to a third wave, a "computational turn" that will examine how shifts in the mediation of culture change our understanding of knowledge and education. This should include a willingness to problematize the disciplinary boundaries and core methodological assumptions of the humanities, Berry suggests, a new reimagining of the task of the university and the student, and a new interest in the cultural significance of software and computer code. According to Leighton Evans and Sian Rees (2012: 29), this computational turn is a search for new methods, concerns, and concepts for the humanities, in response to the new landscape of information access created by digital media, and must include a new willingness to "illuminate and challenge the sciences."

These proposed waves can be loosely mapped onto the themes in religious studies already identified above. Academics and religious communities have applied digital tools to the analysis of primary resources, developing an impressive array of digital literature and modes of reading that is beginning to fulfill Schnapp and Presner's criteria for "second-wave" research. Scholars have begun attending to religious uses of digital media, even though their methodological and conceptual approaches have remained conservative. The computational turn seems closest to some of the work being undertaken by theologians and religious leaders, often outside the academy, to reimagine religious pedagogy, explore the role of digital technology in society, and critique popular understandings of the digital. Written publications in the area of digital theology remain plagued by poor understanding of technology and culture and a reluctance to rethink established narratives, but here, too, there are encouraging signs; recent books by Andy Byers (2011) and Jana Margueritte Bennett (2012) represent considerable advances in their willingness to engage seriously with academic studies of media as conversation partners for Christian theology. Valuable work has been undertaken in each of these areas, but there is much still to be done.

Religion and Text

We now turn to the first of three more focused case studies: the use of digital media to promote engagement with religious texts. This is a project with a long history; indeed the very first humanities computing project was undertaken by a scholar of religion. In 1949 Father Roberto Busa started work on a vast enterprise, using punch cards to record data on every individual word written by the prolific medieval scholar Saint Thomas Aquinas. The Index Thomisticus was finally completed on magnetic tape in 1980, transferred to CD-Rom in 1992 and the Web in 2005, and was still being expanded when Busa died in 2011. The dream of humanities computing, Busa (2004) has argued, is "precisely the automation of every possible analysis of human expression."

Many religious traditions have been engaged in the rigorous study of texts for centuries or even millennia, and have been understandably keen to exploit the power of computational analysis "to take on the tedium of word counts and comparisons" (Ess 2004). The first Bible concordances appeared in the thirteenth century, detailing the location of words in the Latin Vulgate, and the first English versions appeared three centuries later. This laborious work was an obvious candidate for computerization. Only a few years after Busa began his Index, the publishers of the New Revised Standard Version commissioned John Ellison to lead a team to create a new concordance by computer, a task that took six years (Harbin 1999).

Christians were not alone in their desire to analyze religious texts computationally. Aviezer Frankel and Jacob Shoika at Bar-Ilan University in Israel began digitizing Jewish legal decisions in 1963. One hundred thousand of these Sheiltot were collected and made available to the public, accessible through the first full-text search system to operate in Hebrew and Aramaic. Yoel Cohen (2011: 66) reports that this database has been "warmly embraced" by the Haredi ultra-orthodox community, and that the database "provided access to Sheiltot manuscripts—which had previously been limited to a few libraries or private manuscript collections—for Jews and non-Jews worldwide."

Contemporary archiving projects are continuing this tradition, collecting hard-to-find texts and other resources and making them available to believers, scholars, and the wider public through digital platforms that facilitate new ways to find and organize information. The *Ad Fontes* libraries offer hundreds of Catholic and Protestant sources from the Reformation period and allow the user to follow the development of ideas over time, across a geographical region, or within a specific religious group (Alexander Street Press 2012). The "Salem Witch Trials Documentary Archive" has published primary sources, court records, and maps, allowing visitors to search for specific names and to map the spread of accusations chronologically and spatially (Ray 2002). The digitization of archives can also be used to promote education and interfaith dialogue. The British project Jainpedia is currently digitizing thousands of pages of manuscripts, with an additional library of multimedia content and a program of offline events intended to encourage schools to learn about the Jain faith (Jainpedia 2011).

Digital tools can also be used to analyze religious texts and archives in new ways. The world's largest society for biblical studies, the *Society of Biblical Literature*, now includes a Program Unit on "Digital Humanities in Biblical, Early Jewish and Christian Studies." Digital Humanities sessions at the 2014 Annual Meeting include reports on the use of digital technology to decipher, transcribe and publish papyri, attempts to annotate intertextual allusions and trace textual developments, an application of Franco Moretti's work on distant reading to the study of the Bible, and a proposal for a new form of quantitative theology (abstracts for these sessions can be found in SBL 2014; search for "digital humanities"). Eric Atwell is currently leading an international research project applying Natural Language Processing to the study of the Qur'an, collaborating with Arabic linguists and reciters of the Qur'an to analyze the relationship between sound and meaning (Dukes, Atwell and Habash 2013) and "give non-Arabic-speaking Muslims more direct access to and understanding of the verbatim source text" (Cheruvalil-Contractor and Shakkour 2012: abstract 8). "Imitatio Mariae," a multidisciplinary digital humanities project based at Umeå University in Sweden, uses an innovative digital research platform to help articulate and visualize the relationship between words and images in medieval texts about the Virgin Mary and promises new insights into the construction of femininity (Carlquist 2011, Lindhé 2014).

The remainder of this section will focus on the digitization of the Bible beyond the academy, a particularly creative and experimental area of Christian media use (Hutchings 2015a). Non-academic Bible projects have frequently achieved much greater public impact than academic initiatives, and in some cases arguably represent more innovative uses of digital media. Attention to the ways in which Christian practitioners are publishing and engaging with their sacred texts is necessary for academic efforts to understand the contemporary landscape of electronic literature, because these Christian projects diverge in some interesting ways from non-religious e-reading.

The Digital Bible Society and the Bible League are both archiving projects, but their goals are quite different from the academic archives discussed above. DBS and BL publish collections of Christian texts intended for distribution in dangerous international missionary work. According to Bible League (2013), "we bring Akses™ Digital Bible Libraries into countries that we can't name here because of the dangers even that would pose to believers." DBS and BL now store their libraries on SD cards, which are much easier to smuggle and conceal than books or CD-Roms. Web developer and theologian John Dyer (2012) has blogged about the Bible software he created for DBS, explaining that his design prioritizes security—his reader does not connect to the Internet, for example—while still offering digital tools that facilitate detailed textual analysis.

In Western contexts, much more open and high-profile efforts at digital proselytism and education have been attempted. The United Bible Societies have repeatedly attempted to use digital technology to promote engagement with the Bible through music videos, drama, interactive comics and social media, but these efforts have achieved only limited commercial success. Robert Hodgson, then head of the American Bible Society's Research Center for Scripture and Media, explained to journalist Pamela Schaeffer (1999) that one early project had failed because it was too innovative for its audience: "young people love the MTV style, especially inner-city kids," but "parents and pastors weren't comfortable" with the choice of media and the presentation. The Australian Bible Society attempted a different approach in the early 2000s, publishing the whole Bible on CD-Rom in what purports to be the language used by senders of SMS messages: "In da Bginnin God cre8d da heavens & da earth" (Bell 2006: 147; BBC 2005).

United Bible Societies achieved much greater success with their Facebook page, The Bible, originally founded as a personal project by former New Zealand Bible Society CEO Mark Brown (facebook.com/DigitalBible). The Bible posts Bible verses, updates on the work of UBS and prayer requests, and has achieved quite remarkable levels of engagement (Ward 2012). The page now has 9.5 million fans, and each status update receives tens of thousands of likes, thousands of shares and hundreds of comments.

These remediations of the Bible transform the materiality, structure, and dissemination of the text. UBS's Facebook page does not promise a new translation of the Bible, but it completely changes how the text is accessed. A Facebook Bible—even one that simply republishes quotes—shifts experience away from the voluntaristic act of setting aside time for concentrated reading toward serendipitous encounters with unexpected words inserted into the flow of everyday communication. The success of the page may partly be attributed to good content selection, but the primary appeal of the Facebook Bible is the way it shares texts with the reader. An innovation in distribution has taken place here that resonates with the interests of millions of Christian Facebook users.

Any discussion of digital religious publishing must also consider the immense success of YouVersion, which has been downloaded 150 million times since its launch as a free mobile app in 2008 (Hutchings 2014, 2015b). The YouVersion reader can choose from 1000 different versions of the Bible in more than 700 languages (YouVersion 2014a), highlight, bookmark, and comment on passages, share verses with social media networks, or sign up for reading plans that automatically track progress and issue reminders. Badges can be earned for successful reading. In 2014, YouVersion launched new social software that allows users to share their bookmarks, comments and other updates with up to 150 of "your closest friends," "a community of people you know and trust" (YouVersion 2014b). By making reading activity visible to friends, this new update aims to encourage intensified commitment through mutual surveillance, encouragement and accountability. YouVersion has also expanded into the education and entertainment markets, launching "The Bible App for Kids" in 2013. The Bible App for Kids introduces children to key Bible stories through interactive animations, promising to "help your kids fall in love with God's Word" (YouVersion 2013).

The Internet "empowers people like never before," YouVersion claims: "With the ability to share, contribute, create, broadcast and communicate, it's easy to express who we are and what we believe with the rest of the world" (YouVersion n.d.). YouVersion is always accessible, perfect for a mobile lifestyle, packed with alternative versions, easy-to-use plans and calendars to help the user maintain their reading, and it is "engaging people into relationships with God as they discover the relevance the Bible has for their lives"—even people whose print Bibles lie unread. According to YouVersion's own statistics, 77 percent of an unspecified sample of users say they now "turn to the Bible more" because it's available on their mobile devices (YouVersion 2013).

Studying Digital Religion

So far, engagement with digital methods among scholars of digital religion has been limited. Researchers have relied heavily on traditional ethnographic methodologies, adapted to incorporate Skype calls, chatroom interviews, and participation in virtual communities (Hutchings 2011). Some examples of more innovative approaches have been mentioned already, and more will be included in Cheruvalil-Contractor and Shakkour's edited volume (2015)—a publication which may serve in turn as a catalyst for future experimentation.

Quantitative studies of digital religion remain rare, but some examples have been undertaken. Lorenzo Cantoni and his colleagues on the PICTURE project have collected responses from 5000 Catholic priests worldwide (Cantoni et al. 2012), while in the United Kingdom the Resonate project has assembled a panel of 15,000 Christians willing to participate in ongoing studies (http://www.christian-research.org/resonate/). Paul Teusner has examined 30 Christian blogs, using NVivo to help code content and NodeXL to generate a visual network map of the connections between them (Teusner 2010). Pauline Cheong, Alexander Halavais, and Kyounghee Kwon analyzed mentions of Christianity in 200 blogs, coding 800 posts by hand, extracting 18,000 hyperlinks to study the networks of attention formed between them, and using Technorati to compare their influence (Cheong, Halavais, and Kwon 2008).

The digital Bibles introduced in the previous section offer new opportunities for the study of religion, but also pose major methodological challenges for researchers. These projects are engaging millions of people around the world who are sharing many of their responses openly through publicly accessible comments. Areas currently understudied include the processes through which Bible apps become part of church congregations and households, the degree to which such software shapes the interpretive work of the reader, and the impact of products like the Bible App for Kids on the religious formation of children. In the case of YouVersion, at least, every interaction with the app is recorded, a vast dataset that YouVersion has begun to publicize through real-time statistics and a global activity map (now.youversion.com). If researchers can gain access to this flow of data and find adequate ways to analyze it, they could learn a great deal about how Christians in different parts of the world read and respond to the Bible (Hutchings 2015b). Digital humanists could be valuable conversation partners in this enterprise, bringing their expertise in methods of distant reading and data visualization to bear on these streams of communication.

My own research has considered the ways in which YouVersion encodes evangelical Christian theology and attempts to guide users through persuasive techniques (Hutchings 2014). Evangelical Christian groups promote particular patterns of exposure to the Bible, including regular reading both individually and socially, as the key to successful character formation and divine guidance. YouVersion and UBS try to promote Bible reading explicitly through their public campaigns and the interpretive content they provide, but they also promote particular reading choices through their software. Both platforms seek to make the Bible a constant presence in the life of the believer, for example, and encourage users to share brief quotations as a form of proselytism—ideas with a long history in evangelical Christian media. Researchers exploring this complex situation can interview users, request time diaries and conduct ethnographic studies of online reading networks, but we also need to draw on software and platform studies to examine the ideologies built into these Bibles.

Researchers interested in digital religion have begun to interrogate the relationship between ideology and design. Heidi Campbell (2010) has used interviews and participant observation in her study of the "kosher cellphones" invented by ultra-Orthodox Jews in Israel. These phones deliberately reject common features—Internet access, SMS messaging—to reinforce the shared religious values of this consumer group, specifically the boundaries that regulate communication with those outside of the ultra-Orthodox community. Paul Teusner and Ryan Torma (2011) collaborated on a study of religious mobile technology, focusing on the details of design to show that the mobile device itself can frame religious experience. Susan Wyche (2009) has used interviews and guided tours to study the place of technologies in religious homes, postulating a style of "extraordinary computing" that can be used to reflect and support family values in domestic settings. Wyche (2008) also engages in design prototyping as a research method to help explore the values and expectations of religious communities, an approach very rarely considered by scholars of religion.

Digital humanists can also offer valuable critical tools to help our analysis of digital religion. The five challenges addressed to data visualization by Bernhard Rieder and Theo Röhle (2012) are a good example. Rieder and Röhle invite us to question the objectivity claimed by computation, the rhetorical power of images, the "black-boxing" of analysis by software, the impact of new methods on traditional scholarly institutions, and the kinds of knowledge that are obscured. One of the best-known digital Bible software packages is Glo, which offers a wealth of visual and multimedia resources for biblical interpretation. Glo locates every incident in the biblical narrative at an exact point on an atlas or a timeline, complete with photographs, tours, and virtual reconstructions (Immersion Digital 2012)—replacing a complex and uncertain history laden with theological and political disputes with simple visual certainty, and inviting all five of Rieder and Röhle's questions.

Theologies of the Digital

For our third and final case study, we will consider the emergence of digital theology. The relationship between technology and religious belief can be explored in at least three ways. Academic researchers and religious practitioners have sought to trace the influence of spiritual ideas on the development of technology; to evaluate the impact of technologies on religious groups; and to use concepts and metaphors drawn from technology to explore and problematize common understandings of religion.

Following the first approach, a number of scholars have explored the religious roots of contemporary technology, arguing that the way society imagines certain media and machines reflects hidden spiritual beliefs. According to David Noble (1997) and Bronislaw Szerszynski (2005), Western fascination with technology is intimately related to religious visions, grounded in a search for divine knowledge, earthly perfection, and transcendence. Margaret Wertheim (1999: 17) identified spiritual ideas in the early visions of cyberspace, which she saw as a new secular attempt "to realize a technological substitute for the Christian space of Heaven," freed from the limitations of embodiment. "In one form or another," she argued, "a 'religious' attitude has been voiced by almost all the leading champions of cyberspace" (1999: 253).

This theme did not end with the new millennium. Karen Pärna updated Wertheim's argument in 2006, claiming that the Internet was regarded as "the long-awaited vessel of earthly salvation" even after its promises began to be discredited (Pärna 2006: 181). For Pärna, the Internet—like the telegraph before it—came to be invested with spiritual significance and perceived as sacred, a radically new creation promising transcendence, wealth, freedom, and harmony. According to Robert Geraci (2010), a Judeo-Christian tradition that looks forward to post-human transcendence of the world's failings is still influential among leading robotics, AI researchers, and online gamers. For Noreen Herzfeld (2009: 130), it is the distinction between Christian and Shinto attitudes to the nonhuman world that explains the different attitudes of American and Japanese society to robotics. These approaches are in keeping with the vision for the digital humanities offered by Leighton Evans and Sian Rees, among others, in which the humanities disciplines move beyond dependence on tools and learn to challenge the assumptions of the sciences (Evans and Rees 2012: 29).

Demands for pragmatic and theological evaluations of technology have generated a healthy conference and publication industry aimed at religious practitioners, exploring the second and third approaches outlined above. As should be expected, evangelical Christians—traditionally among the early adopters of new communications technologies—have taken the lead here. Technical websites, magazines, and conventions (like the Christian New Media Conference in the United Kingdom and the BibleTech and Technologies for Worship conferences in the United States) are flourishing, seeking to ensure that technology is used efficiently and effectively in worship and ministry. Christian authors have encouraged believers to embrace social media (Heim and Birdsong 2010) and warned that new technologies affect how users think and act (Hipps 2009).

Christian theologies of digital media have been less common within the academy, but this is an emerging field of work. Eric Stoddart has developed a Christian theology of care and visibility to challenge sociological interpretations of surveillance technology (Stoddart 2011), Andy Byers has used the concept of media to interpret the Bible's stories of divine communication (2011), and Jana Margueritte Bennett (2012) has drawn on classical Christian theologies to critically evaluate the Web's challenge to traditional Christian thought and practice.

The Catholic Church has also sought to offer advice, praising the Internet as a resource for education and proselytism while setting clear boundaries around its use in ritual (Pontifical Council for Social Communications 2002: 9). The annual World Communications Day first considered digital media in 1990, when John Paul II delivered a message titled "The Christian Message in a Computer Culture" (John Paul II 1990), and the significance of computers, the internet and social networking for Catholics have been discussed with increasing frequency in subsequent papal addresses. Catholic authors have also considered these issues, and Brandon Vogt's edited volume *The Church and New Media* (2012) and Antonio Spadaro's recently-translated *Cybertheology* (2014) both merit close attention.

According to a number of Christian commentators, digital media are bringing about a relational shift in contemporary society that will force the Church to adopt new theological ideas (Hutchings 2012b). Leonard Sweet's *Viral: How Social Networking is Poised to Ignite Revival* (2012) is a particularly high-profile example of this genre, arguing that the mind-set of the new "Google generation" is perfectly suited to Christian theology. Elizabeth Drescher's (2012a) theological work is much more subtle and sociologically-informed, but she also sees social media as a "reformation" for the Christian church. Drescher is also one of the few academic researchers to date to apply a religious studies perspective to the study of online bereavement, arguing that social media—used to integrate the dying into everyday life and to communicate with the dead—are bringing about a shift in understandings of the afterlife (Drescher 2012b).

In Rachel Wagner's (2012b) multidisciplinary writing, digital media take a different role. Wagner compares religious rituals, games, and stories, and argues that these three categories cannot ultimately be distinguished (2012b). In her theological work, Wagner builds on this analogy to critique conservative Christianity as a kind of first-person shooter game, a violent conflict between winners and losers (Wagner 2012a). The ritual-game-story analogy is only intended as a series of suggestive parallels, and is not entirely convincing, but Wagner's approach is interesting from a digital humanities perspective because she uses media studies to challenge the basic assumptions and methodological approaches of the study of religion.

Conclusion

This chapter began by asserting that religion is still a marginal presence in the digital humanities. Digital tools are being used to analyze primary sources, and digital media are being used to share academic ideas with global audiences, but these initiatives are still relatively rare. Religious communities outside the academy are engaging with digital media, but the research methods used by researchers to investigate these emerging phenomena are not yet equal to the task.

This situation is beginning to change. Evangelical Christians and other religious groups are adopting digital media in innovative ways that echo the interests of digital humanists, producing digital texts, designing new technologies, using digital tools to record, analyze, and visualize data and developing new models of pedagogy. Collaboration with these developer communities may be the best way for researchers to access some of their data and learn more about their approach to design, but we also need new methods and conceptual tools to critically investigate the ways in which religious media shape knowledge, networks, and subjectivities. Theologians and historians are beginning to work with concepts drawn from the study of religion, media, and society to question the core assumptions and boundaries of each of these disciplines, a project that has come into its own in recent years as digital theology has grown in sophistication. In each of these areas, collaboration with the digital humanities can help identify the concepts, methods and models needed to push religious studies research in intriguing new directions.

22 CYBER ARCHAEOLOGY: A POST-VIRTUAL PERSPECTIVE

Maurizio Forte

Although archeology has a long history of engagement with technology, it is typically not considered a core member of the digital humanities. The reason may be exactly this strong engagement in technology and discipline-internal self-sufficiency as well as the view of archeology as placed between science and the humanities. The story of archeology's shift from presentational to interpretative use of digital technologies is highly relevant to the digital humanities and so to making a case for archeology as a humanistic discipline.

I frame this discussion in terms of contrasting cyber archaeology (CA) with its predecessor virtual archaeology (VA). I use the archeological site Çatalhöyük in Turkey as a post-virtual CA case study. Cyber archaeology represents the postmodern evolution of virtual archaeology, its cybernetic code in an ecological sense (Bateson 1972). By "cybernetic code," I mean the capacity to interact digitally with an entire environment and its affordances, not just with single models.

If VA has been "model oriented," aimed at the 3D reconstruction of models of the past, then CA is instead aimed at the development of interactive and immersive cyber worlds in the domain of hyperreality. I use "hyperreality" in its cybernetic sense, that is, a domain where information is increased, interactive, and simulated. I argue that even if digital technologies are tools mediating empirical analyses, they are able to create an enormous amount of information, almost in real time, and can, in principle, be experienced in simulation processes. The fast evolution of digital technologies and techniques of data recording have had a great impact on archeological research. The first effect of this revolution was an increased number of strongly technologically oriented projects and applications; the theoretically driven attempt to raise new research questions followed later. I am arguing that we now can see archeological work located in between empiricism and simulation characterized by multivocality, collaborative interpretation, networking, and open digital narratives. This kind of (cyber) archeology is based on the capacity to interpret and learn through interaction in a specific sociocultural context: an interaction that involves relations in the past and from the past. Among other things, these led to discussions about the validation processes of the reconstruction, the transparency and authenticity of data, accuracy, and the demonstration of scientific methods (Frischer et al. 2002; Beacham and Niccolucci 2006; Ryan 2001).

Background

Virtual archaeology was born without an adequate theoretical background, as early archeological modeling merely demonstrated the great potential of computer renderings and computer graphics. Still, given the data-rich opportunities archeological simulations provide, archeology became one of the leading fields in developing applications; in fact several corporations invested in archeological modeling in the 1990s. These corporations included Taisei Corporation in Japan, EDF in France, and ENEL in Italy (Forte and Siliotti 1997). In the early days of VA, each 3D, high-resolution model demonstrated extreme computing power. These first photorealistic, digital replicas of archeological models were truly technologically impressive; however, they only focused on architectural elements and not on life as lived in the past (Forte 2010). For example, almost no human activities were represented in the reconstructions, and very few avatars populated the digital landscapes making the empty models appear to depict an artificial past. In the VA dominated 1990s, applications were mostly very technologically oriented without adequate interdisciplinary engagement in terms of archeological research. In short, the virtual reconstruction process was initially separated from archeological interpretation (Forte 2000).

The diffusion of virtual archaeological models increased meaningfully worldwide because of the invention and popularization of personal computers. The unconstrained publication and dissemination of archeological models without critical background or scientific verification of the reconstruction process created a variety of problems within the archeological community (Ryan 1996, 2001). Thus the promising potential of virtual reality and computer models was seen with some skepticism (Ryan 2001) in the absence of a rigorous philological approach to archeological validation. One response to these issues was the attempt to find a way to "authenticate" the virtual reconstruction by trying to introduce guidelines and methods common in art history and art collections, which require the validation and authentication of a piece artwork. However, the assumptions of authenticity that undergird the status of the art object are not easily analogizable to digital models.

In response to this skepticism and the authentication debates that followed, several international initiatives were created which defined the validation of virtual models at different levels: the Ename Charter for the Interpretation and Presentation of Heritage Sites (ICOMOS

2007; Callebaut et al. 2002), CURO (Cultural Virtual Reality Organization, Frischer et al. 2001), the London Chapter (Beacham et al. 2006; Bentkowska-Kafel et al. 2012), and, lastly, the Principles of Seville (http://cipa.icomos.org/fileadmin/template/doc/PRAGUE/096.pdf). All of these initiatives began in the 2000s, roughly ten years after the publication of the first book on Virtual Archaeology (Forte and Siliotti 1997), which depicted the state of VA as a field in the early 1990s. The temporal nearness of the attempts to theorize and legislate the field shows scholars' awareness of VA as a site of academic research, but one whose virtual reconstructions needed scientific validation.

Each of the legislating initiatives was aimed at finding a way to validate and standardize virtual reconstruction, a way to verify methodologies and technologies and to create specific "scientific" standards in VA for research and education. Each responds in a slightly different way to an urgent need to validate the entire process of virtual reconstruction and to make it more transparent, standardized, and "scientific." On October 4, 2008, the ICOMOS General Assembly formally ratified the ICOMOS Charter for the Interpretation and Presentation of Cultural Heritage Sites, known as the Ename Charter. Article 2.4 states:

> Visual reconstructions, whether by artists, architects, or computer modelers, should be based upon detailed and systematic analysis of environmental, archaeological, architectural, and historical data, including analysis of written, oral, and iconographic sources, and photography. The information sources on which such visual renderings are based should be clearly documented and alternative reconstructions based on the same evidence, when available, should be provided for comparison. ("Ename Charter 2008")

Similarly, in the "Introduction to the London Charter," Beacham, Denard, and Niccolucci call for clear standards of validation arguing that "while 3-dimensional visualisation methods are now employed in a wide range of humanities contexts to assist in the research, communication and preservation of cultural heritage, it is increasingly recognized that, to ensure that such work is intellectually and technically rigorous, and for its potential to be realised, there is a need both to establish standards responsive to the particular properties of 3D visualisation, and to identify those that it should share with other methods" (Beacham et al. 2006: 1). Frischer, Nicolucci, Ryan, and Barcelo tried to launch another initiative, CURO (Cultural Virtual Reality Organization). The authors correctly point out that the charter of CURO seeks to establish "an explanation [that] can be presented as a visual model, that is, as a virtual dynamic environment, where the user asks questions in the same way a scientist uses a theory to understand the empirical world. A virtual world should be, then, a *model*, a set of concepts, laws, tested hypotheses and hypotheses waiting for testing" (Frischer et al. 2002: 9). Although CURO as an organization had a very weak impact in the international community of digital archeologists, it started to advance some critical methodological questions about the validation of virtual models. According to CURO, "there is a clear need for a theoretical debate with practical implications to enable heritage managers to use the best that new technology can offer them in this area while minimizing its most controversial applications. In short, some basic principles must be established to govern practices in this growing field" (Frischer et al. 2002). However, good selection and standardization of these basic principles are very difficult to achieve (Lopez-Menchero and Grande 2010).

Each charter or initiative calls for standardized methods of validation for archeological modeling, but is this really possible? Is it really possible to find a methodology capable of analyzing a virtual reconstruction step by step to prove its reliability? Can the communities of archeologists, modelers, designers, and computer scientists rethink this kind of digital workflow?

It is quite difficult to find a balance between standardized and controlled workflow and the multivocality and openness of archeological reconstruction. In fact it is impossible to define a digital reconstruction as "authentic," but it is surely possible to make the full process of model creation transparent. Without this transparency, it was impossible to offer alternative simulations of the virtual models in VA; once the graphic rendering was complete, the digital pictures or animations were not modifiable.

In the last decade there has been a great deal of attention from a variety of scholars on the evaluation and transmission of virtual models. In fact, there has been much more emphasis on the digital workflow of virtual archeology than in any other archeological domain or sub-field. Why has there been so much emphasis on these methodological and theoretical discussions in VA?

The Hyperreal, Simulacra, and Digital Worlds

One explanation for these discussions would seem to be that the great potential and seemingly endless capacities of virtual reality and computer graphics are intimidating for many archeologists, heritage managers, and museum experts. Because the hyperreal can reduce the relevance of empirical experiences in the physical world and because it is easy to change established interpretations of empirical data simply by suggesting a virtual reconstruction, Virtual Archaeology is concerning to large portions of the archeological community. What counts as accepted modes of knowledge production is another important factor here. And the more impressive and photorealistic the model is, the more difficult it is to validate it or to offer alternative interpretations of the initial data. Thus the virtual model constitutes a mediated experience between datum and interpretation; it can be seen as a simulacrum of the empirical past. VA is challenging because its core goal is to digitally render the past. The virtual simulation of reality is able to hybridize reality itself; in archeology, one of the biggest risks is that a virtual reconstruction can produce a too peremptory view of the past closed off from scientific, interpretative research.

In a networked, postmodern society, the hyperreal also "avatarizes" information, disseminating digital replicas, models, and virtual experiences. In its immateriality, the hyperreal is like a virus in the sense that it multiplies its code in order to transmit content in endless forms. I use the verb "avatarize" for indicating the transformation of empirical data in narrative experiences through virtual interactions in the cyber world. Hyperreality necessarily suggests simulation, which is importantly not a copy of reality but a potential reality, in cybernetic terms.

While the hyperreal-inflected concerns about archeological modeling (including validation, transparency, and authenticity) and the discussion and attempts at legislation that follow are understandable, the above-mentioned charters, guidelines and principles are difficult to apply. Their shared assumption, that an essential premise for validating authenticity should be a comparison with the original, is particularly problematic. For example, how can we evaluate authenticity in a reconstruction if we don't know the original exactly? I do think that any serious attempts to validate virtual archeological models fail in terms of authenticity. What is possible, however, is the continuous attempt to validate the methodological approach, making all the data and information used transparent. We may not be able to prove authenticity, but we can study the validation of the methodological work. If we translate Baudrillard's idea of simulacrum in the archeological context, we could say that digital data are "signs," markers of the past, new and cybernetic codes re-processable during a simulation. In practice, digitally constructed worlds are endless and once the digital simulation starts, the cybernetic outcome (which is unpredictable) is in the digital performance, human–environment, interaction–enaction (Bateson 1972; Maturana and Varela 1980), and not so much in the models. I argue that if in the 1990s this emphasis on the analysis of virtual models was justifiable today, the most important revolutionary aspect of cyber archaeology is in the simulation/interaction/performance. We do not reconstruct the past anymore; we perform the digital past (Forte 2010).

From VA to CA

Virtual archaeology was born in the 1990s as a methodology for "the reconstructive process for communication and interpretation of the past." This archeology was mainly "reconstructive" because of a deep involvement with computer graphics and high-resolution renderings in the generation of virtual worlds (Forte and Siliotti 1997). By contrast, cyber archaeology typifies mostly the last decade and is influenced by the advanced use of interaction design and the global dissemination of online computer games, serious games, immersive environments, and

Figure 22.1
TeleArch—Teleimmersive System for Archaeology (UC Merced, UC Berkeley): a kinaesthetic and collaborative interaction with a virtual model

virtual communities (Forte 2010). Importantly, CA is much more focused on the human factor, on human interaction, including a deep engagement with digital interactors in the role of avatars. Thus, if VA was more centered on models, CA is centered on models' interactions and environment. In other words, avatars in CA interact primarily with the virtual environment rather than single models.

Moreover in CA the interaction between user and cyberspace can be ruled by a kinesthetic approach where the users exchange information with the environment by embodiment, physical engagement, and collaborative efforts (Forte and Pietroni 2009; Forte and Kurillo 2010; Forte and Dell'Unto 2010; figure 22.1). In fact the frame separating cyberspace and reality is disappearing and involves a total hybridization of devices and visual content (in augmented reality, holograms, etc.).

In table 22.1 I have attempted to delineate a general scheme differentiating VA and CA according to their specific characteristics and tasks. The most important differences concern the role of humans and virtual environments: in VA the interaction is very limited; there is no collaboration, and the focus is mainly on models and visualization. In CA, interaction, engagement, and feedback are the core; the virtual environment is a simulation space. In terms of usability of digital data, it is important to point out that in VA a large amount of data is born analog and secondarily digitalized. In CA, because of the wide use of laser scanners, digital photogrammetry, computer vision, and remote sensing technologies, the data

Table 22.1

Virtual archaeology	Cyber archaeology
Visualization process	Simulation process
Basic interaction	Feedback, behaviors, embodiment
Passive users	Content providers
Models engagement	Users' engagement
Individual environments	Collaborative environments
Desktop	Immersive
Analog to digital	Digital to digital
Models	Enaction/interaction
Computer renderings	Cyberspace
Individual users	Virtual communities
Animations	Real time
Flythrough	Serious games

are born digital and usually captured by indirect survey or remote instruments (Forte et al. 2012). The passage from digital to digital cancels any manual phase of data acquisition, such as hand-drawing, draft maps, and paper notebooks; everything is digitally captured (almost in real time) and then exported or implemented for virtual environments or 3D models. Until now, very little attention has been paid to this epochal aspect: digital control of archeological information is remote. Digital data are at a distance; the human interpretation is in between the real datum (on site, by artifacts, etc.) and its virtual ontology. The passage from digital-to-digital pushes the interpretation to another territory: the simulation, which is the endless perspective to generate multiple models at different resolutions and with different content.

Problems of Mediation

This distance between digital data and real datum recalls another well-known and quite old discussion in archeology about the mediated experience of digital tools, recently delineated by Sue Hamilton: "The product of embodied enquiry is expressed by Tilley as being reliant on detailed penning of experiences on paper/notepad and is contrasted with the evils of using highly technical landscape recording and investigating equipment (unspecified but by implication total stations, geophysical prospecting equipment, compasses and the like), which mediates and dulls experience, or the abstracted outside perspective of statistics and computation" (Hamilton 2011: 271; Tilley 2008). Against Tilley, I argue that every experience is per se a mediated and subjective experience. Whether it is digital or not, as soon as we experience a piece of

Figure 22.2
Laser scanning model of the Building 77 of the Neolithic site of Catalhoyuk (Turkey): about 500 millions of points

information, it is mediated by our subjective and individual analysis. It seems a false distinction to suggest that archeologists with just pen, pencils, and notebooks are able to interpret a non mediated reality, while scientific and technological tools are somehow "mediated systems." Who validates the "penning" of the archeologist and his or her notes? Is that an objective interpretation? Should we only pay attention to "empirical data"? However, even empirical data are mediated by the human mind and as such cannot be considered "objective."

The fundamental idea behind the discussion above is the belief that only a nonmediated reality can offer up archeological evidence, not data provided by mediated instruments (figure 22.2) and digital tools such as laser scanners, remote sensing, and computer vision. I think that the real issue here is the relation between data in situ and remote indirect data capturing (laser scanning, geophysics, etc.). Mediated/remote tools of investigation operate almost in real time, meaning that, for example, when a laser scanner records millions of point clouds (figure 22.2) or a stratigraphy is generated by computer vision in a tablet PC, the operator, the archeologist, is still there, comparing digital data on a screen and "real" data on site. In short, it is now unnecessary to wait until post processing in the laboratory (although this is, of course, still important) for interpreting and evaluating data and models. The digital performance acts on site, on landscape, in different operative contexts. All the definitions on the right side of

Figure 22.3
Virtual reconstruction of the imperial villa of Livia (Rome, first century AD): bottom-up (archaeological evidence) and top-down (reconstruction) models

table 22.1 characterize CA as a nonpredetermined simulation: the action/interaction designs the informational process.

Is Archaeology Different Now?

If our premise is correct, there is a sort of hybridization in the process of reconstructing the past: cyber archaeology embodies multiple interpretations through virtual simulations embedding real and digital data. Therefore interpretation comes from the dialectic between real data (sites, objects, landscape, museums, etc.) and virtual data. In this second category we should include data captured by digital tools (e.g., laser scanners, georadars, and remote sensing), which generate 3D models of archeological evidence, and archeological 3D models reconstructed for virtual environments. The first ones virtualize an existing ontology, the site or the object; the second ones virtualize a hypothetical reconstruction (figure 22.3). These two categories represent different phases of the interpretation process, the bottom-up (data recording) and the top-down (reconstruction, simulation, figure 22.3).

Thus the apparent bias in archeology—that the use of advanced digital technologies in the field represents a mediated experience insufficient for a correct interpretation and not understandable without post-processing (or not understandable at all)—is false. In actuality the first digital data produced on site by "mediated" tools are already interpretable, sharable, and comparable with their authentic source: the excavation context. In fact it is because of an interaction between bottom-up (data capturing) and top-down (hypothesis/reconstruction) that it is possible to achieve a multivocal interpretation (Hodder 2008; figure 22.3).

The main difference between this archeology and past archeologies is that all of this process is completely digital and interactive. Today instruments of archeological data recording produce an enormous amount of information in a very short time (figure 22.2). For example, in my work at the archeological site of Çatalhöyük during the 2012 fieldwork season, one georadar was able to collect 54 GB of data in three working days and a laser scanner (Faro Focus 3D, Trimble FX) was able to generate up to 976,000 points per second and hundreds of millions of points for an excavation trench (e.g., a Neolithic house, figure 22.4). At these rates a site could be mapped by remote sensing imagery, geophysical prospections, laser scanning, digital photogrammetry, computer vision, total stations, GISs 3D modeling, and other technologies in just one day.

All of these digital data are not redundant even if they overlap in the same geospace (figure 22.4) because they represent different ontologies and processing of the same case study. Every single digital layer is a piece of the information puzzle (figure 22.4), a view, a different perspective, or analysis. In this way the digital interaction with the past becomes more challenging and stimulates new research questions and hypotheses. Ultimately this archeology creates data almost in real time, posting digital information from the field to cyberspace. In this case, hyperreal and real coexist: for example, we can see a 3D model of a stratigraphy on site during the excavation or laser scanning data simultaneously (figure 22.4).

Thus this archeology, in between empiricism and simulation, is different: it pursues multivocality, collaborative interpretation, networking, and open digital narratives. This kind of (cyber) archeology is based on the capacity to interpret and learn through interaction in a specific sociocultural context: an interaction that involves relations in the past and from the past. The study of interaction in the past regards the relations among structures/activities, objects, artifacts, and humans: what Donald called "*symbolic storage*" (Donald 1991). According to Renfrew and Scarre (Renfrew and Scarre 1999: 1–7), "the term 'symbolic storage'...is not a static concept, but one which refers to the interaction between humans and artefacts in a general sense." These relations are studied in the process of reconstructing the past. For example, a stratigraphy has relations and affordances with its original context (a postdepositional phase or the domestic life) but also with the area of excavation where

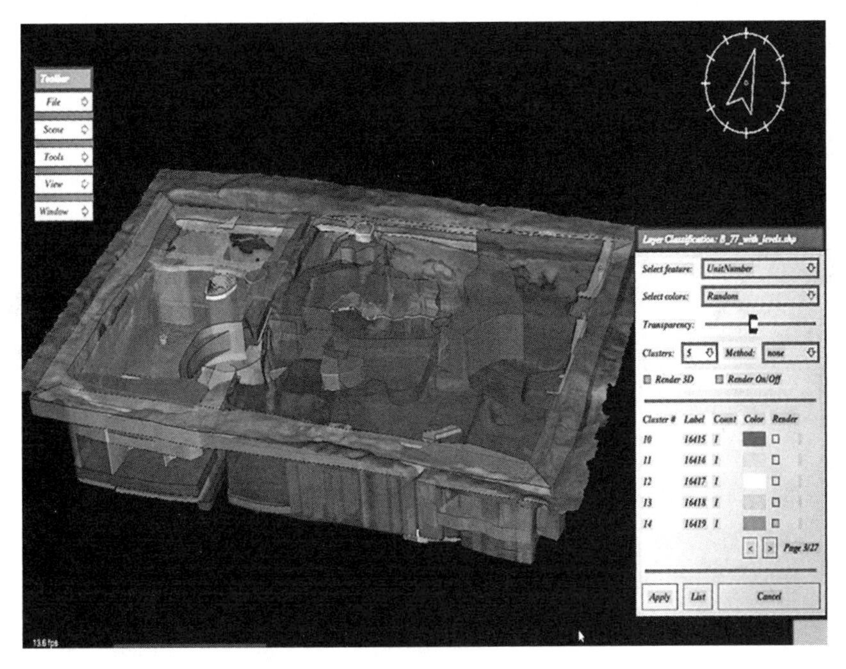

Figure 22.4
Neolithic house at Catalhoyuk (Building 77): virtual reconstruction by laser scanning with the superimposition of 3D GIS layers and artifacts

archeologists operate. The first set of affordances/relations (Forte and Pescarin 2012) can be understood through a reconstruction process (*what was the original context?*), while the second ones, concerning the archeological evidence, are interpretable by the process of data capturing as mediated experience (e.g., laser scanners, remote sensing, and digital photogrammetry). In short, these last metadata are directly post-processed by the human mind and not by technological systems.

3D Archaeology—3D Digging: A Post-virtual Experiment

Since 2010 I have been directing a digital experiment of 3D virtual representation of an archeological excavation by units, objects, and layers at the archeological site Çatalhöyük in Turkey (Forte et al. 2012; figure 22.4). The methodology involves 3D data recording by laser scanning, photogrammetry, and computer vision of any stratigraphic unit once it is identified. The unit could be a deposit, a structure, an infilling, or a negative unit. All these data are captured during the excavation and processed the same day. At the end of the day the data are visualized

Figure 22.5
D sketching of the building 89 at Catalhoyuk during the excavation (season 2012)

with 3D projectors in the lab and shared among the team members for an initial discussion on interpretation. In addition the models are also exported in 3D (pdf) format for a daily 3D sketching (figure 22.5). 3D sketching is an interactive way to annotate thoughts and draft interpretations and comments before the final definitive identification and classification once the layers are mapped and excavated. This digital approach in the use of different 3D technologies transforms the excavation and, in general, digital data recording/processing, into an immersive experience where data entry, interaction/embodiment with the data, and (first) interpretations coexist in the same real space but also in cyberspace (Forte and Dell'Unto 2010). The last processing of the work involves the implementation of all the data for a tele-immersive system (Forte and Kurillo 2010; figure 22.6): a collaborative virtual environment able to connect in real time and remotely to different university labs (in our case UC Merced, UC Berkeley, and UC Davis)

The integrated use of various methodologies of systematic 3D data recording on site allows us to simultaneously gather and represent the information. When the data appear in a tablet PC onsite, in a laptop computer, or in a 3D projector, they are available also for real time interaction and comparison with all the other data in situ. This co-sharing of 3D spatial data in the same context (the excavation) but in different domains (real and virtual)

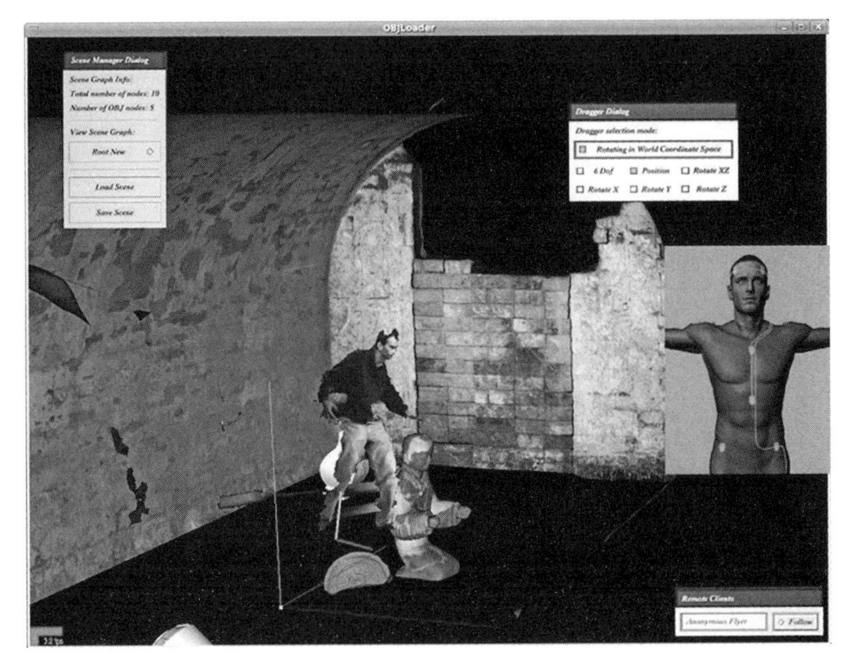

Figure 22.6
TeleArch: a collaborative session of digital archaeology inside a Western Han Dynasty tomb (late first century BC): human avatars studying funeral objects

moves the interpretation forward toward a new archeological hermeneutics: a direct, real time interaction between observation and data simulation (figure 22.6).

Conclusion

In this chapter I have attempted to analyze and discuss recent advances in the field of digital archeology after the "big bang" of VA and in relation to the large-scale, new phenomenology of information coming from digital simulations and cyber worlds. With this goal in mind, I have compared definitions and domains of VA and CA, identifying in the first domain the virtual reconstruction of the past by post-processed computer graphics, and in the second, the enactive processes determined by real time interactions between users, avatars, virtual communities, and cyber environments. The transformation of users into actors/avatars makes the role of virtual environments more focused on affordances of data (their relations with the environment) and narrative and digital performance rather than the semiotics of the models (Gibson 1950).

In VA, the visual attention is on the background of the application, in CA, on the foreground: interaction, enaction, narrative, and cultural presence generate the simulation. Cultural presence plays an important role in a virtual environment, as Champion and Bharat point out:

> The major way of creating a virtual space is where cyberspace enables and constrains human interaction in ways similar to physical space. A feeling of place is dynamically impacted by environmental constraints. We place or site and center ourselves optimally inside a flux of forces that affect our task efficiency (e.g. path of least resistance), our social standing, and our feelings of comfort. Place is also artifactually defined. We place artifacts in relation to our perception of how we appreciate or dislike environmental features. (Champion and Bharat 2002: 7)

In other words, cultural presence designs the social enaction in cyberspace, determining how cyberspace can become a cyberplace (Harrison and Dourish 1996).

Thus, although we are not able to authenticate a reconstruction the past, in CA we can simulate it, and thereby, we "perform the past." The authenticity of the digital past is, of course, a false dilemma because we do not know enough information about the past to validate its reconstruction. Therefore any attempt at validation would need to analyze the whole digital hermeneutic cycle from the bottom-up (data gathering) to the top-down phase (interpretation). However, nowadays CA plays with real time generation of data, from the field up to the cyberspace where individual avatars or entire communities can collaborate and interact.

I have also briefly analyzed several charters that have attempted to discuss the process of virtual reconstruction in archeology in light of methodological issues like standardization, validation, and transparency. Although the theoretical principles expressed by the Ename and London Charters, CURO, and the Principles of Seville are inevitable for archeological modeling, they are too utopian to be practical and do not address the new challenges of CA (interaction, embodiment, simulation, collaborative communities, cultural presence, and serious games).

Deleuze has argued that the fact "that the Same and the Similar may be simulated does not mean that they are appearances or illusions. Simulation designates the power of producing an *effect*. . . ." (Deleuze 1990: 261). In other words, a simulation produces feedback through observation, affordances, and interaction: "The simulation changes the way that we view a work of art or experience a sensation, disposing with an earlier hierarchy that valued the original work highest, and what we are left with is exactly what Plato condemned, a system in which the viewer and his manipulation become more important than any underlying ideas." (Sandoz 2003; Deleuze and Krauss 1983). For example, digital data recording, virtual communities, and post-human and avatar interactions are hyperreal. In a post-virtual perspective we are *avatarizing* the past: in fact the creation of multiple and multivocal realities, and the interaction with these digital environments, have stimulated a new knowledge acquirable by avatars and digital objects (and their affordances) as minds in action in the cyber world.

Finally, I presented the 3D-Digging Project at Çatalhöyük as an example of post-virtual CA. The outcome of this process is the generation of an enactive archeology, where the datum itself is part of a simulation process. The 3D modeling, the real time visualization, the daily 3D interactions and enactments with archeological models push the discussion forward in a new and still unexplored ontological field, the archeological hyperreal.

23 LITERATURE, NEUROSCIENCE, AND DIGITAL HUMANITIES

Natalie Phillips and Stephen Rachman

In Laurence Olivier's 1948 film of *Hamlet*, the melancholy Dane's popular soliloquy begins with a reverse angle shot of Hamlet contemplating his own suicide, looking into the cruel Elsinore surf hundreds of feet below. The camera, however, bores in on the back of Olivier's head. Zooming further in, it penetrates his skull, revealing the brain itself. The wish of that camera angle to let us into Hamlet's mind, through the brain to the tissue of thought itself, expresses perhaps the central dilemma of the relationship between neuroscience and literature. In this adaptation of *Hamlet*, the repressed mind yields to yet another level of penetration, an image of the human brain in the act of thinking—one in which the human form dissolves into little more than the medulla oblongata and the sea. In this moment of cinematic suturing, the figure of Olivier disappears: we hear his voice intoning the "to be or not to be" soliloquy; we see a pulsating image of the cerebellum. For a fleeting almost subliminal moment of screen time, the divides between mind and brain, between consciousness and the organic basis on which it relies, appear to be bridged or fused. And yet the screen illusion of this fusion poses another set of questions: Is the brain merely synonymous with the mind? Does the brain, or visualizing the brain, shed any light on the nature of cognition? Does the brain-image reveal anything more, or more important, than Shakespeare's words delivered by one of the great thespians of the twentieth century?

We raise these questions, in part, because they are so deeply familiar to us. We face them almost daily: first, as literary scholars working in the history of science and mind (eighteenth-century British and nineteenth-century American literature, respectively); second, and perhaps more intriguingly, through our recent collaborations in literature, neuroscience, and digital humanities. The last ten years have seen a wave of scholarship seeking to integrate cognitive science—the modern disciplinary haven of brain images—and literary studies, the usual guardians of the Shakespearean mind. Much of this work draws on broad frameworks from cognitive psychology to explore literary trends across historical periods. Lisa Zunshine (2006: 7–8), for example, moves between modern cognitive psychology and

Samuel Richardson's *Clarissa* to illustrate literary aspects of what psychologists call "Theory of Mind," or the "tendency to attribute observed behavior in terms of underlying mental states." Similarly, Blakey Vermeule (2011: 21–48) reads works from John Milton's *Paradise Lost* to Mark Haddon's *A Curious Incident of the Dog in the Night Time* through an array of cognitive mechanisms (for ascribing animation, agency, and shared feelings) in order to establish connections between reading, brain function, and the strange ways we care about literary characters. Mark Turner (1996: 57) moves seamlessly from "Shahrazad" in *The Thousand and One Nights* to the neurobiology of "narrative imagining," claiming that meaning is best conceived of not as "mental objects bounded in conceptual places but rather [as] complex operations of projection, binding, linking, blending, and integration." With all of these assertions about the fundamentally cognitive nature of literary activity, questions about the power (and limits) of the brain have become increasingly pertinent.

Here we offer a different model of cross-field engagement between literature and neuroscience—one framed around a radical interdisciplinarity that seeks to integrate humanist questions at the level of the experiment itself. This methodology grows out of our experience conducting concrete, neuroscientific research: a brain scan of students reading Jane Austen. This neuroimaging study on literary attention has moreover branched into a series of new projects: experiments (led and designed by literary scholars) on the cognitive and temporal rhythms of poetry reading, on empathetic responses to trauma narratives, and on differences in distraction while reading novels on iPad, Kindle, and the traditional book. Whereas previous models in cognitive approaches to literature have tended to bring theories more unilaterally from cognitive science to literature (often relying, by necessity, on studies of nonliterary activities, or tests of single words and short phrases), our experiments bring together literary critics, cognitive psychologists, and experts in neuroimaging to highlight how literature influences cognition. Rather than relying on metaphors of cognition as they might be applied to literary contexts, we refine these mental models in two ways: first, by being more cautious—historically, experimentally, and theoretically—about analogizing between literary activity and cognitive function and, second, by designing cognitive experiments on literary materials themselves.[1] Instead of hypothesizing about literary cognition based on other studies, we ask, why not study the brain in the fundamental act of reading? Here cognitive science is more than a metaphor; it becomes the prompt for theorizing new literary methodologies and experiments—using neuroscience to ask, and answer, intensely literary questions. This reorientation also opens up new ground for concrete intersections among literature, neuroscience, and digital humanities. Digital tools—particularly those used to map patterns in and across literary and artistic texts—may provide the lynchpin for bringing literature and neuroscience together in their full complexity. Simultaneously, tools from literary neuroscience have the potential to radically extend the impact and reach of digital humanities by reconnecting it to the physicality of reading and to the mind.

Digital Humanities: A Radical Interdisciplinarity

In proposing this approach, we are aware that there is considerable resistance to integrating neuroscience and literature. While interdisciplinarity has become the watchword of the humanities, to mention literature alongside the brain seems to invoke the ultimate kind of scientific essentialism, as well as a tacit critique of qualitative approaches. The questions raised by our opening image of Hamlet's brain thus have typically functioned as straw men for diagnosing distinctions between humanists and scientists. Now that both sides of the university are discussing such topics, however, it is increasingly important to view neuroscience, not as a disciplinary divider, but as a tool—one that can be used in a number of fields to achieve ends as varied as the people who decide to employ it.[2] As a recent call for papers on literature and neuroscience notes, "whether we witness art historians finding fault with neuro-enthusiastic colleagues, linguists warning of a "new biologism," ethicists, science policy strategists and anthropologists pondering the future impacts of neuroscience, literary critics and artists dabbling in mirror-neurons, or media-savvy neuroscientists forming a new kind of public intellectual," we frequently forget to move beyond debates over problems in neuroscience and into questions of how to solve them. "It has become [so] routine," the call concludes, "to celebrate or alternatively, to castigate, the ... recent expansions of the neurosciences" that these problematizations rarely move beyond "partisan polemics" into the methodological engagement and reworking they deserve (Choudhury 2011: 1).

Objections to literary neuroscience tend to fall into three areas. First, it is nothing more than a fashionable trend—the "next big thing"—or a ploy for cash-strapped humanities programs to gain funds and attention for themselves. Second, there is a fundamental divide between "literary" language and scientific language and never the twain shall meet. Any attempt to fuse them will inevitably be reductive or denaturing to the "literary," and risks reinforcing empiricism as the only reliable avenue to "truth." Third, science is a pseudo-apolitical discourse that deliberately avoids the more relevant frames of history, race, class, gender, and aesthetics. Beneath objection number one is a belief that the god of science rules in lordly fashion over the humble domain of literature and any attempt to pose questions that move between these domains is a feeble attempt to use the power of the stronger to support the weaker. This appears to threaten the autonomy of literary study, and those who reject neurological questions in this way often accuse practitioners of being merely self-promoting and trendy (or naively complicit with the military-industrial-scientific grant-getting complex).

Beneath objection two lies an assumption that no mediation can exist between "literary" modes and "scientific" modes without denaturing them both or decontextualizing, dehistoricizing, or de-humanizing the literary. Beneath the third objection lies a suspicion that neuroliterary study is prima facie politically reactionary or at best insensitive to the primacy of social/historical modes of analysis.

Digital Brain Images and Disciplinary Reciprocity

While some of the aforementioned objections have merit, the idea that literature and neuroscience have nothing to say to one another seems misguided, wasting a powerful opportunity to genuinely advance knowledge about the brain and its engagement with literature. The imagined communication barrier between the humanities and sciences, moreover—C.P. Snow's famed "unbridgeable gap"—emerges from a reified (and increasingly outdated) model of absolute competition. Supposedly, we must choose: one or the other, humanist or scientific. But why not both? Or, to appreciate the complex subdisciplines and frameworks within the humanities and sciences: why not many? "We are living in a time," writes Suzanne Keen (2006: 2), "when the activation of mirror neuron areas in the brains of onlookers can be recorded as they witness another's actions and emotional responses.... For the first time we might investigate whether [the brain] can be altered by exposure to art, to teaching, to literature." If so, we can return to our questions about Olivier's *Hamlet*, and reframe them slightly. Instead of asking whether the brain-image reveals anything "more important" than Shakespeare's description of mind (or a more traditional approach to it), we can ask what visualizing a brain engaged with literature—someone reading *Hamlet*, or watching Olivier's film adaptation—might add to our understanding of these works, and the modes of reading they inspire. We need not stop here, with science "adding" knowledge to literature. Literary neuroscience can model a profound reciprocity, a series of horizontal exchanges, translations, and moments of productive dissonance rather than a one-way traffic from science into the humanities, and vice versa. Bringing literary questions about poetry, novels, drama, and film (etc.) to cognitive science has moreover the power to reshape traditional methods and experiments in neuroscience for testing advanced cognition, adding a crucial qualitative dimensionality and cultural-historical richness, producing by consequence more complex models for understanding aesthetic engagement, narrative processing, empathy, and reading. Indeed, approaches from the humanities have the power to qualify our understanding of brain images and the kinds of arguments that are made on their behalf. As Bernd Huppauf and Peter Weingart (2008: 12–13) note, the results of fMRI, or what they call "science images," cannot be "understood in terms of a relationship to a given reality but also must be read as implicated in science models and theories." Literary neuroscience, by consequence, calls precisely for a mode of analysis that is also a mode of critique that seeks to clarify the manifold ways that experimental practices, techniques for visualizing neuroscientific results, and the popular tendency to read brain images as facts require careful framing and interpretation.

This enterprise need not embrace approaches that take a reductionist or overly universalist view of cognition. Literary neuroscience, in the form we propose, has nothing essentially to do with searches for a "narrative gene," which, like hunts for the "God gene" or the "gay gene," risk reproducing a tired biological determinism or "brain-essentialism." If we resist such models in the humanities because they smack of a kind of neo-social Darwinism, neuroscientists tend to find them frustrating for another reason: their tendency to downplay neural plasticity and to reduce intricate neural processes to singular locales. Literary criticism has long resisted integrating scientific approaches to cognition out of a suspicion that linking thought and brain necessarily entails a prioritization of the physical, and a reduction of art's infinite complexity to an ironclad mechanism of neural function. Yet this doesn't have to be the case for neuroliterary theory. Leading scholars on brain imaging, in fact, emphasize the importance of *interlinked* regions in advanced cognition: "Unlike phrenologists who believed that complex behaviors or personality traits were associated with brain regions," reads one seminal work on neuroimaging, "modern researchers recognize that many functions rely on distributed networks, and that a single brain region may participate in more than one function" (Huettel 2009: 3). Brains moreover are highly plastic. "Years ago," Ira A. Black (2004: 107) notes in *The Cognitive Neurosciences*, we recognized that "plasticity, the collective mechanisms underlying brain adaptability, emerges at multiple levels of the neuraxis." In the ensuing half decade "a newly recognized source of plasticity has swept through neuroscience: the genesis of new neurons and glia throughout life."

The implications of these points are profound both for our understanding of the brain and for cognitive studies of literature. If older models of cognition emphasized the static nature of brain tissue (we all get a certain amount of gray matter and no more), newer models recognize the brain's ability to create not only new neural connections, but also new tissue, new growth, new myelination. These nexuses are furthermore generated by external stimuli and our engagement with them, and thus are profoundly responsive to the environment, which opens up new questions in literary neuroscience. (Brains are more responsive to place and time than had ever been supposed; history then is profoundly important, as are our many acts of reading over time.[3]) It is in this more historically sensitive spirit that we envision connecting the traditionally polarized views of the mind and brain, suggesting methods from literary studies and neuroscience can be powerfully (if carefully) integrated. For why *wouldn't* humanists be interested in the brain and studies exploring how we engage with art? In a 1979 interview, cultural critic par excellence, Raymond Williams, argued for just such a venture.

He describes an attempt to conduct some experiments that investigate our physical responses to literature, or what he called "the deep material bond between language and body":

> Years ago I tried to set up some actual experiments of what happened to physical rhythms in certain reading contexts, but such was the atmosphere of specialization that the work was never done. I believe, however, that we've got to move towards active cooperation with the many scientists who are especially interested in the relations in language use and human physical organization.... What is needed is ... an introduction of literary practice to the quite different practice of experimental observation. (341)

This is precisely what we are trying to do.

Literary Attention: An fMRI Study of Reading Jane Austen

Our experiment on Jane Austen raises a key question about literary attention: what happens in the brain when readers focus on a novel? We use two neuroscientific technologies, in particular: functional magnetic resonance imaging (fMRI) and fMRI-compatible eye tracking, to explore the cognitive dynamics of the different kinds of focus we can bring to novel reading. When we began designing our fMRI of literary attention, our central question was: What is the difference between the light pleasure reading we do at a bookstore and the intensely focused reading of literary scholarship?[4] This question prompted several others. Are there significant cognitive differences between the levels of attention we bring to reading? What distinctive brain regions do pleasure reading and close literary analysis engage, and how do they work in concert? What does it mean, cognitively speaking, to read Jane Austen?

Our foundational premise was that attentive reading was not simply the opposite of distraction, but a cognitive mode that moves through a range of intensities and levels of involvement. We believed that close reading, as a form of heightened attention, would create more neural activity than pleasure reading, particularly in areas (e.g., the pre-frontal cortex) associated with attentional control. Rather than seeing this increase as occurring in just one area of the brain, however, we also hypothesized that close reading and pleasure reading would activate overlapping but distinctive sets of brain regions, producing what cognitive scientists call discrete "neural signatures."[5] If so, teaching close reading (i.e., attention to literary form) could serve—quite literally—as a kind of cognitive training, teaching us to modulate our concentration and use new brain regions as we move flexibly between modes of focus. In examining our results, we discovered an implicit bias in our elevation of close reading as inherently more rigorous than reading for pleasure. Pleasure reading, we found, had its own cognitive benefits; close reading, its own pleasures.

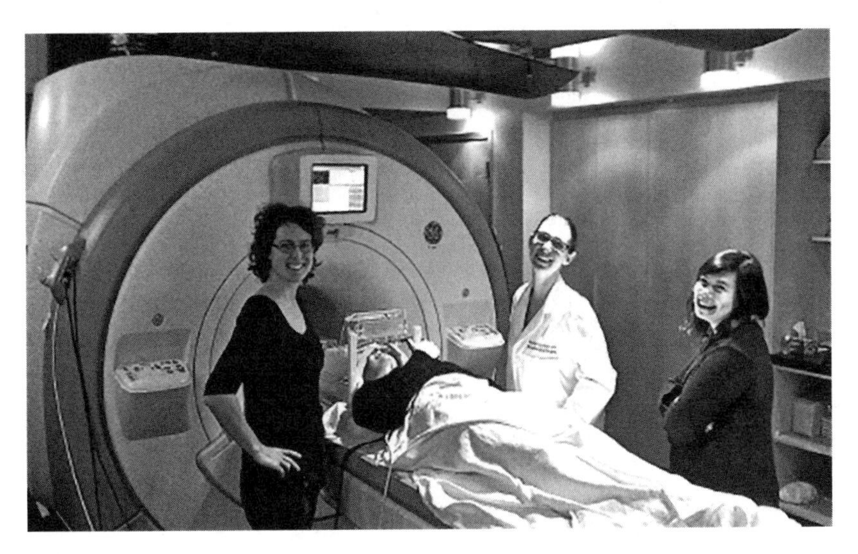

Figure 23.1
A subject entering the MRI scanner to begin reading chapter one of Jane Austen's *Persuasion.* Participants read the text presented through a mirror on a screen directly above their eyes. They proceed through the chapter, paragraph-by-paragraph, at their own speed, pushing a button to advance the text.

Using fMRI gave us a dynamic picture of blood flow in the brain—basically, where neurons are firing, and (with a slight delay) when. Neurons need blood flow to fire, which allows us, as leading neuroscientist Scott Huettel (2009) notes, "to identify different parts of the brain where particular mental processes occur, characteriz[e] the patterns of brain activation associated with those processes," and build "maps that link brain activation to mental function." fMRI's strength is its spatial specificity; it can "localize brain activity on a second-by-second basis, and within millimeters of its origin" (Huettel 2009: 4). In addition to the brain scans, we also used fMRI-compatible eye tracking, which allowed us to see how people's eyes were moving as they read in these two ways. These patterns of saccades, or the micro-jumps our eyes make while reading, provide a more detailed view of how we shift between pleasure reading and literary analysis, which can be aligned with the temporal maps of neural activity in different regions of the brain.

In our experiment, graduate students in literature read a full chapter from *Persuasion* (pilot) or *Mansfield Park* (experiment) inside an MRI scanner as we gathered images of their brain activation, recorded their eye movements, and monitored heart rate and respiration (figure. 23.1).[6]

As they read, each participant completed two sections of pleasure reading and two sections of close reading, with each section broken down into eight smaller blocks, or "paragraphs,"

Sir Walter Elliot, of Kellynch Hall, in Somersetshire, was a man who, for his own amusement, never took up any book but the Baronetage; there he found occupation for an idle hour, and consolation in a distressed one; there his faculties were roused into admiration and respect, by contemplating the limited remnant of the earliest patents; there any unwelcome sensations, arising from domestic affairs changed naturally into pity and contempt as he turned over the almost endless creations of the last century; and there, if every other leaf were powerless, he could read his own history with an interest which never failed.

Figure 23.2
Example of pleasure reading text from *Persuasion*. In our experiment, this was framed with a green border.

--Three girls, the two eldest sixteen and fourteen, was an awful legacy for a mother to bequeath, an awful charge rather, to confide to the authority and guidance of a conceited, silly father. She had, however, one very intimate friend, a sensible, deserving woman, who had been brought, by strong attachment to herself, to settle close by her, in the village of Kellynch; and on her kindness and advice, Lady Elliot mainly relied for the best help and maintenance of the good principles and instruction which she had been anxiously giving her daughters.

Figure 23.3
Close reading text from *Persuasion*. In our experiment, this was framed with a red border.

presented on a screen. Reading this digital text through a mirror above their eyes, they moved sequentially through the narrative at their own speed, pushing a button to advance from one paragraph to the next. Together with a verbal cue, we used color-coding to alert participants that they should move to the next style of attention: (1) reading for pleasure (green) and (2) reading with a heightened attention to literary form, or close reading (red) (figures 23.2 and 23.3).[7] After reading, the students left the scanner and wrote a short literary essay on the sections they examined closely.

Full data analysis for the experiment is in progress, with new results emerging even as we write this article. Results for the group as a whole are still to come. Yet what we can say at this point is that individuals from both our pilot study and the final experiment are demonstrating astonishingly strong cognitive differences between close reading and pleasure reading—differences that proved far more widespread than ever expected. In many cases, moreover, we

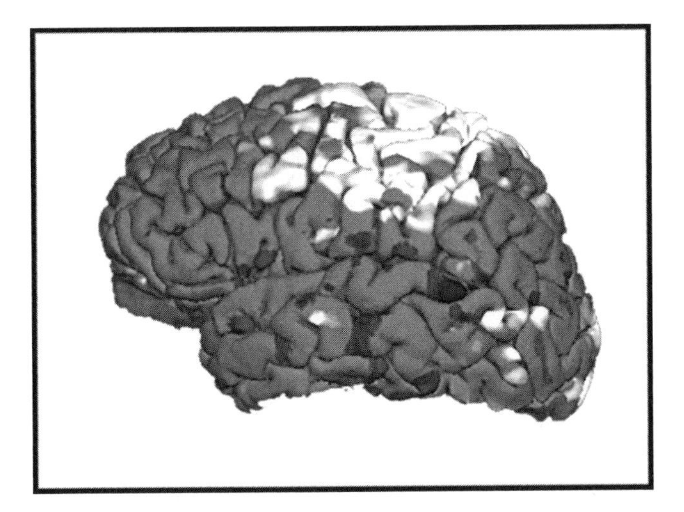

Figure 23.4

Brain image of a single subject (s4) from our pilot study of Persuasion that shows close reading and pleasure reading activating a number of distinct brain regions. Sections of the brain (in white) represent regions of heightened activity for close reading, while darker regions (in black) identify increased neural activity for pleasure reading. See note on interpreting individual results.

are seeing distinctive regions of heightened neural activity, not just for close reading but also for pleasure reading (figure 23.4).[8]

If such results hold across subjects, this suggests that close reading is not just a state of "heightened" attention to literature, nor is pleasure reading merely a lax, or more dormant, mode of cognition and focus. Instead, each style of reading seems to pose its own cognitive demands and create its own distinct patterns in the brain—and patterns far more complex than just "work" and "play." In the face of these results, the traditional metaphorics of attention—often relying on words like *focus* and *concentration,* and their optical analogues—are no longer adequate to describe the cognitive complexity of literary reading.

As Barbara Johnson (1995: 140) asserts: "Teaching literature is teaching how to read. How to notice things in a text that a speed-reading culture is trained to disregard, overcome, edit out, or explain away; how to read what the language is doing... how to take in evidence from a page.... This is the only teaching that can properly be called literary." Rather than using literary neuroscience to critique traditional humanistic practices, such as close reading, our experiment explores its complexity, finding new ways to demonstrate the cognitive intricacy and value of teaching literary analysis. In the process our experiment re-theorizes close reading, not merely as an interpretive act, or an act of writing, but as a cognitive mode of attention in reading—a style of focus, observant of form and formal properties, that serves as a necessary prequel to subsequent interpretative or analytical work.

This re-orientation of close reading parallels the ways in which digital humanities has come to re-imagine reading practices, especially with respect to attention. In the last decade, scholarship on digital reading by Cathy N. Davidson, Katherine Hayles, and Alan Liu has begun to articulate the terms on which "the brain science of attention will transform," as the title of Davidson's (2011b) study puts it, "the way we live, work, and learn." Our research explores the last of these areas, learning—particularly, learning literary analysis. Instead of separating modes of reading into the digital and the analog, this research explores the science of attention by inviting readers to voluntarily shift modes while reading the same kind of text. Our experiment explores the basis of attention that undergirds reading as a whole by studying the literary activities that regulate how, what, and when we notice things in a text: a mode of mind, eye, and body that attunes us to narrative patterns.

Digital Tools for Literary Neuroscience

What can tools from digital humanities add to such a study? And what do technologies from neuroscience add to the arsenal normally included in the digital archive? Bringing together questions about reading and attention, our experiment is already conceptually aligned with digital humanities; however, we also intend to increasingly align the two experimentally.

Recent collaborations between computer science and cognitive psychology have demonstrated the potential that lies in combining digital and neuroscientific tools, in particular, through the development of a new method for processing fMRI brain data known as *functional connectivity*. Rather than localizing cognitive experiences to a predefined brain region, such studies are interested in the *synchronous* patterns that unfold and evolve as we engage with a stimulus—or, for us, a work of literature. These experiments do not map single regions, but the activations and rhythms that emerge in parallel across different areas of the brain. Functional connectivity studies have produced cutting-edge work in narrative studies, storytelling, and the neuroscience of film, all of which push the boundaries of fMRI to map the complex temporality of cognitive processes.

One 2010 fMRI study, for example, showed that rhythmic brain patterns in storytellers and story-listeners in fact become aligned over time, demonstrating that narrative processing doesn't just activate a single region, or set of regions, but in fact synchronizes the brain activity of both the storyteller and the listener, and it does this across multiple regions of the brain (Stephens et al. 2010). Neuroscientific studies of film and moving images have taken this further. The Gallant lab, for example, uses computational technologies for reconstructing fMRI data to create dynamic brain images of movie clips people have just viewed (figure 23.5). The

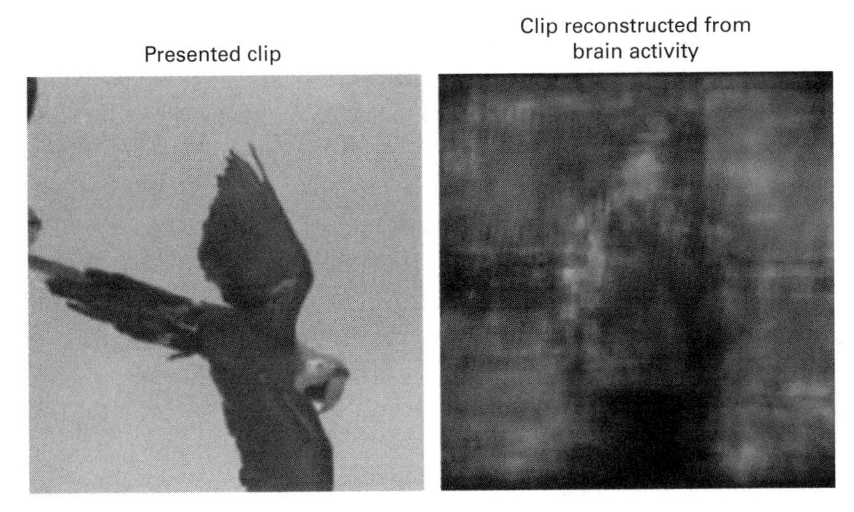

Figure 23.5
Still-frame from the Gallant lab that shows the possibility to use fMRI brain data about blood oxygenation levels (BOLD) to reproduce, or visually reconstruct, what subjects are viewing inside the scanner. Here, a movie clip showing a tropical bird in flight.

reconstructed images of the brain activation match the original images to an uncanny degree, in color, in shape, and in motion, and adjust their mirroring to keep pace with the film's shifting images (Shinji, Nishimoto, et al. 2011: 1641).

It is this model of multiplicity and dynamism that motivates our next steps for the fMRI study of Austen as well as the new experiments emerging in our *Literary Cognition and Digital Humanities* lab at Michigan State. First, we hope to add this kind of complexity to our analysis of readers' engagement with *Persuasion* and *Mansfield Park*. As one fMRI study in Japan found, neuroimaging technology has become fine-grained enough that we can predict just from the brain images whether an art student had been looking at a work by Picasso or by Dali (Yamamura 2009: 1632). If we can navigate still-frame images of deeply complex artistic works, the next step is to begin mapping these "still-frames" of textual engagement over time—a textual version of the Gallant study above. The broader cognitive implications of these connections are immense, providing an unprecedented view of what it means to read and think about a literary work in time. As we revise our old concept of attention to embrace these more complex notions of focus, so, too, will previous ideas of localized brain function move toward a more dynamic (and accurate) model of readerly attention and aesthetic engagement.

Tools from digital humanities will be crucial to this endeavor. Fortunately for our fMRI study of Austen, these interdisciplinary connections have been present from the start. The kernel of the experiment began, in fact, in conversation with one of the leading figures in the field of literary-digital mapping, the oft-cited champion of "distant reading," Franco Moretti. As we move forward, the same tools that enable him to scour 7,000 texts at a time in search of unifying patterns will also enable us to examine micro-moments of literary attention to an unprecedented degree. Digital software such as TextArc, Diction, and Natural Language Processing, among others, can help us create maps of lexical ambiguity, syntactic difficulty, and word frequency in Austen's chapters, forming the crucial underpinnings for a new kind of cognitive-narrative "thick description." Brain images from our pilot study of *Persuasion* (figure 23.4) reveal striking differences between close reading and pleasure reading as a whole (usually 10–12 minute blocks of reading). The next task, however, is to devise a textual instrument that can help us assess more sensitive and subtle temporal changes in the brain—word by word, paragraph by paragraph—working to match them in a precise, meaningful way with the text being read.

Some of the study's most interesting questions arise from thinking about what people are noticing *within* and *across* each paragraph. Here digital tools can provide crucial links between reading and the brain images, as well as between the eye tracking and the literary text. The eye tracker records subjects' physical patterns of attention as they read (e.g., places in the text where their gaze returns, patterns of eye movement, and moments when we tune out or look away).[9] In this way we can obtain a record of a given subject's neurological reaction, adding crucial detail regarding their distinctive responses to each line and paragraph. By correlating the eye tracking with the literary work, we hope to get a moment-by-moment sense of each individual reader's patterns of engagement.

The final essays students write about *Persuasion* or *Mansfield Park* provide a fascinating picture of consciously recalled experience—a written record of what was paid attention to. Part of this picture will be about discovering what responses readers share; another part will be about individual response. Take for instance, these two examples from close readings of *Mansfield Park*.

Subject A

> Fanny's position as a member of the lower classes is marked primarily by her lack of education, or, as the text calls it, her "ignorance." The narrative emphasizes that Fanny can read and write, but does not have many skills beyond this, while her cousins are able to speak French, play music, and paint, making them "accomplished" young women.

Subject B

> It's Edmund who [first] notes to himself, after their heartwarming encounter upon the stairs, that Fanny had become "an object of interest." It is as an object, particularly a class object-lesson, that she has become useful to the family. Lady B's curious and equivocating defense of the girl "she was stupid, [but] must take more pains" but Lady B had always found

> "her useful in" suggests a barbed comparison between Fanny's value and that of Pug [the Bertrams' dog], also good at "fetching" things, presumably, and "carrying."...

Both readers focused on the theme of class and on the Bertrams' slights of the orphaned Fanny in regard to her intelligence and education, a theme one might trace across the essays. Yet these general patterns also reveal individual differences in reading and attention. Subject A, for instance, concentrated on the Bertram children playing with "artificial flowers"; subject B on Austen's use of sibilant words in describing the visit of Fanny's nautically minded brother: "(sea, sister, sailor, serious)." As we gather general information about blood flow, brain regions, and eye movements, we will also be gathering these individual idiosyncrasies surrounding literary attention. Though each reader may have shared patterns in eye tracking and brain activation—each too will be distinct; none will be exactly the same.

Remarkably, given that the subjects no longer have the text in front of them, many commentaries actually quote words and phrases from what they have read. Subject A recalls a specific section where Fanny is described as "ignorant" in comparison to the Miss Bertrams' "accomplishments" (Austen [1814] 2003: 17). Subject B details a particular scene where Lady Bertram says of Fanny that she was "stupid, [but] must take more pains," and compares the girl to her pug (16). Here again, technologies from digital humanities are of aid. Using word-frequency software, as well as tools for linguistic analysis and pattern recognition, we can begin to determine if the words people notice and quote are the ones most often repeated, or foregrounded, by Austen, exploring how such textual repetitions might prime the reader to single out themes for close reading in their final essay (e.g., "stupid" appears six times in chapter 2, twice close together; "ignorance" or "ignorant" appears four times, and once in chapter 1, which subjects read outside the scanner just before entering). A carefully tagged text coordinated with eye tracking should allow us to reconstruct moments of readerly attention and connect them back to literary patterns noticed in the essay. Deciding which aspects of the text to tag—here, both in *Mansfield Park* and the literary essays—as well as what software to use and develop becomes central in bringing literary neuroscience closer to our real-time experience of reading and writing about fiction.

In choosing to have subjects read a novel and write a literary essay as part of an fMRI experiment, our study takes an interdisciplinary stance that reclaims qualitative, "natural" response as innately valuable. To use neuroscientific tools, we show, does not inherently mean elevating the empirical and numerical over the qualitative and experiential. Instead of privileging humanist methods *over* scientific, however, our project intentionally took up an "everything-and-the-kitchen-sink" model of data collection and visualization that integrates complex literary stimuli and neuroscientific results, written essays, and brain-imaging software. (Statistical algorithms for processing brain activity exist alongside open-ended prompts;

social-scientific surveys asking subjects to rate their ease of close reading and pleasure reading on a scale of "one to ten" adjoin rich literary essays on Austen; our brain images for close reading, in the end, will be recalibrated in ways that are deeply responsive to quotes from *Mansfield Park* in the essays students write.) In this back-and-forth between the numerical and the narrative, digital tools offer a key connection between qualitative and quantitative, providing a humanistic version of what Peter Galison (1997) calls "trading zones," or moments of "local coordination" and translation between specialists in complex moments of scientific exchange (46–48).[10] For our literary fMRI, the ability to link patterns in the final essays with the data yielded by the brain scans and eye tracking promises to offer new insights not only into this novel (and the essays written on it) but also into the fundamental—and fundamentally individual—nature of the reading acts that precede literary interpretive work. As this reminds us, literary neuroscience, in its intersections with digital humanities, can be a field committed not to quantitative reduction, but to ever-expanding complexity: to dynamic mappings of aesthetics and mind.

As mentioned, our fMRI of literary attention has inspired a series of new projects, including an experiment on empathy and trauma narratives (Duke), poetry and cognitive rhythm (MSU), and patterns of distraction while reading fiction on digital readers as opposed to traditional codex (MSU, Stanford, and Lund). Algorithms for macro-level text searching will be essential to developing these new projects, letting us trace similar literary formulations through the digital vista of the database to find the best samples, isolate key variables, and produce richer, more rigorous experiments. As Matt Jockers pointed out in a recent conversation, one might begin by looking for nineteenth-century novels (and chapters) with a similar narrative structure to those being tested in *Persuasion* and *Mansfield Park*. The digital affords the opportunity to begin tracking structures that extend beyond one book by a canonized author into many nineteenth-century texts, as well as to compare readers' responses to a fresh work (never seen before) and those created by well-known and culturally significant authors. In all these cases, bringing together new tools and technologies from literary neuroscience and digital humanities can energize both fields, adding a crucial piece to the larger puzzle we know as literary reading.

In "Humanistic Theory and Digital Scholarship," Johanna Drucker (2012) discusses the value of using digital tools in the humanities: "using new platforms and networked environments," she writes, "humanists entering the digital arena learned a great deal. . . . [T]he tasks of creating metadata, doing markup, and making classification schemes or information architectures forced humanists to make explicit many assumptions often left implicit in our work" (85). This effect certainly held true for our fMRI experiment, whose design urged us to consider—and rethink—a number of assumptions about close reading, pleasure reading, and literary analysis even as we worked to demonstrate anew their cognitive value. Yet Drucker poses a second—

and crucial—question about the reciprocity of exchanges between the humanist and the digital: "Can we create graphical interfaces and digital platforms from humanistic methods?" (85)

These questions and challenges are particularly pertinent as literary neuroscience develops as a field. Fortunately, the critical gaze of digital humanities scholarship and science studies in the last fifteen years have framed our approach to intersections between humanities and neuroscience from the start. The attempt to ground literary neuroscience's tools, methods of visualization, and interpretive practices in humanist methods has shaped our experiment design, our collection and interpretation of data, and our analysis of brain images. While our basic equipment (MRI) traditionally has been used for scientific purposes, and we use preexisting neuroscientific algorithms for interpreting the data gathered about blood flow in the brain, our experiment's larger investment in literary questions have necessitated a departure from previous cognitive studies of attention and reading.[11] These choices, fundamentally, required the technological integration of fMRI and eye-tracking and the creation of a subject-based design. This yielded truly complex data sets that combine the qualitative and the quantitative. In our new humanities lab, *Digital Humanities and Literary Cognition* (DHLC), faculty and students from literature, neuroscience, psychology, linguistics, and education are working together to develop a unique set of cross-disciplinary tools to analyze patterns across individual brain scans, Jane Austen's chapter, and the literary essays.

Many scientific studies today—particularly experiments that use fMRI—are interested in generalizable principles about the brain across subjects. In most cases, individual differences are erased in cross-group analysis, producing an image of the brain and its activation that integrates all subjects together. Our project allows us to track traditional cross-group data while simultaneously exploring what is unique for each participant; we take into account idiosyncratic patterns in brain activation and in reading, as well as distinctive aspects of individual brain structure and shape. In so doing, we take another approach to the goal Lisa Cartwright and Morana Alač (2007) laid out in an article on laboratory practices in clinical magnetic resonance imaging, making the whole research subject "visible or present in the process of interpreting his or her MR data." Rather than presenting brain images as scientific facts that "solve" questions of literary reading and attention, we recognize that understanding all such visualizations requires "not only the investigation of images as an instrument in scientific discourse but also . . . a reconstitution of theories of knowledge production . . . including the material conditions of their production" (Hüppauf and Weingart 2008: 5).

To translate these questions and methodological frameworks from digital humanities to literary neuroscience: can we use fMRI to produce *humanist* brain images? For us, this has become a strange question. The brain images that will come out of our lab at MSU—whether derived from reading Jane Austen, William Shakespeare, or Olivier's *Hamlet*—are produced, designed, and run by literary scholars, radiologists, and psychologists all together, in collaboration throughout, testing literary questions. In the literary neuroscience we are calling for, the kind of brain imaging these studies produce is in many ways much more complexly framed from a literary point of view than the composite of Hamlet's brain with which we began this essay. For if there is a universalizing tendency in the imagery of a generic brain superimposed on Olivier reciting Shakespeare's words—one that suggests through a synchronization of brain and words that *this* is the human condition—then these studies of Jane Austen's readers are much more careful not to accept such images as mere "objective entities" or cognitive facts.

Stephen Ramsay and Geoffrey Rockwell argue that digital artifacts are tools, "hermeneutical instruments through which we can interpret other phenomena.... 'telescopes for the mind' that show us something in a new light" (79). Similarly we see neuroscientific images not as scientific "answers" but as tools able to prompt new literary insights. In the end, an fMRI image (and the microscopic view it offers of the brain's responses to literature) is another piece of information. It is a new kind, to be sure, with unusual elements to contextualize, integrate, and theorize. Yet like a letter from Austen's sister Cassandra, an eighteenth-century conduct book, a reader's diary, or a Foucauldian theory, it depends on the interpretations we produce. As two scholars interested in literary neuroscience and the history of mind, we return to narrative analysis and historical context—that is, to literature and culture. Instead of viewing our fMRI results as offering some final scientific "truth" about the cognition of literary analysis, or of reading Jane Austen, brain images must be re-interpreted as artifacts of the intersection between a historically (and culturally) embedded text and the brain that receives it. As Jonathan Kramnick (2011) notes, if "cultural artifacts like works of literature are part of the environment our minds build, respond to, and build again," then "the historical and formal methods of our discipline... contribute [deeply] to understanding how the mind works" (347). In light of this, there is something prescient in the superimposition of Hamlet's brain on the tumultuous waves of Elsinore, as well as in the choice to imagine the connection between mind and brain as a pulsating fluctuation between the two. It becomes emblematic of the plasticity of the brain in its responsiveness to an ever-changing environment and the mind engaged in the act of reading. And this is where things get truly interesting.

Notes

The authors would like to thank Erin Beard, Austin Gorsuch, Shannon Sears, Laura McGrath, Katie Grimes, and Bridget Whearty for their assistance with the writing of this chapter, as well as collaborators Bob Dougherty and Laima Balthus at the Stanford Center for Cognitive and Neurobiological Imaging,

Samantha Holdsworth at the Lucas Center, and Franco Moretti. We would also like to thank the Stanford Literary Lab, the Duke Neurohumanities program, and the Stanford Humanities Center for their early feedback on the experiment, as well as the Wallenberg Foundation, *Neuroventures*, and the College of Arts and Letters at Michigan State University for their support.

1. A number of others are working in this vein studying literature by way of neuroscientific experiments, including neuroscientist Raymond A. Mar as well as literary scholars Lisa Zunshine, Gabbi Starr, and Angus Fletcher.

2. While there is division within neuroscience communities (as with any discipline) as to whether its commitments best lie within its own precincts, questions, and protocols or through its applications to other fields, the majority of neuroscientists view the field as fundamentally interdisciplinary and view with excitement potential applications to other fields.

3. This historical recognition about cognitive plasticity has deep implications for how we intend to interpret our study. To do a brain scan of someone reading Jane Austen, as in our experiment, does not mean that we can make concrete claims about the distant past, or the dynamics of a nineteenth-century reading experience. (We can't put nineteenth-century readers in the scanner; they're dead.) While we can access modern readers' cognitive experiences, any speculations about historical readers drawn from this kind of neurological evidence would require profound caution.

4. This experiment, formally begun at Stanford in 2010 with Phillips as primary investigator, is now a collaborative effort involving Bob Doughterty (co-investigator), Samantha Holdsworth (radiology), and Stephen Rachman (digital humanities), and an emerging team of graduate and undergraduate researchers in the *Digital Humanities and Literary Cognition* lab at Michigan State University.

5. By "neural signatures" we mean complex brain patterns of blood flow recorded during the act of reading. Our experiment uses the 3D capacity of fMRI to create full brain maps of activity, gathering data in 2-mm slices for sagittal, axial, and occipital planes. Rather than localizing function to a single region, these maps offer us images of multi-region patterns of activity for individuals. These patterns exist in individual readers and group aggregates.

6. All participants had advanced training in literary study, analysis, and close reading. Some had familiarity with the Austen novels. None had read them in the last two years.

7. Reading a literary work inside a loud MRI scanner—either for pleasure or for critical literary analysis—is, of course, a profoundly different experience than doing so in the everyday world. One way we are attempting to correlate the experience in the scanner with normal reading is by comparing the times for pleasure reading outside the scanner with those done within. That said, any task or experiment performed inside a scanner is intrinsically distorted to some extent by the technological environment, and any attempt to translate fMRI results must take test conditions into account. For the sake of scientific experimentation, the point is that this distortion remains constant for both reading conditions. Close reading and pleasure reading are both challenging in this environment, and in different ways. Results thus far do not suggest a strong contextual bias toward one or the other.

8. Such individual results must be interpreted cautiously as readers' experiences can vary on a day-to-day basis. Only group analysis (or repeated testing of an individual subject) can provide the statistical significance and reliability required for robust cognitive results. For more on interpreting fMRI results, see Lindquist (2008). As literary scholars, however, we found differences in these individual results intriguing. We use them as a prompt, beginning to ask new questions about how maps of brain activity (and patterns in eye tracking) translated into the personal—and distinct—reading experience of each subject.

9. Eye tracking can reveal various types of information about the reading process. As Salvucci and Goldberg note, researchers typically analyze eye movements in terms of "fixations (pauses over informative regions of interest) and saccades (rapid movements between fixations)," where common metrics include "fixation or gaze durations, saccadic velocities, saccadic amplitudes, and various transition-based parameters between fixations and/or regions of interest" (Salvucci and Goldberg 2000: 71). This information about the reader's gaze can be joined with heart rate and respiration measures, which offer additional data about subtler variations in readers' levels of excitement, engagement, and interest, as well as about difficulty (or nervousness) regarding the task at hand.

10. Digital tools also can widen the scope of fMRI experiments. With an hour-long experiment run costing $400 to $600 an hour, these studies tend to focus, by necessity, on a single work and a small group. Methods for digital surveying can broaden the group of subjects to give such work more reach. Rather than restricting our final analysis to the 20 to 30 subjects traditionally possible in fMRI, we can crowdsource the essays, posting tested passages from *Mansfield Park* and *Persuasion* online along with the final close reading task to gather hundreds of examples.

11. Instead of using single words or phrases to study reading tasks, as in the majority of fMRI experiments, we used paragraphs. Rather than controlling the pace of reading and presenting words at a set time, we developed a design that could allow each subject to move through paragraphs at his or her own pace. In other words, we did not create our own stimulus, or sentences for participants to read, seeking to unify every aspect of their linguistic, syntactic, and narrative structure to isolate a single variable. Breaking in crucial ways from older models of scientific experiment design, we had students read a *real* literary work by Jane Austen and write intricate literary essays in response.

III KNOWLEDGE PRODUCTION, LEARNING, AND INFRASTRUCTURE

This part brings together a series of chapters that center knowledge production and knowledge infrastructures. Whereas part I focused on the institutional and scholarly conditions for humanities and the digital and part II on disciplinary questions and challenges, this part addresses processes and infrastructures necessary for creating new humanistic knowledge and for making possible the institutional and disciplinary work described in the previous parts of this book. The issue of infrastructure is not just a concern for the digital humanities, but a question for the humanities at large; and it is not just a matter of research, but also of education and, fundamentally, learning. What we are suggesting here, too, is that digital humanities challenges the humanities generally to take issues of infrastructure seriously, as driving conditions—both material and conceptual—of the possibility of humanistic research and teaching.

The cultural, institutional, and material situatedness of the production of knowledge provokes a set of deep humanistic sensibilities and questions about matters such as the epistemic status of documents (Gitelman 2014), the media archeologies of the moving panorama (Huhtamo), psychological relationships between screens and social perceptions (Robles, Nass and Kahn 2009), the visual as a form of reasoning (Galison 2014), the epistemics of visualization (Drucker 2009), critical making (Ratto 2012), and the social and institutional foundations of scholarly publishing (Fitzpatrick 2011). Such work is important in its particularities, but also demonstrates that history, technologically inflected modalities, distribution channels, and infrastructures enable and condition knowledge production. Without critical sensitivity and historical awareness, we cannot even begin to understand the situatedness and conditioning of knowledge production and knowledge infrastructures.

At the same time, emergent expressive modalities, the intertwining of digital materials and scholarly production, the privileging of spatially organized information and evolving artistic practice, encourage humanists to engage with alternative modes and genres of scholarly knowledge production. One key challenge then becomes that of infusing our own knowledge production and practices with the critical sensibility that we typically recruit when engaging with other domains, knowledge communities, and historical strata in our work. On a more fundamental level, this requires us to imagine humanistic knowledge infrastructures that are built on the core challenges of the humanities and connected to the questions at the heart of exploring the human condition. Ideally, such infrastructures can both help us approach such questions and evoke new sets of humanistic inquiries and engagements. While discussions of infrastructure may not immediately attract major excitement among most scholars, infrastructure can be seen both as an enabler, facilitating and supporting academic work, and as a relevant object of critical study. Indeed infrastructure is about imagination and connecting deep conceptual ideas with material manifestations. In this sense, articulating an infrastructural agenda for the humanities is about shaping the future of the humanities and the academy. At the same time, infrastructure embeds institutional, cultural, and political values, and humanists need to bring their critical awareness to making evident the commitments and implications of such values.

Work placed between the digital and the humanities calls for a strong engagement with both research and learning. Essentially, as a liminal zone invested in knowledge production across epistemic traditions with a multiplex and changing engagement with the digital, the digital humanities is best positioned when built on continuous learning. Such zones benefit from being inclusive and including students as well as technologists, experts, and humanists. As an institutional formation, the digital humanities has traditionally been more focused on development and research than education, and this bias or neglect has received criticism. However, with the expansion and leverage of the field, there has been a stronger investment in education. This investment partly comes from an increased interest in courses and programs in the digital humanities and from institutional expansion, but also from a broader engagement in the future of the humanities and their role in the academy. The digital humanities has to some degree become a place for (re)imagining the humanities and the academy, and it is simply not possible to envision the future of academe without beginning with learning and students. Furthermore the digital humanities engages with the construction of knowledge production structures such as encoding ontologies and database tools, and such structures fundamentally condition learning and knowledge production. While this capacity might not have been employed to leverage a strong learning engagement, we can see a change in and conflation of interests. A simple example is the conscious effort to change the gendered, national, and racial inflections and assumptions of resources such as Wikipedia, and its more critical use by students.

Importantly, as learning is mediated one way or another and as technology is often essential to that mediation, it would seem that a humanistic interest in the technological (as well as in materiality, mediation, history, and knowledge production) could serve as a useful platform for considering and developing education at this point in time. Online learning and what we now refer to as massively open online courses (MOOCs) offer a topical example. MOOCs have been much debated, and while certainly suggestive, these platforms are typically based on an information distribution paradigm concerned with size and focused largely on simple feedback and monitoring of students. As online learning is about technology, culture, and knowledge production, it might be assumed that digital humanists would have relevant competency to contribute importantly to this discussion, and possibly even the know-how to build conceptually grounded platforms for online learning. However, given that the digital humanities has not been an enterprise particularly focused on education and learning tools, one viable position is that there are better candidates for taking on this task than the digital humanists. Or it could be argued that there is need for a humanities-based sensibility in thinking deeply about online learning and in devising other ways of carrying out such education. Indeed, if present-day online learning systems are lacking conceptually, intellectually, pedagogically, and technologically, they are in dire need of what a richly conceived digital humanities can bring to the table.

It is at the level of knowledge production that the various aspects of digital humanities come together most distinctly. This is where epistemic positions and tensions are most evident and where disciplinary questions meet infrastructural conditions, as well as where new possibilities for knowledge are enabled or prevented. This is where we need simultaneous intellectual, expressive, and technological engagement.

In chapter 24 Patrik Svensson argues that making humanities infrastructure requires simultaneously drawing on the critical sensibility of the humanities and on our own imagination and ideas. He argues that infrastructure is a matter of social, cultural, and political situatedness as well as a question of material and conceptual situatedness. There is a risk that humanities infrastructure will be based on existing infrastructure or science models rather on the central, contemporary needs of the humanities. Svensson posits an imaginary and material device called the humanistiscope as a means of exploring and framing humanities-based infrastructure.

Lisa Parks (chapter 25) points to the rich material and situational complexity of infrastructural systems in her chapter on media infrastructures. Her analysis of the material objects and sites that define media infrastructures is highly relevant to both academic and media infrastructures. Park's combined material and systemic engagement is demonstrated through her analysis of the representations of mail sorters, power poles, and satellite dishes, and the corresponding infrastructural systems of the postal system, electrical grid, and the global satellite system. Park's key argument is that audiovisual media can be analyzed in infrastructural terms, and that this infrastructural disposition goes beyond a frame or an individual's perception. Infrastructure in this sense includes bodily acts such as rigging power poles and policing audiovisual signals, biophysical resources, and the local and global implications of physical installations.

Matthew Kirschenbaum (chapter 26) performs an infrastructural reading of scholarly knowledge production mainly within the digital humanities community by using two MLA conferences (1996 and 2013) as anchor points and by discussing the platform MLA Commons (launched in 2013). Much like Parks, he addresses both individual infrastructural objects or platforms and infrastructure on the systemic level. There is historical and material sensitivity here as well as a situational reading based on power, access, and time. For instance, Kirschenbaum stresses the power dynamics built into access structures and the importance of paying close attention to the procedural and relational models that are part of a platform such as MLA Commons. He also discusses scholarly platforms more generally, and how they may actually be further from interoperability than ever. Consequently Kirschenbaum recommends that the MLA Commons may want to think about its own future degradation.

One way of looking at infrastructure is to see it as placed somewhere between materials/data and scholarly questions. The community of digital humanists described by Bethany Nowviskie in chapter 27 might be seen as living in this space, encountering the friction embedded in the toolset, and seeking to explore and articulate the relationship between craft and interpretation. Nowviskie discusses three critical factors with regards to this space. First, the ongoing digitalization of cultural heritage materials is a major conditioning factor, especially in combination with new digital materialities such as augmented spaces and 3D printing. Second, Nowviskie points to what is not spoken, expressed or reflected on in processes of tool making and coding, and how the craft side of digital humanities work is pressured to become more introspective. Finally, attention is drawn to the people side of digital humanities infrastructure, and how the field is conditioned by the rise of alternative academic positions and the widespread casualization of academic work.

Amy Earhart (chapter 28) investigates the laboratory model as an institutional structure for the digital humanities. She notes how this model has gained momentum, partly as a result of the interest in the kind of embodied, physical work that Nowviskie addresses in her chapter, while also being a source of contention. One key question concerns to what extent digital humanities labs mimic science models and epistemologies, and how such mapping between knowledge domains may or may not be a problem. Earhart points to a range of parameters at play here including spatial design, power and gender hierarchies, reward systems, mentoring systems, training practices, the level of neutrality associated with the space or operation, and views on the knowledge processes and products.

Maps are high on the list of objects and materials likely to play an important role in humanities laboratories and infrastructure. They are likely candidates for digitization because of their cultural significance and they are central resources in an increasingly geo-oriented world. As Patricia Seed shows in chapter 29, the idea of printed maps as untouched and original is problematic, and indeed digitization processes can help detect flaws in printed maps. For instance, high-quality digital images of large maps (several meters by several meters) allow the comparison between the original version and print reproductions. Detailed comparisons are made possible through overlaying print copies with large-format prints of the digital file on translucent or transparent material. Seed's detailed analysis of the material conditions of maps and digitization processes demonstrates the necessity for standards that can validate the accuracy and quality of printed maps.

Zephyr Frank's chapter 30 moves us to one of the laboratory environments discussed in Earhart's chapter 28 with a deep connection to Seed's (chapter 29) engagement with maps, although the Spatial History Lab at Stanford University is mostly focused on digital maps and visualizations. Lab life and current projects provide a grounding to Frank's chapter, which continues to consider the intellectual lines of spatial history, discussing key works such as Paul Carter's *The Road to Botany Bay* and scholarly directions such as Time of Geography, spatial economics, and historical GIS. Frank sees spatial history not so much as a spatial turn but as a more incremental development in the practice of history. He also demonstrates that as a field, spatial history draws on both humanistic and science-based models of spatial thinking, and argues that there is probably much to be gained from these cultures of scholarship coming closer together.

The close relationship between knowledge production, such as project work with students going on in the Spatial History Lab, and learning is evident from Elizabeth Losh's chapter 31. Losh points to the relative lack of engagement with learning in the digital humanities, at least as evidenced by conferences such as the Digital Humanities 2012 Conference. She also urges the digital humanities community to align more closely with human–computer interaction and values-centered design. Her chapter examines what she calls utopian pedagogies, set apart from administratively driven online platforms and much other digital pedagogy driven by rationalization and a paradigm invested in information distribution. Losh sees potential in such new platforms and initiatives, but she also cautions against the celebration of varieties such as remix and mash-up pedagogies and questions the rigor of many such initiatives. She further problematizes the alleged connection to egalitarianism and democractic inclusion. Unmix pedagogy, as suggested by the author, is about unpacking aggregate materials by using digital tools for surveillance and search engine optimization. Losh argues in favor of renewal rather than revolution, learning from mistakes, and situating learning both in current and emergent infrastructures.

In chapter 32 Jennifer González investigates race and race discourse in relation to digital spaces and digital art in her contribution to the volume. Her chapter is a deep investigation of the racial registers of Internet infrastructure as a relation of public encounter. She explores the face and assumption of facelessness of the Internet and other digital media, finding problematic Mark Hansen's argument that "passing" in online environments can suspend constraints of the body as a visual surface for racial markings. She also critiques the positions of Giorgio Agamben and of artist Nancy Burson in her *Human Race Machine* for inviting us to pay attention to our physical traits and properties in order to transcend them, while simultaneously failing to address the political and social constraints that may hinder this very transcendence. By contrast, in addressing Keith Obadike's controversial work, *Blackness*, in which the artist proposes to sell his blackness on eBay, González finds performativity beyond the question of shifting appearances. In investigating *Human Race Machine* and other art works on the level of software and algorithms, she stresses that such software may not only come to mimic restrictions existing in social worlds more broadly, but may also neglect the larger forms of resistance effectively at work there.

Kathleen Fitzpatrick (chapter 33) explores types of scholarly knowledge production mostly far removed from the artistic and experimental modalities analyzed by González. The focus of Fitzpatrick's chapter is scholarly publishing in a digital age, and her work points to the continuities of academic publishing, the dynamics of digital systems, and the complexities of publication systems and their financial, distributional, collegial and epistemological embeddedness. And even as there is continuity across forms, new publication forms—to the extent that they are actually new—develop novel kinds of formal constraints and conventions. Fitzpatrick reminds us that the transition of scholarly communication to new infrastructural digital platforms is not a zero-sum game, and while publishing houses, scholarly societies, and peer review systems will inevitably have to change to accommodate new modes of scholarly communication, they are not likely to disappear. Importantly, Fitzpatrick argues that we need to have a critical engagement not only with the world, but also with our own ways of working. This position aligns well with the combined intellectual and material engagement pursued in several of the chapters in this section.

Mats Dahlström (chapter 34) focuses on a different but closely integrated part of the publishing and content transmission system, namely research libraries. He addresses the question of criticality in relation to content transmission, and looks at scholarly editing and library digitization as separate, but connected endeavors. A starting point is provided by a discussion of tensions in scholarly editing including the division between critical and non-critical scholarly editing and image management. This resonates with Patricia Seed's earlier discussion of maps as images. Dahlström argues the identified tensions are not media specific, but rather a trait of textual transmission as a cultural phenomenon. In looking at library digitization, he contrasts mass digitization and critical digitization as end points along a scale. A limited number of libraries carry out mass digitization and this model does not work with certain artifacts, including fragile manuscripts and texts that are difficult to read. Critical digitization, according to Dahlström, engages with digitization in an intellectual, critical, and manual way. A project may focus on a single object, and Dahlström argues that, here, libraries can be said to be engaged in work close to textual criticism. He ends the chapter by suggesting two possible developments. The first scenario describes the merging of scholarly editing and library digitization. Here scholarly editions and archives would seem to turn into digital libraries and high-quality critical digitization done by libraries may take on the form of "editions." A reverse, somewhat provocative scenario suggests that digitization is seen as a linear and flat affair, nursed by digital humanists, editors and libraries, where the digital product is seen as a reduced offspring. It is obvious that Dahlström is querying his own discipline when he juxtaposes the traditional role of textual criticism, problematizing variation on multiple levels within and between artifactual instances, with the seeming linearity of the digital reproduction.

In the concluding chapter 35 to part III, Tara McPherson argues that, not unlike a century ago when photos and films opened up a space of possibility and resistance in relation to the traditional archive, we can understand contemporary digital configuration as a post-archival moment. Today fantasies about total knowledge are structured around the database, and McPherson claims that the digital both enables and makes possible resistance to those dreams. While computational structures have a tendency to support positivism and control, technologies can be recruited to build other kinds of systems. According to McPherson, humanists need to be involved in developing policies, practices, digitization schemes, and indeed, models of use for the post-archival database. She reflects on several projects along such lines, including Murktu (a database platform for cultural heritage aiming to resist the colonial imperative) and Scalar (a scholarly authoring and publishing platform developed by the author and others as part of the Alliance for Networking Visual Culture). McPherson invokes a vision where technological sensibility is important and where cultural and critical theory play key roles in conceptualizing, designing, and implementing digital scholarship and their supporting infrastructures. She writes, "I argue that only through such an articulation can the digital humanities engage the archival in its richness, its absences, its ambiguities, its pleasures, and its transformations."

24 THE HUMANISTISCOPE—EXPLORING THE SITUATEDNESS OF HUMANITIES INFRASTRUCTURE

Patrik Svensson

The humanities have a complex relation to infrastructure. Humanists engage with infrastructure—the university, digital systems and tools, cultural heritage institutions, seminar rooms, and networked computing—on a daily basis. They also carry out critical work on infrastructures through looking at matters such as the history of scientific instrumentation and the social, cultural, and political situatedness of systems such as national infrastructures. However, when it comes to thinking about the humanities in terms of infrastructure, there seems to be a lack of both everyday systemic awareness and extensive critical work. Humanists do not thus necessarily think of what they do as situated and conditioned in terms of infrastructure, which leads to challenges when it comes to imagining and implementing new infrastructures. There has been a lack of infrastructural engagement from the humanities, and I argue that this has been at our own cost. Such engagement does not necessarily boil down to requesting funding for technology, but to consider our work and future in infrastructural terms. The digital humanities has been helpful in drawing our attention to our infrastructural needs, and I suggest that the field can help us unlock infrastructural thinking and making.

This challenge is also one of moving from critical sensibility to creative, if conditioned, making, which often does not come easy to the humanities. A real risk is therefore that new humanities infrastructures will be based on existing infrastructures, often filtered through the technological side of the humanities or through the predominant models from science and engineering, rather than being based on the core and central needs of the humanities. This is an important concern and possibility for the humanities. This chapter explores the conditions necessary to allow the imagination and implementation of humanities-based infrastructures as conceptually, critically, and materially situated.

While academic infrastructure as constructed by governmental and funding bodies may make reference to the humanities and social sciences, this seems to be mostly a nominal interest. Infrastructural funding given to these areas is very small in comparison, and actual investments made in humanities infrastructure tend to be based on a notion of infrastructure taken from other domains or based on certain qualities of familiar infrastructures, such as the library. Some humanistic fields are more likely than others to be recognized as infrastructure, including cultural heritage, language technology, computational linguistics, and archeology. This does not, however, mean that they should automatically be regarded as the main model for humanities infrastructure.

Moreover, while there is an emerging body of humanities scholarship exploring infrastructure critically, such work does not normally engage with the infrastructure of the humanities. Neither does this critical awareness seem to be applied when the humanities conceptualize and make their own infrastructures. In particular, it seems difficult to infuse intellectual issues and scholarly challenges into infrastructures as articulated in various white papers and reports on humanities research infrastructure. Technology and methodology often get foregrounded at the expense of a strong link to the core of the humanities. At the same time, such articulations of infrastructural visions paradoxically lack a deep engagement with the material level of infrastructure and the associated making.

Even if it is necessary to point out the fact that the humanities as a whole are marginalized in the specification and allocation of infrastructural resources, it is more important that the humanities present far-reaching thinking about what their infrastructure could be than oppose the current state of affairs. Such a strategy involves acknowledging the fact that making a case for infrastructure is one of politics and packaging as well as ideas, people, and equipment. An important note here is that while there is need for much stronger infrastructural support for the humanities, I am not suggesting that the humanities need as much infrastructural funding as science or engineering. These areas are larger to begin with (given current priorities), and the realpolitik of infrastructure and research policy (and common sense) will prevent such a trajectory in any case.

Essentially infrastructure is about situated imagination, and it would seem that the humanities lack a clear and systematic idea about their current and future infrastructure. In this chapter, I argue that we need to see infrastructure as situated culturally, socially, politically, technologically, and spatially. If we are to base our future infrastructure on a science and engineering model, on the library, or on the sense that we do not need any infrastructure, it should be a carefully thought-through decision. Naturally, we are not concerned with one decision or one type of infrastructure, but rather, we need to work toward a humanities-wide perspective and a framework connected to the humanities as a scholarly and educational endeavor. Among other things, this stance requires us to connect the conceptual level to the material level, and neither shy away from the ideational underpinnings nor the material details of implemented infrastructure.

I suggest that that the idea of a humanistiscope, a humanistic infrastructural platform, can be used to free ourselves somewhat from our own infrastructural past and science and technology-driven visions of humanities infrastructure, while also acknowledging the need to incorporate and learn from these perspectives. It might have been possible design humanistic microscopes or telescopes, but such "scopes" come with very particular situatedness, function and scientific heritage. Instead, I propose that the idea of a humanistiscope as a rhetorical and practical device can help us reconnect the humanities and their infrastructures. It is much less laden than the aforementioned devices, although it borrows from a scientific infrastructural logic and the Greek for "to look/examine" (which also points to the commonplace privileging of the visual in the infrastructure turn).

There are several reasons for using the humanistiscope as a "thought piece." For one thing, there is not necessarily a name for the kind of things under discussion here (existing or possible humanistic infrastructures), and the notion of the humanistiscope gives us a way of packaging and imagining humanities infrastructure without being locked into current vocabulary and infrastructures. Additionally this device allows us to appropriate language that belongs to science, technology, and engineering as well as being usefully suggestive and provocative within and outside the humanities. There is also a helpful material push in the idea of humanistiscopes, as they would seem to be material devices one way or another. It should be stressed, however, that the idea is not to exoticize the humanities and humanities infrastructure or to discard previous layers of infrastructure. The humanities need to articulate their own infrastructural visions, but they also need to relate to the larger world of infrastructure and to their own past.

The chapter starts off with an extended background of the infrastructure turn and humanities infrastructure followed by an analysis of a call for proposals for data-intensive projects in the humanities. This leads to a more general discussion of academic infrastructure and a look at two main infrastructural templates for humanities infrastructure. The latter part of the chapter discusses the humanistiscope in more depth, including criteria and potential use, and some examples of humanistiscopes are presented. These examples are based on my experience from HUMlab at Umeå University. I argue that humanities infrastructure often may be multiplex beyond what fits in one humanistiscope, and that platforms such as humanities laboratories can be useful in allowing the alignment of many humanistiscopes.

Background

Over the last fifteen years there has been an increasing interest in academic infrastructure under rubrics such as cyberinfrastructure, research infrastructure, and e-science. The 2003 National Science Foundation report "Revolutionizing Science and Engineering Through Cyberinfrastructure," the introduction of the UK-based e-science program in 2000, and the formation of a European Strategy Forum on Research Infrastructures (ESFRI) in 2002 have been important factors for this development. The infrastructure turn, as Geoffrey Rockwell (2010) describes it, has been characterized by an opening up of academic infrastructure in two senses. First, there has been an interest in moving beyond actual equipment to also including services and people. Second, there have been attempts to include other academic areas than those normally associated with large investments in infrastructure, including the social sciences and the humanities. One interpretation of this more open stance is that it is part of repackaging infrastructure discursively to make a case for new funding.

Hence it is not surprising that the humanities are often included nominally, but not actually, in terms of significant funding allocations or far-reaching thinking about humanities infrastructures. For instance, the "What are RIs" page on the ESFRI website tells us that research infrastructures are "tools for science," that they should be "at the core of research and innovation processes," and that they offer "unique research services to users from different countries, attract young people to science, and help to shape scientific communities" (European Commission 2013). A couple of examples of high-quality, important research infrastructures are given early on in this text: radiation sources, genomics data banks, environmental sciences observatories, imaging systems, clean rooms for developing new materials, and nanoelectronics. As observed by Christine Borgman (2007), such definitions, by example, are common in infrastructure discourse. And although social sciences and libraries are mentioned in a cursory way at a later stage in the text, it seems quite clear that this kind of presentation does not make an effort to include the humanities or to speak to a community outside of science, technology, engineering, and mathematics (STEM).

The report "The Swedish Research Council's Guide to Infrastructures 2012" makes this pattern clear. Crude frequency statistics show that "science" occurs 190 times and "humanities" 36 times, while there are 92 instances of "biology" and no instances of "linguistics" or "archeology." We should be careful not to draw too far-reaching conclusions from this kind of data, of course, but this pattern points to the humanities mainly being dealt with on an aggregate level rather than at a discipline level. There is an effort to mention the humanities, but not very much more, although certain areas such as language technology and digital cultural heritage are given more attention. There is also a focus on databases and data. The report states that the "most urgent need for infrastructures in the humanities and social sciences is that of creating better conditions for research using databases and registers" and that a "fundamental aspect of the infrastructure for research in the humanities is access to digital data." Even including this particular idea of humanities infrastructure, there are typically very few resources allocated to the humanities. Looking at the Swedish Research Council infrastructure grants for 2013, one out of 27 grants was given to the humanities (The Swedish Research Council Project Database). This grant was unsurprisingly given to the Swedish node of CLARIN, which is a European platform for language technology, and is also clearly labeled as a social sciences project. The same year, incidentally, the Swedish Research Council made a principal decision not to continue to support a number of humanities-based research print journals (*Svenska Dagbladet*, November 15, 2013). There were two principal arguments put forward. First, it was argued that other fields than the humanities and social sciences do not have this kind of support. Second, a comment was made that these journals may actually keep the humanities the same rather than develop them. This example shows the effects of trying to have one template for all disciplines and a failure to provide a strong enough infrastructural encapsulation for the journals.

Humanists are not new to academic infrastructure, and indeed, the history of the humanities is embedded in different kinds of infrastructure. The classical concept of *ars memoria* is related to remembering, and more important, to the understanding and interpretation of materials, and as such, is arguably an example of virtual infrastructure for the humanities. The Mouseion in Alexandria, established in the third century BC, is cited by the European Science Foundation report "Research Infrastructures in the Digital Humanities" as an early example of humanities infrastructure. The university is clearly a major infrastructural platform, whose development is intertwined with the evolvement of the humanities (Grendler 2004). At another level of specificity, the domestic study that emerged in the Italian Renaissance is clearly an infrastructural installation connected to humanistic work (Thornton 1997: 8). Humanists were also early adopters of technologies such as the printing press (Burdick et al. 2012: 4).

The relation between humanities affairs and infrastructure can be nicely illustrated by the seminar, which emerged as an institutional activity in the late eighteenth century and early nineteenth century at European universities but, of course, goes back to the Socratic dialogue. The seminar itself could be seen as a type of infrastructure, although it lacks some of the qualities normally associated with the infrastructure turn. For instance, it does not require massive investments, it is conventionally local and not distributed, and it is normally not described as a way of solving major societal challenges. It might be argued that the seminar room is a fairly low-key form of infrastructure that does not necessarily get acknowledged as such. This does not mean that the seminar as organization, space, technology and key intellectual resource cannot be packaged as academic infrastructure or a humanistiscope. If so, it would probably make sense to stress the increasing reliance on remote participants, broadcasting, and shared digital resources. And why should humanists not have enormous touch visualization tables to manage their disparate and complex materials in a collaborative fashion? Another kind of seminar can be exemplified by platforms such as the HASTAC website, which may be easier to package as infrastructure. Such changes would not be the first ones in the history of the seminar. When the seminars were first introduced at German universities, they challenged the strict hierarchy of education because knowledge rather than status was taken into account (Hansson

2007). It would seem quite unlikely that the seminars removed the hierarchies, but they did change the nature of education and they put pressure on infrastructure. For example, the format was not necessarily compatible with traditional lecture halls, and Hansson makes the point that this is why the seminars often took place in the professor's home at this point in time.

Scholarly work on humanities infrastructures in a contemporary context is rare, although there is substantial work on scholarly publishing, libraries, the tools of the trade, and visualization. However, there is a set of work within critical studies and science and technology studies that is highly relevant to the current topic. Some of this work clearly points to the situatedness of infrastructure. As Lisa Parks (chapter 25 in this volume) argues in her discussion of media infrastructure, infrastructure cannot easily be snapshotted. There is complexity, systematicity, and messiness to infrastructure. And as Jo Guldi (2012) shows in her work on Britain as an infrastructure state, the story of infrastructure is one of tension, politics, skyrocketing expenses, and necessity. Chandra Mukerji's (1997) work demonstrates—in wonderful material detail—how the Gardens of Versailles, under Louis XIV's stewardship, became a material site for representing power structures and imaginary worlds. Shannon Mattern (2013) pointedly argues that the idea of big data in the service of urban development is a problematic concept. She discusses the NYU Applied Urban Science and Informatics initiative (placed in Brooklyn) and notes how the "all-white, all-male leadership team, perched high above Brooklyn's MetroTech" sets out to model cities to come up with new solutions and develop new expertise. Mattern's work reminds us of the important work on categorization done by scholars such as Geoffrey Bowker and Leigh Star. Categories are not just "out there," but situated constructs. This is not least true of knowledge infrastructures such as metadata systems. An interesting example is presented by Ursula Heise's (2011) work on biodiversity databases, in which she shows how such databases are subjective constructions full of human decisions, cultural preferences, and storytelling.

Any infrastructure has built-in predispositions. For instance, the Web as a platform is tied to the page as a frame, a predominantly one-screen environment (as opposed to multiple-screen environments), and the encoding structures imposed by HTML. The Text Encoding Initiative offers not only a way to annotate and describe mainly textual materials, but as Dino Buzzetti and Jerome McGann (2006) have pointed out, is also itself an interpretation that among other things imposes a hierarchical structure on texts. The underwater network cables that deliver Web content and other data are intertwined with the content, the industries that use them, and the material and political conditions that govern their placement and routing. The traditional classroom supports a structured learning situation with clear teacher and student positions and expectations embedded in a long cultural and social history. Hence instruments and infrastructures always come hand in hand with ways of perceiving, interrogating, and enacting the world. This is not necessarily a problem and should not keep us from engaging with infrastructure. However, when imagining and implementing humanities infrastructures, we need to incorporate the critical sensibilities that come so naturally when we study other domains.

A Call for Proposals Investigated: Digging into Data

When a science, technology, engineering, and mathematics (STEM) framework is taken to the humanities and social sciences, there will often be attempts at aligning specific tool sets and technologies with the subject areas in question and the starting point will often be the research material or the technology. For instance, it is expected that because of the assumed access to big data, humanists need to engage with it. An example is a 2013 call for project proposals issued by multiple organizations, including the UK-based Arts and Humanities Research Council (AHRC) and the US-based National Endowment for the Humanities (the following analysis is based on the AHRC document "Digital Transformations in the Arts and Humanities: Big Data Research. Call for proposals"). The excerpt below is a description of projects in the most costly funding strand followed by a general comment on what is expected of projects:

> Projects submitted under this strand would need to take a more in-depth approach to their proposed research. They could possibly include visualisations and analysis of big data, creation of new tools and workflows for big data, the assessment of use of high performance computers, creation of artworks and other objects with big data, and may generate new big data. These projects may involve greater collaboration with both academic and non-academic partners and within or between disciplines.

The aim of both strands is to produce innovative, collaborative projects that add value to the digital transformations theme, can potentially have a big impact in the arts and humanities, and raise enthusiasm about the potential of big data to facilitate and support innovative research in the arts and humanities.

On the one hand, this call is quite attractive in that it is open-ended, encourages exploratory work, and seeks to innovate research in the arts and humanities. It is not unlikely that the call may result in some high-quality research projects, and I would be delighted if the Swedish Research Council dared to propose calls of this type. On the other hand, though, there seems to be little substance to the conceptual foundation articulated in the call and the guidance document. There simply does not appear to be an intellectual rationale for why this investment would lead to innovative research or why it is important in the first place. Why do such projects potentially have a big impact? The expectations in relation to the more expensive projects (up to £600,000) as described in the text above seem almost naïve, and there is little focus on the scholarly challenges or a deep conceptual rationale.

Furthermore the call is clearly based on a science model, and the first paragraph of the "Digital Transformations in the Arts and Humanities: Big Data Research. Call for proposals" document, states that some of the most known examples of use of big data come from the sciences. The call cites statistics from the Large Hadron Collider, which is said to produce 15 petabytes of data every year, and points out that a grid consisting of 140 centers in more than 35 countries is used to analyze these data. There is, however, very little discussion of what the collider actually does in terms of facilitating research or tackling research challenges. Corresponding numbers are given for other, humanities-like data sets. The George W. Bush email archive, for instance, consists of 200 million emails (equaling 80 terabytes). It is somewhat troubling that there is no discussion in the call of the perceived objectivity of data or relevant work done in science and technology studies on data. Lisa Gitelman and Virginia Jackson (2013: 4) remind us that objectivity, as situated and historically specific, is the result of "conditions of inquiry, conditions that are at once material, social, and ethical." While it is important to encourage exploratory work and to engage with technology, it seems that the call has significant weaknesses that come from a combination of starting out with the material (generic data), assuming a science model (through a humanities lens), not focusing on research challenges and not incorporating the critical modality that we associate with the humanities. Some of these alleged weaknesses can be explained by the realpolitik of a complex institutional setup across funding agencies and countries and by the attempt to encompass both the humanities and social sciences. Nevertheless, this is where the digital humanities and the humanities more generally should be involved in discussions with funding agencies, and making sure to connect scholarly needs and perspectives with infrastructure.

Academic Infrastructure

Whatever we imagine humanities academic infrastructure to be, we will have to relate to the notion of infrastructure established by the policy makers, funding agencies, and institutions of higher education. I will now discuss this sense of infrastructure and look at some of the assumptions built into the institutional sense of infrastructure.

One assumption is that infrastructure is primarily for research. Research infrastructure, cyberinfrastructure, knowledge infrastructure, and e-science are concepts normally associated with research and particularly science-based research. I use the alternative term academic infrastructure to emphasize that I see infrastructure more broadly. If we believe that education and research are intrinsically connected, we need to think about how infrastructure can serve both needs. This does not apply equally for all infrastructure, of course, but is an important overall sentiment, and something that the humanities should take care to point out when discussing infrastructure. After all, if we look at a key example of humanities infrastructure, the library, it serves both education and research.

Another assumption is that it makes sense to coordinate research infrastructures nationally and internationally. The rationale for this is partly economic. It takes massive resources to fund infrastructure and there is an interest in not duplicating resources unnecessarily. This is not just a question of technological platforms, it also has to do with competence and reaching critical mass. Furthermore coordinated efforts are seen as a way of maximizing the use of expensive equipment, as can be seen in the following EU description ("The European Landscape"):

> Adequate research infrastructures are essential in promoting technological innovation, as they provide the conditions and critical mass required to carry out cutting-edge research. New scientific and technical challenges call for increased performance of research facilities and better knowledge exchange between different disciplines. This increase in capacity and performance can, in part, be achieved through better coordination of existing facilities and the development of simple operational mechanisms. In addition, funding the design and construction of new infrastructures affects the direction of research for many years afterwards.

This text also exemplifies some other key assumptions. Beyond the already discussed notion that research infrastructure relates to science and engineering, it is clear that infrastructure is expected to be expensive, to be used beyond single research groups, and to typically have considerable longevity. Moreover there is an expectation that progress will be made at the intersection of different disciplines and areas, and that infrastructure thus depends on being in between to help facilitate intersectional work. Typically the text also points to how infrastructure is necessary for innovation, producing world-class work and for meeting key challenges. In some ways, infrastructure represents national and pan-national dreams.

If we go back to the examples of research infrastructures listed by ESFRI and discussed earlier, these point to another important assumption about infrastructure. Radiation sources, environmental sciences observatories, imaging systems, genomic data banks, and clean rooms all have a sense of discreteness and unity to them. Infrastructure, in this sense, consists of material installations or platforms that can be described by a name or a phrase. Such platforms are not "given," but the product of packaging and strategic framing as well as function. Arguably, the humanities need to engage with infrastructure on the level of packaging.

Infrastructural Templates

Institutional assumptions about infrastructure as well as our own infrastructural past and our sense of the humanities shape contemporary ideas about humanities infrastructure. This means that humanities infrastructure, as envisioned inside and outside the humanities, relates to certain templates, and I will briefly look at two such templates in this section: a science and engineering model and using existing infrastructure, such as libraries, as a model.

As the earlier discussion of the "Digging for Data" call for proposals demonstrates, there is a strong tendency to relate to and adapt to a scientific and engineering paradigm when conceptualizing humanities infrastructure. Not only are science infrastructure and data sets used as examples, but the call in question draws heavily on ideas associated with science, technology, and engineering, such as big data, instrumentation, and visualization. Another example of this type of alignment is provided by a report from the American Council of Learned Societies (Unsworth et al. 2006):

> Humanities scholars and social scientists will require similar facilities but, obviously, not exactly the same ones: "grids of computational centers" are needed in the humanities and social sciences, but they will have to be staffed with different kinds of subject-area experts; comprehensive and well-curated libraries of digital objects will certainly be needed, but the objects themselves will be different from those used in the sciences; software toolkits for projects involving data-mining and data-visualization could be shared across the sciences, humanities, and social sciences, but only up to the point where the nature of the data begins to shape the nature of the tools. Science and engineering have made great strides in using information technology to understand and shape the world around us. This report is focused on how these same technologies could help advance the study and interpretation of the vastly more messy and idiosyncratic realm of human experience.

There is a clear risk here of adopting a science and engineering based model for humanities infrastructure in such a way that the model significantly constrains and shapes possible research enterprises and directions. Hence it is assumed that much of the structure of infrastructure will be the same, including the technologies, although the objects used by the humanities will be different from the sciences. There is also an assumption in the text excerpt that the objects are static and restricted to specific knowledge domains. However, as Jonathan Sterne points out, "disciplines never fully constitute their objects; they fight over them" (Sterne 2005: 251). Sterne argues that these fights are partly what make disciplines maintain their intellectual vibrancy.

Another relevant question is whether it is at all possible to discern a point like the one referred to by the statement: "only up to the point where the nature of the data begins to shape the nature of the tools." It could be argued that it is naïve to think that there is such a point and that that shaping starts even before one actually starts to use the tool. Geoffrey Bowker argues that databases should be read both materially and discursively as sites of technical, political and ethical work, and that there "can be no a priori attribution of a given question to the technical or the political realms" (Bowker 2005: 123). Furthermore, just like the "Digging for Data" call, the ACLS report advocates a notion of infrastructure very much concerned with data and incorporating as much data as possible. Humanities infrastructure can certainly be about big data, also immensely large data sets, but the science and engineering template seems to strongly suggest a privileging of large data sets in the humanities without adopting a critical perspective and awareness. In her work on data fetishism, Shannon Mattern points to how complex phenomena are reduced to data in an automated fashion, and how these data get reified through visualization and other means. Her position contrasts distinctly with infrastructural visions such as the ACLS report, not least in emphasizing the taintedness of data and associated processes:

> If we gather lots of (mostly well-educated, male) programmers, armed with expensive machinery, and put them in a room with a tank of coffee, their version of "social change" will almost always involve finding the right open data set and hacking the crap out of it. Not only does the hackathon reify the dataset, but the whole form of such events—which emphasize efficiency and presume that the end result, regardless of the challenge at hand, will be an app or another software product—upholds the algorithmic ethos. (Mattern 2013: **https://placesjournal.org/article/methodolatry-and-the-art-of-measure/**)

This is the kind of critical awareness that the humanities need to incorporate into its thinking about humanities infrastructure. For instance, the central role played by the visual in many infrastructural installations must be addressed critically and practically not only in terms of particular installations and current visual predominance, but also in the long-term cultural and historical perspective. Who can do this better than the humanities? And how do we best incorporate critical work along the lines of Mattern's analysis of hackathons to the digital humanities investment in THATcamps and maker labs?

Another important model for humanities infrastructure is existing infrastructure, in particular, libraries and memory institutions. It is clearly evident in the American Council of Learned Societies report "Our Cultural Commonwealth":

> The infrastructure of scholarship was built over centuries. It includes diverse collections of primary sources in libraries, archives, and museums; the bibliographies, searching aids, citation systems, and concordances that make that information retrievable; the standards that are embodied in cataloging and classification systems; the journals and university presses that distribute the information; and the editors, librarians, archivists, and curators who link the operation of this structure to the scholars who use it. All of these elements have extensions or analogues in cyberinfrastructure, at least in the cyberinfrastructure that is required for humanities and social sciences.

While this is probably a fairly valid description of existing infrastructure, it largely leaves out infrastructures outside of libraries, archives, museums, and publication systems. Also it accentuates certain aspects of such institutions and systems. For instance, the library as a situated place for knowledge production is not given any attention.

Thus we are concerned with a library and collection-based model, which admittedly is well in line with a large part of the humanities. It is important to realize, however, that there is a set of epistemic commitments associated with this model—pertaining to structure, delivery, material types, retrieval systems, selection procedures, the relation between researchers and library institution, and other issues basic to the humanities—and that any major new investment in academic infrastructure should not uncritically be based on such existing structures and descriptions. For instance, it would seem that the model presented above makes a fairly strong delineation between the collections (institutions, distribution systems, professional functions involved, etc.) and the researchers and the research community. Johanna Drucker addresses such delineations critically stating, "modelling scholarship is an intellectual challenge, not a technical one. I cannot say this strongly or clearly enough" (Drucker 2009a). In contrast to much of the discourse of academic infrastructure and digital humanities, Drucker importantly focuses on the scholarly challenge and not on the technology or technology-induced visions. And although her point is valid, the challenge is of course both intellectual and technological, especially if we believe that the humanities need to create humanities infrastructure.

The Humanistiscope Revisited

The humanistiscope is a rhetorical device that can help us conceptualize and create humanities infrastructure that meets the combined intellectual and technological challenge referenced above. Needless to say, the idea of the humanistiscope is not an all-in-one-solution, but rather, a tool to help us think about and enact humanities infrastructure. In the sciences the notion of platforms is often used in a similar way, but as we will see, the suggested neologism offers more precision. Platforms tend to refer to set of instruments rather than a single apparatus and notions such as apparatus and instrument already come with much meaning. In any case, I argue that the humanistiscope can push against the humanities in productive ways. The common sensibility that science instruments are materially and conceptually defined can be used to push the humanities to think about their infrastructure in terms of material, technological, and spatial situatedness. At the same time, the humanistiscope as a thought piece can make it easier for the humanities to draw on their own critical work on instruments and apparatuses, notably in science and technology studies, in order to push against their own imagined and implemented infrastructures. Such critical work often stresses the cultural, social, and political situatedness of infrastructure.

Importantly, the humanistiscope should not be seen as independent of current discourse on research infrastructure, but rather, it offers a humanities-based way of productively aligning with and tweaking such discourses. This means that there will necessarily be a relation to national and international research policies and agendas. The point is not to oppose these generically, but rather, to articulate infrastructural models that are built on humanities-based needs and challenges. I will now use an extended example to illustrate some of these points and to discuss what makes a humanistiscope.

During a medievalist conference at the Faculty of Arts at Umeå University in the spring of 2013, a few sessions took place in HUMlab, a digital humanities studio. It turned out that the visual capabilities of the lab environment were quite useful compared with the traditional lecture halls in the humanities building (with relatively small, single, and upfront screens). Research infrastructure was also discussed during the session and, in particular, after one of the presentations. This lecture described a project that made use of a number of discreet screens organized around the space (a screen scape) to enact a pre-print experience of church spaces in relation to the Virgin Mary as a virtuous role model in medieval Sweden (Lindhé, chapter 14 in this volume). One question that came up several times concerned the portability of the installation. When would it be available on the Web? While this is a relevant question, it also demonstrates a lack of awareness of the situatedness of infrastructure. This particular installation was heavily grounded in the particular physical and digital infrastructure of HUMlab, and it cannot foreseeably be moved to a Web environment. Aspects of the installation could certainly be transferred, but that would result in a different sensory experience and knowledge infrastructure. This example points to a tendency to think about infrastructure as placeless, immaterial, and neutral in the humanities. For example, the Web is usually a default platform for digital humanities projects, but there is little critical discussion of the material constraints and conditions associated with this platform and other platforms are rarely considered.

The history of instrumentation has a clear material component, and the humanistiscope borrows a material sensibility from its science counterparts, such as microscopes and sonoscopes, which can help give humanities infrastructure a material and conceptual definition. There arguably has to be a certain degree of configuration and unity to infrastructure. If there are many parts and components, these can be incorporated as long as the heterogeneity is not foregrounded discursively at the expense of the integrity and unity of the infrastructure. Indeed having a fairly discrete definition would seem to be a useful criterion for categorizing something as a humanistiscope. This adds a certain degree of conceptual and material precision, and as such may help the humanities to be both more intellectual and material about their infrastructures and avoid some of the airiness of much of infrastructural discourse. Importantly, building infrastructure is not just a matter of starting out with an intellectual foundation and finding or making infrastructure to implement it (or vice versa), but rather, it is an iterative process. Infrastructure is intellectual, in this sense, and it will shape and condition the research that engages with it.

The preceding discussion and the earlier example of the medievalist conference point to another criterion for humanistiscopes, namely that they cannot to be too specific or too general in terms of function. It would not be possible to have one humanistiscope that covers most research needs in all the humanities disciplines. This piece of instrumentation would be a black box and would not be materially and conceptually well defined. However, humanisticopes must not be too specific or they do not have the applicability and range that we would associate with infrastructure. This is the case with the Virgin Mary installation as it is restricted to one project and associated research questions. It would seem to have too little generalizability to qualify as a humanistiscope in itself. The screen scape, on the other hand, can be used with multiple projects and has a considerably wider range. It also has specificity through having an ideational and material basis. The principal conceptual basis of the screen scape is to allow interrogation of complex cultural and historical matters through multiplex visualization and enactment. In some ways the screen scape challenges predominant visualization infrastructures for science and technology, where there tends to be a focus on immersive environments such as CAVEs as well as very large and uniform display walls (Svensson 2011). There is a humanistic sensibility in the conceptual basis, but also in the way the screen scape has been put together and the underlying technological systems. And beyond the technology, one of the most important parts of the installation is the centrally placed seminar table.

The materiality of infrastructure can naturally be both physical and digital. The screen scape has a strong physical presence but is also digital, whereas an online retrieval system may be mostly digital. Let me now use another example to discuss a mostly digital humanistiscope. Again, I draw on a project based at HUMlab at Umeå University.

The faceted browsing system was developed in a European Union funded project begun in 1997. It was created to manage specific archival data sets and their ontologies, but it came to be expanded over time. In essence, the system allows navigation of rich and complex data sets through selecting and defining facets. These facets determine what will be produced in terms of tables and visualizations in a data view window. Materials can be searched and delimited through the actual data view too and, for instance, map-based navigation can be used to filter data. The facets are shown through small windows or applets in the browser. The basic idea is simple and powerful, and the system has come to be used with many other funded projects and has continued to be developed over time. While it might not have been generic enough to start with, the system would now seem to be a good candidate for being called a humanistiscope. It has a material definition through the Web application and it has a conceptual basis. It is not tied to an individual data set or project anymore, and it is not an all-in-one generic solution.

Beyond a Single Humanistiscope

There is not merely one humanistiscope, but many. And they can be interrelated. For instance, in 2013, HUMlab finished installing a large floor screen and an adjacent triptych screen (a screen with a large central part and two slanting side parts). The triptych screen draws on a long history of triptychs and configurations of three, and the floor screen challenges conventional orientation and placement of screens. The screens are placed on the Umeå Arts Campus, and this situatedness plays an important factor. The exact position, inside a massive glass wall with high visibility for the whole Arts Campus, is meaningful. While the two screens and associated sensoring and interaction technology can be used separately, it would seem to make sense to consider both as one humanistiscope in most cases. They complement each other, and are close enough to be read visually and conceptually as one unit.

Mostly physical infrastructure, such as floor and triptych screens, can also be combined with mostly digital humanistiscopes. The faceted browsing system discussed above has always been a Web application, but in the fall of 2013, it was deployed on the floor-triptych humanistiscope. The idea was that a Web browser client imposes a number of constraints, one of which being that normally content is displayed across only one screen. In the new iteration, the content (data windows) can be shown on the floor screen, while the facets are shown on the triptych screen. Since a very commonly used data view in the system is cartographic visualization, the floor screen seems quite suitable as we often look at maps from above. The triptych screen is used for the facets, and active facets are shown centrally on the screen, while inactive facets are shown in the periphery (using the slanted side screens). In this way we have two humanistiscopes, one mostly physical and one mostly digital, interacting.

Unless the humanities come up with individual humanistiscopes of very large reach and critical mass in terms of researchers, it is likely that there will be a number of separate humanistiscopes. In the sciences these are sometimes subsumed under a platform or a laboratory. I suggest that a particularity of humanities infrastructure is that it is likely to be multiplex to accommodate different scholarly and educational needs. Major science infrastructure, in contrast, tends to be seen as more specific in terms of relating to certain projects, questions, or even certain problems. This may be an issue of scale in the sense of the sciences having a much larger infrastructural and operational footprint, but it may also be a matter of the nature of humanistic issues and endeavors. The humanities deal with ambiguity, complexity, layering, theorizing, and a wealth of perspectives. While a science-oriented infrastructure such as the immersive CAVE is generic, it would seem to be built around a set worldview, one that is not normally problematized and that does not easily lend itself to the theorization of the very premises of the infrastructure.

The humanistiscope can help us package things together, but only up to a certain point. In some cases we also need a larger configuration that allows the bringing together of different humanistiscopes. In the same HUMlab space as the triptych and floors screens, there is also a portal room that allows multiple-party Skype meetings and wall-based visualizations. This is also a humanistiscope, and it would not make much sense to group it too tightly with the other screens. It has its own conceptual grounding and function. This does not mean, however, that these humanistiscopes are unrelated. They are part of the same platform or set of co-aligned humanistiscopes, and the rationale for bringing them together is not least related to management of resources. Just like in the sciences it is important to maximize the use and accessibility of infrastructural resources, have a professional organization around them, and allow for crossover effects.

There are many models for accomplishing this clustering of humanistiscopes, including network organizations, lab environments, and open access platforms, and I will end this chapter by briefly considering the humanities lab or digital humanities lab as such a model. A digital humanities lab or studio would seem too large and heterogeneous to be seen as a single humanistiscope. It can usefully bring together humanistiscopes, competence, scholars (inside and outside the humanities), networks, and an investment in the humanities and the academy. Most labs are seen as physical installations, although they are digital as well. It is quite possible to imagine mostly digital humanities labs, but there is also value in the physical situatedness of physical installations. For one thing, such installations allow both physical and digital humanistiscopes to coexist in the same operation. Such infrastructures can make other things possible, not least, such things that do not work well within current, distributed infrastructures. Localized infrastructure can also be worthwhile at a time when we see increased digitalization of higher education and a distinct pressure on space. It is also a good use of space and a way of increasing the attraction of the physical campus. The website for a new Swedish and European infrastructure, the European Spallation Source (ESS), states "[e]ach year an estimated two to three thousand visiting scientists will come to ESS to perform experiments." Moreover having a lab can be useful in other ways. It can be a place and culture for people and dialogue, house a number of different humanistiscopes (different methodologies and tools), integrate those as well as seminar tables, exhibition space, studio space, and so forth, and allow for multiple points of interaction and for making and critical work (essential to digital humanities). Additionally a digital humanities lab can enable and facilitate new research and challenge the disciplines, manifest the operation (be somewhere to take people and to channel resources), have strong technological engagement, and push technological imagination; it can develop the humanities (knowledge production, work modalities, making, and reaching out), and it can manifest the university as a place for dialogue, intellectual rigor, excitement, and reaching out.

Luckily, the question of humanities infrastructure is not an either-or question, and we will continue to have a mix of different models. What is most important is to connect infrastructures to ideas about the humanities and what intellectual challenges we want to tackle. Infrastructure is a question of social, cultural, and political situatedness as well as a question of material and conceptual situatedness. I have argued that a device such as the humanistiscope can help us do the job. Instead of infrastructure to preserve traditional modes or a wholesale adaptation of infrastructures produced for other communities, we might invite a more nuanced, complex, kind of treatment in order to help us dream up all kinds of yet unseen things for which we are the core.

Note

The humanistiscope as an idea for this chapter came up in my discussions with Erica Robles-Anderson in 2012.

25 "STUFF YOU CAN KICK": TOWARD A THEORY OF MEDIA INFRASTRUCTURES

Lisa Parks

The word "infrastructure" emerged in the early twentieth century as "a collective term for the subordinate parts of an undertaking; substructure, foundation," and first became associated with permanent military installations (OED). Since then the term's meanings have expanded to include electrical grids, telecommunication networks, bridges, subways, dams, sewer systems, and so on, and infrastructures have been the topic of research in fields such as Urban Studies, Communication, Geography, and Science and Technology Studies.[1] In digital humanities scholarship, researchers have explored the topic of "networks" developing important historical and critical studies of networked technologies, institutions, corporations, and cultures.[2] Fewer, however, have investigated the physical infrastructures through which audiovisual signals and data are trafficked.[3] By physical infrastructure I am referring to the material sites and objects that are organized to produce a larger, dispersed yet integrated system for distributing material of value, whether water, electrical currents, or audiovisual signals. Engineers often refer to infrastructures as "the stuff you can kick." Such stuff is typically relegated to the fields of electronic or civil engineering or urban planning and is thought of as irrelevant to or beyond the purview of humanities research.

In an effort to develop a humanities-based approach to the study of infrastructure my recent research has combined approaches from phenomenology, cultural geography, and object studies to explore the sites, objects, and discourses that shape and inform what might be called *infrastructural imaginaries*—ways of thinking about what infrastructures are, where they are located, who controls them, and what they do.[4] By exploring such topics as the endpoints of cable television systems, the locations of cell phone towers, and the territories of satellite and wireless footprints, I have tried to develop a critical methodology for analyzing the significance of specific infrastructural sites and objects in relation to surrounding environmental, socio-economic, and geopolitical conditions.[5] This critical methodology has involved site visits and physical investigations of infrastructural objects using personal observation, photography, maps, video, art, drawings, and other visualizations. These observations and mediations are intended to foster infrastructural intelligibility by breaking infrastructures down into discrete parts and framing them as objects of curiosity, investigation, and/or concern.

While most research on physical infrastructures has taken place beyond the bounds of the digital humanities and media studies, these fields have evolved a rich body of work on screens, interfaces, and networks.[6] To build upon this work, this essay explores "media infrastructures"—-the material sites and objects involved in the local, national, and/or global distribution of audiovisual signals and data. Media infrastructures include phenomena such as broadcast transmitters, transoceanic cables, satellite earth stations, mobile telephone towers, and Internet data centers. I use the term "infrastructure" as opposed to "network" for several reasons. First, the term infrastructure emphasizes materiality and physicality and as such challenges us to consider the specific locations, installations, hardware, and processes through which audiovisual signals are trafficked. Second, the term infrastructure helps to foreground processes of distribution that have taken a back seat in much humanities research on contemporary culture, which has tended to prioritize processes of production and consumption. Third, since it refers to physical sites and objects that are dispersed across vast territories, the concept of infrastructure can encourage digital humanities' further interdisciplinary engagements with fields such as environmental studies, geography, and science and technology studies. Finally, an infrastructure is difficult to visualize in its entirety within a single frame and as such can help to stimulate new ways of conceptualizing and representing what processes of media distribution are, where they are situated, and what kinds of effects they produce. In sum, this concept can bring a renewed focus upon critical issues of materialism, distribution, territoriality, and conceptual visualization.

Broadly, then, the term "media infrastructure" refers to the material resources that are arranged and used to distribute audiovisual content. These resources extend far beyond the studio and the screen, and include raw materials such as the sun, electricity, land, water, petroleum, chemicals, heavy metals, plastics, and spectrum. Without these resources, film and television as we know them would not exist. As Nadia Bozak (2011) observes in her book, *The Cinematic Footprint,* "accessing images at all means tapping into a complex system of resources" (2). She continues, "Images, however intangible or immaterial they might...appear to be, come bearing a...biophysical make up and leave behind a residue—a "cinematic" footprint, as it were" (8). One of the goals of Bozak's book is to "expose the energy requirements, economy of obsolescence, and...lingering afterlife of digital technology concealed behind the crisp, clean infrastructure that supports binary-based images and information" (12).[7] While Bozak focuses on what she calls the "resource-image," I am interested in the resources that are required to distribute audiovisual content around the planet and the layered infrastructural "footprints" (whether carbon or territorial) that emerge as a result.[8]

To begin building a critical vocabulary for the study of media infrastructures, this chapter analyzes media representations of three infrastructural objects—mail sorters, power poles, and satellite dishes—which are part of three media infrastructures: the postal system, electrical grid, and global satellite system. My analysis is intended as a critical provocation rather than a detailed historical study. It draws upon short historical films, a media art project, news media, and film segments, all of which are archived online, as evocative platforms for conceptualizing media infrastructures and demonstrating the kinds of concerns that can emerge through an infrastructural analysis. Each of my examples features human workers as part of infrastructures and, as such, reinforces critical theories of the post-human and historical studies of technologized labor that conceptualize human-technology relations as "integrated circuits" or as part of biotechnical or technosocial formations.[9] I argue that since infrastructures cannot be captured in a single frame, we must read media with an *infrastructural disposition* – that is, when viewing/consuming media we must think not only about what they represent and how they relate to a history of style, genre, or meaning but also think more *elementally* about what they are made of and how they arrived. While the examples I discuss represent infrastructures quite literally, I want to suggest that all images can be read with an infrastructural disposition, that is, with questions of resources and distribution in mind. Even when infrastructures are not visible at all in the frame, it is possible for them to be inferred and imagined.

Mail Sorters

Media infrastructures are not just a product of the most contemporary technological formations of the digital age; they should be thought about in an historical and intermediale sense. That is, media infrastructures demand a consideration of the ways that distribution processes have emerged, changed, and been layered upon one another over time, how they are part of a media archaeology.[10] To advance this logic, I begin with a short film dated August 7, 1903, produced by American Mutoscope and Biograph Company entitled "Throwing Mail into Bags," which offers a fifty-five second glimpse of activity within a US postal service sorting center in Washington, DC.

Three postal workers can be seen coordinating their labor. One brings in large bundles of mail and drops them off on a table as two others quickly pick up individual parcels, visually scan them, and toss them with astonishing precision into delivery bags arrayed in a semicircle in the center of the room. Another film from the same company entitled "Carriers at Work" dated August 22, 1903, features a similar though slightly longer sequence (one minute and fifty-eight seconds) in which four workers bustle around a postal sorting area, rapidly sort mail

Figure 25.1
Screen capture from the 1903 film "Throwing Mail into Bags" featuring US postal workers sorting mail into geographically zoned delivery bags

into a bank of mailboxes, stack and stamp letters, and move them to a table where a sorter picks them up and flings them into geographically zoned delivery bags.

The tableau perspective and slight high angle of the camera in these films present the postal sorting room as if the side of the building had been sliced open. As cinematography and mise-en-scène spotlight the sorting process with a tableau perspective they also introduce an arrangement of space, objects, and movement that is hardly ever seen by most people who receive mail. The films represent a tiny sliver of the physical infrastructure by which postal mail once moved into individual mailboxes in the United States and invites the viewer to imagine his or her own letters transiting through this same sorting process. In doing so, the sequences address the viewer as a citizen of infrastructure—as one who uses, subsidizes, and recognizes the infrastructure of postal distribution.

While these films represent the practice of mail sorting at the turn of the twentieth century, when viewed retrospectively, they also anticipate the current era of Internet servers and cloud computing, an era in which use of the human mind and muscles to sort mail into delivery bags has been supplanted by automated packet switching and remote data storage and processing. Yet even this more contemporary level of automation requires hands-on engagement as manual laborers regularly service the machines responsible for the routing of information. In this sense these early films serve as an historical metonym for content distribution as they visualize and model the physical movement of media (mail) to a portal through which the mail is scanned and sorted into separate containers that correspond with different zoned areas that are linked to distinct locations. What we do not see or know, however, is where the bundles of mail came from and where exactly they will end up. It would be impossible to present a photorealist view of an entire postal infrastructure within a single frame, so we are given a part of it and invited to use this part to infer and imagine the rest of it. Such a view can be used to activate an *infrastructural disposition*—a disposition toward audiovisual media that approaches what is framed as a starting point for imagining and inferring other infrastructural parts or resources, such as the vehicles, horse power, electricity, and petroleum used to transport bundles of mail, the human labor required to lift and carry parcels, the paper on which the letters were written, or the time it takes for a letter to travel through this system.[11] What I am suggesting is a way of engaging with media that not only involves questions of documentation or representation, but one that fosters *infrastructural intelligibility*—a process by which ordinary people use images, sounds, objects, observations, information, and technological experiences to imagine the existence, shape, or form of an extensive and dispersed media infrastructure that cannot be physically observed by one person in its entirety.

By isolating moments in which content is *in the process of moving from one site to another*, these short films foreground the physicality of distribution and the dynamism of media infrastructure. In doing so, they challenge us to imagine other forms of content distribution in the present. For instance, how would one visualize contemporary phenomena such as packet switching or cloud computing, phenomena that occur at scales and speeds that neither the human eye nor the medium of film can readily bring into view? The closest contemporary analogies to these early films of postal sorting might be videos of technicians working inside Google's or Facebook's data centers that have been uploaded to YouTube ("Inside a Google Data Center" 2009). While it might be tempting to imagine and represent packet switching or cloud computing as fully automated and digitized (as advertisements often do), even the cloud relies on human labor. Cloud computing requires people to organize, install, and maintain equipment at huge data centers, to develop applications that manage and monitor transactions that occur "in the cloud," and to secure and clean the buildings in which data centers are located.[12] Though millions of people on the planet now receive email in their inboxes each day, few have ever seen inside the enormous data centers that host servers and clouds. Not only are most consumers socialized to have little interest in the "back ends" of the infrastructures that they use each day, these data centers are so thoroughly privatized or militarized that they are typically secured away from public view. Within this context, these historical films are instructive.

Figure 25.2
"Vertical classroom" as explored in Michael Parker's 2009 multimedia installation, "Lineman." Photo by Michael Parker, courtesy of Michael Parker.

Besides providing opportunities for intermediale analysis, they prompt further consideration of the kinds of infrastructural views and details that we, as citizens of infrastructure, should demand access to.

Power Poles

While mail sorters can be thought of as an historical metonym for content distribution in the digital age, power poles mark an infrastructure of equal significance. My second example focuses on contemporary electrical infrastructure in the United States. To explore it, I engage with the work of media and performance artist, Michael Parker, whose multimedia project, "Lineman," features the technicians who install and rig the electrical power poles that are strung throughout urban and rural landscapes across the country. In 2009 Parker trained with forty-nine electrical linemen at the Los Angeles Trade-Technical College during a sixteen-week course and collaborated with them to produce videos, photographs, and a newspaper-like yearbook, some of which are archived on the Internet.[13]

Parker's project set out to capture the linemen's weekday training exercises from 6:45 am to 2:30 pm, five days per week, in a pole yard known as the "vertical classroom." The linemen's

Figure 25.3
Linemen trainees carrying a heavy wooden pole to a site so that it can be installed into the ground and mounted with electrical equipment. Photo by Michael Parker, courtesy of Michael Parker.

training included technical pole climbing, gloved and blindfolded knot tying, removal and replacement of poles, mounting of cross-arms, and wiring setups. As one lineman aptly put it, this job is about being "connected to all that is out here that we can't see.... It's power" (*Lineman* 2009: 21). Many of the trainees with whom Parker collaborated were Latino and African American men who grew up in South Central or East Los Angeles, and some had been recently laid off from other jobs during the US recession that began in 2008. Parker documented their work using photography and video and featured the linemen working both together and individually.

One photo shows eighteen trainees in hard hats carrying a log pole to its destination before it is turned upright and lodged in the ground for rigging. Another photo reveals a trainee at the top of a pole, dangling backward and outstretched in his harness while attempting to install an insulator on the crossbow. While most of the photographs documented training exercises, one photo appears to have been carefully staged by the group at the end of the course and shows the forty-eight linemen vertically arrayed along the sides of three poles. Each lineman is leaning out from the pole wearing a safety harness, spiked boots, gloves, and a hard hat while waving to the camera. The photo transforms an infrastructural object and class of workers that usually go unnoticed into a lively spectacle, inviting the viewer to recognize and celebrate the human resources that support the electrical grid.[14]

Figure 25.4

Staged photo of linemen trainees, that is, the human capital supporting the construction and maintenance of the electrical grid. Photo by Michael Parker, courtesy of Michael Parker.

Figure 25.5
Linemen pass around a camera designed by Michael Parker to reveal the world as seen from an infrastructural perspective. Photo by Michael Parker, courtesy of Michael Parker.

To enable the linemen to record videos during their training exercises, Parker devised a special camcorder mount that could be carried up the pole and handed off safely from one lineman to another. A video projected as part of Parker's multimedia installation opens with several wide shots of the pole yard where linemen are practicing climbing up and down the poles, and then transitions to several short segments shot by linemen from the top of the poles. A split screen is used to convey several camera hand-offs. As the camera is passed from one lineman to another, the video emphasizes the linemen's connection to and integration with the pole and constructs what might be described as an *infrastructural perspective* in that the viewer can see, and indeed imagine, the world from the vantage point of a power pole. This perspective is characterized by high angles and bird's-eye views, as well as the sounds of safety belts squeaking against the wood when they are raised and lowered, boot spikes piercing the poles, and the atmospheric noise of the wind and environs. One of the linemen eloquently alludes to this perspective in his yearbook comments: "...this atmosphere of electricity wants me there with it.... The electricity wants me with it. It wants me to join with it.... All those voltages, watts; it's power, pure power and I feel like I *am* that, like I'm wrapped up in that" (*Lineman* 2009: 21).

There are approximately 270,000 linemen employed in the United States, but we rarely see the world from the vantage points of the linemen or electrical poles that make our digital culture possible (Bureau of Labor Statistics). As Parker's project visualizes the training and physical labor involved in mounting and rigging electrical poles, it also draws attention to the objects that are so common in the built environment—power poles, cables, and metal fixtures—that we scarcely notice them, and challenges us to think about their position, function, and value. By revealing workers' bodies moving up and down these poles, Parker's project also dramatizes the process of infrastructural formation and encourages the viewer to recognize that the distribution of electrical power is contingent not only on the training of personnel to install and rig electrical poles but also on particular arrangements of resources such as lands, trees, and cables. As much as "Lineman" celebrates the dexterity, strength, and coordination of electrical linemen, it also forces the viewer to confront vital questions about the energy and resource requirements of the digital age. How much energy is used to produce and distribute videos, television shows, and films around the world each year? How much electricity is used to power the television sets, theaters, computers, mobile devices, and networks that people use to access these media? In 2011 CNN published a story with the headline: "The Internet: One Big Power Suck," explaining that the electricity needed to power the millions of servers that support the audiovisual streaming on the Internet has increased 10 percent each year over the last decade (Hargreaves 2011).

Historically, power poles have been strung together across the environment to energize film theaters,[15] radio stations, television networks, and satellite earth stations; today, they are also powering Internet server farms and mobile telephone networks. By honing in on power poles and linemen, Parker's project sharpens our focus on what can be understood as the underbelly of modern media—that is, the extensive, patch-worked, and varied electrical infrastructures that undergird world processes of mediation. Parker's project is particularly meaningful because it brings this process of mediation full circle—that is, his conceptually driven artwork uses electrical energy to transform views of power poles and linemen's labor into digital media that are powered by the very electrical currents that transit through the power poles that the linemen have learned to install, climb, and rig. Parker's scenes of linemen can in fact be understood as representing a kind of below-the-line labor, or as part of "pre"-production cultures, in that they draw attention to the electrical infrastructure that is a necessary precondition of digital media production, distribution, and/or consumption.[16]

Satellite Dishes

Just as quickly as physical infrastructures are built, they can be susceptible to attack, sabotage, or destruction, whether by a computer virus, an air raid, a lightning strike, or a repressive state. My final example of infrastructural analysis draws upon a series of photographs and

Figure 25.6
Iranian special forces removing satellite dishes from an apartment building in Iran. Photo by Hamid Forootan, courtesy of Hamid Forootan.

videos of police confiscating and destroying satellite dishes in Iran and of Iranians manufacturing and installing them. If the case of the mail sorters helps to make infrastructures intelligible and the case of the power poles enables an infrastructural perspective, then the case of satellite dishes in Iran brings *infrastructural contestations* into scrutiny. As Hamid Naficy discusses in his *Social History of Iranian Media* (2012), satellite dishes have been banned in Iran since 1994 because the government claims that they import negative influences from the West. Despite this official ban, dishes, Naficy explains, have "cropped up everywhere" and become a "public obsession" (2012: 345). It is estimated that 65 percent of Tehran's residents use satellite dishes, and 30 to 40 percent of people use them in religious cities such as Qom (Esfandiari 2012). Since 2009 the Iranian government has ramped up efforts to enforce the ban, deploying state special forces and police to destroy, remove, and/or confiscate satellite dishes and receivers throughout the country. In May 2011 police confiscated more than 2,000 dishes in a single day in Tehran, and during October 2011 Iranian police seized more than 6,000 dishes in the Mazandaran province in northern Iran ("Iran Police" 2011). Another aggressive round of dish raids occurred in Tehran in February 2012. Organizations such as Iranian Student News Agency (ISNA), Mission for the Establishment of Human Rights (MEHR), and the Iranian police have photographed and videotaped these satellite dish removals and circulated them online, where they have triggered a range of responses to Iran's restrictive communication policies.

Figure 25.7
Iranian special forces officer stomping on and destroying a satellite dish. Photo by Hamid Forootan, courtesy of Hamid Forootan.

Some of the photos, which have also been published in state newspapers as warnings to satellite dish users, feature Iranian state special forces suspended from ropes while removing satellite dishes from the facades of high-rise apartment buildings, or exiting buildings with confiscated satellite dishes or receivers (Esfandiari 2012; Foroutan 2011). In other photos police can be seen bending, stomping on, or pummeling satellites dishes so that they can no longer be used, exemplifying the brute force the state has applied to terminate citizens' access to satellite infrastructure and international signal traffic. Still other photos reveal ruined dishes scattered in parking lots next to police cars, piled up in the back of trucks being hauled away, or in mounds on the sides of streets, serving as telling reminders of their illegality.[17]

As the most visible part of satellite infrastructure, the satellite dish functions as a peoples' portal to an expansive and complex global media system that is based on the organization of resources ranging from heavy metal deposits to building facades, from bandwidth in the

Figure 25.8

Iranian police officer carrying destroyed satellite dishes to his car. Photo by Hamid Forootan, courtesy of Hamid Forootan.

Figure 25.9

Pile of satellite dishes and media equipment confiscated by Iranian police. Courtesy of Hamid Forootan.

electromagnetic spectrum to slots in the geostationary orbit (Parks 2012: 64–84). Photos of satellite dish removal in Iran are significant because they highlight the kinds of contestations that occur at the edges of this global infrastructure, through the sites and objects where it interfaces with publics. These photos also serve as reminders that media infrastructures are fueled by biopower: that they are dynamic assemblages subject to practices of localization, contestation, intervention, and control that can vary worldwide. When the satellite dish is forcibly destroyed, it becomes an object of even greater attention, investment, and affection.

Just as soon as satellite dishes are destroyed in Iran, they are rapidly replaced. Iranians report buying and installing new dishes right after old ones are removed and trying to position them so that police cannot see them. They pay, on average, USD$150 for a satellite dish and its installation. A 2008 documentary film entitled *The Dish* details the risky job of installing satellite dishes in Iran, which typically occurs on high-rise apartment buildings at night. The film also reveals that satellite dishes are manufactured illicitly within Iran. One sequence features a makeshift operation where old pots and pans are melted down, reshaped in the form of satellite dishes, spray-painted gray, placed in car trunks, and delivered to locations throughout the country. In another sequence, a man on a motorcycle delivers a satellite dish to a customer in a remote location who uses a generator to power his family's television set, satellite receiver, and sound system. After the dish is installed, the man of the household appears overjoyed to be able to receive international music television channels in his home, despite the fact that people caught with satellite dishes in the area are reportedly beaten, whipped, or banished from the community (*The Dish* 2008).

Another film by Iranian filmmaker Saman Salour entitled *Lonely Tunes of Tehran* (2008) features a war veteran and former radio communications operator named Behrooz who finds himself back in Teheran where he encounters a long lost cousin and former telecom engineer named Hamid. Both unemployed, the two decide to start an illegal satellite dish installation business together. They retrieve satellite dishes from hiding spots in the city, climb up high-rise apartment buildings, and bicker or contemplate the meanings of life while working, often in the dark of night.

In one segment the two characters appear standing side by side on a Teheran rooftop with their bodies half occluded by the two satellite dishes, just as integrated with satellite infrastructure as the linemen are with the electrical grid. As the sequence develops, they discuss Behrooz's romantic life and their shadows are projected onto the surfaces of the two satellite dishes as the massive state TV tower looms in the distance. This clever mise-en-scène efficiently brings a constellation of infrastructural tensions into frame, contrasting the state's centralized and highly visible control over media infrastructure with Iranians' clandestine use of satellite dishes to downlink signals from elsewhere. The dish, according

Figure 25.10
Screen capture from *The Lonely Tunes of Teheran* featuring the two lead characters eclipsed by satellite dishes while standing on a rooftop. Reproduced under the Fair Use Doctrine.

to the film's narrative logic, is a way of contending with solitude, a mechanism for making life in Teheran feel less lonely.

These images resonate powerfully with Cristina Venegas's (2010) discussion of what she describes as the "human rooftop antennas" featured in Cuban filmmaker Fernando Perez's 1994 film, *Madagascar*. Venegas suggests this "inspirational image encapsulates an ethos of Cuban identity in the 1990s as one adrift and in search of reinvention and connection" (43–44). This "antenna body," Venegas continues, "provides a metaphorical interface for new pathways of information," and articulates a self that "exceeds national definition" and is "calling out to be heard" (44, 53). Iranian citizens have found themselves in a similar position in recent years in light of Iran's rigid controls over political expression and information flows. In 2009 Iranians too were "calling out to be heard" when they loudly protested the results of their country's presidential election and forged communication pathways via cellphone and Internet to expose state violence and corruption to the world. Understood in this context, the illicit making, mounting, and using of satellite dishes in Iran is part of a broader set of infrastructural contestations in which ordinary people refuse to surrender the technologized power to communicate—to produce, send, and receive audiovisual signals—to the state. As Naficy puts it, "the desire to be in touch with the world and to defy the Islamic Republic's isolation and censorship" is "a key reason for Iranians' love affair with satellite TV" (2012: 347).

Conclusion

In focusing on representations of mail sorters, power poles, and satellite dishes, I have tried to show how audiovisual media can be read *infrastructurally*—that is, to evoke infrastructures that cannot be reduced to the frame or perceived by one person in their entirety. Though we live in a digital age and processes are increasingly technologized, not all infrastructures are fully automated and not all labor is immaterial. The examples I presented corroborate the persistence of bodily acts in media infrastructure, whether tossing mail into bags, climbing and rigging power poles, or hanging from a roof to knock out a satellite dish. A theory of media infrastructure would account not only for the bodies of actors that appear on screen but for those involved in *supporting acts* such as the trafficking of content, the flow of electrical currents, and the policing of audiovisual signals. While in this chapter I used mediations of infrastructure to draw attention to such processes, these processes are often invisible and their relation to digital humanities and media studies research is under analyzed. A theory of media infrastructures would also need to draw attention to the biophysical resources required to make those acts possible, the sites, materials, and objects that have been organized to move signals throughout the world, whether via the fanned arrays of bags in the mail room, the trees of which power poles are made, or the aluminum used to make a satellite dish.

In foregrounding such objects and materials, I hope to suggest the need for further research on media infrastructures in different local, national, transnational, and non-Western settings. While communication scholars have provided historical analyses of the rise and dominance of Western telecommunication networks, we know relatively little about the historical processes by which media infrastructures have emerged in different parts of the world. For instance, when and where were broadcast transmitters, cellphone towers, satellite earth stations, transoceanic cables, or Internet data centers installed in certain regions and why? Where did the labor and materials needed to build those media infrastructures come from? What are the specific local, national, and/or global implications of these physical installations? While pondering such questions, it is important to remember that "infrastructure" means different things across cultures, and infrastructural sites and objects often take on distinct forms or physical characteristics as they are scaled and adapted to local economic, political, cultural, and environmental conditions. Given such considerations, critical studies of media infrastructures should engage with theories of difference, critiques of knowledge/power, analyses of geopolitics and processes of territorialization. My final point is that to fully appreciate media infrastructures, it is important not only to analyze how infrastructures appear in media culture but also to visit infrastructure sites and objects, witness the infrastructural construction processes, interact with infrastructure workers, and get as close as possible to these massive and dispersed things that always feel so unintelligible or so far away. A theory of media infrastructure, in other words, needs to be formulated not only through the frame but also through the body and from the ground up.

Notes

Earlier versions of this chapter were presented at the Epistemic Engines conference at UC Irvine, the Backward Glances conference at Northwestern University, and at the American Studies Association, and Society for Cinema and Media Studies conferences. I am grateful to attendees for their comments and questions. I thank Michael Parker for his willingness to discuss his "Lineman" project with me and for sharing access to project materials, and David Theo Goldberg and Patrik Svensson for their helpful editorial comments.

1. See, for instance, Leigh Star, "The Ethnography of Infrastructure," *American Behavioral Scientist* 43 (1999): 377–91; Stephen Graham and Simon Marvin, *Splintering Urbanism, Networked Infrastructures, Technological Mobilities, and the Urban Condition*, London: Routledge, 2001; Kazys Varnelis, ed., *The Infrastructural City: Networked Ecologies in Los Angeles*, Barcelona: Actar, 2009; Geoffry C. Bowker, Karen Baker, et al, "Toward Information Infrastructure Studies: Ways of Knowing in a Networked Environment," in *International Handbook of Internet Research*. J. Hunsinger, et al, eds. Dordrecht: Springer, 2010, 97–118; and Christian Sandvig, "The Internet as Infrastructure," in *The Oxford Handbook of Internet Studies*, William Dutton, ed., Oxford: Oxford University Press, forthcoming.

2. See, for instance, Manuel Castells, *The Rise of the Network Society*, Hobboken, NJ: Wiley-Blackwell, 2009; Michael Hardt and Antonio Negri, *Empire*, Cambridge: Harvard University Press, 2001; Geert Lovink, *Dark Fiber*, Cambridge: MIT Press, 2003; Tiziana Terranova, *Network Culture: Politics for the Information Age*, London: Pluto Press, 2004; Michael Hardt and Antonio Negri, *The Multitude*, London: Penguin: 2005; Alex Galloway, *Protocol: How Control Exists after Decentralization*, Cambridge: MIT Press, 2006; Alex Galloway and Eugene Thacker, *The Exploit: A Theory of Networks*, Minneapolis: University of Minnesota Press, 2007; Wendy Chun, *Control and Freedom: Power and Paranoia in the Age of Fiber Optics*, Cambridge: MIT Press, 2008; Zizi Papacharissi, *A Networked Self: Identity, Community, and Culture on Social Network Sites*, London: Routledge, 2010.

3. The concept of infrastructure has, however, made inroads in the work of some film and media studies scholars. Jonathan Sterne (1999: 503–30), for instance, has investigated the historical processes by which US television emerged as a system of distribution and perceives its infrastructural formation as a critical problematic, insisting, "In the formation of American television, the creation of a national infrastructure was a problem and a project, not a given." Brian Larkin (2008: 6) has examined media infrastructures in urban Nigeria, using the term to explore the "technical and cultural systems that create institutionalized structures whereby goods of all sorts circulate, connecting and building people into collectivities." And Jonathan Beller (2006: 209) implicitly addresses infrastructural matters

when he observes, "Rather than requiring a state to build the roads that enable the circulation of its commodities, as Ford [Motor Company] did, the cinema builds its *pathways of circulation* directly into the eyes and sensoriums of its viewers." These scholars have adopted the concept of infrastructure to analyze the emergence of broadcast networks, theorize processes of cultural distribution, and critique the cinema's modes of production. See Jonathan Sterne, "Television under Construction: American Television and the Problem of Distribution," *Media, Culture and Society* 21 (1999): 503–30. Brian Larkin, *Signal and Noise: Media, Infrastructure, and Urban Culture in Nigeria*, Durham: Duke University Press, 2008, at 6. Jonathan Beller, *The Cinematic Mode of Production: Attention Economy and the Society of the Spectacle*, Lebanon, NH: Dartmouth College Press, 2006, at 209. Also see *Signal Traffic: Critical Studies of Media Infrastructures*, Lisa Parks and Nicole Starosielski, eds. Champaign-Urbana: University of Illinois Press, 2015..

4. Giuliana Bruno, *Atlas of Emotion: Journeys in Art, Architecture and Film*, London: Verso, 2007; Doreen Massey, *Space, Place, and Gender*, Minneapolis: University of Minnesota Press, 1994; Mike Crang and Nigel Thrift, eds., *Thinking Space (Critical Geographies)*, London: Routledge, 2000; Irit Rogoff, *Terra Infirma: Geography's Visual Culture*, London: Routledge, 2000; Bruno Latour, *Re-assembling the Social: An Introduction to Actor Network Theory*, New York: Oxford University Press, 2007; Fiona Candlin and Raiford Guins, eds. *The Object Reader*, London: Routledge, 2009. My work in this area has also been influenced by a 2005 UC Humanities Research Institute research residency led by Amelie Hastie entitled "The Object of Media Studies." See project online here: http://vectors.usc.edu/projects/index.php?project=65.

5. See my essays, "Where the Cable Ends: Television beyond Fringe Areas," in *Cable Visions: Television beyond Broadcasting*, Sarah Banet-Weiser, Cynthia Chris, and Anthony Freitas, eds. New York: New York University Press, 2007, 103-126; "Around the Antenna Tree: The Politics of Infrastructural Visibility," *Flow*, March 2009, available at http://flowtv.org/?p=2507; and "Postwar Footprints: Satellite and Wireless Stories in Slovenia and Croatia," in *B-Zone: Becoming Europe and Beyond*, Anselm Franke, ed. Barcelona: ACTAR Press, 2005.

6. See, for instance, Ann Friedberg, *The Virtual Window: From Alberti to Microsoft*, Cambridge: MIT Press, 2009; Kate Mondloch, *Screens: Viewing Media Installation Art*, Minneapolis: University of Minnesota Press, 2010; Wendy Chun, *Control and Freedom: Power and Paranoia in the Age of Fiber Optics*, Cambridge: MIT Press, 2008; Alex Galloway, *The Interface Effect*, Cambridge, UK: Polity, 2012.

7. For another book that investigates these issues, see Rick Maxwell and Toby Miller, *Greening the Media*, Oxford: Oxford University Press, 2012.

8. For further discussion of footprints, see "Satellites, Oil and Footprints: Eutelsat, Kazsat, and Post-communist Territories in Central Asia," in *Down to Earth: Satellite Technologies, Industries, and Cultures*, Lisa Parks and James Schwoch, eds., New Brunswick: Rutgers University Press, 2012, 122–40.

9. See, for instance, Donna Haraway, *Simians, Cyborgs, and Women*, London: Routledge, 1990; Gregory Downey, *Telegraph Messenger Boys: Labor, Technology and Geography, 1850–1950*, New York: Routledge, 2002.

10. I made this argument about layered media distribution systems in my essay "Where the Cable Ends: Television beyond Fringe Areas." For further discussion of media archaeology, see Erkki Huhtamo, ed., *Media Archaelogy: Approaches, Applications, and Implications*, Berkeley: University of California Press, 2011.

11. A 1960s documentary about the US postal service entitled *River of Mail* emphasizes the multiple resources required to distribute mail. Available at http://www.youtube.com/watch?v=yR9iOKvDlD0&feature=relmfu, accessed August 10, 2012.

12. For an analytical discussion of the cloud, see Paul T. Jaeger et al, "Where Is the Cloud? Geography, Economics, Environment, and Jurisdiction in Cloud Computing," *First Monday* 14 (5: May 4, 2009), available at http://www.uic.edu/htbin/cgiwrap/bin/ojs/index.php/fm/article/view/2456/2171, accessed August 21, 2012. For an interesting discussion of regulatory issues and cloud computing, see Jennifer Holt, "Platforms, Pipelines, and Policy: Regulating Connected Viewing" in *Connected Viewing*, Jennifer Holt and Kevin Sanson, eds. New York: Routledge, 2014.

13. Information about the "Linemen" project and select photos can be found on Michael Parker's website, www.michaelparker.org/1/lineman.html (accessed August 20, 2012). The full "Linemen" multimedia installation was presented as part of Parker's MFA thesis exhibition at USC in 2009.

14. For another homage to electrical linemen see, Michelle Larson's illustrated children's book, *Lineman: The Unsung Hero*, Bloomington: AuthorHouse, 2012.

15. For a fascinating study of electricity and early film studios in France, see Brian R. Jacobson, "Building a *Cité du Cinéma* in Paris: Film Studios as Urban Industrial Centers," in *Studios before the System: Architecture, Technology, and Early Cinema* (PhD dissertation, University of Southern California, 2011), 231–93.

16. Vicki Mayer, *Below the Line: Producers and Production Studies in the New Television Economy*, Durham: Duke University Press, 2011; John Caldwell, *Production Culture: Industrial Reflexivity and Critical Practice in Film and Television*, Durham: Duke University Press, 2008.

17. In addition to photographing this process, some have posted Youtube videos that reveal police removing dishes or satellite dishes destroyed by Iranian police. See, for instance, "Police raid on satellite dishes, Tehran 2009," posted July 3, 2009, available at http://www.youtube.com/watch?v=7-UE7pcqhT4, accessed, August 20, 2012; "Iran May 2011 – Wall Climber of the Iranian Regime destroying satellite dish," posted May 28, 2011, available at http://www.youtube.com/watch?v=nfB_gkjDE9A&feature=related, accessed August 20, 2012; "Iran Police collect Satellite Dishes from houses...!" posted August 30, 2011, available at http://www.youtube.com/watch?v=KU_nnFiytZM, accessed August 20, 2012; Iran turmoil when police try to dismantle satellite dishes in Tehran, posted August 8, 2011, available at http://www.youtube.com/watch?v=1suFmvGiOSA, accessed August 20, 2012.

26 DISTANT MIRRORS AND THE LAMP

Matthew Kirschenbaum

I attended my first MLA convention in 1996. I was a PhD student at the University of Virginia in Charlottesville at the time, and the MLA was just up the road in Washington, DC. Like many first time attendees, I had earnestly mapped out my convention schedule in the space provided at the front of the program. One session, in particular, stood out to me, promoted with arena rock production values.

"The Canon and the Web: Reconfiguring Romanticism in the Information Age" was organized by Alan Liu and Laura Mandell, two names very familiar to contemporary observers of the digital humanities. The Web presence you see in figure 26.1—one of those distant mirrors of my title—had been placed online months prior to the convention, on May 29 to be exact. [figure 26.1 here]

As Liu commented in a recent email to me, "I am struck (as you are) by how fleshed out that panel site was." Look past the vintage design: there are animated GIFs and gratuitous tables, yes, but there is also an evident will to situate the session amid a thick contextual network (the links to associated projects, related readings, and relevant sites); there is a clear desire for interactivity, as expressed through the live email links and the injunction to initiate correspondence; and there is also a curatorial sensibility that seems very contemporary to me, most notably through the "Canon Dreaming" links that take users to collections of materials assembled by Liu and Mandell's students—certainly prototypes of what we would today realize through, say, an Omeka installation. The participants' listings, complete with links to email and home pages alongside the thumbnail bios, were then an early instantiation of the ubiquitous user profiles we routinely create for social media services.

The other element, of course, was heralded directly in the session's title and instantiated in this-then formidable site: the Web itself, which was being celebrated not just as a convenience or contrivance for delivering content but as a potentially liberating force, a new paradigm, a corrective to the very notion of canonicity. But when I say "the Web itself," what I really mean is the *World-Wide* Web, which we then dutifully spelled out complete with conjoining hyphen

Figure 26.1

Splash page of Alan Liu and Laura Mandell's "The Canon and the Web," MLA special session site, 1996. As of this writing the site is still available online here: http://oldsite.english.ucsb.edu/faculty/ayliu/research/canonweb.html

and glossed with the ghastly descriptor "the graphical portion of the Internet." This *World-Wide* Web was not the same Web as the Web we have today, neither technically nor experientially nor aspirationally. There are continuities to be sure, but fewer than you might think. The Web then (Liu has recently taken to calling it the *reWeb*) was still mostly flatland, the dominant issues being the HTML rendering idiosyncrasies of different browsers (the war between Internet Explorer and Netscape Navigator was only just starting to heat up). The source code for the site invokes the DTD for HTML 3.2. The code itself, according to Liu, was most likely written in an editor called HoTMetaL, from SoftQuad. There is no XML or DHTML here. There is some rudimentary Javascript in the form of an earnest little site navigation widget. There is no commenting facility. There is no feed to subscribe to. There are no Twitter or Facebook sprites. This was 1996. My dearest friends, there was no Google. You Yahooed—there's no shame, we all did—or summoned forth the awesome powers of Alta Vista. Attendees at the 1996 convention would have bragged about their Internet acumen by telling colleagues about ordering a book from something called Amazon.com. Ordering a pizza online you could only do if you were Sandra Bullock in *The Net*, released a year earlier.

The real backbone of scholarly communication at the time remained listserv email. There was a rawness to it. You subscribed and you were on, sometimes pending moderator approval, usually more for spam control than anything else. Once you were on, you posted. Or lurked. Or flamed. Or accidentally hit reply-all when you meant to backchannel. But you didn't have to worry about how many followers you had or if you were popular or pithy enough to be retweeted. You didn't have to ask someone else if you could be their friend in order to converse with them. Strange, down-the-rabbit-hole geographies of influence formed, where the mainstay of a list would turn out to be a graduate student, or an emeritus at an obscure institution in New Zealand.

The starting point for any discussion of email in the digital humanities must be the venerable HUMANIST listserv, whose first substantive message is time-stamped 14 May 1987, 20:17:18 EDT, from one MCCARTY@UTOREPAS. This is, of course, Willard McCarty, who still edits the list to this day. HUMANIST: the name itself reminds us of a time when it was hard to conceive of a need for more than one listserv serving the academic humanities, that most general of titles serving to distinguish it from, say the LINGUIST list, which came along in 1990.

Humanist remains active today, its digests delivered regularly to, I suspect, a number of your inboxes. Some of you may also follow an account on Twitter dubbed @hum_comp. The owner of this account, who wishes to remain anonymous, is chronologically culling the Humanist archives, starting with the earliest entries, for tweet-length tidbits that either seem quaint or remain relevant. "I was immediately struck," said the account owner to me via email, "by how similar they sounded to the conversations I [have] been listening to, and tentatively engaging in in 2009/2010 . . . [T]he early HUMANIST listers saw much clearer than I ever did in the 1980s and 1990s what the challenges and promises of humanities computing and communication really were, right from the beginning of an expansive networked communication."

The longevity of Humanist notwithstanding, by 1998 it was clear that many of the scholarly listservs that had sprung up in the first half of the decade were already living out their use horizon. One of the lists that was important to me at the time was entitled H-CLC, which was part of the H-Net consortium. The CLC stood for Comparative Literature and Computers. In early November of 1997 Nelson Hilton wrote to it: "A growing number of inactive lists, it seems, have folded their tents and disappeared into the electronic night after a ritual call into the void met with resounding stillness. Perhaps it is a sign of advanced maturity in the medium—novelty has long worn off and we return to work at hand. If a list speaks to that work, we pause and read, perhaps respond—otherwise, quite rightly, why bother?"

The next Great Migration was to the blogosphere. Even as the lists were folding, the blogs began to spin up, built on nascent social networking scaffolding in the form of blogrolls, comments, trackbacks, and RSS feeds. Blogs are still very much with us today of course, but these were the salad days. Not WordPress or Blogger but Movable Type. Remember MT? Or maybe you hacked your own. Some bloggers seemed to exist only in the interstices, "comment blogging" as some of us called it, living out their online identities in the long trellis of text that dangled from the bottom of prominent postings. Group blogs were a particularly notable feature of the scholarly landscape, with venues like The Valve, Crooked Timber, Wordherders, and HASTAC becoming daily reads for many wired academics. For me, the most important such experiment was GrandTextAuto, which for five or six years was host to numerous important and intense conversations in digital studies, electronic literature, computer games, procedural literacy, and what we nowadays call critical code studies and software studies. Its "drivers," who included Scott Rettberg, Nick Montfort, Noah Wardrip-Fruin, Andrew Stern, Michael Mateas, and Mary Flanagan, created an active and energetic user community, one with sufficient gravitas and presence to spur the MIT Press to use it as the platform for the open peer review of Wardrip-Fruin's first monograph, *Expressive Processing*. When I emailed Nick, he said this to me: "Those of us who started out as graduate students took on professorial and administrative responsibilities, and there was less time available to play gadfly, perpetrate April Fool's jokes, and explore aspects of projects through online discussion—instead, we had to write grants, teach classes, supervise graduate students, and so on. So, the heavy burden of responsibility, growing up, and so on." For my part, I think the rise of Twitter also had a strong impact on the vitality of the blogosphere, even though, as I've written elsewhere, blogs and Twitter coexist with one another in powerful mutually enabling ways.

This brings me to the new MLA Commons, launching this weekend here at the convention. The MLA announcement states that it is intended "to facilitate active member-to-member communication" and "offer a platform for the publication of scholarship in new formats"—language that echoes, among other early electronic exemplars, the Welcome message from the Humanist listserv so long ago. The MLA Commons is built out from the CUNY Commons in a Box package, which is built on top of the BuddyPress extensions to WordPress, which itself depends on the Linux-Apache-MySQL-PHP LAMP stack of my title, four bedrock open source technologies that all existed, but were in their infancy, at the time of that 1996 Canon and Web panel. Working from the distant and very partially mirrored history of online scholarly communication I've been sketching, there are a few propositions I want to leave you with.

First, access always engenders power. Power dynamics are built in to our social networking services at the most basic level—indeed the ability to define and operationalize various strata of relationship functions—trust, visibility, and reciprocity—is arguably at the heart of the read/write Web. To wit: for the past few days my email inbox has been regularly populated by messages from "MLA Commons" with the subject heading "New contact request." I've accepted them all, including the ones from people I don't really know, because the platform is new and my instinct is to err on the side of openness. But I don't accept all Friend requests on Facebook (nor are all of my earnest solicitations accepted) and I don't follow back everyone who follows me on Twitter even as I do follow people I'd give my eye teeth to have follow me—but alas, they don't. As MLA Commons gains steam, at what point do the filters go up, to an extent replicating existing power dynamics in the profession? Do I accept contacts from everyone in my home department? Does someone with a convention interview send contact requests to members of a search committee? If they do, is it presumptuous? If they don't, are they antisocial in the most literal sense? Do I accept all contact requests from those of a higher professional rank than me? Accept no contact requests from those of lower rank? Anyone who manages their relationships so coarsely is missing the point and very likely has life problems far greater than their facility with online social networking, but given that reputation metrics (whether in the form of services like Klout or scholarly rankings such as the social sciences' H-Index) are operationalized in the very marrow of our online media, including digital publications and citations, we'd be mistaken to believe that these concerns are extraneous or can be entirely sidestepped by any mature scholarly communications network. It's not a reason not to have a Commons, but it is a reason to be mindful of the relational and procedural models built in to its fabric, especially given the varied technologies the end-user experience rests upon.

Access also always entails risk. I have had a home page on the Web since the summer of 1995. In 1997 I began writing my dissertation online, not quite "live," more of a time-shift as I edited and managed different document technologies but always posting the prose in full, not just excerpts. My inspiration here was Harlan Ellison, who regularly wrote short stories seated in the window of community bookstores. In 1977 he did it for a full week, a story a day, in a Los Angeles bookshop. I recognized early on that the Web had the potential to be an even more perfect panopticon; no great stretch, but I was also in touch with enough of my baser instincts to understand that visibility and feedback the project seemed likely to attract would entice me and keep me going. Inevitably I was asked whether I was worried about people plagiarizing my work. Again, this was in the era before blogs, and indeed most middle-state writing online; the equivalent of the blog essay, or "blessay" as Dan Cohen has called it, was mostly confined to listservs and text files distributed via FTP drops—very different from the agora of an open Web indexed by even pre-Google search engine technology. My response to the question about plagiary was to invoke another literary authority, specifically Poe's purloined letter. Hiding my ideas in plain sight, I argued, was the single best way to get them into circulation and ensure the necessity of referencing and citing them (as opposed to merely swiping them). And for the most part, I was right. The experiment paid real professional dividends. It got my work an audience, airtime in front of the eyeballs I most wanted for it, and the work was judged usually, though by no means universally, favorably. More I could not ask for, and I carried the same ethos over into my blogging, which began in early 2003. From the outset I blogged under my own name, and I wrote a response—a blessay, I suppose—explaining why when Ivan Tribble (remember all that trouble with Mr. Tribble?) penned a 2005 piece for the *Chronicle of Higher Ed* called "Bloggers Need Not Apply." In the comments I asked others to weigh in with examples of the professional dividends their online identities and reputations had reaped, and I collected several dozen instances further testifying to the power of, as the title of Phil Agre's classic text has it, "networking on the network." I got onto Twitter in 2008, again perhaps just ahead of its mainstream uptake. Nonetheless, I come before you today to say this: I have not blogged every good idea I have ever had. I have not tweeted every insight or reference or revelation. There's stuff I keep to myself, or better, stuff I release strategically rather than spontaneously, and it will fall to all of us, on MLA Commons and elsewhere, to find our own personal and professional comfort zones regarding what we give out to our contacts and groups, the membership at large, the public at large. Access always entails risk, and while we know scholarship is not a zero-sum game, more tangible and no less sustaining forms of reputation and reward sometimes, even often, are.

Finally, access always requires time. One scenario, I suppose, is that our online interactions will eventually progress to the point that genre and platform dissolve. Our work becomes our discourse stream, and a comment, a tweet, a blessay, or a monograph will all become part of it, winding more or less tightly within and amid the equally interwoven discourse strands of others. Fuzzy mathematics measuring out influence, recognition, and trust will iterate and calculate in real-time as our conversations thrive, hum, surge, and falter in the collective hive. Rivals will wage guerilla flame campaigns across the endlessly reticulating strata of the network, their memes warring with one other, spawning algorithmic eddies and tides that will buffet entire fields and disciplines. Tenure, if it still exists, may be based on cycles of attention rather than the accumulation of publications—propagate or perish. But I don't think so. For one thing, platforms are perhaps further from interoperability than ever; Bruce Sterling, for example, predicted that 2013 would be the year of silos and tactically burned and broken bridges. Genres may indeed fray around the edges, but in the end every comment, every sentence blogged and logged, every essay chapter finished and book published is a withdrawal from the finite bank of productivity that is shaped by our individual intellects, our irreducible bodies with all their vulnerabilities, our jobs, our responsibilities, our loved ones, our hobbies and distractions and vacations and time in the woods. That is to say, that however much attention spans may be redefined and reconfigured—psycho-cognitively, neuro-chemically, and otherwise—most scholarly careers, for now and the foreseeable future, are still going to be measured out in fairly pedestrian ways, in the summer months and three-day weekends and that rare "empty" day during the semester, perhaps on fellowship or sabbatical if we're lucky and privileged, often at the expense of an hour's sleep or a movie with the rest of the family when we're not. Our scholarly communications networks will either cohabitate with the myriad obligations of that ticking temporal complex, or they will lie fallow—most of us are too far out as it is, and not Google Waving but drowning.

So I would leave you with one final urging—that it's not too soon for MLA Commons to be planning for its own planned obsolescence, or what Bethany Nowviskie and Dot Porter have termed "graceful degradation." Our social (and our scholarly) networks are ever more porous, but they have yet to become reliably portable. This is a problem. Relationship economies require enormous amounts of attention and investment, and for those of us who live active online professional lives, a nontrivial amount of time is devoted to cultivating and aggregating those networks in diverse, subtle, and not-so-subtle ways. Many of you have seen your own online networks shape-shift across listservs, blogs, and now the newest social media platforms. Universal avatars, imported contact lists, these are stop-gap measures; more promising for the next phase of scholarly communication may be mature personal unique identifiers, as promised by ORCID, a service that "distinguishes you from every other researcher and, through integration in key research workflows such as manuscript and grant submission, supports automated linkages between you and your professional activities." I'm currently on Twitter, Slideshare, Zotero, Google+, Facebook, Academia.edu, GitHub, and Ello, to name just a few. I want to migrate and port not just my content but also my reputation and relations. Regardless of if or when the MLA Commons folds its tent and decamps into that vast electronic night—hopefully not for a great many attention cycles—what should endure are the relationships it fosters and the work thus performed—the "work at hand" as Nelson Hilton put it on H-CLC on Monday, November 3, 1997, at 10:33 in the decidedly nondigital morning.

Note

This text is a talk I wrote for the 2013 Modern Language Association Convention's Presidential Forum on Avenues of Access organized by Michael Bérubé. The specific session theme was "Digital Humanities and the Future of Scholarly Communication." The text that is printed here is an artifact of its occasion and only very lightly revised from the occasion of that original delivery. A copy with live links and slides is available on MLA Commons.

27 RESISTANCE IN THE MATERIALS

Bethany Nowviskie

Most mornings, these days—especially when I'm the *first* to arrive at the Scholars' Lab—I'll start a little something printing on our Replicator. I do this before I dive into my email, head off for consultations and meetings, or (more rarely) settle in to write. There's a grinding whirr as the machine revs up. A harsh, lilac-colored light clicks on above the golden Kapton tape on the platform. Things become hot to the touch, and I walk away. I don't even bother to stay, now, to see the mechanized arms begin a musical slide along paths I've programmed for them, or to watch how the fine filament gets pushed out, melted, and microns-thin—additive, architectural—building up, from the bottom, the objects of my command.

I'm a lapsed Victorianist and book historian who also trained in archaeology, before gravitating toward the most concrete aspects of digital humanities production—the design of tools and online environments that emphasize the inevitable materiality of texts, and the specific physicality of our every interaction with them. I suppose I print to feel productive, on days when I know I'll otherwise generate more *words* than *things* at the Scholars' Lab, the digital humanities center I direct at the University of Virginia Library. Art objects, little mechanisms and technical experiments, cultural artifacts reproduced for teaching or research—cheap 3D printing is one affirmation that words (those lines of computer code that speak each shape) always readily become things. That they kind of... *want to.* It's like when I learned to set filthy lead type and push the heavy, rolling arm of a Vandercook printing press, when I should have been writing my dissertation.

I peek in, as I can, over the course of a morning. And when the extruders stop extruding, and the whole beast cools down, I'll crack something solid and new off the platform—if a colleague in the lab hasn't done that for me already. (It's a satisfying moment in the process.)

Sometimes, though, I'll come back to a mess—a failed print, looking like a ball of string or a blob of wax. Maybe something was crooked, by a millimeter. Maybe the structure contracted and cracked, no match for a cooling breeze from the open door. Or maybe it's that my code was poor, and the image in my mind and on my screen failed to make contact with the Replicator's sizzling build-plate—so the plastic filament that should have stuck like coral instead spiraled out into the air, and cooled and curled around nothing. Those are the mornings I think about William Morris.

It's not too long ago that we couldn't imagine humanities computing becoming so mainstreamed as to have a cutesy acronym, or cluster hires everywhere, or a dedicated office at the National Endowment for the Humanities and common campus centers, full-time strategists, and digital humanities librarians—much less frustrated outsiders and active (rather than passive) detractors. In those days, as a grad student in the late 1990s, I apprenticed under Jerome McGann at the Rossetti Archive. Jerry had recently been interviewed for *Lingua Franca* by a then-unknown, 26-year-old tech writer (Steven Johnson), and had thrown a little Morris at him, by way of explaining the embodied frictions that become beautifully and revealingly evident when you move scholarly editorial practice, born in book culture, from print to digital media: "You can't have art," Morris, the master craftsman of the Pre-Raphaelites had said, "without resistance in the material."

It's a compelling line—reproduced (somewhat mechanically and often slightly mangled) all over, and only rarely contextualized or traced back to its source. Morris's erstwhile son-in-law, Henry Halliday Sparling, reports it in a 1924 study of the Kelmscott Press, as part of the designer's extended complaint about a newfangled device: the typewriter. For many years, in its precise terms, this seemed to me an odd quarrel to pick.

"Morris condemned the typewriter for creative work," Sparling tells us, saying that "anything that gets between a man's hand and his work, you see, is more or less bad for him. There's a pleasant feel in the paper under one's hand and the pen between one's fingers that has its own part in the work done." Morris goes on to extol a nicely proportioned quill over the steel pen, and to condemn the pneumatic brush, "that thing for blowing ink on to the paper—because they come between the hand and its work, as I've said, and again because they make things too easy. The minute you make the executive part of the work too easy, the less thought there is in the result. *And you can't have art without resistance in the material.* No! The very slowness with which the pen or the brush moves over the paper, or the graver goes through the wood, has its value." So far, so good, but then Morris—whom I believe had never used a typewriter—concludes, a little awkwardly: "And it seems to me, too, that with a machine, one's mind would be apt to be taken off the work at whiles by the machine sticking or what not."

I'm generally with Morris until the final turn. Isn't "the machine sticking or what not" just another kind of maker's resistance? A complication we might identify, make accessible—which is sometimes to say tractable—and overcome? After all, the "executive part of the work" should never be "too easy." Isn't a sticky typewriter something to be worked against, or through—a defamiliarizing and salutary reminder of the material nature of every generative or transformative textual process?

But as I reflected on "Avenues of Access" (our theme for today's presidential panel at the MLA convention), I came to understand. Morris's final, throwaway complaint is not about that positive, inherent resistance—the friction that makes art—which we happily seek *within the humanities material we practice upon*. It's about resistance unhealthily and inaccessibly located in a *tool set*. Twentieth-century pop psychology would see this as a disturbance in "flow." Twenty-first-century interaction design seeks to avoid or repair such UX (or user experience) flaws. And, closer to home, *precisely this kind* of disenfranchising resistance is the one most felt by scholars and students new to the digital humanities. Evidence of friction in the means, rather than the materials, of digital humanities inquiry is everywhere evident in the program of this convention. And it's written in frustration all over the body of proposals and peer reviews for a conference of much greater disciplinary, linguistic, generational, and professional convergence I will chair later this year, the annual meeting of the Alliance of Digital Humanities Organizations, DH 2013.

When established digital humanities practitioners and tool builders are feeling overly generous toward ourselves (as we occasionally do), we diminish our responsibility to address this frustration by naming it the inevitable "learning curve" of humanities computing. Instead, we might confess that among our chief barriers to entry are poorly engineered and ineptly designed research tools and social systems, the creation of which is a sin we perpetrate on our own growing community. It's the kind of sin easily and unwittingly committed by jacks-of-all-trades. (And I'll return to them, to us, in a minute.) But it's worth reflecting that tensions and fractures and glitches of *all* sorts reveal opportunity.

When Morris frets about "the machine sticking or what not," it is with an uncharacteristic voice. He offers the plaint of a passive tool-user—not of the capable artisan we're accustomed to, who might be expected to fashion and refine and forge an intimate relationship with the instruments of his work. The resistance in the typewriter Morris imagines, and the resistance digital humanities novices feel when they pick up fresh tool sets or enter new environments, is different from the positive "resistance in the material" encountered by earlier generations of computing humanists. It's different from that happy resistance *still felt* by hands-on creators of humanities software and encoding systems.

Until quite recently every self-professed digital humanist I knew was deeply engaged in tool-building and in digitization: the most fundamental and direct kinds of humanities re-mediation. The tools we crafted might be algorithmic or procedural—software devices for performing operations on the already-digitized material of our attention—or patently ontological: conceptual tools like database designs and markup schema, for modeling humanities content in the first place. These were frameworks simultaneously lossy and enhancing, all of them (importantly) making and testing hypotheses about human texts and artifacts, and about the phase changes these objects go through as we move them into new media. No matter the type, our tools had one thing in common: overwhelmingly, *their own users* had made 'em, and understood the continual and collective *re-making* of them, in response to various kinds of resistance encountered and discovered, as a natural part of the process of their use. In fact, this constructivist and responsive maker's circle was so easily and unavoidably experienced as *the new, collaborative hermeneutic* of humanities computing, as *the work itself* that—within or beyond our small community—we too rarely bothered to say so.

So much for the prelude. Three crucially important factors, all touching on modes of access, are converging for humanities computing today. I believe we're at the most critical juncture for the welfare of digital research of any in my eighteen years of involvement in the field. The first factor I'll share with you sets unheard-of conditions for real, sustained, and fundamentally new advancements in humanities interpretation. The second de-familiarizes our own practice so thoroughly that we just might all (established and new actors alike) feel levels of "resistance" adequate to allow us to take advantage of the first. But I lose heart when I think about the third. I'll walk through them one by one.

The first of my three factors starts with the massive, rapid, and inexorable conversion of our material cultural inheritance to digital forms. Hand-crafted, boutique digitization by humanities scholars and archivists (in the intrepid, research-oriented, hypothesis-testing mode of the 1990s) was jarred and overwhelmed by the mid-2000s advent of mass digitization, in the form of Google Books. Least-common-denominator commercial digitization has had grave implications not only for our ability to insert humanities voices and perspectives in the process, but also for our collective capacity *and will* to think clearly about, to steward, and to engage with physical archives in its wake. A decade on, as a community of scholars and cultural heritage workers, we have only just begun to grapple with the primary phase change of digitization-at-scale, when we've become (for the most part) bystanders at the scene of a second major technological shift.

I gestured at it in the images with which I began my talk. Momentous cultural and scholarly changes will be brought about not by digitization alone but by the development of ubiquitous digital-to-physical conversion tools and interfaces. What will humanities research and pedagogy do with consumer-accessible 3D fabrication? With embedded or wearable, responsive and tactile physical computing devices? What will we do with locative and augmented reality technologies that can bring our content off the screen and into our embodied, place-based, mobile lives? Our friends in archaeology and public history, recognizing the potential for students and new humanities audiences, are all over this. Writers and artists have begun to engage, as we can see next door at the 2013 MLA Convention e-literature exhibit. And I believe that scholarly editors, paleographers, archivists, and book historians will be the next avid explorers of new digital materialities. But what might other literary scholars do? What new, interpretive research avenues will open up for you, in places of interesting friction and resistance, when you gain access to the fresh, *full circuit* of humanities computing—that is, the loop from the physical to the digital *to the material text and artifact* again?

The second factor I want to address has a twinned potential. It could be dangerously inhibiting or productively defamiliarizing for our field. Currently it's a little of both, resting on the uncomfortable methodological and social axis of embodied inquiry. Without a clear call from people feeling barred from access to the tool-building side of the digital humanities, our software developers' community might not now be talking about things we have long internalized—about *what goes unspoken* or is illegibly expressed in our day-to-day practice. And, frankly, if it weren't for some measure of annoyance at that much-quoted false binary of "hack-vs-yack," we might have remained disinclined—disinclined to voice the ways in which *tacit knowledge exchange* in code-craft and digital humanities collaboration contributes to a new hermeneutic, a new way of performing thoughtful humanities interpretation. You might call it exegesis through stage-setting. It comes into focus as interface and architecture, through our own deliberate acts of communal, mostly non-discursive humanities design. The work we do is graphical and structural and interactive. It's increasingly material and mobile, and it's almost never made alone. Whatever it is, like any humanities theorizing, it opens some doors and shuts others, but it's a style of scholarly communication that differs sharply from the dominant, extravagantly vocal and individualist verbal expressions of academic humanities in the last fifty to sixty years. And, like any craft, it'll always be under-articulated.

The call prompting this new introspection about the nature of our work comes most strongly from women, minority scholars, and other groups underrepresented in software development, responding in their turn to an aggressively male global tech culture that is (on a good day) oblivious to its own exclusionary practices and tone. Now, all this is much more the case *outside of digital humanities* than within it, and in truth, I find the humanities a piss-poor battleground for a war that should be fought in primary and secondary STEM education. But the prompt *to make accessible the unspoken in digital humanities* also comes not only from people who feel they have lacked the basic preparation to engage, but from those who lack the time and tools: from rootless contingent faculty and scholars from under-resourced or largely teaching-focused schools—newly interested in digital humanities but feeling unable to play along with their counterparts from research institutions.

All these people would find the murmurings of the digital humanities developers' community sympatico and sincere. But our conversations are pretty much *sub rosa* now and (part of the problem) are happening in places either technologically inaccessible to most scholars or so coded as "unscholarly" as to be ignored by them. We're doing what we can, from our end, to fix that, including by hosting events such as the Scholars' Lab's NEH-funded summit, "Speaking in Code." But will it matter? Maybe not to *this* discipline. Literary critics and cultural theorists may not (after the current digital humanities bubble bursts) ultimately wish to engage in a brand of scholarly communication that places less premium on argument and narrow, expert discourse, and more on the implicit embodiment of humanities interpretation in *public production* and open source, inter-professional *practice*. For the most part, though, I suspect many of our colleagues *just can't tell*: to them, everyone with direct access to the means of digital humanities production speaks, sometimes literally, in code.

When I'm feeling sad about this stuff, I turn, again, to William Morris. As a self-help strategy, that yields mixed results: "In the Middle Ages," he tells us, in *Art and Labor*, "everything that man made was beautiful [eh.], just as everything that nature makes is always beautiful; [yeah?] and I must again impress upon you the fact that this was because they were made mainly for use, instead of mainly to be bought and sold. . . . [hmm.]. He continues: "the beauty of the handicrafts of the Middle Ages came from this, that the workman had control over his material, tools, and time."

I said there were *three* new conditions at play in this, our late age of the digital humanities. The final one is the rise of casual and alternative academic labor. First, I'll briefly address what has come to be called "alt-ac," the increasing recruitment of humanities PhDs to full-time, hybrid, scholarly professional positions in places like libraries, IT divisions, and digital humanities labs and centers. Real advantages and new opportunities for the humanities are attendant on this development. Properly trained and supported, long-term, "alternative academic" faculty and staff are potential leaders in your institution. They are uniquely positioned to represent and enact the core values of our disciplines; to serve as much-needed translators among scholars, technologists, and administrators; and to build technical and social systems suited to the work we know we must do. Absent their energetic involvement in shaping new structures in higher education, I am convinced that digital humanities will only scale *as commodity tool-use for the classroom*—not as a generative research activity in its own right.

But they (to continue a theme of this conference), like far too many of our teaching faculty, are subject to the increasing casualization of academic labor. Positions in digital humanities centers are especially apt to be filled with soft-money (or short-term, grant-funded) employees. In a field whose native interdisciplinarity verges on inter-professionalism, full-time, long-term digital humanities staff already struggle against the pressure to become jacks of all trades and masters of none. How can grant-funded digital humanities journeymen find the time and feel the stability that leads to institutional commitment, to deep engagement and expertise, and to iterative refinement of their products and research findings? And the situation is worse for more conventionally employed, adjunct academics. If the vast majority of our teaching faculty become contingent, what vanishing *minority* of those will ever transition from being passive digital tool-users to active humanities makers? Who among them will find time to feel a productive resistance in her materials?

Casualized labor begets commodity tool sets, frictionless and uncritical engagement with content, and shallow practices of use. I am not an uncritical booster of the tenure system, nor am I unaware of the economic realities of running a university, but I find it evident that, if we fail to invest at the institutional and national level in full-time, new-model, humanities-trained scholarly communications practitioners, devoted to shepherding and intervening in the conversion of our cultural heritage to digital forms (now *there and back again!*)—and if we permit our institutions to convert a generation of scholars to at-will teaching and digital humanities labor—humanities knowledge workers of all stripes will lose, perhaps forever, control over Morris's crucial triad: our material, our tools, and our time.

We can't allow this to happen at any stage of the game, but most especially today, it seems to me—as I listen to a community struggle to articulate the relationship between interpretation and craft, and as I crack some warm artifact off the printer of a morning. We've come to a moment of unprecedented potential for the material, embodied, and experiential digital humanities.

How do we, all together, intend to experience it?

28 THE DIGITAL HUMANITIES AS A LABORATORY

Amy E. Earhart

Academic institutions interested in growing local digital humanities scholarship are experimenting with a range of institutional structures designed to leverage digital humanities work. Some institutions are launching undergraduate majors in digital humanities, others are developing digital humanities certification programs or PhD tracks, and others have launched digital humanities centers, an increasingly popular approach evidenced by the formation of centerNet, an international network of digital humanities centers (digitalhumanities.org/centernet/). Other institutions have experimented with the development of digital humanities laboratories. Unlike many early versions of digital humanities spaces that were constructed as centers or institutes, the more recent laboratory model implies a scholarly working space, real or virtual.[1] Existing labs include the Stanford Literary Lab (Stanford), Scholar's Lab (University of Virginia), Digital Scholarship Lab (U of Richmond), Humanities and CriticalCode Studies Lab (USC), the Humanities Laboratories (Duke U), Electronic Textual Cultures Lab (U Victoria), HUMLab (Sweden), The CulturePlex Laboratory (Western U), Digital Humanities Lab Denmark (Aarhus Denmark), and AlfaLab (Netherlands), among others. These labs are located in physical spaces designed to support scholarly inquiry by providing meeting space, programming, training, support, and equipment. Other digital humanities entities, such as HASTAC, the Humanities, Arts, Science, and Technology Advanced Collaboratory, define themselves as collaboratories, virtual labs that exist as centers without walls (Siemens and Siemens 2012).[2] Participants in both types of labs may include postdoctoral fellows, librarians, research assistants, administrative staff, technical staff, such as programmers or developers, students, and subject specialists. Digital humanities labs are often multipurpose, with activities ranging from research to pedagogy.

The digital humanities lab has been constructed from various traditions including the design lab, art studio and science lab to meet the distinctive needs of humanities scholars, or as Patrik Svensson has noted, digital humanities labs are fusioning forms from other traditions to develop a lab that serves our unique purposes (Svensson 2011). While the design lab or art studio has had a great influence on many digital humanities labs, the model of the science laboratory is most often invoked by those that imagine the digital humanities labspace. In part, this is because our current moment of digital renaissance mimics the historical moment in which science labs were developed. According to Pamela Smith's work on the history of science, laboratories underwent a dramatic shift in the seventeenth century. A laboratory was no longer an individually run "artisanal workshop" but the "site of science" (Park and Daston 2006, 292) which emphasized exploration through hands on activities and "the observation and manipulation of nature by means of specialized instruments, techniques, and apparatuses that require manual skills as well as conceptual knowledge for their construction and deployment" (Hannaway 1986, 585). This shift to a modern understanding of knowledge produced with the manipulation of tools is in many ways like the current trend in digital humanities where technological tools and methods are both built and applied. Like many science labs, digital humanities labs contain equipment utilized by digital humanists, emphasize collaborative research, and focus on theoretical and applied research. Digital humanities labs also train and mentor students and faculty, serving as pedagogical and outreach or service arms dedicated to growing work within the field, tasks which are often conducted in centers or offices that are separated from the science laboratory.[3] For example, Umea University's HUMLab is multipurpose: "HUMlab is an environment of innovation which works as a place of study, a research laboratory, a place for project development, as well as a lecture hall or exhibition space" ("About Humlab"). While science labs are often clustered by area, equipment or topic, digital humanities labs are often broader, more multipurpose, and more inclusive by both design and funding limitations. Digital humanities labs are experimental models attempting to fill the various needs of the digital humanities community including the desire to expand the field, support a broad range of projects, and provide training for students and faculty.

The fusion of various traditions is necessary for the humanities lab experiment due to the type of scholarly questions and research methodologies best suited for such questions, which makes the science lab model an imperfect fit within a humanities framework, even a humanities framework that emphasizes technology. Instead, the maker traditions from the arts and architecture have been useful in the construction of laboratory models. HUMLab, Duke's Greater Than Games Lab, and a newly constructed digital humanities lab at North Carolina State University, for example, were developed in close connection with design schools and, because of this influence, outwardly emphasize process oriented approaches and artistic spaces.[4] As digital humanities continue to evolve it seems likely that laboratories will increasingly blend various traditions to meet particular needs. At this particular historical moment, however, the digital humanities lab is primarily imagined as science lab-like by both supporters and detractors.

The potential application of the science model often generates contention. During the DH 2012 conference, Lynne Siemens gave a talk titled, "Notes from the Collaboratory: An informal study of an Academic DH Lab in Transition" to which Eric Johnson tweeted: "I jotted a note to myself earlier: 'Interesting to study: how DH views science.' Seems ... idealized?" Doug Reside responded: "Hmm ... or unfairly vilified."[5] Johnson and Reside capture the current tension in the use of a science model. On the one hand, digital humanists utilize a means of scholarly production more closely related to science than traditional humanities, including the use of high-performance computing and collaboration, and are necessarily interested in adopting models of scholarly production that support such inquiry. Yet the digital humanities are deeply enmeshed with the humanities, of which some strains, such as literary criticism, have been imagined as oppositional to science. The two responses that emerged on the Twitter stream in response to science—idealization versus demonization—are divergent approaches that nonetheless must coexist within the digital humanities framework.

There are those in digital humanities that see the laboratory as the ideal space in which to conduct future research. Franco Moretti (2004) has mused, "'My little dream,' he added wistfully, 'is of a literary class that would look more like a lab than a Platonic academy.'" "Labs are built around the process of discovery," writes Cathy Davidson (1999: B4–5), "and discovery is rooted in the practice of what is already known (past experiments, lab technique). A lab supports work that is new, and it concomitantly requires collaboration across fields and disciplinary subfields, as well as across generations."[6]□ While in many labs this is true, and certainly an ideal to which digital humanities labs aspire, it is important not to romanticize the lab. Science labs emphasize collaboration, but hierarchies may be apparent. Linda and Michael Hutcheon (2001: 1367–68) agree that laboratory science "requires collaboration" but remain cautious of adopting the model wholesale, as there is a "hierarchy implicit in that model, with its 'stratified division of technical and intellectual labor.'" As Lisa Ede and Andrea Lunsford (2001: 363) remind us, "the sciences have a poor record of including women and members of minorities—or their perspectives—in research." John Unsworth concurs, noting, "In humanities, we often emulate what we think the sciences do, but our emulation may not actually bear that much resemblance to the reality of what goes on in science. Often science looks more collaborative because a lot of people get together to write a grant proposal, but that does not mean that they have necessarily figured out how to work together" (Unsworth and Tupman 2012: 232; Patrik Svensson 2011: para. 22) succinctly articulates the problems of uncritical adoption of a science model: "First, existing humanities infrastructure may be disregarded as we do not have a tradition of science-like infrastructure. Second, the science-based and data-driven model may be imposed on the Humanities (sometimes by humanists themselves) without careful discussion of the premises and consequences. Third, there is a risk that infrastructural needs or agendas compatible with the largely science based model will be the ones most likely to be prioritized.

Fourth, new humanities infrastructure may be uncritically based on existing infrastructure and associated epistemic commitments." So, while we might look to the laboratory as a model, we need to be critical about its implementation in our field.

As many working within digital humanities laud the laboratory model, there remains deep suspicion of bringing a science model to humanities work. The split between science and the humanities is longstanding, á la C. P. Snow, and resistance to utilizing anything from science is increasing as university funding and prestige is increasingly seen, by some humanists, to correlate with STEM areas rather than humanities fields. We might also have a historical structural problem in the very emergence of digital humanities that contributes to the tension. Martha Nell Smith (2007) contends that digital humanities developed as a space to which practitioners hoped to flee from the shifts in the profession that arose out of the cultural studies movement. In "The Human Touch: Software of the Highest Order, Revisiting Editing as Interpretation," Smith highlights the digital humanities' retreat into modes of analytics, objective approaches as "safe" alternatives to the messy fluidities found in literary studies. She notes, "It was as if these matters of objective and hard science provided an oasis for folks who did not want to clutter sharp, disciplined, methodical philosophy with considerations of the gender-, race-, and class-determined facts of life . . . Humanities computing seemed to offer a space free from all this messiness and a return to objective questions of representation" (4). My own experience in running a focus group during an NEH seminar on tenure and promotion confirms how deep the science/humanities divide remains. When I suggested that participants consider what constitutes a humanities data set, most responded in a negative, visceral manner to the use of a science term to describe humanities materials. As one participant noted, why apply science terminology to humanities? Science and the humanities are thus often seen as two very separate and distinct fields that have good reason to remain separate and should remain so.

As Christine Borgman (2009: para. 2) has noted, "The humanities need not emulate the sciences, but can learn useful lessons by studying the successes (and limitations) of cyberinfrastructure and eScience initiatives."[7] Certainly the science laboratory is a structure that we might use to examine the way in which equipment could be shared amongst scholars or funding is acquired and distributed, but it might best serve the digital humanities to look beyond the sciences to model how individuals might work collaboratively. A hallmark of digital humanities work is an emphasis on multiple partner participants, including technologists, subject specialists, and librarians, all necessary to complete the wide-ranging projects we imagine. Subsequently, we are developing models that link our faculty to students, community partners, and the greater public. How do we develop a humanistic laboratory "where no solitary thinker—no matter how brilliant or creative—could think through a complex problem as comprehensively as a group of thinkers from different fields, with different areas of expertise, different disciplinary training and biases, and from different intellectual generations?" (Davidson 1999). Rather than emphasize an adopted laboratory model, it might be more useful, and even politically savvy, to emphasize a laboratory model as one that privileges traditional humanistic inquiry through material and spatial construction. Svensson (2011) has described the way that the seminar table serves as an established model of humanistic collaboration and, therefore, a key material piece of any lab. Further, making spaces of architecture, art, and design offer equally interesting ways to consider collaboration and creativity. From town art production studio spaces, such as the small town Concord Art Association, Concord, Massachusetts, that allows members a space in which to work, to community driven open access design studios, such as the Portland, Oregon Radius Community Art Studios, which allow open drop-in studio use, laboratories might become an interface to the localized community. Such a model is very enticing when considered in connection to public digital humanities movements. The lab might exist physically, as is the case with groups like Scholar's Lab, or virtually, as is the case of HASTAC. Nancy Nersessian (2006) conceptualizes the research laboratory as "not simply a physical space existing in the present, but rather a dynamic problem space, constrained by the research program of the laboratory director, that reconfigures itself as the research program moves along in time and takes new directions in response to what occurs both in the laboratory and in the wider community of which the research is a part" (130).

As we develop collaborative approaches, it is imperative that we think carefully about the impact of the institutional structures that we adopt. Lynne Siemens et al. (2009) discovered, in her extended study of digital humanists and collaboration, that "collaborative issues include equitable distribution of work, interpersonal issues, working across disciplinary boundaries and expectations." Bethany Nowviskie (2011b) has written extensively on collaborative practices in digital humanities, giving particular attention to tensions between tenure and tenure track faculty and alt-ac career digital humanists driven by "institutional policies . . . that *codify inequities* among collaborators of differing employment status." Nowviskie has argued that equitably distributed credit will prove beneficial to all partners, leading "to strengthened research-and-development partnerships in DH" and "promoting a sense of *shared ownership of knowledge production*" that "will result in better design decisions and more enthusiastic preservation of our cultural and scholarly record." The sciences provide various models for representing individual contributions to projects, primarily denoted by the author order on publications. The author order represents the contribution level. Some science fields list the graduate students who conduct the research behind the lead faculty member's name, others place the scholar that conducted the most work last in order, and still others allocate the last space in a list to the funder of the lab in which the research was conducted. Each field understands how to interpret the author order on papers. This is not so in the humanities, where a common understanding of author order is not shared. Add to this the differing measure of credit applied to joint authored papers by tenure and promotion committees, and collaboration becomes an extremely fraught problem. Regardless of such challenges, digital humanities scholars report that they produce better work together than alone (Siemens 2012).

While humanists' relative lack of collaborative research experience presents challenges, it also presents an opportunity for digital humanities to actively build, examine and rebuild institutional environments that foster collaboration.[8] It is crucial that we tailor the existing science laboratory model to meet best practices in the digital humanities. We might not yet understand how to most effectively manage collaborative projects, but digital humanists are engaging with tough ethical questions, from how to appropriately credit work to diversity in the field. For example, a Collaborators' Bill of Rights emerged from an Off the Tracks NEH workshop that articulated best practices for crediting participants in a project (mith.umd.edu/offthetracks/recommendations). The authors point to science and the arts as possible models for defining equitable crediting of project participation, but make stringent recommendations that connect to core digital humanities values. The bill of rights and subsequent discussions of equitable project team credit bodes well for the future of digital humanities and sets a standard for the field. The INKE project provides a useful example of credit attribution:

> INKE establishes collective intellectual property provisions, specifying that all research materials generated in the course of the project be deposited in a "research commons" for shared access among team members. Work in the commons is understood to be open to reuse and publication by any INKE collaborator, "with full acknowledgment of that work's origins." For "presentations or papers where [INKE itself] is the main topic," the charter specifies "all team members should be co-authors." It also defines when, how, and where individuals should be listed for "named co-authorship credit" as active participants and defines situations in which an agreed-on corporate authorship notation (i.e., "INKE Research Group") is appropriate. Postdoctoral fellows and student assistants are specifically identified as eligible for equal acknowledgment when making "significant contributions to INKE's research." Project leaders are instructed to pay special attention to mentorship and to the professional growth of such employees. (Nowviskie 2011c: 176)

The emphasis on shared data, appropriate credit for work, and the importance of mentorship are common themes in the formation of field specific guidelines.

A theme in digital humanities distributed credit guidelines is the importance of a neutral space in which collaboration might occur. Star and Griesemer's (1989: 387) concept of boundary objects, where collaboration "requires cooperation—to create common understandings, to ensure reliability across domains and to gather information which retains its integrity across time, space, and local contingencies" that "creates a 'central tension' in science between divergent viewpoints and the need for generalizable finding." Development of a project model that denotes boundary objects provides a way to allow a group to work around some of the conflicts that occur during project development. Many digital humanities projects are co-invented, but it remains difficult to create real use and meaning for all in the participating group and to appropriately give credit where credit is due. Developing a shared territory could be a successful strategy for fostering equal participation and creation of a stronger project that benefits from the shared expertise of all partners. If we think about the boundary object as related to space and place, and of neutral spaces as crucial to such shared work, then the laboratory model emerges as one that could allow us to foster an equitable collaboration. The concept of neutral spaces, of spaces that are not highly regulated, the desire to offer open access and shared space, are more akin to a design lab than a science lab and emphasize the importance of looking to various laboratory traditions to produce a humanities based digital humanities lab.

As the INKE guidelines make clear, the neutral space of the laboratory must not neglect the importance of mentorship. As we utilize the science laboratory model, it is beneficial to carefully review how we are training the future of digital humanities, our graduate students. Many digital humanities practitioners have used traditional research assistantships as a means of immersing students in a project, creating, in effect, an apprenticeship for graduate students. If we return to the idea of the archive as a laboratory, we effectively find bench space for grad students. But while we often give students project tasks, we are less likely to allow the student to carve out a problem that might become the capstone of their PhD work. In other words, we have developed an apprenticeship model that is far more utilitarian then research oriented, tied to specific tasks set by the managing faculty member or the parameters of the project. This is not to suggest that we don't need to teach students basic skills in digital humanities. Our digital humanities students "must first master the relevant aspects of the existing history of an artifact necessary to the research, and then figure ways to alter it to carry out her project as the new research problems demand, thereby adding to its history"□ just as do students who undertake science research. A science laboratory model emphasizes interdependence, shared scholarship and exchange of ideas, and a closer working relationship for faculty, staff and graduate student than the humanities dissertation model currently in place. Newly developed programs, such as The Praxis Program at the Scholars' Lab at the University of Virginia,[9] attempt to model the consolidation of training with graduate student centered research support. The Praxis Program is an exciting experiment that might prove an enticing model for replication in other programs. The project's success is partially based on the decision to locate the program in a neutral laboratory space. Clearly the project is designed to replace traditional research methodology courses with a more current set of skills for graduate students training to become contemporary digital scholars (Nowviskie 2012). The benefits of such training within a neutral lab space are myriad. The students receive cutting edge training from scholars who are doing R&D work on a daily basis. By moving students out of disciplinary departments the lab fosters the type of interdisciplinary work that we hope to achieve. Of utmost importance is the detachment of the training from the stasis of curriculum and degrees that move slower than most digital skills. Yet the program's detachment also raises the problem of what it means to remove such work from where most of our students and faculty reside, where most academic work is conducted, the department. Does such a separation ensure that most of those trained within the traditional humanities will be unaware of and untrained for the digital future? Does this approach exacerbate the split between graduates trained in post-digital humanities and those who were unaware of digital humanities during their formative graduate years? The Scholars' Lab has launched what I suspect will be a model adopted by many digital humanities programs in the years to come. But we might wonder if the move to laboratory based training and research will signal the shift from a digital humanities contained within traditional fields to a separation from the traditional fields of humanities.

In addition to utilizing the laboratory model to develop new training approaches, we might also look to the science laboratory for more robust and effective mentoring approaches. While it is true that science has not solved all of its problems with mentoring, reward structures in sciences tend to enforce a codependency that is not present within the humanities. The lead faculty member in a science lab is responsible for funding and recruiting students for particular projects, most often financed by hard won grant monies. To complete the grant and to publish the paper, the faculty member, or a postdoctoral surrogate, will train the graduate student, who will be supported from the grant and receive some form of publication. While the system is not perfect and not all research labs encourage research groups that are codependent, in the best functioning groups, the faculty member needs the student in a way that a humanities faculty member does not need the humanities graduate student. While I would like to see my graduate students in English publish, my career is minimally impacted if they do not. If a science faculty member does not work with the student to publish, the faculty member's publication rate is diminished, adversely impacting the faculty member's career. Certainly there are abuses of graduate students in the sciences, but the interaction between faculty and student occurs in a symbiotic manner that tends to spur higher quality mentoring and increased scholarly production.

The problem with moving to a symbiotic mentoring system is that most digital humanists are trained and housed within traditional humanities programs, which emphasize products that are produced by individual scholars, such as monographs and individually authored papers, not the collaboratively produced documents common to digital humanities. While the best humanities mentoring emphasizes working closely with students in reading groups, providing feedback to student writing, and discussing student ideas for scholarship, the structures of the disciplines do little to reward faculty for good mentoring. For digital humanists, the laboratory has become a space to challenge the isolated scholar model. Also students that produce individually authored work might congregate in the laboratory to learn skills, share ideas, and receive feedback. To encourage such work, many digital humanities laboratories are supporting scholars with fellowships or internships that emphasize different structures of working. But these are stopgap solutions, where digital humanities scholars must work within seemingly contradictory environments to produce what often times becomes double the work of those positioned within traditional fields. Ultimately we need to be sure that graduate students interested in training as digital humanists are allowed to utilize the form of scholarship that best suits their research projects, and a potential model for such interactions comes from the sciences.

In 1989 R. G. Potter called for a revision of literary studies; "What we need is a principal use of technology and criticism to form a new kind of literary study absolutely comfortable with scientific methods yet completely suffused with the values of the humanities" (xxix).□ Over twenty years later, we still have not adopted a model that selects the best of the two disciplines. "In general, we must acknowledge," says Liu (2004), "the profession of the humanities has been appallingly unimaginative in regard to the organization of its own labor, simply taking it for granted that its restructuring impulse toward 'interdisciplinarity' and 'collaboration' can be managed within the same old divisional, college, departmental, committee, and classroom arrangements supplemented by ad hoc interdisciplinary arrangements" (13).□ We need to work together, in shared spaces, to develop working models that best match our scholarship. We should not merely mimic the current institutional structures of the sciences nor the humanities. If we use the laboratory as a neutral space to foster the type of collaborative work we imagine is possible, we are using the laboratory as more than a space, but a symbol of our hopes. Patrik Svensson (2012: para. 2) notes, "the digital humanities often become a laboratory and means for thinking about the state and future of the humanities, as well as how this visionary discourse shapes the field and what that tells us about the current state of both the field and the humanities." Svensson identifies what seems to be the driving interest in the science laboratory. It is a space into which we can imagine our hopes for new practices.

Notes

1. Notable forerunners to the laboratory model are the Stanford Humanities Lab, launched in 2000, and the even earlier HUMLab, launched in 1998.

2. Lynne Siemens and Ray Siemens, "Notes from the Collaboratory: An Informal Study of an Academic Lab in Transition," Digital Humanities 2012, http://lecture2go.uni-hamburg.de/konferenzen/-/k/13921

3. The separation of pedagogy from research in the science lab is a current invention. Park and Daston note that the early labs emphasized a multiplicity of purposes from "the production of natural knowledge" to "a place of pedagogy, entertainment, and the production of goods." Katherine Park and Lorraine

Daston, eds. *Early Modern Science*, vol. 3, Cambridge: Cambridge University Press, 2006. Print. *The Cambridge History of Science*, 305. While labs would continue to serve multiple purposes for several hundred years, science labs would become increasingly specialized. See Stvilia, Besiki, et al., "Composition of Scientific Teams and Publication Productivity at a National Science Lab," *Journal of the American Society for Information Science and Technology* 62 (2: 2011): 270–83. Print.

4. See Patrik Svensson, "From Optical Fiber to Conceptual Cyberinfrastructure," *DHQ* 5 (1: 2011): no page. Print. Svensson offers an important meditation on the design of studio space might be articulated within a digital humanities lab.

5. Twitter feed, July 18, 2012.

6. Jonathan Arac also calls for work that is "laboratory" based. Jonathan Arac, "Shop Window or Laboratory: Collection, Collaboration, and the Humanities," in *The Politics of Research*, ed. E. Ann Kaplan and George Levine, New Brunswick: Rutgers University Press, 1997.

7. Borgman lists collaboration as one of the primary areas of exploration that the humanities might examine.

8. Humanities scholars have some forms of trained collaboration. Peer review offers a form of collaboration and co-authored scholarship occurs. However, the norm of the lone scholar continues to dominate training and practice.

9. For additional information about the Praxis program, see **http://praxis.scholarslab.org/**

29 A MAP IS NOT A PICTURE: HOW THE DIGITAL WORLD THREATENS THE VALIDITY OF PRINTED MAPS

Patricia Seed

When you see a printed map in a book, a calendar, or poster, you might well be looking at a picture significantly altered from the original map. As the general public has been increasingly aware, fashion magazines considerably transform front-page pictures of celebrities, leading to calls, and in some countries such as the United Kingdom, laws against any but the most minor retouching of advertisements.[1] As it turns out, unbeknown to many, printed maps have been subject to the very same process of airbrushing with one major difference.

While the editors and publishers of magazine covers of famous people have deliberately modified the images, museums and libraries have rarely played such a role in the alterations. Instead, many librarians and curators are only discovering now, to their great dismay, that printed reproductions of their holdings are inadequate, distorted, or in the most extreme cases, just plain false.

Print has long been the gold standard for reproduced maps. Images of maps in books have been judged more reliable, more trustworthy, and more accurate than those reproduced on microfilm or even in digital media. Part of this respect stems from the widely held conviction that print materials are inherently trustworthy. That conviction rests upon knowledge that one or more experienced or knowledgeable individuals have vetted the book prior to publication. Additonally authors presumably have closely scrutinized the final version. As a result printed maps have shared in the generally higher repute accorded published texts.

As the digital world has struggled with means of validating knowledge, it has suffered from a generally held belief that digital and online information are inherently less reliable. In a twist of fate, however, digital technologies have recently exposed previously unrecognized flaws in print maps; deficiencies that fundamentally challenge the reliability of maps in print. The reasons for this challenge reside in the history of map reproduction.

Prior to the era of high-resolution digital scanners, cameras, and high-powered software, images of maps were relatively hard to acquire. Most maps were either transported to photographic studios where a professional photographer created a print or a slide of the map. Often libraries and museums would wait until they had a significant number of requests before sending the images over to the studio. In cases where the maps were extremely large, the photographer would bring his equipment to the museum or library. However, that visit had to coincide with a day when the museum would be closed. As a result, through the end of the 1990s, photographs were expensive and slow to be created and transmitted to the individual scholar, library, or publisher.

Furthermore to compare a printed version to an original was a daunting task. Printed map books are often large, very heavy, printed on thick, coated paper, and highly expensive. Carrying around a twenty-pound book in order to compare one map to the original remains impractical. Most libraries and archives holding rare maps refuse admittance to anyone carrying or holding a book for fear the book could be used to conceal stolen maps. Cutting an image out of such an expensive book for comparison with an original remains equally unrealistic. Even then, comparison of a removed book page with an original would remain impressionistic and imprecise since the map could only be placed alongside the original.

Ironically, earlier, nondigital methods of reproduction (Xeroxes or printed copies) failed to raise any of the challenges introduced by digitization. Some copiers introduce subtle changes in the size of objects, which might not disturb the overall impact of an image but might alter the location of a place on vertical and horizontal axes. For example, a copy might bring Umeå closer to Stockholm (vertical axis) or move London closer to the English Channel (horizontal axis).

Colors have also been integral to the composition of the map in many traditions. For example, a red color signified south in traditional Chinese directional maps based on the heavens, and a red–green contrast became critical for Mediterranean nautical charts over six hundred years ago. While color copiers have improved, they vary widely in the ways they represent colors' hues, chroma, tints, and shades. Grays may be bleached out; high contrast reds muddied, and so on. Without the red for orientation on the Chinese map or the red–green contrast for Mediterranean charts, a scholar might wrongly interpret the map. To lose the correct color would alter both understanding and interpretation.

Digitization, however, has profoundly altered the study of maps on many levels. Very high-resolution digital images allow for more detailed study than possible with print or even conventional magnification. Customarily stored as well as laid out on a flat surface, huge maps—2 by 4 meters, for example—remain very difficult for a human to physically examine beyond a narrow border. With a digitized version, lines, shapes, colors, and names in the center of the map can be scrutinized in a way not previously possible.

Comparing digital pictures to printed maps has led to increasing recognition of the differences between the original maps and their representation in books. For purposes of comparison with an original, digitization has primarily vastly improved the portability of maps. The low cost and widespread availability of high-quality digital cameras has made it possible for researchers to photograph printed maps from impossibly heavy as well as from rare or remotely located books and bring their digital images of print maps to compare to the original. Instead of a weighty tome, a one-ounce thumb drive allows researchers and curators to transport an image of even a very large printed map. The digital image of the print on a screen, projector, or even a printed version of the digital image on the same scale as the original permits comparisons not previously possible. For example, if you take a digital photograph of the map in a book, calendar, or poster, you can print that image out using a translucent or transparent material. If you overlay the image you have printed out on top of the book, poster, or calendar, you can adjust the proportions so that the digital print exactly matches the map in print. When printing to archivally safe paper such as vellum or mylar, you can safely place the image of the reproduction on top of the original. Since the original will remain visible through the translucent material, you can compare and measure major differences as well as smaller distinctions between the map in the book, and the original.

In addition to cameras, a second digital technology has dramatically improved the quality of map reproductions. This second change rests upon the vast improvement in scanners. Beginning in the mid-1990s, scanners became capable of reproducing higher quality images from books. As scanners became increasingly more complex, their manufacturers began to include features such as retaining the *x* and *y* axes and adjusting the color saturation or hue (Allman et al. 2007). More precise than any form of photography, digital scans can be calibrated to allow more accurate reproductions of the maps.

Suspicions that not all scanners performed equally well led to a critical reevaluation of machines used to digitize maps. One of the pioneers was the Catalan Institute of Cartography (Institut Cartogràfic de Catalunya). Near the end of 2005, technical staff members realized that the scanners were producing map images of widely varying color hues, chroma, and resolutions. Recognizing the importance of accuracy and consistency, members of the Institute conferred over the course of the next year and agreed upon standards for a new map scanner. In 2007, they held an open international competition inviting manufacturers to scan the same maps and compared the resulting digital images with their originals. Eventually they discovered, to their surprise, that the best scanner came from a boutique Italian firm that provided superior results to those of better-known finalists from Germany and Japan (Roset 2012).

All of this digitization and improved access to maps, however, has resulted in an unanticipated discovery—that the image on a printed page may not actually reflect the map.

In 2011, the publisher of a popular calendar of maps in a European Union country asked a relatively obscure library in that country to include a photograph of a map in their collection for a popular calendar. Although the cartographer of the map was widely esteemed and had composed a variety of maps of the area, this calendar would mark the first time that the version in this library would appear in print. The library consented, the photographer took the picture, and the calendar subsequently sold with great success. Viewing the image in the calendar, a well-known historian concluded that the copy in the more remote library belonged to a certain style of maps that this cartographer had designed. When this scholar arrived at the distant library expecting to see the map pictured in the calendar, she was dismayed to discover that the library's original was a significantly different variation of the map than the librarians had trusted the printer to publish. The librarian's faith in the photographer's integrity was shaken when the scholar requested the original from the collection and placed it alongside her digital enlargement of the calendar image. Only then did the curators become aware of the substitution.[2]

While a digital image enabled the discovery, it required comparison with the library's original to complete the evaluation. Thus the digital representation created an affordance, the quality of an object that allows an individual to perform an action, in this case, comparison done by a skilled human hand and eye. Digital technology's role as an affordance in this and similar cases remains an often overlooked aspect of its function.

As it turned out, the calendar had simply published an already existing and widely circulated image of a map from another library. Whether the mistake was deliberate or inadvertent, the substitution of a photo of one map for that of another showed the absence of prior inspection of the printed map. Scholar, the general public, as well as the librarians themselves were misled.

Not all such alterations result in the complete substitution of one image for another. More subtle changes, similar to those performed on magazine covers are sometimes deliberately made.

The logic behind these deliberate actions taken behind the scenes remains impeccable from the standpoint of marketing. In the early 1990s, on the occasion of the quincentenary of Columbus's arrival in the New World, the imaging department of the French National Library (Bibliotèque Nationale de France), on its own initiative, produced a very attractive poster of a map reportedly created by Columbus, which it sold in the library's gift shop.[3] Seven years later, I had the opportunity to see the original for the first time and was struck by the very different background color as well as the overwhelming number of deep creases, disconnected lines, and stains obvious on the original. Without any great expectations, I visited the library's imaging department to see if I could acquire a digital picture of the original, furrows and all. The official immediately called up a remarkably well-scanned image of the stained and crinkled original. When I asked why that image differed so much from the one appearing in the poster that they had been selling only a few years before, he looked at me oddly. If we were selling an image of the obviously damaged original, he replied, how on earth would we have been able to sell it? Of course, they straightened the lines, evened out the colors, and eliminated the creases. Only a carefully photoshopped image would prove popular with purchasers. He was right. Few, if any, visitors to the exhibition would have found the original appealing. The shop was selling poster maps to commemorate the show. In short, the same rationales for editing photographs on the cover pages of magazines were used to sell copies of maps.

However, any serious students of maps would have been led astray, had they mistaken the poster for the original. What appeared as a straight line between two different places on the map turned out to be nothing of the sort. One place had been slightly shifted so that it lay in a pleasingly straight line from another larger feature on the map. While the change was minor in an artistic sense, it would have led to erroneous conclusions about the original.

One solution for such publicity-worthy goals may be for the publisher to acknowledge that the poster is "based on" or "derived from" the original. While this step may deprive the bookstore of sales, because visitors think that they want a picture of the real map, this small acknowledgment would allow libraries and museums to sell publicity materials while at the same time acknowledging the re-composition or artistry exercised on the image offered for sale. It would also alert scholars that they would need to contact the library or museum before embarking upon a study of the map.

Although digitization can vastly improve the quality of reproductions, the same structural problems remain. Just as in the days of simple photographic reproduction, major libraries and archives on both sides of the Atlantic continue to grant a monopoly concession to an outside company to produce images requested by researchers, television, and newspapers. The company holding the contract follows its own customary policies and practices, usually without input from the guardians and those knowledgeable about the object. Institutional rules sometimes constrain map curators and staff even at several major EU national libraries from supervising imaging process within their own organizations. All they are able to do is complain, but the distorted and airbrushed images keep reappearing with the seal of their institutions. The need for such review is imperative as the following fortunate case illustrates.

In Switzerland two years ago, a map curator discovered by accident that a very expensive scan of a valuable map in the university's collection was being altered by the scanning company, just as the company was putting together the CD. Near the scheduled completion date, the map curator decided to go to check on the status of the scan. He returned to the reading room ashen faced. The original map consisted of three entirely separate sheets, but when he arrived at the imaging company, he discovered to his great dismay that the scanning organization, on its own initiative, was in the process of joining the separate scans together, so the parts would align perfectly. He was able to call an immediate halt to the proceedings, informing the imaging department that for many centuries the original had been kept in three separate sheets for good reason—and that the scans needed to reproduce the sections. The imaging department recovered the original scans and released the CD the next day. But for his checking the status in person, the final scan would not have shown the original map but a distorted one. This curator, renowned for his attention to detail, would have recognized the problem as soon as he looked at the final image. Had he uncovered the alterations later, the scanning department might have simply separated their photoshopped images from the original ones. But by visiting the scanning department in advance, he managed to prevent either a delay or an expensive disaster requiring rescanning of all three large sheets.

Severe repercussions have resulted from the publication of altered reproductions, and the unrealized distortions have misled even prominent writers on maps.[4] Over the past fifty years, a reproduction of a world-renowned atlas purportedly reproduced on the size of the original has widely circulated to libraries around the world. The print version, introduced and published by prominent European map scholars and printers, has served dozens of academics and researchers in lieu of the relatively inaccessible original.

Despite its proclaimed fidelity to the original, this widely circulated reproduction reduced the atlas's pages by varying degrees. Some were reduced by 3 percent, some by 7 percent; others by 12 percent, and a final striking example reduced the original sheet by 33 percent. However, an opportunity to study the original or even a reproduction on the size of the original would have shown that the reproduction had significantly altered the map. The status of the author and publisher misled a number of unsuspecting prominent map scholars into publishing mistaken arguments about the map and its controversial method of composition. While no one knows exactly how this alteration in scale came about, the process probably developed as follows. The well-known historian of maps wrote an introduction, preface, or careful explanation of the group of maps. He studied the original with great care, and took meticulous notes on his observations. When the printed text was sent to him for revising to catch any typographical errors or other mistakes, he reviewed the text carefully. He might or might not have seen the reproductions of the map prior to their printing, and if he did, he probably gave it a perfunctory look (relying on the reputation of the photographer to have performed his task properly) or he might have forgone this step entirely because he placed the image as secondary to his textual exegesis. As a result the printed map failed to undergo the kind of scrutiny usually accorded the printed word and appeared in print without having been subject to further scrutiny or evaluation.

Passing over the visual material in a print medium still occurs frequently today, since no recognized standards or methods exist to evaluate the accuracy of a printed map. While the proofs of words are carefully inspected by the author, no similar scrutiny is accorded the map. Such examination is most definitely needed at two stages in book production.

From the photograph of the original map to its layout and printing, opportunities for alterations multiply. The reasons reside in the fact that imaging departments, printers, photographers, and scanners have adopted the same approach to maps as they have to other images. In other words, they have treated maps as pictures and strived to create the most aesthetically pleasing object to place in a book. In both processes—creation of the image and its placement for print—they may correct the color or the contrast, straighten out crooked lines, and eliminate bumps and wrinkles so that the printed object will look attractive in a publication.

During the preparation of the book for its printing, alterations may creep in, just as they do in straight forward text. While getting the manuscript ready, press staff may crop an image, increase the contrast, or change the colors or proportions so that the image better fits the flow of the text. Such alterations can be checked if caught when reviewing the proofs.

A second stage in which alterations may slide in occurs during printing. Printers may correct lines, heighten color contrast, or brighten or darken underlying colors, sometimes unbeknown to the publisher. In interviews with printers who work for major US academic and commercial publishers, all admitted that they had frequently altered or "fine-tuned" images in the course of printing. While major color shifts would be noticed, none of the printer's finer adjustments had been detected, nor had any of them ever had their rectified images returned. Both additional sources of potential distortion in the map suggest that maps deserve the same scrutiny accorded the text prior to appearing in print.

Minor alterations to colors, boundaries may be suitable for artwork, since paintings, drawings, and sculptures often convey an emotional rather than a literal understanding of the scene, person, or object. So the longstanding practice of brightening and sharpening images to make them clearer or more attractive is understandable, especially when seeking to reproduce the emotional message of the artwork effectively

Often a single photograph or scan may leave an important area of an artwork obscure or shaded, despite help from lighting experts. Because of the expense of color, several photographs lit from different angles are sometimes joined (photoshopped) together so that the reproduced image shows all parts discussed by the author, instead of what a viewer would see gazing at the actual painting, drawing, or sculpture.[5]

But maps are not pictures. Although they are also visual images, they depict spatial relationships. Hence any effort to visually enhance the original image may alter the relationship of space—size, distance, or direction—between objects depicted on the map. However imperfectly the map-maker might have represented space, changing those relationships in the process of printing distorts what the original drafter was trying to convey in haphazard and random fashion.

Even minor alterations can distort a map. As noted earlier, altered colors can significantly mislead readers of a map. If a photographer or image processor changes colors, the identity of the original feature is lost. Beyond this element lie additional reasons why maps cannot be treated as illustrations.

Let us take-maps that represent the earth as an example.[6] Since the world is a three-dimensional object, it has to be flattened into a two-dimensional one in order to become a smooth surface, or a map. Taking a simple paper globe and smashing it would yield torn pieces of paper and a two-sided shredded object—that could only be seen from a single side at a time. In order to create an orderly solution to this problem, mathematicians and astronomers since Hipparchus in the second century BCE have devised the many systematic means of flattening the earth.

Since any three-dimensional object cannot be translated into a two-dimensional one without some type of distortion,[7] professional map creators choose a particular representation to conserve specific features such as area or direction, depending on the purpose of their maps. Distorting a line here or a shape there would alter spatial relationships and render the retouched printed map unsuited for its objectives. If a map was created to compare areas of decimated rainforest, a prettied map could lead to targeting the wrong region for reform. If the map were showing the direction of a river, the loss of fidelity could result in poor planning for irrigation projects. The consequences for altering maps are, potentially, highly significant.

In short, unlike texts, there exists no standard process for vetting the quality or accuracy of maps reproduced in print. Nor is there any indication that such standards existed in the past. However, clearly standards are needed. First, curators and librarians need to be brought into the process of evaluating the digital image of a map rather than being excluded. Visiting the scanning department, as the curator did in Switzerland, is one such opportunity to provide better digital images from the outset.

Second, when a map is altered for publicity purposes, the copyright notice should state clearly that the image is "after the original," thus acknowledging the changes. If the publisher seeks to create fictional, visually lush images for coffee table production, that objective (and the techniques used) needs to be acknowledged in the beginning. When accuracy is the goal, publishers' production teams need to reframe their practices prior to their printing, verifying the proportions and colors of originals rather than taking the liberty to create sumptuous illustrations, unless that is their openly stated objective.

A strategy for so doing might adopt the standards outlined earlier for the evaluation of a print map at a museum or library. If the image is blown up so that it is the size of the original, and printed in grayscale on a translucent sheet of paper such as vellum or mylar in the fashion described above for verifying the print reproduction of an original, the production staff as well as the author would be able to judge whether the image producer (photographer, scanner, company, or department) had altered the content of the map in any significant way. Ironically, the digital age has brought any previously unknown retouching of maps into the open, making it easier for scholars and curators to recognize changes made to maps appearing in print and challenging the validity of the one-time gold standard for reproduction.

Notes

The author wishes to thank the editors of this volume for their helpful comments. Patrik Svensson provided a wonderful opportunity to visit the Digital Humanities Lab at Umeå University where I enjoyed his and his colleagues enlightening discussions. Additional thanks are due to Kim Ricker, Jean Aroom, Francis Hebert, the staff of the Institut Cartogràfic de Catalunya, and the audience at the 7th International Workshop on Digital Approaches to Cartographic Heritage in Barcelona for their support and helpful commentary.

1. The United States Better Business Bureau's National Advertising Division banned a misleading photoshopped advertisement in 2011: Jim Edwards, 2011, "US Moves toward Banning Photoshop in Cosmetics Ads," *Business Insider*, December 16; Christine Haughney, 2012, "Who Can Improve on Nature? Magazine Editors," *New York Times*, July 20; Jessica, Seigel, 2012, "The Lash Stand Will New Attitudes and Regulatory Oversight Hit Delete on Some Photo Retouching in Print Ads?" *Adweek*, May 29. The United Kingdom's Advertising Standards Authority has made similar bans. Fashion and Apparel Team: Sheppard, Mullin, Richter & Hampton LLP, 2011, "About Face: Lancôme's Airbrushed Makeup Ads Banned in the UK," *The National Law Review*, August 21. The UK's advertising authority also maintains a website where the public can post complaints: http://www.asa.org.uk/.

2. Greek map historian, personal interview, Barcelona, 2012.

3. "Carte dite Carte de Christophe Colomb," Bibliothèque Nationale de France, Paris.

4. Mercator, G., & Hoff, B., 1961, Gerard Mercator's Map of the world (1569) in the form of an atlas in the Maritiem Museum "Prins Hendrik" at Rotterdam Reproduced on the Scale of the Original. Rotterdam: Maritiem Museum.

5. John Berger pointed out the selectivity of reproductions. "Photographs are not, as is often imagined, a mechanical record. Every time we look at a photograph, we are aware, however slightly, of the photographer selecting that sight from an infinity of other sights." John Berger, 1972, "Ways of Seeing," London: British Broadcasting Corporation, 10.

6. Celestial and galactic maps also use the same techniques for producing two-dimensional images.

7. Carl Gauss's (1777–1855) "Theorema Egregium" or "Remarkable Theorem" provides mathematical proof of the impossibility of simultaneously retaining direction and area. Carl Friedrich Gauss, 2007, *General Investigations of Curved Surfaces*, trans. Adam Hiltebeitel and James Morehead, Bel Air, CA: Wexford College Press.

30 SPATIAL HISTORY AS SCHOLARLY PRACTICE

Zephyr Frank

It is a summer morning and the Spatial History Lab at Stanford University is running at full throttle. Large tables configured in rectangles bristle with screens and keyboards. Students, staff, and faculty hunch or slouch at nearly every available workstation. The coffee mugs are half-empty and the sound of half-a-dozen conversations, carried on in hushed voices, which rise from time to time into animated exchanges, filter through the fourth floor of Wallenberg Hall. This is what practicing spatial history in the lab looks like at first glance. It is a collective enterprise, diverse in objects and methods, bound together by a common desire to work in teams, to learn new technologies, and to conduct exploratory research through the creation of models of spatial relations and movement over time.

This chapter aims to describe the practice of spatial history in one laboratory at one particular point in time. In doing so, by way of contextualization and contrast, it also addresses the broader field of spatial history (in necessarily truncated and general terms) and assesses the development of the field over the past decades. In short, the purpose here is to show what it is that we are doing in the Spatial History Lab at Stanford and to explore how what we do is related to broader historiographical trends.

What this chapter does not intend is to provide specific operating instructions. A "how to" spatial history essay would necessarily read, in ten years' time, a bit like the early quantitative history how to essays, with their mentions of antiquated IBM punch cards and the like. Computer programs and their routines and quirks will come and go. The value of space as a concept for historical analysis and the opportunities afforded by new technologies and teamwork will remain. What is more, there is something in the truth-value of common, everyday experience. It is interesting to know something about the place where scholarship is conducted and the people conducting it. Spatial history teaches us to attend to the place-bound aspect of experience; it reminds us that academic knowledge emerges from processes in particular places, not from some abstract space of pure ideas and information, veiled behind the façade of a place-free, process-free abstract neutrality.

A Day in the Life of the Spatial History Lab

Start with an example. It is still the same summer morning and at one of the tables, Scott Saul, a professor of English and American Studies at the University of California, Berkeley, is discussing the creation of a short video introduction to accompany his *Richard Pryor's Peoria* website. Saul's research has led him to delve deeply into the history of Peoria, Illinois: the city of Pryor's childhood and young adulthood; the scene of his formative experiences and his first gigs as a comedian. Pryor was arguably the most influential (not to mention most hilarious and challenging) stand-up comedian of the twentieth century. Pryor's life trajectory, from growing up around the brothels run by his family in the red-light district of Peoria in the 1940s and 1950s, to his triumphant, coruscating routine captured in *Live on the Sunset Strip* (1982), is one in which the deep well of childhood experience in a small river city in Illinois offered the most enduring material for his comedy. There, in Peoria: the outsized characters, the idiosyncratic speech patterns, the odd mixture of insider and outsider status seen in his family and his own youth, all played out against the backdrop of a changing city in the throes of urban renewal, economic change, and the mutual incomprehension of race in 1950s and 1960s America.

Understanding the space of the city offers a way to understand Pryor. But the space of the city is complex; it is layered and changing. The traditional means of dealing with complexity and historical change have been the monograph and the research article. These time-tested forms are excellent in their own way and need not, should not, be abandoned; the idea of researchers like Saul is to supplement and extend historical and literary research through the use of alternative media—namely, in this case, interactive websites filled with links to primary documents and short filmed sequences designed to provide a visually rich narrative pathway into the material for novice users (becomingrichardpryor.com/priors-peoria/). Websites and films offer different points of access and alternative means of showing and telling history.[1]

Building a website of this kind is an exercise in horizontal thinking. There is no clear hierarchy. Rather, there are relationships, among people, among people and places. The advantages of this approach over traditional forms of scholarship are relatively easy to spot and tend to derive from the affordances of digital media. Among the advantages are (1) the possibility of creating multiple interlinked narratives; (2) the possible integration of images, maps, commentary, and primary sources in the same field of vision; and (3) the ability to curate and shape the reader/viewer's experience, with nudges as well as explicit direction, allowing for a hybrid experience in which exploration of the material is conducted with the aid of an unobtrusive but effective guide.[2] These advantages are all present in Saul's early version of the website. The site will offer a menu of categories, each linked behind the scenes to the others, allowing for just such a curated exploration of the spaces and the people of Peoria, which include images,

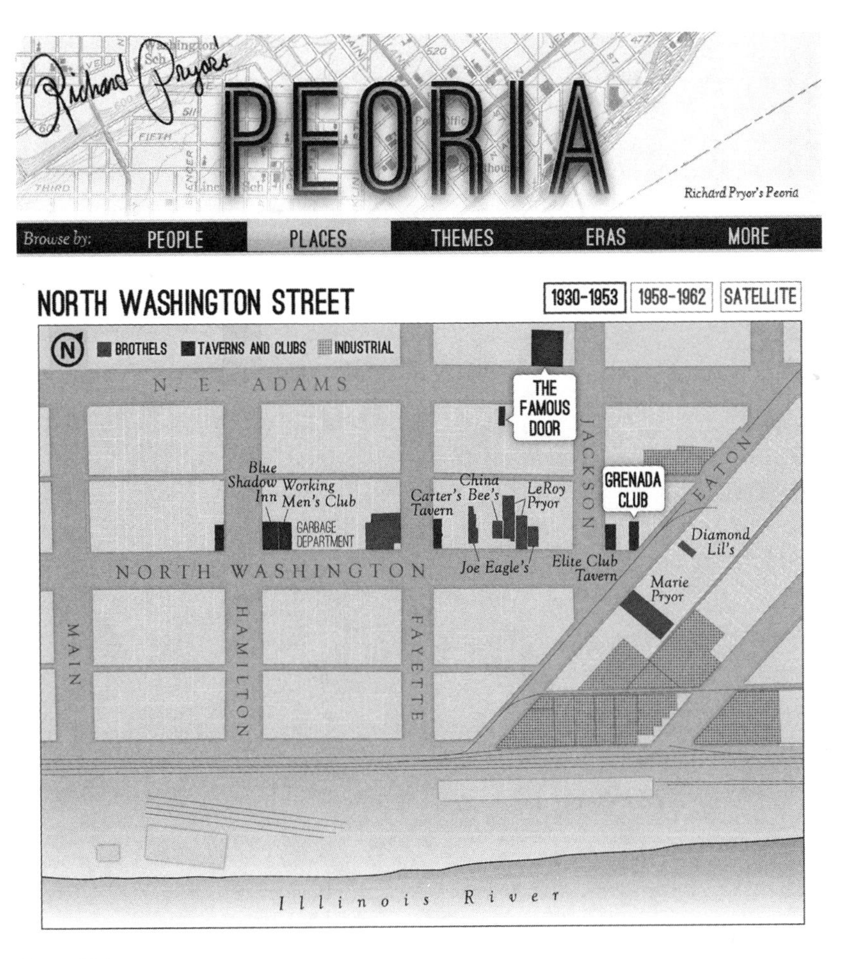

Figure 30.1
Places in Richard Pryor's childhood

commentary, and primary documents. Maps (figures 30.1, 30.2, and 30.3) are being created to allow for the display of detailed spatial information, locations of key venues and places of residence in Pryor's childhood, all linked back to primary sources and to the other themes. The neighborhood, reconstructed through the use of fire insurance maps, is shown in figure 30.1, the locations of two key venues where Pryor honed his comedic talents are highlighted in figure 30.2, and the result of urban "renewal" is shown in figure 30.3, as Pryor's old neighborhood gives way to a freeway interchange and bridge.

The disadvantages of websites built upon such horizontal and open thinking include, perhaps most critically, (1) the danger of getting lost, of missing major aspects of the larger

Figure 30.2
Coming of age in Peoria: First gigs

argument owing to the lack of a single unifying narrative line (the kind of thing a book provides) (Otter and Johnson 2000), and (2) the risk of simple cognitive overload or dissipation of energy in clicking through various strings of linked material on the site.[3] These problems have led Saul to create an additional resource to complement the website and the book that he is writing—short films. With a short clip, he can provide a single, compelling narrative voice (addressing the first problem), and he can at the same time provide a simplified and coherent pathway into the material (responding to the second problem).

A morning's work complete, a rough cut of the clip is ready to view. Working with research assistants and professional staff in the lab, Saul has selected the images and maps that will go

Figure 30.3
Lost worlds: The urban fabric transformed

together in this particular telling of spatial history. The resulting clip, of approximately five minutes duration, will eventually reach a level of polish and completeness adequate for "publication" linked to Saul's *Richard Pryor's Peoria* website.

Practicing spatial history, in this instance, means the creation of what might now be termed conventional digital media (an interactive website), albeit filled with carefully arranged content, including vectorized, interactive, historical maps and images keyed to locations along specific streets. It also means the creation of narrated clips designed to introduce key themes and to explain the functionality and conceptual underpinnings of the website. What is distinctive about Saul's project is its consistent emphasis on the spatial dimension of Pryor's childhood experiences and its creative use of the vernacular of the Web: interactive sites with tree and branch organization of linked primary material and short video clips, to provide readers/viewers alternative resources for thinking about Pryor and Peoria.

What difference does all this work make? Drawing on William James ([1907] 2000), we might ask: What is this information's value in *experiential* terms? What concrete difference, granting that the maps and clips are "true"? Does taking this approach to scholarship make?[4] It seems the answers, if affirmative, would have to be one of two things: either the maps and clips tell us something in a better, more efficient ways, or they show or reveal to us something that cannot be otherwise seen or revealed. In any case, spatial history aims at something much more than illustration when it elects to use maps and other visual representations of space. It makes arguments rather than illustrations.

Movement and Relational Space: Building Spatial Historical Models

Richard White (2010) suggests that a basic category of analysis in spatial history is movement. Space and time, brought together, become visible as co-present elements in history through the analysis of movement. Doing this kind of history invites the use of digital tools—more specifically, visualizations of historical data that capture something of the sense of movement. Making dynamic visualizations is time-consuming and requires specialized knowledge that historians generally lack. Spatial history, as practiced in the Spatial History Lab, is of necessity collaborative. Working in teams, with diverse skills in programming, database management, cartography, and visual storytelling, historians have at their disposal an expanded set of tools for the exploration, interpretation, and presentation of spatial data.

An example from White's project, "Shaping the West" will help show what is meant by exploration, interpretation, and presentation. The first term is perhaps the most important. The point of building digital models of spatiotemporal relationships is not necessarily to publish a visualization or static map. Rather, the model serves as a powerful tool for exploring data, for identifying patterns and discontinuities (White 2010). One early example of a model that works primarily as a research tool, even though it is also presented as a finished visualization, is "Seeing Space in Terms of Track Length and Cost of Shipping" (figure 30.4). The viewer is presented with a dashboard interface, with an abstract map of geographical distance and "cost space" composed of nothing more than the lines of the railroad and concentric rings indicating distances at regular intervals. To the right is a map of California highlighting, again, the railway lines as well as the topography and major hydrology of the state (figure 30.5). At the bottom of both of these rectangles, there are tabs allowing the user to toggle between physical

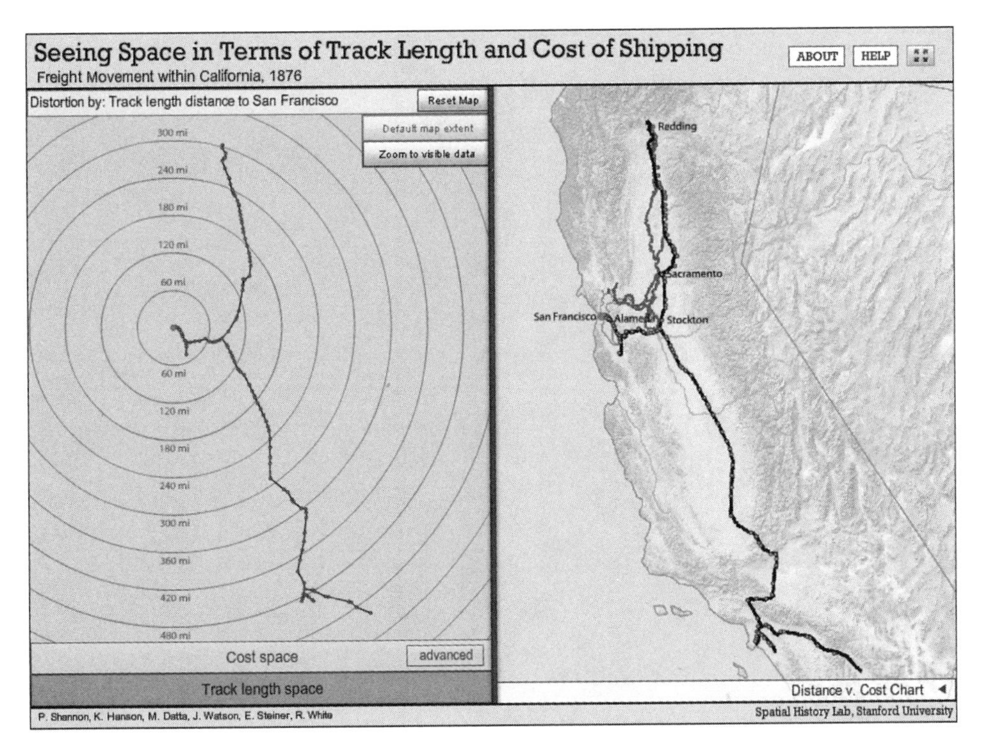

Figure 30.4
Railroad space measured by track distance

distance (length of track between points) and cost of shipping distance (reflecting the cost of shipping between points).

The visualization is disorienting at first. It is meant to be so. There is something inherently strange about using maps to distort space. The viewer expects the map to answer, not ask, questions. Exploring the spaces of distance and cost serves at least two basic scholarly purposes. First, experientially, the cost space toggle calls attention to the shifting and constructed nature of the space of railroads. The viewer sees the difference made by categories of freight, first class, second class, and so forth; interior towns move closer or farther from San Francisco. As the text accompanying the visualization suggests, this manipulation of space by the railroads, setting differential rates, helps us understand how economic power was experienced through spatial relationships. Second, as an exploratory research tool, the visualization calls attention to patterns and occasional ruptures in the fabric of spatial relations created by the railroads. In this sense the visualization can help formulate new research questions framed in a data-rich environment wherein specific places and relations can be explored without erasing the surrounding patterns.

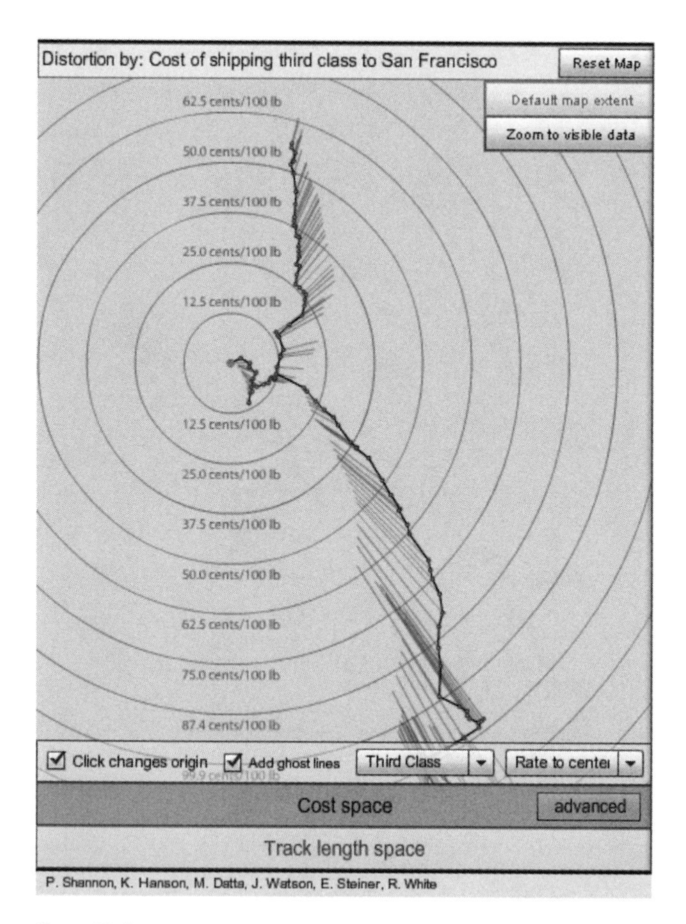

Figure 30.5
Railroad space distorted by cost of shipping

A similar project, developed independently from the Spatial History Lab by the classicist and economic historian Walter Scheidel, extends the idea of the construction (distortion) of space through the estimation of cost distances for the Roman Empire circa 200 CE. In the "Stanford Geospatial Network Model of the Roman World," Scheidel and his collaborators Elijah Meeks and Karl Grossner use a distortion cartogram to model distances according to season and origin. In a related text, the authors point out how:

> This perspective captures the structural properties of the imperial system as a whole by identifying the relative position of particular elements of the network and illustrating the impact of travel speed and especially transport prices on overall connectivity. Distance cartograms show that due to massive cost differences between aquatic and terrestrial modes of transport,

peripheries were far more remote from the center in terms of price than in terms of time. (orbis.stanford.edu)

Visualizations, then, are more than maps. They provide interactivity and the possibility to explore, to look for patterns, to experience, visually, the distortions and *differences in perspective* generated by humans through their spatial relationships.[5] This might best be seen as spatial history practiced through model building and data visualization. The tools and purposes differ here from the linked web content and film clips created in Saul's Pryor project. In common, however, is the use of space as a category of analysis—the insistence on the importance of spatial practices and relationships in historical explanation, whether of the trajectory of an individual through a changing city or the systemic and place-specific (contingent) relations between towns in geographic and economic space.

These descriptions of spatial history projects at Stanford are meant to show how this kind of research is being practiced in a particular time and place. The idea has been to introduce the range of projects and research questions that have emerged from this work, showing how spatial history can operate in both humanistic and social scientific fashion. What, however, is the state of the field at large? Where does the work described above come from and how does it fit in? The second part of this chapter attempts to trace out, in the briefest of terms, some of the major lines of scholarship in spatial history over the past century. It then concludes with an assessment of the future possibilities of the field in light of changing technologies and scholarly practices.

It is worth thinking about the origins and fortunes of the term "spatial history" as it is both older, in a sense, than much of the work in vogue today, and younger, as a widely circulated term, than its actual practice suggests. Judging by the use of the term according to citation indexes and Google N-grams, the approach has seen at least two "booms" in interest since the 1980s. The initial impetus came with the publication of several path-breaking theoretical works. First, perhaps, was Paul Carter's influential book, *The Road to Botany Bay,* which, at the time of its publication in 1987, was subtitled "an essay in spatial history." Henri Lefebvre's work, *The Production of Space*, and Michel de Certeau's theoretically sophisticated treatment of spatial practices in *The Practice of Everyday Life*, gave further impetus to the growth of spatial analysis in a theoretical vein throughout the 1990s.[6]

Something has happened since the 1980s to shift the ground under the term spatial history, to extend it in the direction of the transdisciplinary field of historical GIS, and at the same time to foster its development along the historico-linguistic path suggested by Carter's intervention.[7] This shift, and the diverging paths of research along these lines, makes the definition of spatial history, as practiced today, particularly difficult. Spatial history, then, is a cluster of distinctive research methodologies and historiographical concerns that cannot be subsumed easily into a single historical methodology or subfield—in this regard, the thematic and

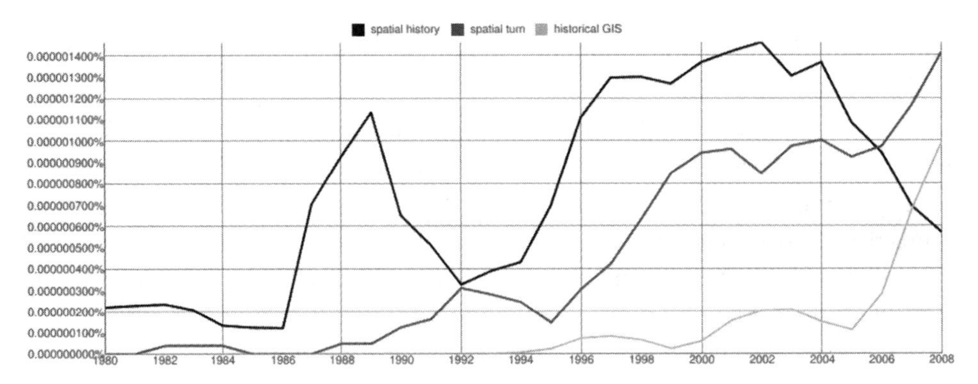

Figure 30.6

Dating the (relative) rise of spatial history, 1970 to 2000. Source: Google N-grams, one-year smoothing.

methodological heterogeneity in the Stanford Spatial History Lab reflects the state of the field as it has evolved.

This difficulty of definition is illustrated by the confusion regarding how best to categorize Paul Carter, one of the founders of the contemporary field, and his foundational text. Although *The Road to Botany Bay* is almost impossible to describe in a few lines, for our purposes it may suffice to note two crucial concepts of space that underpin the book. First, Carter defines space as a field constructed by cultural practices and social power rather than as a natural entity. Second, he develops an analysis of the linguistic construction of landscape, which is tied to struggles over power and knowledge—the language of the state and the language of local practice, the language of ascription and the language of fantasy. These conceptual elements resonate with the practice of historical geography itself, as this field of geography has moved further from its positivist-deterministic origins toward a critical engagement with the cultural construction of space and the articulations of power inscribed in spatial patterns and practices.[8]

The subtitle of the *Road to Botany Bay* has changed from "an essay in spatial history" to "an exploration of landscape and history," in the newly reissued version published in 2010 by the University of Minnesota Press. There also appears to be some confusion in the categorization of the book as "Geography" and the author as a "historian, writer, philosopher, and artist." The rise of historical GIS, I suggest, is part of the explanation. An increasing number of practitioners of spatial history have migrated toward historical GIS, just as a strong current in the discipline of geography has moved away from GIS and toward Carter's linguistic and cultural orientation. Notwithstanding these changing labels, I consider *The Road to Botany Bay* an exemplary work of spatial history at the same time that recent work in historical GIS is also spatial history of a different sort. The shared commitment to interpreting the past with reference to space and spatial meanings is what draws the two approaches together and, perhaps in the right hands, makes them compatible.[9]

I would like to suggest, leaving aside the problem of defining precisely what the term means, that spatial history as practiced today marks an incremental change in the practice of history (instead of a "turn," perhaps a series of more modest engagements). These engagements with space as a category for historical analysis are potentially significant but hardly revolutionary. Spatial history, particularly in the sense of Carter or Lefebvre, is more a general sensibility about the role of space in explaining historical structure and change as well as the importance of the historical construction of certain kinds of space. It is not, in this sense, best thought of as a technical term or a specific methodology, despite Lefebvre's claims in this regard. There is, of course, another sense of the term, which connotes spatial analysis in the coordinate space of two- or three-dimensional computer-generated maps, as part of the historian's toolkit. In this sense spatial history essentially means historical GIS. I would argue that all historians can improve their own practice with a stronger engagement with spatial history in the first sense; the utility of the second sense of spatial history is necessarily more limited, as is any tool-based technical method.[10] It applies to cases where the questions being asked are best answered with the specific tools of historical GIS—that is, generally, questions amenable to quantification, placement in real geographic space as coordinates of longitude and latitude, and thus "spatial" in an abstract Cartesian sense and in a unique localized sense.

Geographers, whether labeled historical or not, have, of course, contributed more than any other scholarly community to our understanding of space in the context of time—spatial history, in other words. Of particular interest for historians, the work of Hägerstrand and his followers in the mode of "Time Geography" introduced sophisticated new ideas and forms of visualization in the analysis of space–time interactions and processes. In this literature it is particularly noteworthy that the visualization of space and time is often depicted in creative graphics rather than maps meant to represent the reality of physical space. When space is represented graphically, it is usually paired with time in a simultaneous visualization of temporal change over a physical surface—such as in the famous time–space aquarium model.[11] Reaching back to a much earlier literature, one is drawn to the fertile ideas of von Thünen (1966) regarding central places and spatial economics. In particular, these insights have proved useful for the study of the articulation of cities and hinterlands, and the changing dynamics of this interaction over time with the development of new forms of urbanization and transportation systems coupled with changing practices in agriculture.[12] These ideas are picked up time and again in historical work, as we will see in a moment.[13]

Spatial history is not new in this broad sense,[14] but its practice, in mainstream history, has evolved in the more recent past, especially in the hands of environmental historians, for

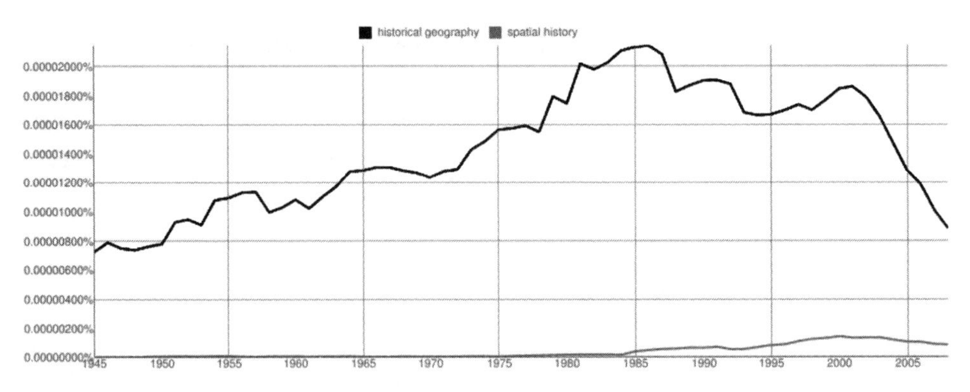

Figure 30.7
Reversal of fortune? The rise and fall of historical geography

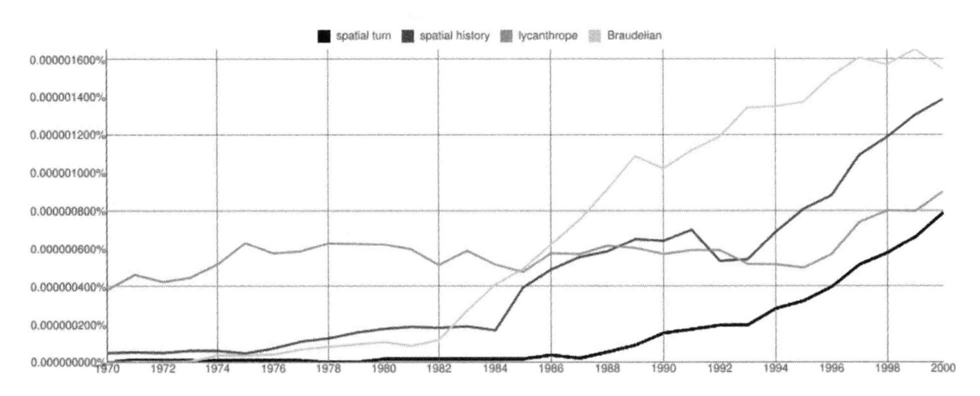

Figure 30.8
Order of things: From Braudelian, to spatial history, to the spatial turn

whom a heightened sense of the importance of space flowed quite naturally from the kinds of questions they wished to answer. For these scholars, as in the case of Paul Carter, the utility of spatial analysis was in its capacity to break apart traditional historical narratives, imperial or otherwise, that tended to posit space as given and thus essentially irrelevant to the more important considerations of human actors in the political, social, or economic realms of life.

During the 1980s and 1990s William Cronon published two major works in which space and movement of people and products played a key part in explaining historical change. Cronon's (1983: 21) first book, *Changes in the Land*, begins with a nuanced description of the way Europeans saw New England landscapes "in terms of commodities."[15] The similarity to Carter's sense of spatial history in *The Road to Botany Bay* is patent. In effect, what is seen is spatially and ecologically distorted: something, in time, to demarcate and simplify—in short, something that changes in historically significant ways and that is best understood as a spatial process of seeing, occupying, and modifying (Cronon 1983: 33). Cronon's masterful study of Chicago and its hinterland, *Nature's Metropolis* (1991), is Braudelian in ambition and in the consistent application of spatial categories of analysis, with chapters such as "Annihilating Space: Meat," in which the production and marketing of meat is understood as a complex spatial system (logistics) embedded in the political economy of nineteenth-century American capitalism. In constructing his argument, Cronon draws directly from Von Thünen's central place theory—showing, I argue, the importance of attending to much older strands in the literature when working in the vein of spatial history.[16]

If the works of the authors discussed thus far show the depth and sophistication of spatial history as practiced in the best of mainstream historical scholarship, then the work of scholars such as Edward Ayres and Anne Knowles represents the spirit of technical innovation associated with the use of GIS and other digital tools in bringing formal spatial analysis into historical practice. As Knowles put it early on: "Historical GIS can support very creative scholarship that raises new questions and offers a challenging alternative to existing methodologies" (Knowles 2000: 451–70, 464).[17] It can also call forth important questions, such as: "What is a meaningful period of time? What is a meaningful boundary in space? These questions lie at the heart of historians' current interest in regions and in our recurring need to understand how we parse the past" (2000: 465).

Ed Ayers and William Thomas's (2003) pioneering digital history project, entitled "The Valley of the Shadow," explores the histories of two communities in the Shenandoah Valley, one in Pennsylvania and the other in Virginia, around the time of the Civil War. What makes this project so significant is that it appeared in print, in the *American Historical Review*, as well as online in an interactive digital archive maintained by the University of Virginia's Center for Digital History. The online portion included maps created with GIS software portraying important environmental, demographic, social, and economic patterns in the two communities. By placing the Augusta and Franklin counties side-by-side, the Valley of the Shadow project used GIS to put forward an argument about the fundamental commonalities found in these two communities divided by the institution of slavery. The argument about the common patterns of life, except for slavery, was substantially strengthened with the addition of detailed spatial information: seeing these patterns, with one's own eyes in the carefully constructed pairings of maps provided by the authors, provides the kind of empirical confirmation that text alone would struggle to achieve, allowing for a layering and juxtaposition of information in spatial context.[18]

The sheer scale of the Valley of the Shadow project called out for extended collaboration, not just between Ayers and Thomas but also with the professional staff at the University of

Virginia Center for Digital Humanities. Indeed, part of what makes historical GIS potentially transformative in practical terms is the high degree to which it demands collaboration. This has led, in the case of Virginia but also in other institutions, including Harvard and Stanford to name just two, to the creation of centers for spatial history with considerable investment in technical capacity in GIS applications.[19]

Although geographers in particular have long used spatial statistics in their work, the application of these tools to history really only took off during the 1990s. Knowles, trained as a geographer but working primarily in the realm of spatial history, has helped take the use of GIS in spatial history to new levels of sophistication.[20] Over the past decade, she has published a series of articles as well as two edited books (sponsored, it should be noted, by the software maker ESRI), *Placing History* (2008), and *Past Time, Past Place* (2002) in which the computing power of GIS programs such as ESRI's ArcMap suite is highlighted and the significance to spatial historical research is examined from various perspectives. The approach taken by the diverse authors of the essays collected in these two volumes seeks to use quantitative methods in order to study the role of spatial variables and their distributions in historical processes and events.

An example from Knowles's (2000) research will help illustrate the way the computational tools provided by GIS can change the way we ask and answer historical questions as part of a general trend of using GIS to reconstruct past landscapes. Take the seemingly simple question: "What could Robert E. Lee see at Gettysburg?" An important question indeed, particularly if we are trying to assess Lee's decisions in light of what he could see and thereby act upon—that is to say, from his geographically specific and thereby limiting point of view. This question can be answered with a good deal of precision through the use of historical GIS and the toolkit available in GIS programs such as ArcMap. Knowles digitized the terrain surrounding Gettysburg and then employed viewshed analysis, a common tool in GIS applications, in order to delineate the space Lee could see from his known position on the battlefield.[21] The result of this study exemplifies the way historians can make creative use of tools in GIS in order to ask new kinds of questions with greater precision and to better explain a subject that might otherwise seem exhausted, such as the battle of Gettysburg. Knowles points out "viewshed analysis indicates [the spaces] Lee would not have been able to see, which were precisely where the greatest threats to his army were accumulating" (Knowles 2000: 255).

Spatial distributions alone do not, for the most part, tell the kinds of stories historians wish to narrate. They are like frozen slices of time where no change is discernible. What makes historical GIS *history* is more often than not the depiction and analysis of movement or change in spatial patterns, which, after all, is nearly the same thing—as movement is presupposed in a change of pattern. Knowles's viewshed analysis only seems static at first blush. What Lee could see according to his location on the battlefield intersected, over time, with the *movements* of the Confederate and Union troops. The battle comes in and out of view and historically significant insight is gained by this method.

Conclusion

This chapter is an experiment. In the first half, I have attempted to describe the practice of spatial history in a particular time and place; I did so mostly without recourse to jargon or reference to specific technologies. My sense is that what is really distinctive about how we practice spatial history (and how it is being practiced in many other places) is its collaborative and exploratory nature. By emphasizing everyday interactions and the process of research, my intention has been to pass along to readers the most important findings of our project. These are not to be found in specific research output, whether digital or paper, but rather in the loose but coherent community of scholarship built around the exploration of a common historical concept from multiple angles and for a multitude of ends.

The second half of the chapter is intended to read more like a traditional if brief and necessarily incomplete "state of the field" overview. Spatial history, much less historical geography, is a vast field. Notwithstanding these limitations, I seek to show how spatial history draws upon two major strands of spatial thinking—one humanistic, the other social scientific. In this regard, if nothing else, the point has been to suggest that space as a category of analysis, approached inductively and collaboratively, might provide grounds for reconciliation between these two cultures of scholarship. Historical GIS + *The Road to Botany Bay*. Not necessarily in the same project but, ideally, in the same space of scholarly work.

Notes

1. For a recent discussion of the use of film as a medium for academic history, see Vincent Brown's recent documentary, "Herskovitz at the Heart of Blackness," PBS (2009), wherein the role of the historian is not merely to provide commentary (talking head) but to conduct original research and shape the manner in which the film itself is constructed.

2. For a now classic statement on the use of digital methods in history and the advantages of hyperlinks and other modes of dynamic organization of historical materials of varying type (photos, maps, manuscripts), see William Thomas III and Edward Ayers, *The Differences Slavery Made*, also known as the "Valley of the Shadow" project. Their statement regarding hypertext, referencing the pioneering work

of Vannevar Bush, can be found at: http://www2.vcdh.virginia.edu/xslt/servlet/XSLTServlet?xml=/xml_docs/ahr/article.xml&xsl=/xml_docs/ahr/article.xsl§ion=text&area=intro&piece=presentation&list=&item=

3. For an overview of the concept of cognitive load and its relation to hypertext, see Peter Gerjets and Katharina Scheiter, "Goal configurations and processing strategies as moderators between instructional design and cognitive load: Evidence from hypertext-based instruction," *Educational Psychologist* 38 (1: 2003): 33–41. There is some dispute regarding the effects of "mouse-click fatigue." A common rule of thumb is that users should be able to find anything by the third click of the mouse. This idea is challenged empirically by Josh Porter in "Testing the Three-Click Rule," *User Interface Engineering*, (2003) http://www.uie.com/articles/three_click_rule/.

4. In James's famous formulation: "Grant an idea or belief to be true," it says, "what concrete difference will its being true make in anyone's actual life? How will the truth be realized? What experiences will be different from those which would obtain if the belief were false? What, in short, is the truth's cash-value in experiential terms?" James also famously suggested that "theories become instruments, not answers to enigmas." (James [1907] 2000). Spatial history as practiced in our lab hews to this stance. It tends to be inductive rather than deductive. It explores and modifies its premises according to the emerging patterns revealed by thinking about subjects in terms of spatial relationships. In this regard, our practice is closer to the stance taken in *The Road to Botany Bay* than, for instance, *Nature's Metropolis*, which is informed, though not determined, by an abstract model based on central place theory.

5. The distance cartogram is by no means a new idea. Braudel, for instance, uses a similar device to map out the space of the Mediterranean. The difference is in the inclusion of multiple variables, such as seasons and modes of transport, and the ability to shift the perspective of the map, from, say, Rome to London.

6. Citation counts derived from Google Scholar indicate that both Carter and Lefebvre became foundational authors in the realms of a linguistically oriented spatial history in the case of the former and a theoretically-oriented conceptualization of the production of space in the latter. Carter's book has garnered over 500 citations since publication in 1987; the translation of Lefebvre's *La Production de l'espace* has garnered over 6,000 citations since 1991. These numbers are approximations according to counts in Google Scholar. The Google N-gram depicted in figure 30.6 provides suggestive evidence of the timing of the booms in spatial history as well as the rise of the so-called spatial turn and historical GIS.

7. In fairness, it should be pointed out that historical geographers, such as Donald Meinig, were making similar moves toward linguistic/cultural interpretation of space a decade before Carter's book was published.

8. See, for example, the work of David Harvey, especially, for historians, his magnificent *Paris: Capital of Modernity* (New York: Routledge, 2003). The classic work of Donald Meinig helped lay the foundation for historical geography with a rich cultural dimension. See, for example, D. Meinig and J. B. Jackson,

The Interpretation of Ordinary Landscapes (New York: Oxford University Press, 1979), for an edited volume showcasing this approach.

9. David Bodenhamer, for example, suggests that historical GIS offers "the potential for a unique post-modern scholarship...that embraces multiplicity, simultaneity, complexity, and subjectivity" in his essay, "History and GIS: Implications for the Discipline," in Knowles, *Placing History*, 230.

10. My colleague Richard White quips: "historians sometimes write history as if it took place on the head of a pin." Spatial history would insist that different spaces and spatial-historical questions require distinctive vocabulary and writing/visualization strategies.

11. For an overview of Hägerstrand and the work of his associates, see the two volume collection by T. Carlstein, D. Parkes, and N. Thrift, eds., *Timing Space and Spacing Time*, vol. 2, *Human Activity and Time Geography* (New York: Halsted Press, 1978).

12. For a good online resource regarding the theory, see http://faculty.washington.edu/krumme/450/thunen.html. In the context of the field of historical geography in the United States, Carl Sauer's integrated regional approach was particularly influential.

13. The fortunes of the field of historical geography appear as the inverse in figure 30.7, albeit representing a much larger scholarly footprint, of spatial history and historical GIS. The former shows signs of impending collapse, the latter are stable or growing from a much lower and more recently established base.

14. Indeed, as figure 30.8 suggests, Braudelian has been a more common term than spatial history, at least through the year 2000. The influence of Braudel's approach, with its emphasis on geographic variables remains significant to this day and reference to his work is by no means merely an antiquarian interest. The unusual word lycanthrope is added in the (fun) spirit of keeping the meaning of "important" or "major" trends in some kind of perspective.

15. A few years earlier, Donald Worster's *Dust Bowl* presented an integrated spatial-cultural-economic framework for understanding environmental change. Interestingly, Worster's argument was revisited and revised a quarter-century later in Geoff Cunfer's historical GIS studies, "Causes of the Dust Bowl" and "Scaling the Dust Bowl," published in 2002 and 2008 respectively as part of the pair of volumes on historical GIS edited by Anne Knowles.

16. Cronon writes, for instance, "When one adds to the abstract models of central place theory this more historical perspective on capitalist expansion and colonization, one can read the bankruptcy court records in a new way" (284). Economic historians have also long used spatial analysis in their work. Distance and related costs are of paramount importance in understanding the development of markets and the patterns of trade. In this vein, we often find work inspired by historical geography. Concepts such as core and periphery or central place help orient research on the distribution of economic activity. Exemplary works in economic history with a strong spatial historical component begin with Robert Fogel's study of social savings and railroads and extend to more recent studies

such as Winifred Rothenberg's *From Market-Places to a Market Economy* (Chicago: University of Chicago Press, 1992).

17. Quoting Miller, 1999, to wit:

1. Do not use GIS to perform a mapping task that will be performed only once. Tracing paper still works fine.

2. Use GIS only with data that can be represented by points, lines, or areas.

3. GIS is most useful for representing and analyzing large populations; iterative or exploratory mapping; testing hypotheses that relate to mappable social entities; investigating inherently spatial patterns and processes; and integrating data from various sources.

18. The online version of the AHR article is located at: http://www2.vcdh.virginia.edu/xslt/servlet/XSLTServlet?xml=/xml_docs/ahr ; The digital archive of the Valley of the Shadow project can be accessed at: http://valley.lib.virginia.edu/)

19. At Harvard, the Center for Geographic Analysis, founded in 2006 and directed by Peter Bol, has developed an impressive roster of projects supported by professional technical staff. http://gis.harvard.edu/icb/icb.do?keyword=k235&tabgroupid=icb.tabgroup53821 At Stanford, the Spatial History Project, founded by Richard White in 2007, has evolved a model for project-based research and teaching supported by professional GIS and visualization experts. http://www.stanford.edu/group/spatialhistory/cgi-bin/site/index.php

20. For a good overview of the field of Historical GIS, see Knowles, "Emerging Trends in Historical GIS," *Historical Geography* 33 (2005): 7–13, the framing essay in a guest-edited volume.

21. Viewshed analysis has been used, as well, by archaeologists and architects to reconstruct sight lines and systems of visible/hidden spaces in both rural and urban settings. For a striking example of the use of viewsheds to reconstruct views of eighteenth-century Rome, see James Tice and Erik Steiner's "Imago Urbis: Giuseppe Vasi's Grand Tour of Rome," a web-based interactive mapping and rendering project hosted at: http://vasi.uoregon.edu/index.htm. This project is of particular interest inasmuch as it combines a map, by Noli, in plan view (from above) with contemporary eighteenth-century etchings, by Vasi, of buildings and their surroundings. The project also includes modern photographs of the same locations from the same point of view, allowing users to explore the sense of continuity and change in the cityscape.

31 UTOPIAN PEDAGOGIES: TEACHING FROM THE MARGINS OF THE DIGITAL HUMANITIES

Elizabeth Losh

In 2012 the international conference for the Digital Humanities met at the University of Hamburg, where the portico entrance of the main dome is emblazoned with the following motto: DER FORSCHUNG DER LEHRE DER BILDUNG.

The words roughly translate into English as a statement set in stone of the institution's equal commitment to "research," to "teaching," and to "education." The last term, *Bildung*, it should be noted, means something very different from education as indoctrination through a standardized curriculum, since the word has strong associations with individual character development, self-construction, and the philosophy of German idealism, thanks to Hegel and famed university founder Wilhelm von Humboldt.[1]

Yet to a disinterested observer watching the proceedings of Digital Humanities 2012, it was obvious that the specific vision of the digital humanities that the conference was promoting emphasized developing tools for research rather than tools for teaching or learning. In notes kept on a collective Google doc by DH 2012 participants, the word "research" appeared twenty-one times, while the word "teaching" appeared not at all. [2] The official conference program documents an event that included papers delivered on "Research Methods," "Facilitating Research," "Research Infrastructure," "eResearch," the "Virtual Research Environment," and "Academic Research in the Blogosphere." In contrast, "Teaching" appeared in the titles of absolutely no conference papers in the published program, although the word "Learning" did at least appear twice in connection with language learning and learning to "play like a programmer."

In a 2005 report about digital sharing practices with pedagogical resources for higher education in California, I complained that "trends favor the consolidation of traditional boundaries between research and teaching borrowed from 'real' bricks-and-mortar universities, which are reconstructed in the 'virtual' universities of the near future." I bemoaned the separation of "text encoding projects for elite groups of scholars" from "distance learning for the masses" (2005: 5). By 2012, I was even more pessimistic about the role of integrated approaches in the building of new online institutions in the United States, as the divide between the research propagated by promotion and tenure guidelines that is funded by prestigious organizations and the teaching outsourced to distance learning became even more marked:

The term "hybrid learning" has recently been appropriated by the distance learning movement to delineate the features of a specific type of educational experience that blends traditional lecture and Socratic discussion with online computer-mediated instruction. In many ways, however, this "hybridity" only reinforces traditional boundaries between learner and teacher, learner and learner, and teacher and teacher, because this kind of courseware-driven pedagogy ultimately only reifies certain ideologies of late capitalism oriented around efficiency, modularity, linearity, and surveillance in which the interfaces of so-called learning management systems are structured like a conventional teacher's grade-book, and spontaneous forms of improvisation made possible by the unexpected connections facilitated by a course syllabus and particular aggregations of students are constrained by highly scripted interactions. (2012e: 86–87)

Although forward-thinking instructional technologists have been working to develop more flexible and extensible alternative systems for the delivery of course content, the basic architecture of "course management" or "learning management" adopted for administrative convenience has very obviously been dictated by the philosophy of the scientific management movement of the last century. Of course, it is ironic that such scientific management thinking maintains its grip in digital pedagogy, despite the fact that other technology and design sectors of the economy long ago jettisoned its overly rigid labor model in favor of fostering more egalitarian, collaborative, and creative workplaces. Even "flipped" classrooms in which live instruction is devoted to time spent mentoring the hands-on practice that was once demoted to homework still may depend on some of the old paradigms of de-skilling at work in the assembly line.

Among other faculty in the University of California system, skepticism continues to be strong about the benefits of online learning. Many worry that the regimes of rationalization that digitization and distribution enforce would endanger basic academic freedoms. For example, the 2012 "Choices Report" puts forward several substantive objections to schemes for a virtual UC campus solely devoted to teaching and able to grant degrees from the same system that includes top research universities. The committee points to a long history of UC failure with remote delivery systems in the previous century, such as correspondence courses and televised lectures, and the fact that the costs associated with good online courses are frequently substantially higher than those of their equally effective face-to-face counterparts. The report also expresses reactionary anxieties from faculty and includes worries that the primacy of research would be eroded if teaching were to be recognized as a more utilitarian aim and fears that faculty would lose valuable personal intellectual property to others.

I examine here a much smaller subculture of pedagogical experimentation that has progressed independently from the larger distance learning industry, as well as from the more recent model of the MOOC ("massive open online course"). Rather than merely delivering, publicizing, or popularizing the offerings of faculty in higher education using coursecasts of traditional lectures as the basic unit of instruction, these pedagogical initiatives highlight the labor of other kinds of cultural curators and often put the onus upon students to produce the bulk of the actual course content. These utopian pedagogies aim to bring research, teaching, and *Bildung* together as integrated practices in the digital humanities and to highlight how new ideas around "critical making" change the fundamentals of the production and consumption of knowledge. Although I call them "utopian" pedagogies, they are often pedagogies profoundly about *topos* in the classical sense, which capitalize on geography and regional advantage in technological innovation, new forms of place-based teaching, and shared rhetorical commonplaces.

These are also pedagogies that draw attention to the procedural character of learning in ways that are potentially deeply subversive. Rather than present such procedurality as a set of predictable linear and hierarchical processes in which learning outcomes and assessment methodologies emphasize prescribed prerequisites and goals that are hard-wired into paths of sequential steps to success or failure, such pedagogies invite questions about rule-making and rule-breaking in general by facilitating activities that bring operational logics to the forefront, make explicit how particular forms of otherwise black-boxed code are executed, and unpack emergent phenomena in complex and heavily trafficked assemblages of users and platforms. At the same time these pedagogies are designed so that learners can be redirected from solely pursuing optimization or devoting themselves exclusively to autonomous leveling-up, badge-collecting, or other goals characteristic of what James Paul Gee has called the "school of one" paradigm.[3] Many of these pedagogical efforts use technology to facilitate new forms of dialogue that undermine the conventional sage-on-the-stage model of patriarchal mastery as well.

As admirable as many of these new teaching philosophies might be, should the combinatory freedoms of "remix" and "mash-up" always be uniformly or uncritically celebrated? I argue, by contrast, for the value of analysis, articulation, taxonomy, and what I call an "unmix pedagogy," which is now made possible with new computational tools for media visualization and signal or pattern fingerprinting. Furthermore I assert that pedagogy in the digital humanities needs to expand the information science model developed in the earlier era of humanities computing to integrate new insights from the information arts and that these pedagogies should borrow more from other informatics disciplines, particularly values-centered design and human-computer interaction.

Place, Space, and Pedagogy: Hypercities and Looking for Whitman

In another essay in this volume, Todd Presner describes the work of the UCLA/USC digital humanities project Hypercities as an exemplary case study that exposes key features of the current debate over the proper role both of "making" and of critical theory in the digital humanities and the possibilities for bringing practice and theory together in a sustained effort to critique present conditions in economics, politics, and the cultures of geosynchronous daily life. He calls for acknowledging the "mangle of practices, performances, constructions, social relations, and disciplinary powers" (Presner, chapter 4 in this volume) at work in digital humanities projects, for recognition of the fact of likely resistance or even failure, and the messiness of kludge-filled successes. Although Presner's essay does not explicitly present a thesis about pedagogy, many of the digital collections of Hypercities were created for classroom use, including material appropriate for "literature, history, culture and civilization as well as upper-level language classes" (Hypercities 2011).[4] Archived materials tied to particular locations on the coordinates of Google maps can be linked to historical renderings of the plans of cities from ancient Rome to Weimar Berlin. Since the project is designed as a mash-up that encourages further appropriation of the Hypercities software platform, Presner describes how graduate students took Hypercities in new directions that document real-time street activism in Egypt and Iran and diverge from the path of the more traditional urban history scholarship of their faculty mentors. Hypercities participants were also encouraged to literally take to the streets in bus-tours of Los Angeles's historic Filipinotown.

In *Divining a Digital Future*, Paul Dourish and Genevieve Bell observe that there are limitations to the "cartographic, Cartesian" "traditional approach" in which "space is understood as a manifold that can be indexed by a coordinate system" (2011: 80).[5] Dourish and Bell argue that an understanding of space, like an understanding of place, actually "arises from within different social practices." Ethnographers might argue that people experience urban history much as they experience stops on a subway system, from the subjective perspective of episodic spectatorship as each station comes into view rather than from the objective framing of actual distance traversed through geography on a 2D plane. In representing distance, subway maps ignore the ratios of such real aerial proportions in order to make the user interface of the colorful map legible and meaningful with evenly spaced stations unrelated to real distances. Similarly, in his earlier work, Presner has acknowledged the difficulty of presenting the sensibilities of the urban flâneur with the same Google tools of command and control created initially for the military and its goals of spacewar.

Matthew K. Gold and his team of pedagogical experimenters developed a somewhat different model of "place-based teaching." They used social media to teach a series of linked courses collaboratively that focused on American poet and rhapsode of the site-specific Walt

Whitman and the geographical locations that were significant in Whitman's life, including New York City, Washington, DC, and the battlefields of the Civil War. The Wordpress platform that Gold appropriated probably suggested different pedagogical approaches from the Google suite of mapping tools that Presner deployed. Gold's team also focused on tagging nearby historical sites that would facilitate experiences of educational co-presence. Yet both initiatives similarly utilized Web-based technologies to connect the college classroom to locations that might be remote temporally if not necessarily spatially, and both encouraged students to situate their own participation in the classroom critically in constructing the space and place of the historical witness. As Presner put it, such collaboratively built digital archives encourage "participation without condition" and spur discussion with broader publics, particularly if participants interrogate "various kinds of agency: human, material, computational, conceptual, and disciplinary."

Like Presner's Hypercities team and many other collaborative efforts in the digital humanities, the group behind the NEH-funded Whitman initiative also benefitted from what AnnaLee Saxenian has called "regional advantage." Much as there has been a large cluster of related digital humanities efforts—including Hypercities—sustained by collaborating and competing campuses connected by freeways in Southern California, there is a hub of innovation in the New York area in reach of a common transit system covering Fordham, NYU, The New School, and CUNY, where Gold teaches, where individual campuses serve as nodes in a local intellectual network sustained by geographical proximity.

Gold not only insists emphatically that the digital humanities must theorize its teaching as well as its research; he also argues that digital humanities pedagogy needs to be more inclusive and to embrace the community college student with as much enthusiasm as the student at the elite research university. He notes that "Looking for Whitman" addresses many types of students often overlooked by the digital humanities, including students on vocational education tracks or English Language Learners. Gold observes that "amid this talk of revolution, it seems worthwhile to consider not just what academic values and practices are being reshaped by digital humanities, but also what values and practices are being preserved by it." As Gold asks, "In what senses does the digital humanities preserve the social and economic status quo of the academy even as it claims to reshape it?" (2012,"Whose Revolution?").

"Looking for Whitman" foregrounds the labor of *Bildung* in classrooms that emphasize personal expression, individual subjectivity, and developmental growth. Gold's group piloted a pedagogical experiment that also seems to privilege what he calls the "model of the 'personal learning environment' and a 'domain of one's own'" in which "the project asked each student to create a personal blog for the course and to post all course work in it" (Gold 2012, "Looking for Whitman"). Although Gold draws on John Dewey's ideas about "collateral learning," collaborative work occurs in the context of particularized authorship, so that private and public activities coincide. As Gold explained, the model was of "small pieces loosely joined" to bring together "a number of different platforms and social networking applications into a confederated learning environment" where "loose connections between tools allowed students to take more control over their online learning environments and to mold those environments to their particular learning styles" (2012 "Looking for Whitman"). In a recent essay, Gold describes how students in "Looking for Whitman" created individual annotations, glosses for unfamiliar terms, and blog reflections, and even selected personal frontispieces to introduce their own work. Students also created a material culture museum, participated in tracking down information about Whitman's street addresses, wrote scripts for a visitor center, and produced translations into other languages of Whitman's work. Final projects consisted of online videos in which students could synthesize multiple elements of the course. By mixing students from many different types of institutions at many different levels, Gold and his collaborators claimed that they were putting into action many of the democratic principles that characterize Whitman's life's work.

Grounds for Dissent: Toward Unmix Pedagogies

The "Looking for Whitman" video projects that Gold (2012) describes as "mash ups and cinepoems" seem to have encouraged strong student engagement in this group of linked digital humanities courses and to have created products in which learners could play an active role as apparent co-constructors of knowledge. Nevertheless, it may be worth raising some questions about remix pedagogy at this point and the problems that can emerge when multimedia sources are inadequately curated with meaningful metadata or combined haphazardly in imitation of the aesthetics of advertising in popular culture. In other words, more incoherent forms of database cinema produced by students that borrow conventions from music videos may lack sustained and rigorous critique. It may also be a mistake to assume that supposedly subversive remix pedagogies automatically spur practices of democratic inclusion and egalitarianism, or that the aesthetic forms generated by participatory video editing and compositing practices are necessarily unproblematic. It may not always be appropriate to accept the assertion of Lawrence Lessig that a video remix "can't help but make its argument, at least in our culture, far more effectively than could words" (2008: 74) or to celebrate video remix unconditionally, as Henry Jenkins does, as an expression of "access, participation, reciprocity, and peer-to-peer rather than one-to-many communication" (2006: 208).

In looking at examples of remix video assignments posted on the Web, I am often surprised by the lack of specificity in the prompts about audience, purpose, format, and genre, especially since clear direction and concrete suggestions about appropriate process have long been

Figure 31.1
Example of database cinema from Al Jazeera news

recognized as valuable for students grappling with more traditional writing assignments. Faced with a vacuum in instruction, students improvise, which can stimulate expression that is just as likely to be banal as it is to be liberating, if not more so, given how our visual culture is dominated by database cinema. TV commercials, news broadcasts, viral videos, political attack ads, public relations films, video family albums, and K–12 class projects all may draw from heterogeneous source files culled from among the results of search engines, which are often driven by the metrics of targeted advertising and user popularity. Critics like Alexandra Juhasz who have taught courses where students watch and make video remixes have asked questions about the practices of commodification and consumerism that students might otherwise take for granted.

So I would like to make a radical suggestion: let's use digital humanities tools not only for remix pedagogy but also for unmix pedagogy. Faculty could present students with materials that juxtapose remixed content without explanation and ask them to unpack all of the visual references that the footage in aggregate contains. For example, in the "Official Prison Version" of Michael Jackson's rock video "They Don't Care about Us," there are clips from the Rodney King police beating and from protests by Chinese political dissidents in Tiananmen Square. Could students locate precisely where the original video was shot, when, and by whom? A variety of free and open source tools on the Web make it easier to identify the digital signatures of video and audio sources, and shot detection and facial recognition software might also be helpful in isolating discrete components in sequences or finding similar framings of specific historical actors. Even some commercial products can be helpful. Soundtracks can sometimes be identified by the popular iPhone app Shazam, which may provide useful clues about origin and authorship, and individual frames from videos can be fed into a Google image search to enable rough matches to similar shots.

The other digital humanities approach that could be deployed in student unmix projects is crowdsourcing, as students learn to supplement their own knowledge bases about the contents of remixed digital collections with input from others who might be more expert in identifying images, footage, soundtracks, or texts. As more digital humanities tools are developed that allow timeline-based annotation of video or frame-specific metadata, classrooms could be used to unmix rather than remix multimedia cultural artifacts, much as traditional forms of explication, close reading, word study, and lexical analysis allow scholars of print culture to trace significant patters of appropriation and reuse by untangling particular strands in the text.

My call for including this pedagogical approach is not intended to be reactionary nor to nullify the important forms of critical thinking facilitated by bringing hip-hop music, queer mash up videos, or other oppositional forms of remix into the classroom. As the pedagogical case study from my own classroom indicates, remediation, translation, and adaptation are central to my own teaching practice. Ironically, unmix pedagogy actually presents an invitation to appropriate tools that were often initially designed to further agendas of search engine optimization, status checking, surveillance, and the protection of intellectual property and re-task them for more liberatory pedagogical purposes.

Codes of Conduct: Considering My Own Classroom Practice

In recent years another rancorous debate about pedagogy in the digital humanities has developed over whether or not students should actually learn to write code in specific computer programming languages (Vee 2012). As discussions about the role of "building" or "making" in the digital humanities have become more heated, strong opinions about classroom practice emphasize either core competencies or the need for greater inclusiveness, with the programming camp arguing for the need for richer literacy practices in the digital age, and the anti-programming camp expressing fears that demanding specialized technical virtuosity would only magnify the differences between the digital haves and have-nots. However, those who argue for the importance of "procedural literacy" for all majors (Mateas 2005) sound very persuasive as more professions and intellectual pursuits require programming abilities, even if it is not always clear which department or group of instructional stakeholders—computer scientists, digital designers, librarians, or writing instructors—should be charged with teaching the basics to students (Alexander and Losh 2010).

Fostering an environment of inclusion matters when teaching students the computer literacy and programming skills commonly associated with success in science, technology, engineering, and math (STEM) disciplines. Young women and underrepresented minorities often find themselves on the wrong side of what Henry Jenkins (2009) has called "the participation gap." Many innovative programs have been developed for K–12 students to feel more strongly connected to competence with computational media (DiSalvo and Bruckman 2011; Harrell 2009; Resnick 2009; Gee 2009), but less attention has been paid to addressing the participation gap in higher education, where students often self-select into either technical or non-technical majors, and first-year mandated programming classes may generate ambiguous results (Simon 2009).

A creative writing class, particularly a class in poetry, might seem to be an unlikely venue for teaching principles of programming and interactive design to a diverse group of students, but that is precisely what the Digital Poetics course in the Department of Literature at the University of California, San Diego attempts to do. Taught in both a traditional seminar setting and in hands-on tutorials in a computer lab, Digital Poetics allows students to participate in lively exchanges about trial and error in a fault-tolerant, low-risk environment.

The metaphors of adaptation and translation (Ramsay 2007; Davidson 2010) are central to the course, because every other assignment asks students to adapt an existing poetic work created for the page with a new software tool. Working with Photoshop, Flash, Audacity, Dreamweaver, and Processing, students are encouraged to interrogate the definition of poetry itself, as they work with different fonts, screen layouts, sounds, and rules and consider the analogies between writing poems and writing computer code. Using UCSD's Archive for New Poetry, students also see examples of experimental work in the twentieth century that use poetic effects from the concrete form of the text on the page, sound, nonlinear composition, and chance.

Students eventually have to turn in a project created with Processing, a free and open source computer programming language popular among artists and DIY enthusiasts, which can serve as a stepping stone to learning languages like C++ and Java. However, my digital poetics class is designed around lowering barriers as well as raising standards. Students could turn in paper prototypes as part of approaching electronic literature much like other digital design processes. For example, one student brought in a selection of origami paper sculptures and another prototyped a project with index cards and string. Paper prototyping has been recognized as an important strategy for teaching digital design, and it helps many students visualize programs, systems, or interfaces more effectively. Barriers were also lowered by allowing submissions created with common computational media platforms such as blogs, YouTube, and even PowerPoint as students gained confidence with software that required more expertise or time peer learning or consulting DIY resources online. All students turned in at least one programming project, but programming was presented in the context of many different phatic, poetic, and metalingual forms of expression fostered by the class, and writing computer code was never presented as intrinsically better than creating work for print media or oral performance.

Like many other digital humanists, my own pedagogy uses the model of the laboratory and supports a paradigm in which access to computational technology should take place in an environment that fosters collaboration and experimentation. Hands-on access allows many parties to "drive" digital artifacts other than the instructor, and students can take part in acts of creative navigation, try out the modding, or even test systems by trying to break them. Yet many classrooms today configured as computer laboratories often constrain creativity and experiential learning rather than encourage it, particularly if students are lined up behind monitors in rows that isolate and alienate. Furthermore expensive technologies are often rapidly outmoded and poorly able to adapt to the new norms of ubiquitous computing. This approach probably reinforces an appealing set of assumptions about the impact and cost-effectiveness of technologically enhanced flagship spaces that can be displayed to the public, but I have also seen the benefits of a flipped classroom in which instructors can actively mentor and advise during periods of hands-on work.

My digital poetics workshop class uses a number of hands-on sessions to instruct students in how to use media production software packages, but sessions in the computer lab are only part of the larger structure of the course. Just as important is the final showcase, which includes an actual public performance in front of invited guests and a permanent link to works that are particularly effective. One student in the digital poetics class was actually chosen for the "Future Writers" section of the 2012 Electronic Literature Exhibit at the annual convention of the Modern Language Association. The emphasis of the course on what Garnet Hertz, following Matt Ratto, calls "critical making" also invites emulating other pedagogical situations borrowed from the discipline of iterative design including the "crit session" in which all students participate in making constructive suggestions and asking questions about the creator's artistic vision and intended audience.

Teaching the Information Arts in the Digital Humanities

Moving away from the information science paradigm so often associated with the earlier era of humanities computing to the information arts paradigm that many are now promoting doesn't mean abandoning existing collaborations with computer scientists and librarians or ceasing to recognize the importance of the common standards that took many years to develop. But it does mean broadening collaborations to initiate new kinds of partnerships that solicit input from those who recognize that user-centered design often resists one-size-fits-all uniform solutions and that people interact with computers emotionally as well as rationally. At Digital Humanities 2012 the lack of citation from the growing body of scholarship in the field of human-computer interaction was surprising, particularly given that so many digital humanists intend to build user interfaces for members of the public.

After all, making choices about code, platforms, and infrastructures is an expression of particular values, whether the intended tool is for teaching or—as it was in many of the presentations at Digital Humanities 2012—a tool for research. A new pedagogical initiative from the feminist collective FemTechNet, Dialogues on Feminism and Technology, is intended to make discussion of those values more explicit in a DOCC or "distributed open collaborative course" in which over a hundred feminist scholars of technology on several continents have attempted a huge team-teaching experiment by using an archive of shared resources and engaging an extremely diverse population of learners. The project brings together those working in science and technology studies with those working in media arts practice on many campuses and from many departments.

In undertaking this initiative in the spirit of feminist consciousness raising, co-founder Alexandra Juhasz reminds other digital humanists that "we already have a model of how people work well with each other: the classroom. We already know how to motivate people to engage there." Although traditional classrooms may be under attack by those who want to remake the university as a more open place for learning, Juhasz points out that not only do traditional classrooms benefit from shared experiences of embodiment and cohabitation in the same physical space and place, they also provide a "structure around which community can form, and where ideas can build, and records are kept." Juhasz insists that it is important not to lose "institutional support" when jettisoning conventional hierarchy. To adapt the classroom model to a less patriarchal design, "we could use technology to move from an isolated but dynamic space (the classroom) to a networked environment, making the embodied experience of the class and a distributed experience of the class simultaneously possible" (Losh 2012a).

As fellow FemTechNet co-founder Anne Balsamo explains, it is also important not to lose the traditional classroom's fault tolerance and its abilities to recover quickly from moments of failure. Balsamo insists that "failure is an absolutely necessary part of the learning experience." The ability to learn from failure—and to help students learn *how* to learn from failure—is a teacher's highest responsibility, she claims, both online and in the face-to-face learning environment where the courses will be piloted. "One of the features that learning online needs to replicate from the traditional classroom is the recognition of the social contract between student and teacher; students have to trust the teacher to create a space of learning that will be safe enough for failure." Classroom errors cannot have "life or death consequences" so that students can be comfortable enough to fumble and take risks. Unfortunately, she notes, "the Internet doesn't forget easily" (Losh 2012d). As beneficiaries of the rise of gender and race studies in the academy, Balsamo and Juhasz are reluctant to jettison the valuable legacies of decades of critical pedagogy that emphasizes the importance of recognizing affect, identity, transgression, oppression, and the co-construction of knowledge.

As revolutionary pedagogies evolve in the digital humanities, Balsamo and Juhasz's desire to preserve some of the values of classroom teaching while interrogating others should serve as a reminder of the ethical demands placed upon instructors to serve the needs of individual students appropriately. In balancing research, professionalizing practices of responsible teaching, and fostering the personal development of individual learners in the digital humanities, utopian pedagogies may actually be strongly grounded in place, space, text, and code.

Notes

1. For a more complete discussion of how the concept of *Bildung* relates to pedagogical theory, including Rorty's views on "edification," see Krassimir Stojanov. "The Concept of Bildung and Its Moral Implications" (New College, Oxford, 2012).

2. The word "learning" appeared three times, twice with "machine learning: and once with "embodied learning."

3. See James Paul Gee. *The Anti-Education Era* (New York: Palgrave MacMillan, 2013), 114–15.

4. Hypercities was also promoted as a platform for rich media scholarly publication. Successor products such as Neatline similarly advertise dual use in research and teaching. For example, Neatline claims that its "suite of add-on tools for Omeka" will allow "scholars, students, and curators to tell stories with maps and timelines." However, promotional materials for Hypercities foreground pedagogical uses much more explicitly than Neatline. See "Neatline.org | Plot Your Course in Space & Time," n.d. http://neatline.org/.

5. For a more complete discussion of the problem of Cartesian epistemology, see Anna Munster. *Materializing New Media: Embodiment in Information Aesthetics* (2006).

32 THE FACE AND THE PUBLIC: RACE, SECRECY, AND DIGITAL ART PRACTICE[1]

Jennifer González

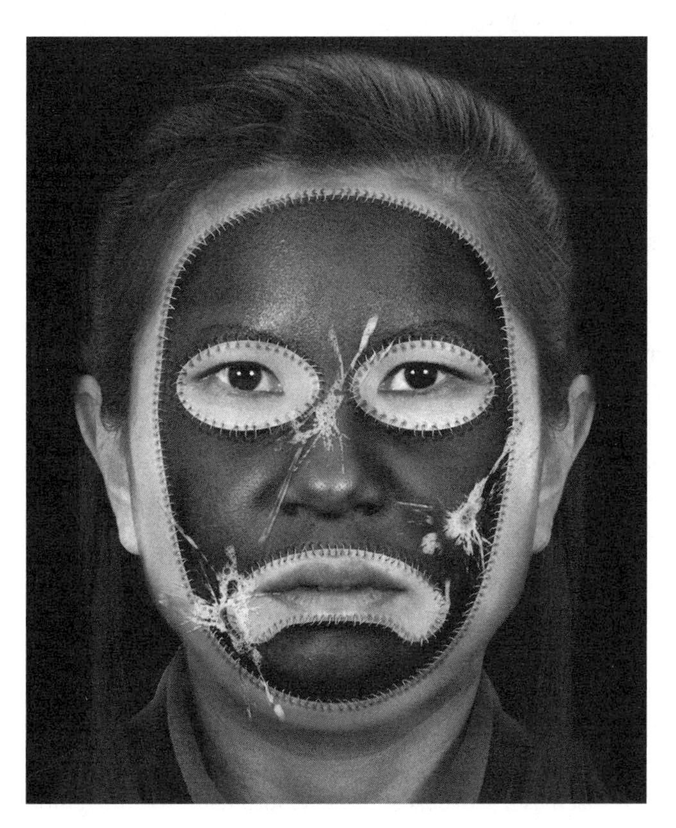

Figure 32.1
Colour Separation (Mongrel 1997), image courtesy of the artists.

The function and importance of race and race discourse in online digital spaces and in contemporary digital art revolve around an apparent paradox. On the one hand, there is a recurring desire to see online digital spaces as sites of universal subjectivity that can escape the limitations of race. This desire tends to intersect with assumptions about public space and systems of ethics that valorize the neutralization of cultural, racial, and sexual difference, as well as historical specificity. The apparently neutral space of the Internet is viewed as a potentially progressive domain for overcoming barriers that otherwise obstruct or restrict ideal forms of participation in the public sphere. On the other hand, a proliferation of racially marked avatars and experimental hybrids (human and nonhuman) increasingly populate artificial worlds and online chat spaces. Race, as a set of visual cues operating in graphical interfaces, has literally become a fashion accessory to be bought, sold, traded, and toyed with experimentally and experientially online (Nakamura 2002: 57). This proliferation of typologies and pseudoidentities provides the opportunity for the expanded display of difference, and this display seems directly and actively to undermine the prospect of the neutral, universal, online subject.

It is not a real paradox, of course, because both conditions operate in parallel to reduce cultural and racial difference to a question of appearance: the domain of visual signs. Online identity, participation, and power have become tethered to images (or their elision) for social and political ends. Questions arise, however, concerning how race discourse actually intersects with the Internet and with digital culture. [2] What are the conditions for ethical relations that entail encounters with racial difference? How do theoretical explorations of the face and the public bear on the subject? If vision and visibility are central to the operative dynamics of race, as has been argued by not only Franz Fanon but many others subsequently, then might it be possible to undo the power of race discourse as an oppressive regime by decoupling it from vision or the visible (Fanon 1967)? Or, alternately, might it be that visual culture is the very place where contemporary race discourse might be most powerfully critiqued and transformed?

These questions are central to recent theories of digital art practice that directly engage race as a dominant and pervasive visual discourse in an emerging public sphere. Technoculture is often praised for the ways it enhances democracy by realizing an ideal public sphere. But this view is generally inattentive to the fact that the experience of the technocultural public sphere can also be one of aggression, exclusion, and invisibility (Jakubowicz 2003: 219). Taking the writings of the media theorist Mark Hansen as a provocative and symptomatic starting point, this chapter explores how the desire for racial neutrality can lead to the unintentional repression of important forms of cultural difference. Two models of ethics, grounded in the writings of Giorgio Agamben and Emmanuel Levinas, respectively, are posed as alternatives in the quest for understanding the importance of the face as a device for the unfolding, or unmaking, of race in the public space of the Internet.

Universal Address

In 2004 Hansen published an essay titled "Digitizing the Racialized Body, or, The Politics of Universal Address," which was later expanded and substantially revised as a chapter titled "Digitizing the Racialized Body, or the Politics of Common Impropriety," in his 2006 book *Bodies in Code: Interfaces with Digital Media*. In both versions of this essay, Hansen argues that the Internet provides an unprecedented possibility for a new ethical encounter between humans, in part because it can render them invisible to each other (Hansen 2004a, 2006). Hansen observes that digital art can produce affective states in the user that might ultimately lead to recognizing incongruities or incommensurabilities between categories of identity and embodied singularity. Race becomes a lens for Hansen's thinking about online identification as making possible community beyond identity:

> Because race has always been plagued by a certain disembodiment (the fact that race, unlike gender, *is* so clearly a construction, since racial traits are not reducible to organic, i.e., genetic, organization), it will prove especially useful for exposing the limitations of the Internet as a new machinic assemblage for producing selves. For this reason, deploying the lens of race to develop our thinking about online identification will help us to exploit the potential offered by the new media for experiencing community beyond identity. (Hansen 2006: 140–41)

Hansen's use of contemporary art and discourses of racial (dis)embodiment to illustrate his argument are worth further analysis precisely because they signal a set of consistent, symptomatic desires in media theory regarding the potential of the Internet. While I applaud Hansen's antiracist goals, the general framework of both essays risks returning us to an overly utopian, universalizing understanding of human relations that leaves little room for more subtle analyses of the concrete effects of cultural, racial, and sexual difference operative online today.

Hansen's argument is engaging and nuanced, but it reveals a certain racial and cultural privilege. For example, he finds that "passing" in online environments "suspends the constraint exercised by the body as a *visible* signifier—as a receptive surface for the markings of racial (and gendered) particularity" (Hansen 2006: 144, emphasis in original). In other words, since we are all theoretically invisible online (webcams notwithstanding) and cannot be marked or mapped visually, we can all pass. Hansen hopes that by celebrating the ubiquity of passing online (we all are equally subjected to the condition of *having* to pass), cultural signifiers (of race or gender) will be shown to have no natural correlation to any particular body and will thus be revealed as no more than social codings (Hansen 2006: 147). Hansen presents this vision of cyberspace as not merely experimental but also pedagogical: through the transcendence of visibility, those who are engaged in passing online will, of necessity, learn the very bankruptcy of categories of identity.

Yet social codings are precisely the forms of ideology most resistant to transformation. If race is revealed to be (or has scientifically been proven to be) a social code, rather than a natural or biological condition, this revelation has yet to transform the social function of race in the maintenance of uneven power relations. The claim that "online self-invention effectively places everyone in the position previously reserved for certain raced subjects" ignores the many ways in which cultural privilege and hierarchy exist online in terms of literacy, access, social networks, and even forms of self-invention (Nerad 2003; Nishime 2005).[3] Hansen suggests that because race is performative and not ontological, online performances of blackness, for example, are all essentially equivalent to the degree that they all are equally subject to the available cultural meanings of blackness. Hansen equates online self-invention, blackface, and racial passing as forms of "imitation of an imitation; a purely disembodied simulacrum" (Hansen 2006: 146).

While some aspects of race, gender, and sexuality are performative, as Judith Butler has so convincingly argued, it must also be observed that not all forms of performance are equal, nor do they have equal effects (Butler 1990: 25). Lisa Nakamura (2002) effectively reveals in her book *Cybertypes* that online passing frequently produces stereotypes of race that become solidified through their repeated performance via a kind of "identity tourism." Nakamura writes, "identity tourism is a type of non-reflective relationship that actually widens the gap between the other and the one who only performs itself *as the other*" (57, emphasis added). While Hansen philosophically hopes performative repetition will render stereotypes void of meaning, Nakamura observes that it appears merely to reinforce narrow conceptions of race. Her argument is echoed in sociological studies showing that racial "identities" may be more immutable, fixed, and shallow in online interactions than offline (Burkhalter 1999: 63).

Passing in the real world, or online, entails more than visually choreographing one's appearance; it constitutes a complex *psychic* activity that foregrounds precisely the ways in which subjects are generally fixed by racial typologies. Anyone who has racially passed, or who has worn blackface, knows that there is nothing, truly nothing, disembodied about it (Riggs 1986). Indeed every element of existence as an embodied subject comes to the fore in real-life moments of racial passing. Every nuance of skin tone, every glance and gesture, might betray the subject's secret. Passing always presumes conditions of unequal power. The *need* to pass (historically to avoid racial discrimination) and the *desire* to pass (to experiment with subjectivity online) are limited in Hansen's argument to an enforced *condition* of passing.

Stuart Hall has argued that race is best understood as a discourse, constructed by thought and language, that responds to real, concrete conditions of cultural difference (Hall 1998: 298–99). If the complexity of race discourse is grasped in the fullness of its multiple articulations, then it is not possible to discount processes of identification, fantasy, and dominance that racial difference elicits simply because an online image may or may not have a "real-world" referent. Race is always an embodied discourse that acts on and through living human beings at the level of corporeal practices, movements, gestures, and gazes, ultimately constructing and deconstructing the psychological states of individual subjects. In her essay "Cyberfeminism, Racism, Embodiment," Maria Fernandez makes a parallel argument, suggesting that unspoken anxieties that attend the conception of race and racial difference produce a kind of physical haunting that emerges as a set of frequently unconscious and involuntary rote behavioral habits (Fernandez 2002). Drawing on earlier feminist analyses of embodiment, Fernandez suggests that although much has been written about race as an ideological construct, the *performance* of racism in everyday physical and social interactions is of fundamental concern for understanding its continued reproduction. Race as a set of embodied practices supports Michael Omi and Howard Winant's conception of race as a *social formation* that is constantly under revision. What they call "racial formations" can be found both in small moments (at the microlevel) of racist encounters and in systemic (or macro-level) epistemological approaches to cultural and ontological understandings of human being (Omi and Winant 1994: 60). Taken together, these theorists provide a framework for understanding race as a complex and nuanced discourse functioning at every level of individual and collective representation, consciousness, behavior, and organization. Online passing is never free from the social, historical, linguistic and psychological constraints and conditions that also shape racial discourse offline. The invisibility of "real" bodies cannot, alone, produce a racially neutral space or even racially neutral subjects.

Race as a discourse is not an unchanging historical framework that limits identities to fixed taxonomies; it is rather a dynamic system of social and cultural *techniques* carefully calibrated to constrain, define, and develop a nexus of human activity where the ontology of the human, the representation of the body, and the social position of the subject intersect. At this intersection, the invention and perpetuation of various forms of race discourse can be understood effectively to employ the human organism as an *experimental* object of signification. The domains of law, commerce, and medicine have participated and continue to participate in this experiment. Thus the Internet might be better understood as, among other things, a new opportunity for such experiments in signification to play out, rather than as the condition for their disappearance.

The Face

Underlying Hansen's basic argument is a hopeful interest in the possibility that some kind of unprecedented ethical relation might emerge from the anonymity—the facelessness—of the Internet and other forms of new media. He turns to the notions of the "improper" and the "whatever body" from the writings of Agamben to argue for digital media's potential to produce the conditions for the emergence of an identityless, subjectless singularity, citing the following passage from *The Coming Community*: "If humans could, that is, not be thus in this or that particular biography, but be only *the* thus, their singular exteriority and their face, then they would for the first time enter into a community without presuppositions and without subjects, into a communication without the incommunicable" (quoted in Hansen 2004a: 110). Agamben suggests, in essence, the utopian possibility of human encounter that relies on a kind of purity of presence, where all else (history, memory, gender, race, and class) falls away. Counterintuitively, for Agamben, the face is not the human visage in its material presence, but rather what he calls an opening to communicability. He writes, "there is a face wherever something reaches the level of exposition and tries to grasp its own being exposed, wherever a being that appears sinks in that appearance and has to find a way out of it. (Thus art can give a face even to an inanimate object . . . and it may be that nowadays the entire Earth, which has been transformed into a desert by humankind's blind will, might become one single face)" (Agamben 2000: 63). For Agamben, the face is a restless power, a threshold, a simultaneity and being-together of the manifold visages constituting it; it is the duality of communication and communicability, of potential and act. It seems therefore to be both the form and the function of signification. Yet it is also an ontological or existential state. He writes, "in the face I exist with all of my properties (my being brown, tall, pale, proud, emotional . . .); but this happens without any of these properties essentially identifying me or belonging to me" (Agamben 1993: 98). Agamben wants us to be able to imagine the unique character of each human subject without limiting this uniqueness to surface representations, to the limits of particular resemblances between people, to the frameworks of socially defined characteristics. He not only wants us to be able to imagine this state but also to somehow voluntarily achieve it. He writes in the imperative: "Be only your face. Go to the threshold. Do not remain the subject of your properties or faculties, do not stay beneath them; rather, go with them, in them, beyond them" (Agamben 1993: 99).

The artist Nancy Burson's *Human Race Machine* echoes Agamben's call, but it replaces the universal singularity of the subject with universal sameness, emphasizing the physical and racial properties of humans in an effort precisely to erase or transcend their significance. The artwork combines a complicated viewing-booth apparatus with a patented morphing technology that will transform a snapshot portrait of the user into a series of racially distinct replicas. A digital algorithm adjusts bone structure, skin tone, and eye shape, automatically reproducing the same face with a range of facial features, which is then displayed on the computer screen as a row of uncanny doppelgängers. Burson claims that the *Human Race Machine* is her "prayer for racial equality" and suggests that "there is only one race, the human one." "The more we recognize ourselves in others," Burson writes, "the more we can connect to the human race."[24] Her work adheres to the same conception of race as primarily a concern with visual appearance found in Hansen, but she reverses the importance of the image in the production of a universal subject. The power of visual representation, for Burson, lies in its ability to produce forms of cross-racial identification, whereas, for Hansen, visual representations of race are always already corrupted by their ideological history and therefore cannot be used productively as sites of identification.

Burson also claims that "the *Human Race Machine* allows us to move beyond differences and arrive at sameness." Despite her progressive intentions, Burson's desire to "move beyond differences and arrive at sameness" seems strangely undone by the artwork itself. Instead of Burson's promise of greater human sameness, the *Human Race Machine* appears to offer a thinly veiled fantasy of *difference*. Presenting the argument that "there is no gene for race," the *Human Race Machine* allows the user to engage in what Nakamura might call "identity tourism." As a form of temporary racial tourism, Burson's machine may make the process of cross-racial identification appear plausible, but its artificiality does nothing to reveal how people live their lives, or even how they engage with cyberspace. To be more specific, the *Human Race Machine* does not offer users any insight into the privileges or discriminations that attend racial difference, such as the experience of being ignored by taxis or denied housing, being harassed by the police, receiving unfair legal representation, or having one's very life threatened. Instead, it offers users a kind of false promise of universality through the visual mechanics of race. By using the face as a device that is ultimately mutable and theoretically nonidentitarian, she shows how any face (this time the actual visage) might become like any other face, any *whatever* face, and by doing so implies that the racial discourses attached to those signs will fall away. Like Agamben, Burson invites us to attend to our physical traits, our "properties," so that we might transcend them. Yet both fail to attend to the social and political constraints that might impede this transcendence.

In contrast, Fanon (1967) has eloquently theorized the involuntary condition of epidermalization that precisely interrupts the concrete possibility of being *only* one's face (in Agamben's sense) because of one's racially defined, physical visage (11). Fanon describes the moment at which he realized that his own "properties" were in fact created by others, writing: "Below the corporeal schema I had sketched a historico-racial schema. The elements that I used had been provided for me ... by the other, the white man, who had woven me out of a thousand details, anecdotes, stories" (111). As scholar Delan Mahendran (2007) nicely summarizes, for Fanon "the racial-epidermal schema is the interior horizon of self and others in immediate perceptual experience of the world. The racial epidermal schema impacts a black person's tacit sense of self. The racial epidermal schema immediately in play is the phenomenon of appearing or showing up as black in an anti-black world" (198). When Agamben (2000) suggests that "there is a face wherever something reaches the level of exposition and tries to grasp its own being exposed, wherever a being that appears sinks in that appearance and has to find a way out of it," he reveals the very fact of a subject who is undergoing the process of exposition—that is, of being defined, of being explained, framed, delimited, and exposed as an appearance—and who is trying to grasp this exposition. One might say that this is an insightful description of the very process of racial formation, of epidermalization, or of subjection per se. But for those human subjects constantly enclosed in these properties or faculties by others, Agamben's call to "go with them, in them, beyond them" seems not only utopian (literally appropriate for a space that does not exist) but also blind to the conditions by which human subjects are, indeed, produced through elaborately constructed discourses and relations with other humans. These discourses and relations are designed to prevent precisely this voluntary opening of the face, to prevent any movement beyond racial particularity. Perhaps this is why Agamben, to his credit, frames his argument as a conditional statement that marks the edge of the possible: if humans *could* be only their face—that is, exist in a state of utter openness and nonidentity—then they might for the first time enter into a "community without presuppositions." Agamben's approach to ethics is ultimately privileged in origin and messianic in structure, working toward a future point of unknowable possibility without attending in any depth to the material conditions of difference in the present.

Writing before Agamben, Levinas (1969) elaborated the face as the critical site of human ethical encounter. For Levinas the absolute infinity of the other, legible in the physical presence of the face, simultaneously manages to appear within and exceed this material frame. Levinas foregrounds his ambivalence concerning visual knowledge by opening his discussion of "Ethics and the Face" in *Totality and Infinity: An Essay on Exteriority* by stating, "inasmuch as the access to beings concerns vision, it dominates those beings, exercises a power over them" (194). He goes on to explain how the face is the condition for the visibility of the other *as* other and the origin for the opportunity to enter into speech and discourse. He writes, "the idea of infinity is produced in the opposition of conversation, in sociality. The relation with the face, with the other absolutely other which I cannot contain, the other in this sense infinite, is nonetheless my Idea, a commerce" (Levinas 1969: 197).

We can see clear parallels with Agamben's theorizing of the face, which is certainly indebted to Levinas, but the latter seems to be more attuned to the involuntary nature of this coming into relation via the face-to-face encounter and to the responsibility and possible fraternity that emerges from this. He writes, "one has to respond to one's right to be, not by referring to some abstract and anonymous law, or judicial entity, but because of one's fear for the Other. My being-in-the-world or my "place in the sun," my being at home, have these not also been the usurpation of spaces belonging to the other man whom I have already oppressed or starved, or driven out into a third world; are they not acts of repulsing, excluding, exiling, stripping, killing?" (Levinas 1989: 82). Even given this somber revelation that the encounter with the other, with the face, is not a pure state of abstracted unity but also always grounded in the conditions of history and contingency, Levinas is not without hope that the radical and uncontainable otherness that appears in face-to-face encounters can nevertheless be maintained "without violence, in peace with this absolute alterity. The resistance of the other does not do violence to me, does not act negatively; it has a positive structure: ethical" (Levinas 1969: 197). While Agamben grounds the possibility of ethical encounters through an *erasure* of difference, Levinas grounds it *through* difference, writing that "the face resists possession, resists my powers" (Levinas 1969: 193). It is this very resistance that allows us to recognize the infinity of the other that always exists beyond and in excess of the mechanisms (whether visual or discursive, historical or taxonomic) that we might use to frame or delimit it. More to the point, our own historicity depends on the other, our situatedness becomes defined by having to answer to and for histories that we may not have previously conceived as our own.

In contrast to Burson's *Human Race Machine*, which works to produce a form of seamless identification in her audience through the visual production of racial *equivalence*, the British-Jamaican artist collective Mongrel (Graham Harwood, Mervin Jarman, Matsuko Yokokoji, Richard Pierre Davis, and Matthew Fuller) leverages the iconicity of the face to elicit a structure *of ambivalence*. Its print and online project *Colour Separation* (1997) offered users the opportunity to encounter masked subjects who signify as imaginary projections of racial types. Each of the composite images consist of a simple frontal head shot of a man or woman on which a smaller photographic mask of a different racial type was apparently sewn, revealing the eyes and mouth of the subject underneath. Produced with the collective's own morphing software, strategically named Heritage Gold, the images were compiled from over one hundred photographs of people who were somehow connected to the core members of the art group into eight racial stereotypes. Echoing the processes of composite photography used in the early twentieth century to define criminal and racial types, the images emerge as the sign of the impossible referent—that is, they signify subjects who do not exist except in digital form and in the imagination of those who created them. The phrase "color separation" also refers to an image processing technique that entails creating separate screens (magenta, cyan, black, and yellow) for color image printing—an artificial and mechanical process not unlike racial categorization.

The layering of a racially distinct mask on top of the face implies not one but two subjects defined both by difference and intimacy, by their mutual interdependence and potential interchange. These double portraits reappeared in Mongrel's installation *National Heritage* (1999) with a dynamic, interactive element: by clicking on individual faces, the user added another layer—of spit. These unexpected marks, not immediately legible as saliva, marred the surface of the face. At the same time a voice recounted in some detail a personal narrative of everyday racial abuse, of which the spit was a visual sign. In drawing out the complexity of human race relations—their microviolence and the inescapable complicity of every viewer—the work functions as a disruptive device in the ongoing experiments of race discourse. By naming its specialized morphing software Heritage Gold, Mongrel plays off the rather insidious euphemistic term *heritage,* used in British culture typically to signify the preservation of a white, English patrimony. Rachel Green (2000) observes, "based on the ubiquitous graphics software Adobe Photoshop, Heritage Gold replaces its banal tools and commands ('Enlarge,' 'Flatten') with terms pregnant with racial and class significance ('Define Breed,' 'Paste into Host Skin,' 'Rotate World View')" (163). Pull-down menus allow users to transform photographic images according to racial types such as East Indian, Chinese, and Caucasian.

Such designations reveal the strange equation of national identities with racial identities and seem to parallel the kind of morphing fantasies and identity tourism found in Burson's *Human Race Machine.* One crucial difference is that Heritage Gold is free, unpatented shareware that allows users to produce these visual manipulations and transformations themselves, rather than imposing a homogenizing algorithm on all participants. Both *Colour Separation* and Heritage Gold software engage not merely the question of racism as a complex, multiparticipant event without immediate remedy; they also emphasize the ways in which this condition is mediated by visibility and invisibility. Graham Harwood (2008) writes, "in this work as in the rest of society we perceive the demonic phantoms of other 'races.'" But these characters never existed just like the nigger bogeyman never existed. But sometimes…reluctantly we have to depict the invisible in order to make it disappear." As Nakamura has observed, "women and racial and ethnic minorities create visual cultures on the popular Internet that speak to and against existing graphical environments and interfaces online. Surveys of race and the 'digital divide' that fail to measure digital production in favor of measuring access or consumption cannot tell the whole story, or even part of it" (Nakamura 2007: 172).

Colour Separation has received attention from a number of scholars, including Hansen and Wendy Chun. Hansen's book *New Philosophy for New Media* includes an analysis of *Colour Separation* that suggests quite rightly that the work "compels the viewer to confront the power of racial stereotypes at a more fundamental level than that of representation; it aims to get under the viewers skin, to catalyze a reaction that might possibly lead to a loosening up of the sedimented layers of habitual, embodied racism" (Hansen 2004b: 198). Chun's (2006) book *Control and Freedom: Power and Paranoia in the Age of Fiber Optics* raises some important and provocative questions about *National Heritage*, pointing out that "making users spit may expose our relation to another's pain, but it also flattens differences between users. Also, making the 'faces' speak after being spit on exposes the ways in which the other speaks its truth in response to the demands of the would-be user/subject, but it forecloses the possibility of silence and refusal" (167). This important observation reminds us that the lack of freedom, flexibility, and choice in the software may not only mimic similar restrictions that exist in the world at large; this lack can also repress the forms of resistance existing there as well. Yet the fact that solutions and reconciliations are not presented in *Colour Separation* should not be read as a form of cynicism or simple *ressentiment*. In drawing out the impasses and intersections of human race relations, the work functions as a salutary disruptive device that rather closely approximates a Levinasian ethics in which the resistance to possession takes place in the public domain of cyberspace.

Common to all these examples is the logic of the face as a visible threshold to the domain of communication, and ultimately to a practice of ethics. In the long tradition of portraiture, so thoroughly theorized in the history of art, the face is the object of public encounter, a device that mediates the historicity of the subject and its interior character. As many scholars have argued, the portrait and the face are primarily rhetorical, functioning like speech acts in both argument and address (Gavalli-Björkman 2002: 141). Sharing an etymology with *façade*, the face is architectural in its features and potentially false in its design. This is the lure and disappointment of the face, both for the early twentieth-century eugenicist, who hopes to discover in the features of the face the proof of racial superiority, and for the artist, who hopes to capture in a glance or a profile the essence of identity. At the bureaucratic level, however, the face guarantees legal status, defines passport control, and provides the focus of most surveillance and security technologies. As Sandy Narine (2006: 15) observes, "In a future presumed by many thinkers to involve digital enhancement, electronic recording and constant surveillance, the technology of recognition (attributed to increased security pressures) promises to make the science of the face an arena for further work and development."

As the most reproduced visual sign on the Internet, the face continues to operate as the threshold to public space. Facebook, the largest social networking site on the Internet with more than 500 million registered members, has uploaded many billions of images since its founding (Ellison, Steinfield, and Lampe 2007: 1153). Ninety percent of the profiles on Facebook contain an image; most are faces. Each face is presented as one point in a nexus of other faces, each with its own extending network, creating vast pools of tenuous social links that grow exponentially. Unlike the portraits of previous eras, depicting wealth or fame, the faces on Facebook depict anyone who can follow the simple uploading directions on the website. More important, the face is no longer presented as singular and isolated but becomes the ultimate origin of other faces, always defined by, surrounded by, and in some way guaranteed by the visual presence of others. The meaning of the Facebook face is not limited to facial features, to the facade, but extends to the other faces to which it is linked. Within multiple trajectories of signification, the face enlivens and mobilizes social connections that become much more significant than the photographic representation of individuals. Yet race and class still play a role in the way Facebook and other sites construct networks of inclusion and exclusion, such that membership and a sense of belonging are already circumscribed via categories existing in the culture at large (Tufekci 2008; Claburn 2007).

The Public Secret

The desire to locate a universal quality in human subjects or the allure of forms of universal address (the two are not the same, but the latter frequently presupposes the former) are probably tied to a will to eradicate not merely individual differences but any difference that is believed to create an impediment to public action, public consensus, or communication. Race has traditionally been thought of as a quality of individuals, therefore reducible by Agamben and other theorists, like Hansen, to a property or mere set of appearances that one can theoretically "move beyond." But race is not a *property*; it is a *relation of public encounter*.

These relations of encounter were the subject of the artist Keith Obadike's performance and conceptualization of blackness in his playful and well-known project *Keith Obadike's Blackness* (2001), wherein he proposed to sell his blackness to the highest bidder on eBay. While the work clearly referenced the history of slavery when black bodies stood on the public auction block, Obadike was nevertheless careful not to equate his cultural "Blackness" (with a capital B) with a black body made visible. By not including a photograph of himself, Obadike thwarted the common expectation that objects for sale on eBay will be visible online—further underscoring the difference between the concept of blackness and physical traits assigned to the term, specifically skin color. On an actual eBay page, the artist described the object for sale, stating that this "heirloom has been in the possession of the seller for twenty-eight years" and that it "may be used for creating black art," "writing critical essays or scholarship about other blacks," "dating a black person without fear of public scrutiny," and, among other rights, "securing the right to use the terms 'sista,' 'brotha,' or 'nigga' in reference to black people." Certain warnings also applied: for example, the seller recommended that this Blackness not be used "during legal proceedings of any sort," "while making intellectual claims," "while voting in the United States or Florida," or "by whites looking for a wild weekend" (Obadike 2001).

Obadike here toys with the idea that blackness is a commodity that can be bought and sold for the purpose of cultural passing, tapping into a long-standing fantasy in the history of race politics of crossing the color line. But the artist also wrote, "this Blackness may be used to augment the blackness of those already black, especially for purposes of playing 'blacker-than-thou.'" Structured around the perceived desires of others to occupy or "own" blackness even if they are already black, Obadike's project brings out the hierarchies operative in cultural conceptions of racial identities while revealing the social inequities that always attend blackness in the United States. The artist uses humor to reveal the daily pain and the ubiquity of racism that revolves around the concept of blackness, yet he also demonstrates the impossibility of selling oneself out of being black, with all of its attendant advantages and disadvantages, both personal and systemic.

Rather than presenting the Internet as an ideal place to racially pass, *Keith Obadike's Blackness* addresses relations of commodification, wherein aspects of performativity are not simply a question of shifting appearances but a set of cultural expectations that inflect ethical, political, and social relations with others online and off. If any user can join Second Life and pay to accessorize their avatar with racially specific visual signs, Obadike's project reminds us of the purely phantasmatic nature of this commodity relation to race that takes place in the public sphere of the Internet. It also reveals the involuntary (i.e., inherited) relations of discursive inclusion and exclusion attending the concept of Blackness as a set of obstacles and choices for those who are perceived as black and, by implication, for those who perceive others as black.

The use of eBay as the quintessential marketplace, as the site of the public or of publicity, demonstrates not the demos of the Internet as public sphere, but as a platform for what Jodi Dean (2002) (borrowing from Paul Passavant) calls "communicative capitalism," which is the condition by which technoculture works in the interest of capital growth while appearing to enhance public access to information and communication (117). Communicative capitalism leverages the public space of the Internet for its own ends, while advertising this space as a site of democratic potential. Obadike's work draws attention to the Internet as a site of communicative possibility, while simultaneously leveraging its publicity to display private—that is, "individual" —experiences of race, reminding us that Blackness and the race politics associated with it are precisely *not* individual, but entirely public, relational, and important elements of today's communicative capitalism.

Dean observes that our widespread differences in culture, practice, language, information, race, status, religion, and education in the world (and especially in online digital culture) preclude the possibility that "the public" can refer to "all of us." Why, then, does the idea of the public persist? For Dean, the public is symbolic; it may not exist in fact, but it still has real social effects both in political thought and in law. For these discourses, "the public" is a central organizing trope commonly contrasted with the private, such that the borders of this demarcation are the subject of theory, debate, and controversy. Dean shifts this opposition by proposing another: that between "the public" and "the secret." She writes, "few contemporary accounts of publicity acknowledge the secret. Instead they adopt a spatial model of a social world divided between public and private spheres. For the most part, the accounts claim either priority of the one or the other, ignoring the system of distrust, the circuit of concealment and revelation, that actively generates the public. To this extent they seem unable to theorize the power of publicity, the compulsion to disclose and the drive to survey" (Dean 2002: 44). The other of the public is not the private but the hidden, the unknown, even the unknowable. The secret is both the object of desire and fascination and the threat to the coherence of the public as a homogeneous, open, knowable condition of universal participation.

Publicity requires secrets, for Dean, insofar as the secret maps the limit of public discourse. Secrecy is always a public fact. Revealing secrets is one of the goals of publicity, but producing secrets is another one. Power resides in what people conceal as well as what they reveal, whether as part of the hegemony or of the subaltern classes. Race and other forms of cultural difference have been historically presented as secret unknowns that require definition, mapping, measuring, and legislating by those in power in order to render them public. Race both constitutes and is constituted by the public. Race produces a form of resistance to ideals of the public because it stands as a marker of difference that stubbornly resists transformation or incorporation. Race serves as an aspect of secrecy in the logic of publicity, but as an already publicly constructed discourse, its secrets are plainly evident. This is its fundamental contradiction. Racial schemas work to hide or mask not only individuals *as* individuals but also their real and imagined historical conditions. As David Marriot (2007) observes, the fearful projections accompanying the gaze that produces the raced subject are always haunted by the past, but "what haunts is not so much the imago spun through with myths, anecdotes, stories, but the shadow or stain that is sensed behind it and that disturbs well-being" (2).

The philosophical imperative for a homogeneous universal subject, without racial or cultural specificity, who might therefore properly participate in a neutral public sphere can be seen as a demand for subjects not only to reveal their secrets but also to find ways to live without them, in other words, to find ways not to be disturbing. Dean argues that while the Internet may indeed provide one site for democratic politics, it does not constitute a public sphere (particularly in the Habermasian sense of equal access and homogeneous participation). In fact she suggests that the public sphere, with all of its structure of spectacle, suspicion, or celebrity, is the wrong model for understanding political process or democracy, especially within technoculture; rather, she suggests that we conceive of the Web as an intersecting nexus of "issue networks" that produce "neo-democracies," borrowing these terms from Richard Rogers and Noorjte Marres. For Dean, traditional public sphere models rely on the nation as a site, consensus as a goal, rationality as a means, and individual actors as a vehicle, whereas the "neo-democracy" model relies on the Web as a kind of neutral institution with contestation as a goal, networked conflict as a means, and the issues themselves (rather than individual actors) as a vehicle (Dean 2002: 170).

We can conclude that it is not yet possible to decouple race discourse as an oppressive regime from vision or the visible, and that visual culture (both online and off) is the very place where contemporary race discourse might be most powerfully critiqued and transformed. As Butler has written,

> It is possible to see how dominant forms of representation can and must be disrupted for something about the precariousness of life to be apprehended. This has implications, once again, for the boundaries that constitute what will and will not appear within public life, the limits of a publicly acknowledged field of appearance. Those who remain faceless or whose faces are presented to us as so many symbols of evil, authorize us to become senseless before those lives we have eradicated, and whose grievablity is indefinitely postponed. Certain faces must be admitted into public view, must be seen and heard for some keener sense of the value of life, all life, to take hold. (Butler 2004: xviii)

The idea of a neo-democracy, with its emphasis on contestation and conflict centered on political issues, rather than a consensus model addressing universal subjects, might be a valuable model, not only for the interactions of "cyberspace" but also for the lived politics of our everyday lives.

Figure 32.2
Caught Like a Nigger in Cyberspace, digital game, 1997.

Notes

1. This is a shortened version of an essay originally appearing in *Camera Obscura* 70, vol. 4, no. 1, 2009, published by Duke University Press.

2. See Jennifer González, "Electronic *Habitus*: Agit-Prop in an Imaginary World," in *Visual Worlds*, ed. John R. Hall, Blake Stimson, and Lisa Tamiris Becker (New York: Routledge, 2005) 117–38; González, "Morphologies: Race as Visual Technology," in *Only Skin Deep: Changing Visions of the American Self*, ed. Coco Fusco and Brian Wallis (New York: International Center of Photography, 2003), 379–93; and González, "The Appended Subject: Race and Identity as Digital Assemblage," in *Race in Cyberspace*, ed. Beth Kolko, Lisa Nakamura, and Gil Rodman (New York: Routledge, 2000), 27–50.

3. For discussions of the psychological and material conditions of racial passing, see Julie Cary Nerad, "Slippery Language and False Dilemmas: The Passing Novels of Child, Howells, and Harper," *American Literature* 75 (2003): 813–41. For an interesting discussion of racial passing and cinematic cyborgs, see LeiLani Nishime, "The Mulatto Cyborg: Imagining a Multiracial Future," *Cinema Journal* 44 (2005): 34–49.

33 SCHOLARLY PUBLISHING IN THE DIGITAL AGE

Kathleen Fitzpatrick

Attempting to cover the breadth and complexity of the field of scholarly publishing in the digital age in one brief chapter is an all but impossible task; the experiments and possibilities in development today are as diffuse as is the relationship between the digital and the humanities in general. What follows cannot for that reason pretend to any kind of completeness; it can instead look at a few of the major issues that the digital raises for the future of scholarly publishing, and at a few of the possibilities that digital work in the humanities presents for the development of new modes of communication. In other words, I am interested, in this chapter, in thinking in a somewhat open-ended fashion about what the digital can do for scholarly publishing, or perhaps, what it can help scholarly publishing to do, rather than what digital scholarly publishing is or will be.

Digital processes and platforms have unquestionably transformed the systems of journal production and distribution, as platforms like JSTOR and Project MUSE have dramatically changed the ways that scholars access journals, and software packages such as Open Journal Systems have significantly lowered the technological barriers to starting a new journal or moving an existing one online. Similarly networked structures are beginning to present ground for innovation in book publishing, even beyond the development of ebooks; within the digital humanities, one might see, for instance, the highly visible experiments conducted online with McKenzie Wark's *Gamer Theory* (2007), Noah Wardrip-Fruin's *Expressive Processing* (2009), Matt Gold's edited volume, *Debates in the Digital Humanities* (2012), Dan Cohen and Tom Scheinfeldt's anthology, *Hacking the Academy* (2013), and the collaboratively authored volume *10 PRINT* (Montfort et al 2012). Each of these volumes, like my own *Planned Obsolescence* (2011), made use of networked environments in the production, review, editing, and discussion of work in the course of its production.

The relationship between publishing and the digital, however, particularly within the already peculiar space of scholarly publishing, is more complex than it may on the surface appear. "Digital scholarly publishing" is not simply a matter of distributing traditional forms of scholarly publications via electronic networks—and even if it were, this matter is in itself not simple. Despite the successful experiments listed above, substantial challenges remain for new modes of producing and disseminating scholarly work. Financial challenges, for instance: many consumers believe that eliminating the expenses of printing, warehousing, and shipping publications should radically reduce their production costs, and thus the prices at which these texts are sold. For this reason, among others, readers are often unwilling to pay prices for digital texts that are comparable to those for printed objects; for many people, the ephemerality of bits renders them much less valuable, in a financial sense, than the permanence of paper. However, the expenses attributable to the physical form of a print publication make up less than a third of its production costs—and, of course, digital texts require substantial investments in new technologies and technical expertises in order to design, produce, host, distribute, and preserve the new form. As a result publishers and readers have not yet found a mutually agreeable price point for digital publications, a situation that's further exacerbated by the rise of a small number of distributors (e.g., Amazon and Apple in the realm of ebooks, or corporate-owned platforms in the case of journals) whose terms are often disadvantageous for both producers and consumers of scholarly work. And then there are challenges and conflicts surrounding the role of digital rights management, the tension between the first sale doctrine and the licensing of digital content, the proliferation of platforms and formats, and so forth.

Already the situation is complicated, and we have not yet even begun to consider whether the book and the journal article should remain the primary forms that scholarly production takes in the digital age. We are accustomed to these forms, as both authors and readers, and we have a feeling for their shapes and for the work that they accomplish. We are taught, little by little, to produce these forms, at every stage of our educations, from the first five-paragraph essays we write as children through the seminar papers and dissertations we labor over in graduate school, such that they come to seem to us utterly natural, the shapes of thought itself. It should be said, of course, that the constraints presented by the forms of the book and the journal article have in many cases been productive, giving structure to the analysis and exploration that we undertake. However, their shapes are also significant limitations. There has, for instance, long been nothing in the large space between the journal article and the book, a space that might have been occupied by the pamphlet or the chapbook but never was, because that inbetweenness of shape made them literally undistributable. (Books need to reach a minimum length in order to have their title and author information printed on the spine; that information is necessary in order for books to be displayed spine-out on shelves; no physical bookseller is willing to display any but the most popular books face-out; ergo, all books must be longer than that minimum).[1] On the other hand, now that this matter of the printed form that our texts take is no longer quite such a determining factor in their distribution, we might take the opportunity to step back and consider not the shape that scholarly texts have had, but the shapes that they might most productively adopt.[2]

These shapes will undoubtedly differ from field to field. In media studies, for instance, it seems apparent that we might want to produce work that takes forms that use or even resemble the objects we traditionally study. This is not simply a matter of our texts including images, audio, video, or other media forms, though we will certainly want the ability to quote from and work with these forms in new ways. Instead, we may find our work actually *becoming* those forms, incorporating their modes of representation and argumentation into our own. *Kairos*, for instance, a journal of rhetoric, technology, and pedagogy founded in 1996, focuses on the publication of "webtexts," texts that employ the structures of the Web in their processes of argumentation; similarly *Vectors*, a journal focused on culture and technology, pairs authors with designers in producing dynamic, multimodal essays (kairos.technorhetoric.net; vectors.usc.edu). Long-form projects experimenting with multimodal argumentation are similarly beginning to emerge, including Alex Juhasz's *Learning from YouTube* (2011) and Nicholas Mirzoeff's *We Are All Children of Algeria* (2012). These new structures for scholarly discourse, of course, come with their own conventions (however emergent) and constraints (however flexible), and they present challenges for preservation and access as our computer systems continue to advance. These new forms thus are best thought of not as *solutions*, in any conclusive fashion, to the issues we face in scholarly communication today, but rather "essays," in the sense derived from Montaigne: attempts to think anew about what scholarship could become.

Further, as Tara McPherson (2010) has explored, the digital can enable not only new forms of multimodal writing and publishing but also new kinds of engagements with the vast stores of archival material to which we have access, opening that material to analysis of both a large-scale and an extremely intimate variety. The Dynamic Variorum Editions project, which emerged from the National Endowment for the Humanities' Digging into Data Challenge, for instance, seeks to "identify and track topics about the Greco-Roman world as they appear in more than a million documents produced across thousands of years and in several languages," while Fred Gibbs and Dan Cohen (2011) used the text-mining potential of Google's N-gram Viewer as the starting point for a closer engagement with the presumed Victorian crisis of faith than had previously been possible.

Many other shifts and changes are beginning to surface in the kinds and shapes of texts that we will be able to produce and work with in genuinely digital forms. As I've explored elsewhere, it's likely that new forms of digital publishing will allow authors to focus more on the *process* of creating scholarly work, with a little less fixation on the end goal of the final *product*.[3] This is not to say that there won't be a final product, or a moment at which work on a project will end, but it is to note that new kinds of scholarly texts produced in a more dynamic environment can continue to grow and develop, rather than being sealed into fixity once cast in print. Stephen Duncombe's edition of Thomas More's *Utopia* (theopenutopia.org), for instance, is imagined as a project that will develop in concert with its community; similarly the MediaCommons project *#Alt-Academy*, which was originally conceived of as an edited volume, has emerged online as an ongoing, expanding digital community. Projects like these can be released to the public over time, in a serialized fashion, allowing development and exploration of related areas, even unexpected areas, rather than taking on a predetermined, immutable structure. And these new kinds of texts can allow for discussion among authors, or between authors and readers, or among readers, whether as comments on the original text or as direct links and responses across texts, facilitating increased communication within scholarly fields.

All of this is to say that, whatever scholarly publishing is becoming in the digital age, it is not a matter of simply transforming the networks through which scholarly work is distributed, or the formats in which it is produced, or even the shapes and modes of the work itself. Genuinely digital forms of scholarly publishing will also transform the *agents* involved in the publishing process and their engagements with one another. Yet in each of these transformations, I believe, we will find not an abandonment of the values that we have brought to scholarship, but rather a reaffirmation of those values and the new means that we have of manifesting them. In this sense, rather than changing the nature of scholarship per se, the affordances that the digital can lend to scholarly publishing may support our abilities to do some of the things we meant to do all along.

For instance, it's apparent that new modes of networked publishing can help to facilitate more fluid exchanges among peers working in particular fields, as they share work, discuss texts, and collaborate on projects. But these interconnections among peers can also exist across fields, and such interconnections present the potential for richer forms of interdisciplinary exchange, as scholars working on related issues in different areas might similarly find and collaborate with one another. Moreover those scholars whose work prioritizes engagement with a range of more broadly conceived publics will have the ability to connect and work directly with those publics in rich ways.[4] And we may increasingly find that these connections enable us to develop a more capacious sense of the audience for scholarship, including researchers working in contexts outside the university, students who have graduated but remain invested in the fields they studied, and general readers interested in looking further into the cultural texts or issues they care about. Networked platforms for scholarly communication, in other words, may enable us to bring the core goals of the university, including the production and dissemination of new knowledge and the education of new generations of engaged participants in public life, to bear in connecting with a broadened audience—a connection that appears increasingly necessary to ensure the survival of the university in a time of budget cuts and austerity measures.

Of course, doing so will require that this broadened audience has access to scholarly work with which to engage, making the adoption of appropriate open access dissemination models crucial for academic publishing's future. It is little wonder, after all, that the presses housed at state universities in the United States seem safe targets for closure; too few legislators, not to mention taxpayers, have a clear sense of what it is that their university presses do, or why they matter.[5] Similarly federal granting agencies are easy targets for budget cuts, and humanities subjects themselves come to seem unnecessary luxuries in a time of austerity.[6] Making the work that is done by these presses and agencies and the scholars who work with them more broadly available is required in order to make their value apparent, but the business models that can sustainably support open access remain unclear.

Many appear to believe that the transition to what is called "gold" open access publishing can be accomplished quite easily by shifting the costs of publishing from the consumer (via subscription fees) to the producer (via author fees).[7] The Finch Report in the United Kingdom, for instance, argues that "publication in open access or hybrid journals, funded by APCs," or article-processing charges, should be "the main vehicle for the publication of research" (Finch 2012). As the SPARC-Europe response points out, however, the transition to a pay-to-publish system will be extremely expensive (Wellander 2012). Furthermore, while such a shift might be feasible in STEM fields, given the nature of grant support for scientific research, there's much less grant support available in the humanities, and typically this funding cannot be used to defray publication charges. Open access publishing in the humanities will thus require that publishers begin to consider entirely different business models—business models that might in fact change the ways that we think about the purpose of publishing itself. Publishers might, for instance, experiment with "freemium" models, making the content they disseminate readily discoverable and readable online, while reserving access to premium formats, sites for interaction, and ancillary materials for paying customers.[8] In this way publishers may find themselves shifting their focus from producing individual chunks of content for sale to providing more robust services for discovering and interacting with that content.

Publishers can continue to play a series of important roles in the process of preparing and distributing content online, of course: a good development editor can be invaluable in helping an author shape an idea for an audience; a line or copyeditor can transform the quality and clarity of a writer's prose; a designer can create a readable, visually appealing format; a marketer can bring the right audience to the text. But it is likely that scholars will increasingly be called upon to do—or will choose to take on for themselves—a fair bit of the production for the work that they publish in digital environments. With the advent of blog engines and other highly user-friendly mechanisms for self-publishing online, a growing number of authors have been able to become their own publishers. In fact, for many authors, these platforms have allowed the mechanics of publishing to begin to fuse with the mechanics of writing itself, in the same way that the rise of the word processor fused the mechanical act of typing with the intellectual act of writing. If the window in which we write onscreen increasingly becomes not just a space for the creation of a manuscript but a space for the creation of publications, traditional publishers' roles in the production process may diminish. Again, this is not to say that publishers will be less necessary—it may simply be that their contribution shifts from preparing a text in order to get it in front of readers to instead getting those readers in front of a text, building the community that engages with scholarship.

One of the key aspects of getting traditional scholarly publications to their readers is, of course, peer review; the evaluation of scholarship by a select number of the author's peers prior to its acceptance for publication has long been a definitional aspect of scholarship itself—so much so that conducting peer review is a key requirement for membership in the Association of American University Presses. Publishers have thus understood peer review as part of the value that they add to scholarly work, a core component of the editorial process. In recent years, however, some scholars have begun to balk at the notion that *presses* do peer review; while presses may manage the process, it's, of course, other scholars—volunteers who are rarely compensated, and then only a fraction of the value of their time—who do the actual reviewing. As a result some of these scholars have begun discussing other means through which peer review might take place: in ambient forms such as exchanges during the drafting process, in collaborative revision, in discussions that take place at the point of reception. Several notable experiments have been launched in open peer review in the humanities, including two experiments jointly conducted by the journal *Shakespeare Quarterly* and MediaCommons; the journal's editors found the process useful enough to have it become a regular part of their process.[9] But these open processes have not been universally accepted; other scholars have raised important questions about the possibility that these processes might replicate in-group conversations or suppress substantive critique, especially when a reviewer occupies a lower status or less secure position in the academic hierarchy. And then there is the extremely pressing question of labor: if review processes, turned into postpublication discussions, become open to more readers than the two-to-three scholars whose responses have traditionally been solicited, and if these discussions become increasingly important to assessing the impact that a given piece of scholarship has, who is going to do all of that reviewing?[10] Of course, this same question, and perhaps the same answer, applies to the traditional peer review process: by and large, a small percentage of scholars do the bulk of the work of reviewing today. In an open process, at least that disparity might be made visible, opening a path for its correction.

In other words, just as introducing new forms for digital publishing will not remove all constraints on scholarly discourse, changing the point in the scholarly publishing process at which peer review takes place will not, in and of itself, "fix" things that might be wrong with the systems within which we work today. And it must be acknowledged that the current system has produced some very real benefits for scholarship: scrutiny by experts often catches problems in scholarly work before it is released, and frequently improves the work; the introduction of double-blind review processes allowed work by women, by scholars from underrepresented groups, and by scholars from non-elite institutions to enter into the mainstream of academic exchange. The hope, though, is that by broadening participation in these review processes, and by permitting reviewers to respond not just to the work but to one another, communities of practice might be able to transform publishing from a means of producing a series of relatively static constative utterances (which respond to one another, but often after a lapse of years) into a dynamic, performative dialogue that works to create knowledge in the act. An open, postpublication review process such as this would of course require careful management, and it's this management—gathering a community, inspiring participation in its discussions, helping to reinforce the community's established rules of engagement—that might be one key role for publishers to play in the review of digital scholarship.

In particular, this is a vital role for scholarly societies, as the work that their members do and the forms that their communications take increasingly move into networked environments. After all, facilitating scholarly communication has been the main business of scholarly societies since their inception, whether that communication has taken the form of meetings or of circulated publications.[11] Universities in the United States came, since the late nineteenth century, to take on a significant percentage of the function of publishing scholarship, by establishing university presses and subsidizing journals edited on campus; more recently, as scholarly work has become increasingly digital, universities have established institutional repositories to archive, preserve, and (at least potentially) disseminate the work done by their faculties. Such repositories, on average, remain significantly under-used, however, at least in part because of one key misunderstanding about faculty members in their roles as scholarly authors: such scholars' primary point of identification is not with their institution but with their fields. It is their colleagues stretched horizontally across many institutions that scholars wish their work to reach, rather than their vertically affiliated colleagues in their home institution. Connecting scholars across a field is precisely the role of scholarly societies, and so it may be that institutional repositories should give way to, or be complemented by, disciplinary repositories operated by those societies; in such disciplinary networks, scholars' work can be connected to that of their peers, and there it stands the best chance of being found by those doing research in the field.

Unfortunately, the position of many scholarly societies with respect to supporting communication within the membership is already severely attenuated. Many have given over their journals to commercial publishers to operate, as those publishers can work at a scale that none but the largest societies can afford; these publishers provide technical and production expertise, access to distribution channels, and options for digital preservation and access—not to mention vital income—that few societies can obtain on their own. But these services come at a cost: the publisher's goals are not always aligned with those of the society and its members, and the subscription fees that such publishers charge to university libraries—which is how those publishers can afford to pay the society so much for its journals—are causing serious damage to those libraries' budgets, and thus to the libraries' ability to maintain other necessary services. As a result societies today risk giving the impression of working at cross-purposes with scholars, narrowing the channels of communication rather than opening them up, restricting the dissemination of members' work rather than facilitating it.

Perhaps for this reason, a number of societies in the humanities, including groups as diverse as the American Folklore Society, the Society of Architectural Historians, and my own employer, the Modern Language Association, are beginning to shift their focus from "scholarly publishing" to "scholarly communication," signaling their investment in facilitating a broad range of ongoing exchanges among scholars. Societies such as these are developing open platforms to support new kinds of communication among their members, allowing them to produce work not just in article or book form, but instead in a much more diverse array of weights and time signatures. These societies demonstrate through these platforms that they can best fulfill their missions by supporting the free and open distribution of the work done by their members. Such open platforms likewise present the possibility of creating new forms of engagement and collaboration across associations, as well as between association members and a broader range of publics.[12]

However, these societies also have a crucial role to play in supporting the scholars who are working in these new ways—producing digital work that takes innovative shapes—by helping them to demonstrate the value of such work to the institutional colleagues and their administrations. Disciplinary organizations can be invaluable in helping to set the terms for the evaluation of digital work, as has been seen in the MLA's "Guidelines for Evaluating Work in Digital Humanities and Digital Media" (Modern Language Association 2012) and the College Art Association's "Guidelines for Faculty Teaching in New-Media Arts" (Davis et al. 2007). Guidelines such as these can help scholars working in new forms ensure that as pressures build from outside the academy for ever more quantified measures of "impact," an emphasis on direct engagement with the work—what John Guillory (2005: 22) has described as the "immanent evaluative scene" conducted by experts within a field—remains. As Don Brenneis (2009) has pointed out, quantifiable measures of impact such as those provided through bibliometrics (including citation counts, journal rankings, etc.) provide an apparent objectivity, but in reducing assessment to numerical indexes, we run the risk of losing the element of qualitative judgment that is most necessary to the evaluation of complex information. Scholarly societies will be crucial in helping new generations of scholars articulate the criteria for qualitative judgment that will best help the profession encounter and understand new forms of digital scholarship.

This concern has apparently taken us somewhat far afield from this chapter's focus on what the digital can do for scholarly publishing, to thinking instead about what scholars, institutions, and societies will need to do for scholarly communication in the digital age. But as the digital enables communication that is more dynamic, more fluid, more participatory, and more open-ended, it demands that scholars and organizations understand publishing as an ongoing process of conversation rather than a mode of production for static objects. As the digital allows published work to reach more scholars, as well as readers from a broader array of interested publics, it requires that scholars engage with their readers in new ways. As the digital permits new means of exploring questions about our objects of study, it also asks us to consider new kinds of questions that couldn't have been asked without contemporary technologies. The digital changes the forms with which we communicate, but it also inevitably changes us in the encounter.

All that having been said, it is important to note that this transition to digital scholarly communication is not a zero-sum game. Open platforms need not replace presses; multi-modal scholarship need not replace journal articles; blogs need not replace books. Instead, these new forms and formats present us with a greater range of options for communicating our work—but also a new set of questions and challenges. Today's modes of digital publishing introduce new kinds of formal conventions and constraints, which are no less conventional or constraining for being new or networked. Rather than settling down into a new array of acceptable forms and modes for scholarship, scholars and institutions must commit themselves to a continual process of exploration and experimentation, to ongoing reassessment and rethinking, to critical engagement not just with the world by which we are surrounded but also with our own ways of working. Digital technologies will not effect transformations in our ways of communicating if we simply trade one set of limitations for another, nor will they effect those transformations on their own. Scholars must agree not only to work in new ways, but also to a never-finished process of renewal—and their institutions organizations must agree to support that work—in order to activate the full range of affordances presented by the digital.

Notes

1. Thanks to Monica McCormick for this explanation of that puzzling gap.

2. Dan Cohen has argued in numerous presentations, including Cohen (2011), for the necessity of "right-sizing" scholarship.

3. See Fitzpatrick (2011: 66–72).

4. See, for instance, the work being done by the consortium of universities participating in the project *Imagining America.*

5. This failure in understanding became painfully apparent during the Summer 2012 closure and reconfiguration of the University of Missouri Press; see, for instance, Elkin (2012).

6. See, for instance, Cohan (2012).

7. "Gold" open access refers to publications that are made freely and openly available by their publishers; this is in contrast with "green" open access, in which publications may remain closed-access, but authors deposit their accepted manuscripts in publicly accessible institutional repositories.

8. See, for instance, the model of the National Academies Press, discussed in Fitzpatrick (2011: 162).

9. The MediaCommons–*Shakespeare Quarterly* open review experiments remain available at MediaCommons Press; the journal's more recent open review project is available at *Shakespeare Quarterly* online. For more examples of open review experiments in the humanities, see Fitzpatrick and Santo (2012).

10. See comments made on Fitzpatrick and Santo (2012), particularly Harley (2012).

11. See, for instance, the history of The Royal Society, which was founded in the 1640s in order that its members might "discuss the new philosophy of promoting knowledge of the natural world through observation and experiment, which we now call science" (The Royal Society). This discussion led to other means of communication among the members, including most notably the journal *Philosophical Transactions.*

12. The obvious tension between this desire to provide a platform for free and open communication among society members and between members and other publics and the society's need to produce revenue remains, of course. New potential avenues for developing new streams of revenue through these open platforms, such as mechanisms for producing EPUB or print-on-demand compilations of material made available through their networks, remain to be explored.

34 CRITICAL TRANSMISSION

Mats Dahlström

Agents in the literary and textual fields are currently shifting mutual relations and reallocating labor, responsibilities, and tasks as their domains become increasingly digital. A field where I have had reason to observe this at close range is that of memory institutions, in particular research libraries. In their engagement with digitizing printed materials, for instance, making written artifacts from their holdings available on the Web, such libraries have, without much reflection, entered the field of producers, publishers, and IPR holders. This means that they not only uphold the task of managing and preserving written heritage but help shape it, even create it anew.

Of increasing importance and interest therefore is the manner in which libraries digitize and subsequently publish the digital reproductions—in terms of, for example, selection, exhaustiveness, level of detail, transparency, and authenticity. And during the last decades, libraries have indeed been experimenting with different levels, sample sizes, and strategies when digitizing collections and artifacts, on their own as well as in various forms of public or private partnerships. A whole terrain of different kinds of digitization endeavors is beginning to be discernible. Our map of this terrain has nevertheless so far been poor, often reduced to equating digitization to the large book scanning project conducted by Google and its partner libraries, sometimes even combined with a trivialization of the digitization process as a flat affair, with a linear derivative relation between a "source" analogue document and a digital reproduction. Likewise the intended users often seem to exhibit a lack of critical distance toward the digitized collections—the digital reproductions are taken, as it were, at (inter)face value.

At the same time much of the experimental work increasingly performed by digital humanities scholars uses empirical testbeds consisting of precisely such digital artifacts that come out of memory institution digitization and digital library initiatives. A hallmark of humanities scholars is an awareness and critical analysis of the inevitable interpretative dimension of reproductions and representational artifacts. But is there a risk that in the digital domain, and, in particular, the domain of mass digitization and big data, humanists lose some of their critical distance toward the digital representational artifact, approaching it as a linear surrogate, as raw data, which can be mined in search of truths about the analogue source of which it purports to be a representation?

On the other side, the partly new role for the library as a producer and distributor of potential research data means that libraries will encounter firsthand some of the problems and challenges that are familiar to publishing houses and distributors as well as to scholars engaged in critically transmitting "contents" between documents and media. Libraries do not often appreciate such a role for themselves, nor do they even acknowledge the kinship between what they are doing when they digitize and what, say, a textual scholar is doing when he or she composes a critical, scholarly edition of a literary work. Rather, libraries are depicted both by their own staff and by scholars as objective copyists or porters of information in the service of "proper" scholars. Such a view risks turning a blind eye to the possibilities and responsibilities inherent in the media transmission of cultural heritage content performed in a deliberate, scholarly based, and critical manner.

Admittedly, the field of digitization within libraries and other memory institutions is largely pragmatically concerned with standards, practices, and manuals. Much attention is devoted to mass digitization efforts such as Google Book Search, and there are very few critical and analytical studies of digitization around.[1] Nevertheless, there is a growing awareness within libraries that digitization does indeed entail considerable theoretical and problematic challenges, and an increased concern that there are more digitizing opportunities and strategies than just mass digitization.

This chapter explores the degree to which research libraries engage in a content transmission that we might refer to as "critical."[2] Consequently it will be relevant to discuss how libraries approach the role traditionally assigned to scholarly fields occupied with that kind of transmission such as art restoration or rare book conservation. In this chapter, however, my choice of comparison is scholarly editing based on textual criticism.

I will start by discussing what kind of activity scholarly editing is, primarily by providing examples of scholarly ideals and issues that are deemed important in the field and that are also frequent sources of conflicts between schools of textual scholars. This will help to draw a crude map of tensions and stakes in the field. I will then proceed by having a closer look at library digitization. This is often equated to mass digitization, where quality resides in quantity. By positing mass digitization at one end of the scale, however, it is possible to assume quite different approaches to digitization, and hence to begin to draw a more nuanced map of the terrain. Last, I will make a tentative comparison between these two fields. I suggest that their respective patterns are similar because they are both instances of institutionally based, bibliographic transmission activities that share a history and a sociocultural legacy.[3]

Scholarly Editing and Transmission Ideals

Scholarly editing is a form of textual transmission in the sense that the editor reproduces existing documents by making a new document that also embodies a documentation of the work's textual history and of the editorial process. Scholarly editions are also hermeneutical documents, expressing subjective editorial interpretations. Nevertheless, editions, by tradition, pretend to convey a sense of value-free objectivity, a mere recording of facts. This twin character of subjective interpretation and the objective reporting of facts is one of many interesting tensions and potential conflicts within scholarly editing. In fact, when studying the history of editing and textual criticism, one can easily come up with a long list of such tensions and conflicts. I list a few of them in table 34.1, and discuss them in the text that follows.

A long-established distinction, first, is the one between scholarly editing that is *critical* and scholarly editing that is *noncritical*—the latter exemplified by documentary editing or facsimile versioning.[4] At times noncritical editing is even looked upon as a more or less mechanical and trivial transmission. One might argue, however, that digital scholarly archives displaying full-text versions in parallel with no single established critical text in the center (as gateway to the complex of versions) threaten to break down this distinction. As I argue later in this chapter, the boundary is further blurred between critical editions and some of the digital facsimiles that libraries produce and that might arguably be considered "critical."

Image management is perhaps the digital editing phase where the conflict between presumed objectivity and situated subjectivity becomes most apparent. Facsimile production, in particular, has traditionally been regarded by textual critics as a noncritical activity, where the editor supposedly recedes into the background and where the user is brought closer to the source documents by having "direct" access, as it were, to the originals. Yet digitization and

Table 34.1
Tensions in scholarly editing

Critical	Noncritical
Interventionist	Non-interventionist
Interpretative	Factual
Facts *as* interpretation	Facts separable from interpretation
Ambiguous	Disambiguable
Idiographic	Nomothetic
Contingent tools	Universal tools
Material document	Abstract text
One text: discriminatory	Many texts: comprehensive

the subsequent editing of images has, more than any other editing phase, made us attentive to the fact that virtually all parameters in the process (image size, color, granularity, bleed-through, contrast, layers, resolution, etc.) require critical intellectual choices, interpretation, and manipulation.[5]

A related distinction is therefore that which exists between the acknowledged presence and the presumed absence of the editor. This is the fundamental issue of *intervening* or *not intervening* in the text, something Greetham (1999) refers to as the Alexandrian and the Pergamanian editorial ideals (50–51). If the former points to eclectic editing, the latter suggests facsimile and best-text editing, strategies where editors might appear to invite readers to assume the authoritative editorial function.

I mentioned earlier the difference between regarding the edition as fact or as interpretation. Similarly there is a tension between the notion that facts *are* interpretations and the idea that we can *separate* facts from interpretation. In the digital humanities this has been a recurring discussion in the fields of XML encoding (between form and content), of stand-off markup, as well as of synoptic full versions of facsimiles that represent no tampering and provide "raw" material. So editions are in this respect either stores of scholarly raw material that support future reusability by other editors and scholars or argumentative and context-bound statements.

Another tension between scholarly ideals is the one between viewing editing as a primarily *nomothetical* or *idiographical* affair. The former maps patterns of common, regular, and predictable traits in large amounts of texts, while the latter highlights that which is unique, different, or contingent. These ideals result in respective strategies to design either *universal, project-general* or *contingent, project-specific* tools (Rockwell 2003: 215). This is also related to a distinction between different conceptions of the core empirical object of editing, on the one hand the ideal text that can be abstracted from one or several witnesses, on the other hand the material and graphical text as inscribed or printed into the individual witnesses. This is, in other words, a difference in perspective between *text* and *document*, manifested in the distinction between text-oriented and image-oriented editing.[6] Text-oriented editing works mostly with text transcriptions, image-oriented editing mostly with facsimile images. According to much text-oriented editing of, for example, intentionalist descent, text is an immaterial, abstract, ideal, copy-independent phenomenon, while to much image-oriented editing in, say, sociology of texts or material philology, text is a material, physical, concrete, copy-dependent phenomenon. Depending on which school of thought you subscribe to, the editing, its tools, and the resulting edition will turn out to be very different.

Finally, a significant tension has emerged between displaying the uniformity or multiformity of the edited work—what Peter Robinson (2000) has referred to as the *one text* or the *many texts*. The former ideal strives for choosing or constructing single copy-texts, whereas the latter ideal turns the edition into an archive. This is the difference between, on the one hand, selection and discrimination and, on the other hand, more or less total exhaustiveness.

Some of these conflicts have been around for the entire history of scholarly editing, while others have emerged during the last decades. Digital editing does not seem to do away with this pattern of conflicts at all, but rather accentuates some of them. For instance, the ability of digital editions to house full-text representation of all versions of the edited work and to support the modularization of documents into movable fragments *across varying contexts* seems to boost the idealist strand in editing. This trend is even further supported by text encoding, where form is separated from content, and where fact is often conceived of as separable from interpretation.

But these notions have forerunners within printed scholarly editing as well. So-called definitive editions attempt to be matter of fact, exhaustive, and final. Parallel and synoptic editions are attempts to accommodate versionality and inclusivity within the covers of the codex book. The *Archiv-Ausgabe* concept in German editorial theory takes this idea even further (Kanzog 1970). Admittedly, those forerunners in the print domain are different in scale, and digital editions such as *The William Blake Archive, The Wife of Bath's Prologue,* or *The Rossetti Hypermedia Archive* can partly be seen as attempts to not only embody but prolong these notions. I would suggest that an extrapolated realization of the *Archiv-Ausgabe* brings you to the digital library, much in the manner that is currently envisioned by many digital humanities archival projects.

Returning to the notion of editions as arguments, there is also a sociocultural function in the scholarly edition. The fact that a work has been the object of scholarly editing is a seal that it has been raised to literary nobility and invited into the inner rooms of the literary salons. Burman (1999) is on the mark when he refers to scholarly editions as the "cathedrals" of literature (85). Frohmann (2004) dresses up a similar thought in a sophism: "A text does not belong to the scriptures because its content is holy; rather, its content is holy because it belongs to the scriptures" (153).

Library Digitization and Transmission Ideals

Let us, however, return to the pattern of conflicts within scholarly editing as depicted in table 34.1. I suggest that this pattern is primarily not media specific to either printed editing or digital editing. It is rather *a general trait of textual transmission* as a cultural phenomenon. We might therefore expect more or less the same pattern in other cases where textual transmission has been similarly stabilized by institutionalization. The discussion below goes more deeply into a particular case of institutionalized transmission.

One new aspect of digital editing is the division of labor and media that surrounds the field of editing and connects it with neighboring activities. I am thinking, in particular, of the changing relation between scholarly editing and the ongoing digitization within libraries. And, interestingly enough, digitization within libraries is developing a pattern of conflicts and tensions between ideals and transmission strategies that *is* in many ways similar to the pattern within scholarly editing.

Libraries and other so-called memory institutions throughout history have developed a range of methods and tools for transmitting full texts between material carriers and across media family borders. In this sense library digitization belongs to the same tradition as twentieth-century microfilming and the transcribing of manuscripts performed by ancient libraries and medieval copyists. During the Gutenberg era libraries devoted more time to producing bibliographical meta-labels for documents rather than to reproducing the full documents themselves. With digital reproduction technologies however, libraries are yet again dedicating much energy and attention to the full-text transmission of their holdings. In so doing, they take on a much more explicit role of producing and shaping the digitized cultural heritage in addition to the accustomed role of preserving it and making it available.

Digitization Strategies

Let us bear in mind that digitization within libraries is much more than the mere technical capture of some content in analogue documents. It is rather a large and complex chain of affairs, from planning, budgeting and selection, via content capture, metadata production and publishing, over to documentation, marketing, and archival maintenance. The links in the chain overlap, cooperate, and support one another. In principle, every link is a factor that might affect and delimit the nature and quality of the final digital resource. This includes to what extent, at what level, and in what form the users are granted access to the resource. How the different links are implemented and work together is, of course, dependent on the overall strategy for the digitization project.

For instance, library digitization works with two modal strategies: *text digitization* and *image digitization*—similar to scholarly editing. In text digitization, documents are primarily interesting as carriers of the linguistic text rather than as graphical and material artifacts. So the task is to create a machine-readable and (usually XML-compatible) encoded transcription of the text. Image digitization, on the other hand, wants to capture the source documents as two-dimensional images (digital facsimiles), using scanning or a digital camera. Needless to say, the two approaches are often combined. There are, of course, other distinctions of digitization approaches, such as that between proactive (i.e., just-in-case) and reactive (i.e., just-in-time) digitization, or between conservational (non-intervening) and restaurational (intervening) digitization. Again, the latter distinction is similar to the one we noticed earlier within scholarly editing.

Mass Digitization

Currently the most spoken-of strategy is, no doubt, mass digitization. It aims to digitize massive amounts of documents (thus an all-inclusive strategy) using automated means, in a relatively short period of time (Coyle 2006), such as Google Book Search, Internet Archive, Europeana, the Norwegian digital national library, or the late Microsoft Book Search. It operates on an industrial scale and with as many chain links as possible fully automated. Mass digitization systematically digitizes whole, large collections, document by document, with no particular means of discrimination. The projects might assume more or less ambitious totality claims: from projects limited—by copyright, politics or administration—to a particular subset of a collection, to the grand supercollection schemas mentioned above; the idea is to digitize "everything" within the collection or sets of collections.

For practical reasons mass digitization has to minimize manual and labor-intensive work and cannot include intellectual aspects such as textual ambiguity, interpretation, descriptive text encoding, and manual proof reading. Neither can it afford to have too much metadata and information about the source document accompanying the digital representation. In mass digitization, transmission has been flattened out into a linear, streamlined affair. Mass digitization has its ambitions and its value in *scale*, not in depth.[7] Vast amounts of books are made available and their texts searchable. As expected, projects such as these have been met with no shortage of critical remarks.[8] One might, for example, question whether the libraries involved in Google Book Search will be able to turn to other digitizing agents with the same source documents in the future, should Google for any reason change its activity or cease altogether as an enterprise.

Despite the many critical remarks, one can definitely see strengths and the positive effects of mass digitization projects.[9] They combine, on the one hand, commercial agents who are strong in financial resources but in need of content with, on the other hand, public libraries who are inversely strong in content but in need of financial resources. A *marriage made in heaven*, it would seem. The result: a gigantic, growing bank of digital texts that can be searched free-text, used as localizing tool, and—perhaps more important—form the technical base for many kinds of future software development and implementation.

But mass digitization is not the only strategy around. Only a limited number of large libraries have the interest, competence, and resources to implement it. Further it suits some objects and collections and not others, such as fragile books, manuscripts, and other unique objects, and also documents whose texts are difficult to read or interpret. Such cases require considerably more resource consumption and manual labor, and so cannot reasonably be referred to as mass digitization. We find in fact a number of digitization projects in libraries worldwide that are anything but "mass." What then do we call them? I suggest we refer to them as *critical* digitization.

Critical Digitization

Critical digitization implements several of the links in the long digitization chain in a manual, intellectual, critical way. At every step one can make choices, deselect, and interpret. Mass digitization turns a blind eye to most of these choices, whereas critical digitization acknowledges and makes active use of them. The project might, for example, focus on a single document or a limited set of documents. It may need to perform a deliberate and strategic selection from a number of possible and more or less complete source documents. Perhaps the source document has text that is difficult to decipher and decode. The text or image may need to be edited and manipulated to make sense or provide context. Perhaps it is vital not to destroy the source document during the digitization process (as mass digitization often does) but rather subject it to careful preservational or conservational measures. The project may wish to manually and critically produce a representation that is as faithful and exhaustive *as possible* in relation to the source document and its text, its graphics, and perhaps artifactual materiality. The digital object may need to be enhanced with large amounts of metadata, indexing, descriptive encoding, paratexts, and bibliographical information, which is to say, bibliographical and other scholarly research may need to be *embedded* in the objects. Critical digitization is qualitative (or idiographic) in the sense that it concentrates on what is unique and contingent in the documents, whereas mass digitization is quantitative in its design to capture what are common, regular, foreseeable traits in large amounts of documents (i.e., nomothetic). In consequence critical digitization normally has to develop project-contingent procedures and tools and tailor them to the nature of the documents in the particular collection. In mass digitization, the single documents in the digitized collection, on the contrary, are subordinated (or "tailored") to more general, perhaps even universal, procedures and tools.

Critical digitization is, in other words, a more exclusive strategy—in more senses than one. Let us briefly sum up the differences between the two approaches using table 34.2 (while also bearing in mind table 34.1).

I am aware that table 34.2 could be interpreted as painting a flattering picture of critical digitization, while mass digitization is attributed a negative role. That is, however, not the aim

Table 34.2
Tensions in library digitization: critical versus mass digitization

Critical digitization	Mass digitization
Is primarily manual	Is primarily automated
Critically recognizes the distortion digitization brings about	Treats digitization as a cloning process
Undertakes a well-informed selective analysis of source copies	Picks whatever source copy praxis or chance happens to present
Maximizes interpretation and metadata	Minimizes interpretation and metadata
Is idiographic	Is nomothetic
Uses contingent tools	Uses universal tools
Treats graphical and material document as artifact	Treats linguistic text as fact
Discriminates	Is exhaustive
Works in depth	Works on scale

here. Indeed I confess to not being perfectly content with the distinction between the mass and the critical mode of digitization.[10] It might suggest that a digitization project is either a mass or a critical mode project. Instead, I would like to think of the two modes as poles at each end of a scale—most digitization activities contain features from both strategies and place themselves somewhere along the scale. A more problematic aspect, one that suggests that the "horizontal" scale should be supplemented with a vertical scale along more nuanced parameters of, for example, collection size and project scale, is that the two labels might lend themselves to ideas of mutual exclusivity. That is, to suggesting that a digitization project cannot comprise huge quantities of objects while still being performed with critical inquiry and high levels of standards and quality. Current development within technology points to increased automation but also to maintaining high standards and requirements for such things as preparation and management of source documents or high-quality digital facsimile production. Many memory institutions are obviously interested in finding ways to satisfy both such demands.[11] A final potential problem with the distinction, or rather the description of it that I have provided above, is that it might suggest that one is objectively superior to the other.

In fact the advantages as well as the disadvantages of mass digitization can be reversed in the case of critical digitization. Critical digitization is slow and very costly in relation to the number of produced objects. It addresses a small audience. It may require rare skills in textual and bibliographical scholarship, among other things. It often has an image-oriented ideal where the facsimiles are left without accompanying machine-readable transcriptions. The result risks quickly being more or less forgotten after the project is completed, and many digitizers neglect to publicize and market the project in proportion to the labor invested. Manual labor, such as interpretations and tailored technical solutions, are seldom properly documented (if at all) and run the risk of becoming silent knowledge locked in the minds of the digitizing persons, and therefore available to the institution only as long as the persons remain employed by it. Mass digitization, on the other hand, requires such an industrial scale that its strategies, practices, and technologies need to be documented in order to be properly implemented by many different people and machines over long periods of time.[12]

There is also another consequence of the way critical digitization intervenes in the documents. Someone consulting a critically digitized collection faces a material that is, in a sense, already encoded, manipulated, labeled, and explicitly interpreted. The more this has been done explicitly, the more it turns into a sort of *comment* on the source document that might work counter to how flexibly the material can be reused in new contexts. Mass digitization, on the other hand, conveys an aura of objectivity around the objects and a lack of manipulation—but that is, of course, due to the cloning ideal of mass digitization (an objectivity that is nothing more than a chimera).

So opinions as to how usable and reusable the products of different digitization projects really are to scholars and scientists can radically differ. On the one hand, critical digitization enriches its objects with intellectual added value and applies some kind of quality seal with regard to selection, textual quality, resolution, proofreading, comments, and bibliographical information. On the other hand, not all scholars may be interested in that particular metainformation and added value but may be in need of quite different aspects than those that happened to catch the interest of the digitizing institution.

One can even turn it all around and claim that critical digitization risks falling for another kind of cloning ideal than the one previously identified as typical for mass digitization. This other cloning ideal might work on the assumption that as long as the digitization process inscribes in the digital representations large and deep enough metainformation, we will obviate any future need of new digitization efforts, since all material and all possible aspects already exist in the digital archive that we have created. We would, in other words, be facing the "definitive" digital representation, *once and for all*. Mass digitization, on the contrary, might be thought of as more advantageous precisely because it does *not* select, provide metadata and explicit text encoding and interpretation, but rather constructs reservoirs of source documents that scholars ideally can use, reuse, and enrich the way it suits them best. We obviously recognize this pattern from scholarly editing and its tension between Alexandrian and Pergamanian ideals. Again, however, we need to remember that the products of mass digitization can also be thought of as dependent on interpretations and selections—but that these are generally ignored and silenced, which leaves the user helplessly dependent on the unknown choices that praxis forced upon the mass digitizing institution. We should also readily admit that the frequent lack of textual quality and proofreading in, for instance, Google Book Search hardly makes any scholar particularly happy, regardless of his or her disciplinary affiliation.

It would seem clear that both strategies have their advantages and disadvantages to different audiences. Whether a digitizing library adheres to an ideal that is closer to critical digitization or to mass digitization, however, should take into account which kind of digital material is being produced, what metainformation and added values should be attached to the material, to what extent it should be able to be used and reused, and to which user group it should prove to be of interest. In that way the digitization strategy chosen legitimizes certain kinds and levels of material at the expense of others, and favors certain user groups over others—all a question of symbolic power. There are, of course, mechanisms in digitization strategies that might be designated as constitutive and perhaps canonizing. Again, we recognize this tension from our previous discussion about the sociocultural functions of scholarly editing. All in all, however, the library community is increasingly favoring the ideals of mass digitization and its pragmatic notion of transmission as a relatively simple, linear, content-capturing affair. The critical digitization activity is much smaller and is in many cases currently threatened to become extinct as an ineffective, costly luxury.

And it could be described as a mere logical consequence that the library community is more prone to mass digitization than to critical digitization. Kjellman (2006) observes that whereas museums discriminate, select and deselect as part of their joint collecting task, national libraries, in particular, display an ambition of comprehensiveness in their collecting activity (239f). Historically this has created an ideal of objectivity within the national library communities that tends to hide the discriminating mechanisms of the institution. Such an ideal is certainly expressed and fueled by mass digitization. The library collections furthermore are largely made up of printed documents that are mass-produced to begin with. Given that the many copies of a published book are normally thought of as identical, it is consequently thought of as more or less indifferent whether the one copy or the other is picked (rather than selected) as source document, that is, as "ideal copy." Archives and museums, on the contrary, manage unique objects to a much higher degree, and their digitization in consequence regularly concerns single document artifacts rather than the text as a presumed commonality in, say, a book edition.

Scholarly Editing and Library Digitization

So to some extent the area of library digitization seems to develop into a map of tensions and conflicting ideals (table 34.2) that bear considerable intellectual similarities to that of scholarly editing (table 34.1). As scholarly editing does, library digitization can make deliberate selections and discriminations, interpret, analytically compare source document candidates, and seek to establish something common, sometimes even ideal, in such a collection of candidates.

They likewise edit, optimize, document, comment, and produce metatexts. They are both examples of transmission practices that are stabilized and legitimized by sociocultural prestige institutions and that are based on agreed upon, publicly declared, and fairly documented principles. In their exclusive character, they constitute, consecrate, and perhaps even canonize works of culture. If scholarly editions are cathedrals of literature, one might certainly argue that ambitious critical digitization projects within libraries turn the selected documents into national monuments—or even testaments.

I would contend that in the case of critical digitization, libraries are in effect engaged in what comes close to textual criticism. Librarians and other employees involved in current digitization projects might feel awkward with such a label of textual criticism or even bibliography put on their work, but the historical connection cannot be denied.

Might we go so far as to say that scholarly editing and critical digitization are one and the same activity? No. The institutional context aside, there are some important differences between them. John Lavagnino (2009) has addressed this issue, formulated as the differences between digital (scholarly) editions and digital libraries (in the sense of collections of objects digitized by memory institutions such as libraries). The former, Lavagnino argues, displays a more profound analytical understanding of the extant source situation, its problems and challenges, for instance, by emending textual errors or making arguments about authorship:

> It is a perfectly respectable thing to provide an accurate transcription or photographic reproduction of one copy of a book, and in many cases those tasks require a considerable amount of scholarly expertise and judgment, but the potential achievements of these tasks are different from those that are possible in a scholarly edition. (Lavagnino 2009: 63)

A library digitizing a collection, on the other hand, "does not involve any analysis; it is devoted to reproducing existing books, but not to any critical or bibliographical analysis." Producing it is largely a clerical task, and it "can be created by workers who have no special knowledge of the material" (ibid.).

This is an accurate description of most mass digitization projects performed by libraries, archives, and partner agents such as Google. But if we broaden our view of digitization in libraries, we see a more complex spectrum of analytical, critical levels in the work performed.

If one looks closely at some of the high-quality digital-imaging projects in memory institutions, one is struck by the degree to which teams of conservators, technicians, and photographic experts constantly make a long series of decisions informed precisely by critical and bibliographical analysis and by a highly specialized knowledge of the bibliographical, graphical, historical, and other research aspects of the material to be digitized. Libraries further produce, with or without the aid of subject specialist scholars, electronic thematic research collections, a task that requires a considerable degree of scholarly, critical skill and deliberation.

I agree that the major distinction one might define between scholarly editing and library digitization concerns the textual level: library digitization does not seek, as scholarly editing does, to establish a text (a copy-text as it were) that perhaps never existed previously and that cannot be literally transmitted from a single source document. But this distinction is not as sharp as one might at first expect. Movements such as versioning, documentary editing, and un-editing push scholarly editions toward becoming documentary, flat archives. There the products of scholarly editing are broken down into small, modular fragments in the manner of a database or "library."[13] Some within the scholarly editing discourse also advocate minimizing editorial interventions, instead providing full-text and "raw" versions, refraining from highlighting or constructing a single uniform, established base text. Scholarly editions are also increasingly based on image-oriented editing, and devote more of their space to harboring digital facsimiles of the source documents.[14] By contrast, critical digitization approaches the domain of scholarly editing and textual criticism, for example, in image-oriented digitization of documents where we have more than one source copy to choose from. There the critical comparison of several source candidates based, for instance, on their condition and completeness results either in the deliberate selection of one candidate or in an eclectic amalgam of fragments from several candidates (e.g., eclectic facsimiles à la Charlton Hinman). In the latter case—at the very least—we are facing, if not the ambition of textual criticism to establish an ideal text, then at least a kind of document criticism that seeks to establish an ideal document. This is the primary aspect where critical digitization and scholarly editing are both different and similar. The former devotes its analytical faculties to the critical transmission of graphical and material documents, and has been criticized for not paying much critical attention to their texts. The latter devotes its analytical faculties to the critical transmission of text, and has been criticized for not paying much critical attention to the graphical and material documents carrying those texts. Understood this way, the two activities not only differ but support and complement each other.

Besides, there is tangible cooperation between the two fields. Libraries normally house the source documents of interest to scholarly editors to begin with, and often perform the technical digitization of them to serve large editing projects. And libraries are arguably best suited to be responsible for the long-term technical and bibliographical maintenance and preservation of the digital files. They are also in a good position to coordinate and manage the intricate web of IPR interests within large editing projects in a way other agents, including scholarly editors, usually cannot do. This is particularly the case with image-oriented projects, where libraries and archives often *are* the very IPR holders themselves.

Increasingly digital humanists in general and scholarly editors in particular are implementing practices, systems, and tools that were developed by the library community for managing, relating, and describing large collections of documents, such as the Metadata Encoding and

Transmission Standard (METS) for metadata, the Functional Requirements for Bibliographic Records (FRBR) for descriptive cataloguing, and Linked Open Data (LOD) for sharing and reusing data through, for example, APIs. Library digitization projects are also valuable to textual scholars and scholarly editors in the way they make large amounts of material available, if nothing else, as facsimiles that, to an increasing degree, can be subjected to OCR and whose text therefore can be turned machine readable and thus reusable in scholarly editing projects. Not to mention the many digitized manuscripts that have previously been unpublished and not subjected to research but that are now becoming identified, cataloged, and made available.

So scholarly editing and library digitization are perhaps approaching a point where the two not only meet but even merge, at least on the project level. We are definitely witnessing a turn in digital scholarly editing, where scholarly editions and archives are turning into digital libraries. Is it likely that we will witness a similar turn in high-quality critical digitization performed by libraries, where the digitized collections are granulated, technically and architecturally sophisticated, based on critical choices and scholarly expertise, and where these digitized collections will increasingly take on the form of "editions"?

Or is the reverse development rather the case? That is to say, memory institutions, scholarly editors, and digital humanists are inclined to conceive of digitization (in the primary sense of image capture) as a flat and linear affair, with the digital reproduction as a reduced, derivate offspring. For centuries, textual criticism has been established as a discipline that documents and problematizes textual, graphical, and material variation within and between artifactual instances (with varying historical trajectories) of a conceived intellectual entity, such as work. This approach, however, seems to be a rare commodity when the representational artifacts are digitized interpretations of the source documents. Are we numbed by the seeming straightforwardness of the reproduction—the scan, the digital photograph, the machine-readable transcription? And if so, what kind of versatility, richness and informative potential will the digitized resources offer when imported into the lab where digital humanists aim to engage and experiment with them?

Notes

1. An example of such a critical study would be Conway 2013.
2. Most of the remainder of this chapter has previously been published as Dahlström (2010).
3. The phrase "critical transmission" might be conceived of as an oxymoron. This is intended.
4. See, for example, Greetham (1992) and Renear (2001).

5. See Patricia Seed's chapter 29 on maps in this volume.

6. See, for example, the distinction between text-based and image-based editing as treated in the thematic issue on image-based humanities computing, *Computers and the Humanities* 36 (1, 2002). See also the sharp essay on the topic by Tanselle (1989).

7. This is, of course, just one aspect of the value of mass digitization. Another is that many mass digitization projects have the economic strength to develop advanced methodologies that remain beyond the reach of many critical digitization projects. And obviously it all depends on how you define attributes such as value or quality. It is perfectly reasonable to state that the main quality of a mass digitization project is its sheer quantity.

8. Paul Duguid (2007) has been an astute critic from the bibliographic field, Paul Conway has performed systematic studies of errors in large-scale digitization projects (e.g. Conway 2013), and Robert Darnton presents his close encounter and critical perspective on the Google Book Search endeavor in his *The Case for Books* (2009).

9. For a broad but still nuanced view on mass digitization projects, see Deegan and Sutherland (2009: ch. 5).

10. This section contains a slightly edited text from Dahlström (2011).

11. For instance, the Münchener Digitalisierungszentrum explicitly states on its website that it does not distinguish between high-quality digitization and mass digitization. Both are possible due to ten years of expertise and the presence of high-quality scanning equipment (digital-collections.de). I am grateful to Lars Björk at the National Library of Sweden for pointing to this example.

12. Whether or not such documentation is being made available to anyone outside of the digitizing agents themselves is of course another matter.

13. "The very existence of variation in the text has become a matter of historical interest (rather than a problem to be removed). . . . We have found that the 20th century models of critical text-plus-apparatus are incapable of answering many of these new questions. The best scholarly environment for addressing these questions would be a digital library of facsimiles and accompanying diplomatic editions." (Blackwell and Smith 2009: 6).

14. There is a tradition within scholarly editing to refer to such facsimile editions as "noncritical." "Critical" would then refer to the established text as either a particular version "critically" chosen among several candidates, or as an eclectic mix of segments from several versions. With that definition as a base, a facsimile edition would be dismissed as noncritical, assuming it is not produced that way. But this is often far from the case. To produce facsimile editions, as well as in critical digitization, a "critical" comparison of several candidates often marks the start, regardless of whether these candidates are treated eclectically. Facsimile editions are parts of an image-based editing strategy, and as such, the crucial issue is really how close the new document (the edition, the reproduction) aims to approach or "mimic" the source document. As strategy, as technology, it can be used to produce both critical and "noncritical" editions.

35 POST-ARCHIVE: THE HUMANITIES, THE ARCHIVE, AND THE DATABASE

Tara McPherson

We live in a culture saturated with image-scapes and data. From the official archives of the Shoah Foundation (containing over 100,000 hours of video testimony) to the ever-expanding vernacular archives of YouTube and Flickr, one is tempted to argue that the archive is everywhere today, from our iPhones to our cities. Given the rapidly expanding hard drives of our laptops (let alone the data clouds promised us by Google or Apple), the archival impulse seems to be flourishing, spurred on by dreams of infinite storage.

However, Diana Taylor (2013) has recently argued that we should exercise care in calling the "proliferation of material in cyberspace" an archive. Certainly the notion of the archive seems to have expanded so completely in the digitized twenty-first century as to lose several shades of nuance and specificity. The online parenting group I belong to has an "archive" of past posts built into its design. The software that backs up my aging PC each night "archives" that day's work. The digital journal I edit has an electronic archive of back issues. The word stretches thin: the archive is a collection is a repository is a database is a folder on my desktop. Less and less does it signal rigorous institutional practices, dedicated spaces, or a devotion to physical (let alone dusty) objects. In fact Taylor goes so far as to claim that "the digital is anti-archival." Wolfgang Ernst has called it "anarchival" (2013: 139). Perhaps. But I want to take a slightly different tack here and sketch another possible relation of the digital to the archive both conceptually and practically.

Kenneth Price (2009: para. 2) reminds us that attention to such terms as "archive" and "database" should be of central concern to the digital humanities. In "Edition, Project, Database, Archive, Thematic Research Collection: What's in a Name?" he argues that the "shorthand we invoke when explaining our work to others shapes how we conceive of and also how we position digital scholarship." His piece explores a variety of terms to categorize online digital collections and the projects they generate, particularly in terms of digital scholarly editions, through an examination of the Walt Whitman Archive he helped to create. While he recognizes the difficulty of settling on a single term that seems adequate to the many forms

USC Shoah Foundation

The Institute for Visual History and Education

HOME | WATCH | TEACH & LEARN | RESEARCH | ABOUT THE VISUAL HISTORY ARCHIVE

Home » Watch

--Choose a Language Portal--

Videos by Topic

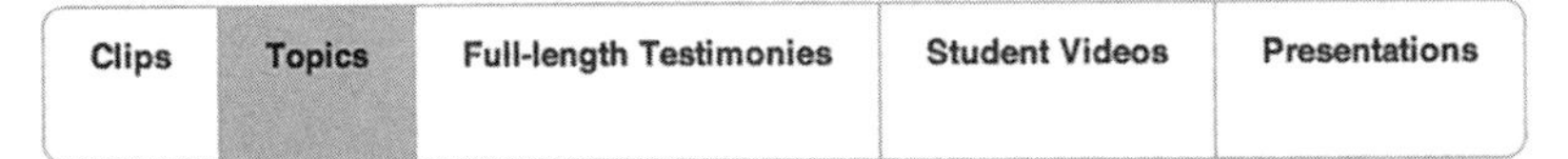

Love During the Holocaust

Clips of survivors recalling times in their lives during the Holocaust when they still managed to find love.

Figure 35.1

Visual History Archive of the USC Shoah Foundation: a database of over 51,000 testimonies from survivors of genocide of online data that we interact with today, the article does direct our critical energy to the value of attending with some care to our use of terms such as "archive" and "database" when dealing with digital data. Of course, following Foucault and Derrida, humanities scholarship has been conceptually querying the archive for quite some time now. This chapter brings post-structuralist encounters with the concept of the archive into dialogue with digital endeavors that seek to reactivate the archive along several registers. I especially focus on a series of recent projects that foreground the value of critical or cultural theory for the digital humanities, including two with which I have been involved. I do so in order to highlight the ways in which critical and cultural theory might inform the conceptualization, design, and implementation of digital scholarship. Indeed I argue that only through such an articulation can the digital humanities engage the archival in its richness, its absences, its ambiguities, its pleasures, and its transformations.

Archival Histories/Database Futures

Theories and critiques of the archive are well known to scholars of cultural and critical studies. From Michel Foucault's framing of the archive as a space of enunciation, as the site for what can be said and known, to Jacques Derrida's fragmentary and feverish archive, scholars have delineated the archive as a complex and contradictory site of power, ideology, and control. The scholarship of the archive is rich and far-reaching if also multi-vocal and sometimes abstract. We have learned of the archive's deep complicity with empire and the colonial (Stoler 2010; Richards 1993), of its many erasures (Baucom 2005; Best 2011; Greetham 1999; Taylor 2003), and even of its joys (Steedman 2002; Hastie 2007). Such work has sometimes generated lively feedback loops with custodians of the archive, as a subset of archivists has argued that the "refusal of the archival profession to acknowledge the power relations embedded in the archival enterprise carries a concomitant abdication of responsibility for the consequences of the exercise of that power, and, in turn, serious consequences for understanding and carrying out the role of archives in an ever-changing present, or for using archives with subtlety and reflection in a more distant future" (Schwartz 2002: 5–6.).[1] Certainly the digital humanities has also turned to the archive as a central locus of activity, expanding our concept of the archive to include collections of digital surrogates that might be marked up, interacted with, and analyzed in new ways. While this work has not always explicitly engaged critical theories of the archive, early archival projects certainly understood that the porting of the archive into the digital realm was a fraught and complex process. The reasons for this complexity were not merely technological. This is a point we would do well to remember.

To what might we owe this broad archival engagement across the humanities, in libraries, in software design, and beyond? How has the archive come to occupy our collective critical imaginations with such plentitude and force? This critical, scholarly turn toward the archival (and the advent of poststructuralism more generally) parallel the emergence and solidification of computation as a key site of cultural power and authority in modern life. Perhaps this is more than a coincidence. Perhaps the archive's privileged role in ordering memory, enunciation, and experience was more easily discerned as its mechanisms and power were coming to be supplanted by a rising electronic technological regime.

In her recent book *Counter-Archive*, film scholar Paula Amad (2010) offers a lively examination of Albert Kahn's "Archives of the Planet," a vast and ambitious attempt initiated early in the 1900s to catalog the cultural diversity of the world through visual media. The collection grew to over 72,000 color photographs and 183,000 meters of film and documented life in more than fifty countries. Amad argues that Kahn's archive (particularly his interest in film) presented "a 'counter archival' challenge to the positivist archive's sacred myths of order, exhaustiveness, and objective neutrality." She maintains that these early twentieth century films were able to activate a rediscovery of the everyday and to push back against positivism, observing that, "once translated into the age of cinema, the archive thus mutated into the counter-archive, a supplementary realm where the modern conditions of disorder, fragmentation and contingency came to haunt the already unstable positivist utopia of order, synthesis, and totality" (21). Thus a new technology brought pressure to bear upon the imperial logics of the archive, opening up a space of possibility and of chance.

We inhabit a similar but different moment today, one in which emergent forms of technological order again restructure our dreams of totality and complete perfection. If Amad argues that cinema circa 1910 instantiated a kind of "counterarchive," we might understand today's configuration of the digital as a post-archival moment. As such, our extensive investments in the archive and the archival—a kind of frenzy of the archival—may actually be responding to the archive's waning hold on culture as we migrate from the archive to the database. But rather than suggest that the "post archival" somehow represents the death of memory or of the historical (the tenor of much popular and scholarly writing on the digital) or a submission to regimes of total mathematical order and control, I argue that our post-archival moment is a moment of dialectical possibility not unlike that moment of a century ago investigated by Amad. Kahn's archive was both a site for the oppressive classifications and omissions of the imperial archive and also harbored the potential for the everyday, the counterarchival, to destabilize that very order through the untamed potentialities of cinema. Our post-archival moment simultaneously encompasses algorithmic regimes of control that push far beyond classification but also set the stage for a "generative, participative form of archival reading"

(Ernst 2013: 82). It is up to us as theorists of the (post-)archive and as practitioners of the digital to seize this possibility.

The object that is supplanting the archive is the database, particularly if we understand the database, following Price, in both a technical and a metaphorical sense. While physical archives obviously continue to exist (and to decay much faster than they can be digitized), the database ingests, supersedes, and obsolesces the archive. If Kahn's dream of capturing everything was consciously meant to build an archive, today's fantasies of total knowledge depend on the database. Take, for instance, DataOne. Funded by the National Science Foundation, DataOne is meant to fuel big and complete science, the science of the grand challenge and of the massive dataset. An early online description of the project described its goals: "DataOne will transcend domain boundaries and make biological data available from the genome to the ecosystem. . . . [It will] provide secure preservation and logical access" to "multiscale, multidiscipline and multinational scientific data." Obviously these claims are grand, recalling the positivist archive familiar from Foucault and others. But in very key ways, what these scientists are constructing is not an archive in any meaningful or traditional sense of the term. This is not (as Diana Taylor describes YouTube) an archive that is erasing its objects. It is an archive without objects.

The database depends on the loss of the thing itself. At one level, this loss is literal, as the "thing"—say a photo or film—is digitized and transformed into data. At another level, this loss is also a key aspect of the functioning of the technology. New media theorist Matthew Fuller has argued that in the development of the relational database, categories come to trump things. Databases work by "normalizing" data, a process that in effect privileges abstract relations among data while also stripping "things" of context. Elements in a database get sorted by a set of structuring relations (like metadata) that radically limit what can be seen. If the card catalog and the filing cabinet of the late nineteenth-century archive also defined relations, they still bore a connection to the object that was archived. The relationality of the database takes this abstraction to a very different level. (Here it might be noted that while we might as humanities scholars think of relationality as a good thing, perhaps related to concepts such as sociability, as is often the case in computer jargon, a relational database is less about sketching rich and detailed relations between objects but instead about rigid categorizations and formal logic).[2]

I have written elsewhere about the ways in which the development of computer operating systems at midcentury installed an extreme logic of modularity that "black-boxed" knowledge. An operating system like UNIX (an OS that drives most of our computation, directly or indirectly) works by removing context and decreasing complexity. Early computers from 1940 to 1960 had complex, interdependent designs that were premodular. But the development of databases would depend upon the modularity of UNIX and languages like C and C++. While "generalization" in UNIX and "normalization" in databases have specific technical meanings, we might also see at work here the basic contours of an approach to the world that separates object from context, cause from effect, data from thing. If Kahn meant to put together many small examples in an archive in order to craft a big picture of the diversity of the world (one that Amad calls kaleidoscopic), the logic of computation proceeds quite differently. The drive of computation is not the classifying impulse of the Fordist era but the governmental impulse of the post-Fordist or neoliberal era. It moves from specificity to interchangeability.

If the logic of early film—following Amad—was counterarchival, the logic of the database is post-archival, pushing far beyond the classification systems of the nineteenth century toward an extreme modularity and a set of abstract relations that retains little of the materiality of film or the traditional archive (which is not to say the database is not material, but its relationship to the collected object is different from both the filing cabinet and Kahn's linear film strip). Nonetheless, this post-archival, techno-epistemological fever is dialectical much as were the projects and films Amad surveys. That is, the digital at once dreams of perfect reproduction and of total knowledge (particularly in its data structures) and also embodies the possibility of pushing back against that dream. Amad observes that early French critics of film saw that the "ideological correspondences of film's two-sided affinity for the everyday were still up for grabs." That is, film might dutifully serve realist dreams of perfectly captured reality, but it was also unpredictable and had a tendency to stir things up.

We are in a similar moment today in relation to the digitization of the archive and to the technologies that enable this digitization. If, as Amad notes, the "instability" of the archive became more explicit "under pressure from film" one hundred years ago, we might now say that, under pressure from social and networked media (particularly via a variety of theoretically inflected projects that refuse a total submission to formal logics and normalized data), that the instabilities inherent in governmentality become more explicit, revealing and disrupting, not exhaustive classification, but instead, the relentless processing of an informatic age. Thus, even as the database future wants to "transcend domain boundaries" and provide "logical and complete access" from the genome to the solar system, a variety of digital media projects have the potential to operate as a critique of that very impulse. We must attend to, theorize, and help create such projects (and the data structures that enable them) as thoroughly as we have attended to the archive over the past four decades. Only then can we hope to mobilize the potentiality of the digital.

For instance, even the DataONE project, which aspires to capture big science, also incorporates the rise of amateur science, as is evidenced in the affiliated eBird project. This project allows recreational and professional bird watchers to log their sightings into a rapidly expanding database that aggregates a variety of information about species migration and biodiversity. While, on the one hand, the project might be seen as harvesting immaterial and

Figure 35.2

eBird crowdsources information collected by the birding community providing "rich data sources for basic information on bird abundance and distribution at a variety of spatial and temporal scales"

free labor from its participants, it, and a variety of related endeavors that crowdsource science, also open up the possibility for a deeper engagement with pressing issues around climate change and human impact on the environment. More playful projects like the Metadata Games developed by artist Mary Flanagan and archivist Peter Carini offer everyday users of the archive a chance to engage in tagging digital objects across a variety of collections. Such user-generated tagging could simply default back into the regimented grid of the database, but it might also inject human playfulness and ambiguity into normalized systems.[3] If film circa 1910 bore a privileged relation to the everyday that could tip toward possible freedoms, some forms of sociotechnological praxis within the post-archive seem likewise to be parsing community and sociability, seeking to reinstall the context that our database-driven machines strip away.

Engaging the Database

If the technological structures of computation are weighted in favor of positivism and control, these impulses are not determining. We might wield digital technologies to other ends, building different systems for scholarly engagement. One mode of engagement might be a continued devotion to the physical archive, even as it wanes as an organizing epistemology of culture. But this will not be enough. We must also engage policy and practice along several registers, work that is already ongoing. We must not assume that digitization will adequately capture the richness and diversity of the cultural record or that digital surrogates can simply replace the physical archive. We should participate in and guide decisions about what will get digitized, ensuring that digitization does not simply reinstall the absences and imbalances of our physical archives within digital realms. Today it is our responsibility as humanists not only to *use the archive* but to participate in its transformations in the digital era, not only theorizing what it is, was, or should be, but actively engaging the construction and models of use of the post-archival database.

Ken Price (2009) observes that "a database in not an undifferentiated sea of information ... [A]rgument is always there from the beginning in how those constructing a database chose to categorize information—the initial understanding of the materials governs how more fine-grained views will appear because of the way the objects of attention are shaped by divisions and subdivisions within the database. The process of database creation is not neutral, nor should it be." In this way, the database is similar to the archive, functioning in fundamental ways to shape what can be seen or enunciated. If for Derrida (1998) "effective democratization can always be measured by this essential criterion: the participation in and access to the archive, its constitution, and its interpretation" (4), the same might be said today of the database.

There is no natural fit between the logic of the database and the interpretative methodologies favored by many humanities scholars. However, productive tensions emerge when humanities scholars turn their attention to the database. As Stephen Ramsay (2004) reminds us, "a number of fascinating problems and intellectual opportunities lurk beneath" the seemingly practical issues of database design. I would argue that no one is now better poised to inform and shape technological development than scholars who have devoted years to analyzing text, to interpreting images, and to thinking through complex questions of the emotions, of embodiment, of representation, and of power. We might help technology better handle the complexity of human thought and human feeling. We can participate in designing technological systems

Figure 35.3

Critical Commons, under the leadership of Steve Anderson, allows its users to collect excerpts from commercial media, serving as a rich media archive and fair use advocacy network

that better suit our needs and modes of analysis. Along the way, we will better come to understand how our machines are also redesigning us.

At the level of policy, good work is well underway. This ranges from interventions in the Digital Millennium Copyright Act to bottom-up interventions like Critical Commons, a repository of largely commercial content repurposed for academic use. Conceptualized by media scholar Steve Anderson, the site functions as a user-driven and vernacular database for the viewing, tagging, sharing, and curation of video, images, and sound. Critical Commons also embodies an argument about the necessity of fighting for fair use and fostering rich collaborations. It speaks back to the restrictive limits of copyright while engaging in crowdsourced content creation from a like-minded community of users. It imagines the database as a platform that can be turned against its embedded corporate logics, contributing to the circulation of knowledge and expression beyond the regime of private property.

Humanities scholars are also participating in the design of new digital systems that take scholarly critiques of the archive's colonial complicity to heart. A good example is an emerging interest in new kinds of ontological systems that wield metadata differently, often challenging the presumed universalisms of Western technological design. For instance, Kim Christen (2006) and Ramesh Srinivasan (2011) each argue for new modes of participatory design practice that engage indigenous peoples in the construction of technology systems that better match their ways of knowing. They call not for universal standards but for fluid and particular ontological systems as well as for database designs that question the contemporary urge to access all information all the time. Recalling the longstanding violence of the archive toward indigenous peoples, Christen has worked side by side with indigenous communities to create Mukurtu, a database platform that allows its users to manage their own cultural heritage materials, resisting the colonial imperative that so often underwrote the archival impulse. The project powerfully reminds us that access comes with costs and that notions of "the commons" often mask imperial assumptions. Critical race theorists have often argued that greater visibility for people of color is not inherently a good thing. Visibility comes with costs (all the more so in an era of heightened technological surveillance), and the goals of and desire for visibility must be evaluated within specific contexts. Similar arguments might be extended to notions of access, particularly in relation to cultural heritage materials. Greater access to and visibility of such materials might itself enact a kind of appropriative violence. Mukurtu seeks to add context, texture and intentionality to the digital's dreams of ordered and open access. As the website notes, their technological practice endeavors to design systems that foster "relationships of respect and trust" rather than efficiency, commerce, or speed.

Thus the normalizing logic of databases might be remapped in their very use and implementation: we might participate both in their design and in their usage, melding, in Diana Taylor's (2009) terms, repertoire and archive. To illustrate further these possibilities, I now turn to two such projects I am involved in, modeling paired approaches to working with, against, and through the database. In 2005 several close collaborators (especially Steve Anderson, Craig Dietrich, Raegan Kelley, and Erik Loyer) and I launched the multimodal, online journal *Vectors*. *Vectors* was developed as a space for experimentation in screen languages, open access publishing, and collaborative design and authorship. Unlike a traditional print journal that reviews largely "finished" articles, we initially deployed a fellowship model that paired humanities scholars with design and programming teams to create projects iteratively, orchestrating deep collaborations that typically lasted three or four months. These partnerships pushed back against the siloed knowledges of the university, building a space for shared practice across very different skill sets. (One might understand the contemporary university to be partitioned into isolated departments in a manner that mimics the modularity of the database; this structure makes meaningful collaboration across different methodologies quite difficult.) These projects also utilized a rolling model of peer review, beginning with the review of fellowship applications and continuing throughout the development process.

Vectors' projects bear little resemblance to articles in typical print journals or to PDF-driven online journals. We began the journal motivated by a series of research questions. We were interested in how multimodal expression might allow for different relationships of form to content and wanted to explore the specificities of digital media for scholarly communication. We asked how scholarship might more directly engage the emotions and multiple senses. We investigated how we might play an argument as we play video games or immerse ourselves in scholarship as we do in films. Such questions led us to create projects that mine the intersection of design and the humanities much more than the science–humanities nexus that drives much digital humanities work. Our projects were speculative in the sense that Johanna Drucker describes, committed to pushing back against the cultural authority of rationalism in the digital humanities and in digital design. They were also centered on the critical and theoretical questions that motivated the scholars with whom we worked, humanities scholars interested in questions of memory, race, gender, embodiment, sexuality, perception, temporality, ideology, and power.

Consider the piece, "Public Secrets," by Sharon Daniel with design by our Creative Director Erik Loyer. The piece opens with a voiceover narration by Daniel describing her regulated entry into a prison near Chowchilla, California, the site of the two largest women's prisons in the world. After the opening, the user encounters a fairly minimalist black, white, and gray design comprised of a moving line that horizontally redraws and divides the screen. Hovering over a block of text launches sound. Click through and the screen resolves into a gridded tree map. The project includes hours of audio footage, testimonies from imprisoned women that were collected by Daniel through her work with the activist organization, Justice Now. The user can navigate the piece via a series of themes (inside/outside, bare-life/human-life, and public secret/utopia), through individual women's stories, or in a more random fashion (or through a combination of all of these). The design and structure of the project reinforces its goals, calling our attention to the shifting borders between inside and outside, incarceration and freedom, oppression and resistance, despair and hope. Through navigating the piece, the fine lines demarcating such binaries will morph, shift, and reconfigure, unsettling any easy assumptions about 'us' and 'them' in the carceral state. Rather, inside and outside mutually determine and construct one another, sketching powerful vectors of relation between individual experiences and broader social systems as Daniel argues that we are all complicit with the prison industrial complex.

A project like "Public Secrets" begins to imagine new possibilities for the archive as it mutates into the database. It is a database with a point of view: it very much relies on the database to structure its content (as well as on custom treemap and typographic algorithms

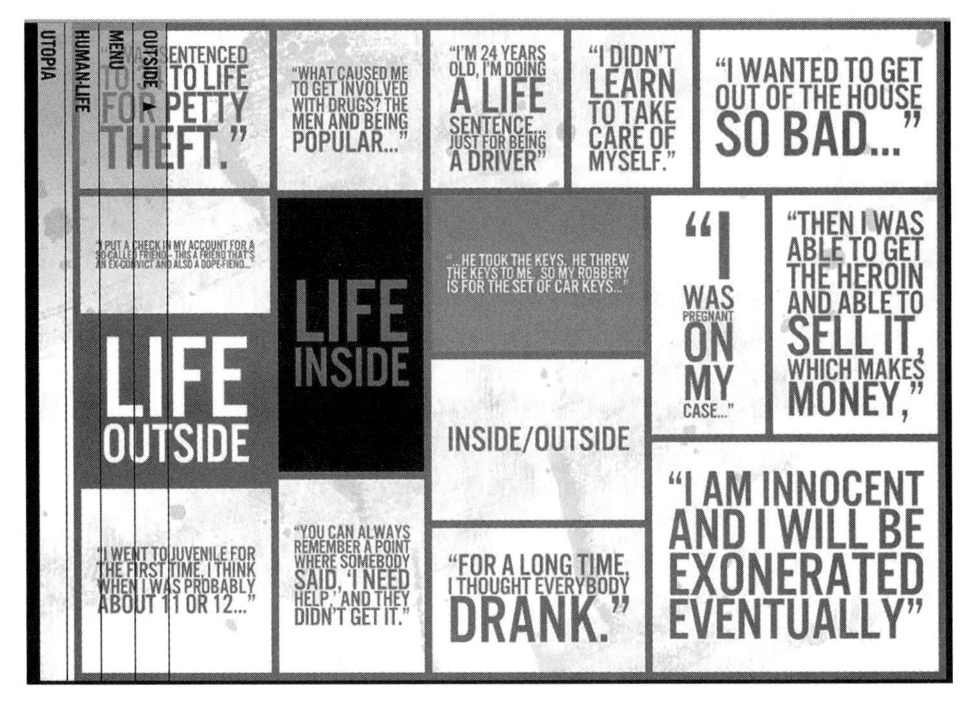

Figure 35.4

Vectors' project, *Public Secrets*, by Sharon Daniel with Erik Loyer, reconfigures the physical, psychological, and ideological spaces of the prison, and allows us to learn about life inside prison along several thematic pathways and from multiple points of view

designed by Loyer), but it also contextualizes the database, guiding the visitor's immersion in the piece, using the logic of the video game to unfold portions of the project at key moments. It plays with linearity but does not entirely abandon it. It encourages exploration more than mastery or completion. Its argument is affective and builds momentum in the temporal unfolding of the piece. If Lev Manovich has (now rather famously) argued that database and narrative are "natural enemies," *Vectors*' projects often insist on recoupling these modalities.[4] In charting the emergence of the "anarchive" or the "dynarchive" (the archive as digital network), Wolfgang Ernst (2013) has written that narrative is being replaced by calculation. Technologically, this may be true in the inner workings of the database and the algorithm, as the dynamic processes of digitization perpetually churn. But, as humans, we still engage our machines via interfaces that offer rich possibilities for reimagining data narratives in and for the digital age.

While *Vectors*' projects began as experiments at the surface of the screen, they soon led us toward building tools and thinking about infrastructure. In particular, we began to grapple with the database as an object to think with and to think against. Many of the scholars we worked with came to us with their own collections of evidence and objects that they wanted to animate and engage in new ways. We found that the constraints of much relational database software were not particularly well suited to the ways in which humanities scholars think and work. Through the guidance and insight of Technical Design Lead, Craig Dietrich, the team developed a customized database tool that allowed more flexibility in how scholars could iteratively work within our middleware, allowing them to modify the database and its tables much more easily. We also began to explore ways in which interface design might mitigate the database's relentless logic, refusing the tyranny of the template, even as we obviously were still working under the sign of computation. In exploring relations of form to content, we privileged particular kinds of content, choosing to work with scholars interested in questions of gender, race, affect, memory, and social justice. These concerns were and are at the core of our research, and they profoundly shaped (and continue to shape) how we use and design technological systems.

For instance, anthropologist Kim Christen (who would go on to develop Mukurtu) worked with the Vectors' team to create "Digital Dynamics Across Cultures," a project that draws upon materials (including photographs and videos) collected during her fieldwork to explore the knowledge protocols of the Warumungu Aboriginal community in Australia. Rather than deploy these materials to illustrate some Western truth about the Warumungu, Christen envisioned a piece that would call into question our assumptions that the Internet should give us access to all that we desire. This project and others like Judith Jackson Fossett's "Slavery's Ephemera" and David Theo Goldberg's "Blue Velvet" subtly draw upon theories of the archive (while utilizing the database and the interface) to underscore the politics of selection and interpretation always implicit in our technologies of memory and storage. Stephen Best writes that the "visual archive of the slave past ... is marked by the clash between the imperative to recover the past and the impossibility of doing so" (2011: 159). In exploring the failed infrastructures of New Orleans or the ghosts of Louisiana's plantation past at the intersection of data structure and interface design, Goldberg and Fossett remind us that our archives and our databases are neither transparent nor innocent.[5]

Database-driven and database-dependent, these are projects for post-archival times. But their impulse is to reanimate the archive—to bring a waning form to new life while blurring the line between form and content, repertoire and archive. They pay attention to specific things and experiences, resisting the decontextualizing logic of the database. They wield technology against its positivist self, foregrounding the work of the interface and refusing an easy transparency and corporate tenets of "good" design via the template. *Vectors*' projects reward slow reading, requiring attention and care from the reader (much like any emerging discipline or new theoretical model). They most often complicate the database at the level of the screen, amping up the auto-critique. They typically deploy the strategies we often associate with the avant-garde or with experimental art.

Like an art or design studio, we were learning by doing, creating prototypes (if not always rapid ones), working by iteration. But, as much as I love the work we do with *Vectors*, there are problems with it as well. It is expensive work to do, it does not scale well, and, at the level of the interface, the projects are mostly one-offs that each demand new reading practices. Many of the early projects were done in Flash, so they will be hard to preserve and don't necessarily play well with the wider world of linked data or with the proprietary world of the iPad. Increasingly we realized that we needed to think about scaling our work and about other parallel approaches for scholarly publishing. One motivation for our research was the realization that many of the ways that knowledge circulates within the academy (through black-boxed books in niche fields speaking largely to the initiated few or behind locked-down corporate paywalls) were themselves complicit with the modular logic of the database.

Over the past four years, concurrent with the work on *Vectors*, our team has also been expanding upon the lessons learned through our ongoing research. With support from the Mellon Foundation and the National Endowment for the Humanities and in close collaboration with many colleagues (including especially Wendy Chun, Brian Goldfarb, Nicholas Mirzoeff, and Joan Saab), we have been forming a larger organization, the Alliance for Networking Visual Culture. Our goal is to formulate new ways of working with digitized archival materials within the humanities and to continue to push for new modes of digital scholarly publishing. In the context of specific research questions in visual studies and media studies, the Alliance focuses on integrating the primary source materials available in online databases more directly into born-digital scholarship. Strategic partnerships with archives (including the Shoah Foundation, Critical Commons, the Hemispheric Institute's Digital Video Library, the Internet Archive, and the Getty) and several university presses (MIT, California, Duke, Michigan, NYU, Oxford, and the Open Humanities Press) provide the testing ground for the investigation of new publishing platforms. We are currently expanding our range of partners to include new archives and presses, as well as scholarly groups like the College Art Association and several humanities centers.

One particular area of interest is to begin to work with images, film, and video in new ways within our scholarly analyses. Obviously, as bandwidth has increased, moving images have exploded on the web. From popular video sites like YouTube to specialized archives like the Hemispheric Institute Digital Video Library and the Shoah Foundation collection, a variety of moving image media is now available for scholarly analysis. Indeed YouTube claims that seventy-two hours of video are uploaded to its servers every minute. These materials are ripe for scholarly interpretation, particularly in a format that positions analysis side by side with archival material, but they also point the way toward new time-based modes of argumentation and scholarship. There are lessons to be learned from our engagement with video about how scholarship itself might unfold temporally in response to a user's engagement with a digital archive. While some projects might take the user along a scholar's carefully pruned and relatively fixed pathway within such a collection, other projects can set the stage for the user to begin navigating her own pathways, adding new layers to the scholarly interpretation of the digital holding. Some of these interpretations could be vigorously peer reviewed and warranted via relationships to established presses; others might encourage more open and bottom-up networks—new, livelier modes of collaboration and curation. Over time we might begin to see how users respond to a variety of materials held with a collection and to produce analyses of how scholarly writing emerges within and across collections. Scholars might alternately track the silences and absences of the archive within the space of the archive itself, haunting it with our own interpretations and poetics, foregrounding the impossibility of total knowledge or complete access. If libraries and museums have now carefully curated and digitized these troves of material, humanities scholars can and should engage these collections in ways that push beyond the linear book or article.

Historically the archive was officially meant to collect, preserve and protect. Selection of, access to, and the use of archival materials were rigorously regulated. The archive cultivated an ethos of the rare and the original. Careful order was imposed. The digitization of archives has upset this careful hierarchy. Digital archival materials might be circulated and shared more freely; the line between archive and library blurs. Amateur and expert might build archives together. We might begin to imagine the post-archive itself as a site of creation, change and emergence. In the past humanities scholars have raided archives in order to capture their treasures for our books and articles. This relationship has often been uni-directional and vampiric, giving little back to the archive. In an era of connected data, our interpretations might live within the digital archive, curating post-archival pathways of analysis through its datasets or framing the archive via multiple points of view.

The Alliance is building both human and technological infrastructure in support of these goals, including a new authoring and publishing platform, Scalar, that was released into open beta in spring 2013. Scalar allows scholars to author with relative ease long-form, multimedia projects that incorporate a variety of digital materials while also connecting to digital collections, utilizing built-in visualizations, exploring nonlinearity, supporting customization, and more. (For a full description of Scalar's capacities, please see our website, which also links to several sample projects, http://scalar.usc.edu/.)

Many of the scholars we have collaborated with are interested in allowing the users or readers of their research to engage with their primary evidence while also exploring the scholar's own interpretation of that evidence. Working in Scalar, scholars are pulling sets of visual materials from digitized collections into a Scalar project or "book," encouraging the project's reader to examine these materials in their own right while also engaging a scholar's analysis of the materials. A good example is a project now underway by media theorist Kara Keeling and filmmaker Thenmozhi Soundararajan. Soundararajan previously founded a nonprofit, Third World Majority, focused on digital storytelling with global youth. Along with Scalar collaborator Micha Cárdenas, she and Keeling have posted materials related to the organization to the Internet Archive, one of the Alliance partners. These materials encompass hundreds of videos and still images, as well as text documents of various kinds. Keeling and Soundararajan have incorporated this work into a Scalar "book" and have invited several collaborators to author pathways through this material. When completed, a reader will be able to explore the collection of primary materials in a fairly open way, but she will also be able to follow one of several pathways through the materials that are authored by filmmakers, nonprofit workers, and several scholars of digital storytelling, media history, and global cinema. Such a project is neither solely a book nor solely an archive, neither simply database nor narrative, but rather a hybrid space between the two that blends an edited collection of essays with an abundant cache of primary materials. Using Scalar's built-in commenting features, the reader of the project can then add her own commentary, providing more context for the collection of primary materials or responses to the scholarly interpretations of the collection.

This type of connected and shared research and writing space has emerged as a key area of interest for Alliance partners, who recognize the capacity of such practices both to reach new audiences (with certain paths more relevant to certain types of readers) and to facilitate close analysis and collaborative methodologies. Within a single project, we glimpse research operating across scales, with scholars able to move from the micro level of a project (perhaps a single image or video annotation) to the structure of the entire project and its integrated media. The researcher can create careful close readings within a project of many components that can also be instantly represented as a whole collection, thus moving beyond the artificial binary of "distant" versus "close" reading that often surfaces in conversations about the digital humanities. The result richly combines narrative interpretation with visualizations that are automatically generated by the semantic elements of the platform. These visualizations allow an author or reader to see the larger structure of a project that may have been built up more organically piece by piece, while also allowing iterative refinement to this structure. This method of researching and writing across scales now predominantly unfolds within a given Scalar project, but the possibility for porting these modes of analysis back to our archival partners' larger holdings represents a key area for future research.

The software that underpins Scalar was born of the frustrations our scholars often experienced working with traditional database tools. *Vectors* engaged intersectional, political, and feminist work at the level of content but also integrated form and content so that the theoretical implications of the work were manifest in the aesthetic and information design. Scalar is now seeking to integrate this methodology at the level of software design. Scalar takes our early experiments at hacking the database for *Vectors*' projects to a very different level by wrapping a relational database in a very particular semantic layer, a process spearheaded by Craig Dietrich and Erik Loyer.

In effect we wanted to build a system that respected the research methodologies of the scholars with whom we work. Scalar resists the modularity and compartmentalized logics of dominant computation design by flattening out the hierarchal structures of platforms such as WordPress. While easy to use, it also moves beyond the template structures that frequently categorize the web, allowing a high degree of customization through its API. Thus it mediates a whole set of binaries: between close and distant reading, user/author, interface/backend, micro/macro, theory/practice, archive/interpretation, text/image, and database/narrative. It respects both the system and the node, or, in the words of our collaborator Nick Mirzoeff, "you can explore an issue and then develop its intersections with other issues" in a very dynamic way. Nick has described working with the platform as a kind of horizontal writing; I think we might also point to its associative possibilities and its capacity to model an in-between (of interface to code; of archive to scholarship; of author to audience).

Vectors and Scalar are about reanimating the archive along several registers even as the archive blends into the database. We are consciously creating technology that allows scholars to be closer to the archive, while also producing new interpretations and new paths out of the archive into other vernacular traditions that depend upon the mutability of the database. Scalar invites the reader in to remix as well, preserving the scholar or curator's voice while also opening up the conversation to other voices. This is the post-archival world.

Following years of training in poststructuralist theory, we have come to see the archive as disciplinary, positivist and ideological (even as we love to hang out there). Yet I have here argued that the official archive is waning, replaced as it is by the database as the preeminent mode of organizing knowledge. The differences between the archive and the database matter, and we'd do well to attend to these differences with care. Of course, this is more easily said than done, as my own writing seems to slip all too easily between the terms. Our tendency to see the archive everywhere makes it hard to use the term with any specificity. It often becomes a modifier (Mukurtu is an "archival platform") or is itself modified (I have often used the term "digital archive"). Thus, in foregrounding the post-archival, my aim is to apply critical (and practical) pressure to these slippages and proliferations. Such pressure helps us recall that the Internet is not an archive in any meaningful sense (even if it seems to be one via common sense).

However, to be post-archival is not simply to cede the archive to the database and to submit to new, more encompassing regimes of order. To be post-archival is to reanimate the latent pleasures and possibilities of the archive for new ends. If early film around 1910 had the capacity both to order knowledge and to reactivate the everyday as a category of experience, digital

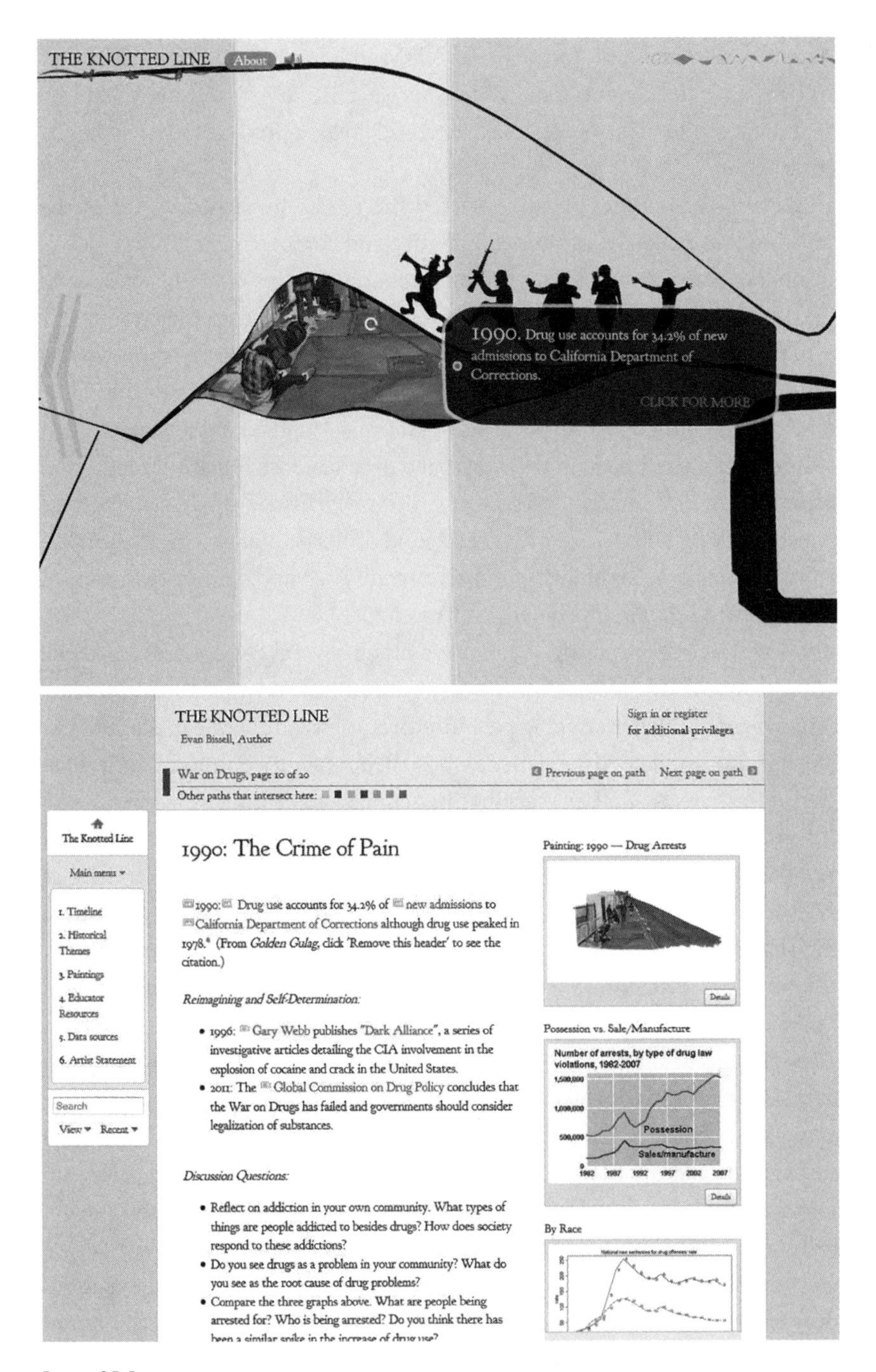

Figure 35.5

The Knotted Line utilizes Scalar's API to combine a tactile, evocative interface (left) with a detailed online resource (right) examining freedom and confinement, pushing far beyond the hierarchal and templated structures of WordPress media today—especially certain strains of social media, linked data, and experimental design—at once are joined to the database and its control fantasies and are able to bring into greater clarity experiences of sociability, tactility, mediation, narrative, temporality, and affect that can inject the human into the network. If database forms tend to divorce object from relation, might a post-archival sensibility remake the very terms of the database, pushing beyond its privileged modularity? The post-archival can reconnect repertoire and archive, content and context, node and network.

The post-archival seeks to retrieve the best of the archive and the digital for new modes of practice, but to truly hone a post-archival sensibility those of us who worry about the archive (both in its transformations and its tyrannies) will need to take seriously the dominance of the database and the algorithm today. Humanities scholars need to become more technologically literate and to move outside of our comfort zones toward broad collaborations. Our insights into culture and aesthetics, our methodologies, and our close readings are deeply relevant to things digital, but we will need to be open to new ways of working and to collaborative forms of creation if we are to help tip the post-archival world more fully toward justice and possibility.

Notes

1. Rick Prelinger was a fellow in our labs at USC in the summer of 2005. At that point our team was well aware of his pioneering efforts to collect and preserve ephemeral film as well as his involvement with the Internet Archive. With his wife, Megan, he also opened the very radical Prelinger Library. From Rick, I learned to flip my thinking about archives, seeing them not as "quiet places" but as places "justified by use." See his *Vectors'* article, "Panorama Ephemera" at http://vectorsjournal.org/projects/index.php?project=58

2. Humanities scholars involved in computation have debated the possibilities and limits of databases for both quantitative and qualitative scholarship for many years. For instance, William Thomas has traced the debates over relational database design for the field of computational history. He notes that the widespread availability and adoption of commercial database platforms in a "third wave" of historical computing had serious implications for the analysis of the "fuzzy" or "incomplete" data often at play in historical research. Such databases were not designed to accommodate the ambiguity at play in much humanities scholarship.

3. A variety of important research has begun to take up the questions of immaterial labor in the digital terrain, building on the work of scholars such as Maurizio Lazzarato. See particularly essays by Terranova, Andrejevic, and Scholz in *Digital Labor: The Internet as Playground and Factory*, edited by Scholz. This scholarship is particularly valuable to the degree that it challenges more utopian readings of participatory

digital culture. While Metadata Games or eBird might be seen to be part and parcel of this tendency toward unpaid and harvested labor, I think that they can also be seen to engage elements of a communal life that might tip digital labor in different directions, especially because neither site is tied directly to flows of global capital.

4. Many scholars have responded to Manovich's formulation. Notably, see Katherine N. Hayles and Marsha Kinder. Kinder is also an early pioneer in creating narrative database forms through her direction of the Labyrinth Project. Our collaborative working process at *Vectors* drew inspiration from the lively atmosphere of Labyrinth.

5. In an exciting essay in *American Literature*, Lauren Klein offers a different take on how we might visualize absence in the archive by bringing together visualization techniques favored within the digital humanities with a careful attention to what data visualization may distort or deform. She aims "to expose the impossibilities of recognition—and of cognition—that remain essential to our understanding of the archive of slavery today" while engaging practices like data-mining and visualization.

36 FINAL COMMENTARY: A PROVOCATION

N. Katherine Hayles

This volume, immensely impressive in its breadth and the range of viewpoints represented within it, clearly shows the vitality of digital humanities, even as it also reveals how many disagreements of various kinds populate the field. If there is an area of general agreement, it is the transformative potential of digital humanities for the humanities and for academic discourse generally. Yet, despite the far-reaching changes envisioned here, I find many of the arguments for transformation surprisingly conservative, in that they see the way forward as revisionist rather than revolutionary, working within departmental structures and academic protocols (Thomas, Liu), re-thinking pedagogy within academic practices (Davidson, Losh) or moving toward a synthesis of critical theory and computational practices (Presner, McPherson, Davidson). Valuable as these perspectives are, to my mind they stop short of articulating the full implications of digital humanities within the landscape of computational media generally.

Johanna Drucker perhaps comes closest to exploring these wider implications when she talks about computational media as submerging the individual voice in the collective. Harkening back to the Romantic ideal of the individual voice as the source of creative inspiration and literary innovation, she suggests that the era of the individual voice may be coming to an end. As Mark Rose has reminded us, the cult of the author originated in the desires of publishers and bookmakers to ensure a reliable source of continuing profit; in this sense, the Romantic ideal of authorial genius was interpenetrated by capitalism from the beginning. But it is not just the individual voice that is at stake in the spread of computational media. The wider landscape referred to above includes one of the most remarkable innovations of the twentieth century: the unprecedented rise of automated cognition, and the re-alignment of human conscious thought as a result.

To appreciate the scope of this change and its implications, we must move beyond thinking only of networked and programmable machines (vast as these networks are) to all of the devices embedded ubiquitously in our environments, including the computational modules perched underneath the driver's seat in cars, the smart paint in buildings, the RFID tags that have revolutionized warehousing and distributing practices, and a host of other intelligent devices and smart technologies. What is remarkable about these computational media is that, although they carry out cognitively sophisticated actions, they are entirely lacking in consciousness and consequently in questions of meaning. We can characterize them, collectively, as the cognitive nonconscious. At present they so completely interpenetrate human systems that, were they suddenly to cease functioning, the result would be catastrophic loss of human lives and social chaos. Think, for example, of what would happen if the computer systems controlling traffic at O'Hare or LAX were to crash; if the software controlling the electric grids were to malfunction; if a malware so effective that it could crash the Internet went on a rampage.

If, as Foucault, Bernard Stiegler, and a host of other theorists have shown, human being is not a transhistorical essence but a mode of living emergent from our engagements with social practices, diffusions of power, and our technologies, there is reason to think that human being may be entering a new era, dominated by and emerging from our interactions with the cognitive nonconscious. In Henry Jenkins' interview with Sherry Turkle, she proposes as an acid test for evaluating technologies, "Does this technology serve human purposes?" To pose the question like this is to risk a false binary, one that positions human purposes as preceding and standing apart from the technologies with which they are enmeshed. A better way to interrogate our present situations is to ask, "How are our engagements with technologies affecting human purposes?"

Our inheritance from the Enlightenment includes not just the individual voice, as Drucker notes, but the individual as the relevant unit of analysis for social contracts, ethical judgments, legal adjudications, and (perhaps importantly) the origins of selfhood and meaning. This assumption is deeply ingrained in the humanities in particular; across a range of disciplines, from literary studies to history to philosophy and art history, questions of meaning are central. If, as I have been suggesting, the cognitive nonconscious poses implicit challenges to the centrality of meaning, then it follows that the humanities, including (better, especially) the digital humanities, should take not the centrality of meaning as a grounding assumption but rather as a site for interrogation, innovation, and revolutionary re-visionings.

Already there are calls to re-examine the costs of meaning—or, if you will, the meaning of meaning. One such inquiry is Ray Brassier's *Nihil Unbound*, in which he draws on the resources offered by such thinkers as Alain Badiou, François Laruelle, and Gilles Deleuze, among others, to argue that the quest for meaning has kept us imprisoned within the Kantian correlationist circle and has prevented us from making contact not just with the "for us" but for the world as "in itself." The costs of meaning, in this view, include an egoistic understanding of the world, in which we view the cosmos as a Heideggerian "standing reserve" to be exploited for our benefit. Breaking out of the correlationist circle, or in Brassier's term, unbinding ourselves from meaning and confronting the void as non-meaning, can be, he argues, as liberating for us in the posthuman present as was the shift to humanism in the Renaissance.

Understood within the context of digital humanities, the implications of our engagements with computational media go far beyond transformations of the academy and humanistic discourse. At stake is the centrality of meaning in the humanities, the interpenetration of human consciousness with the cognitive nonconscious, and the openings these developments create to re-think the role of the humanities in our historical present. The one avenue not open to us, these developments suggest, is to remain complacently rooted in past verities. This is the *revolutionary* potential of digital humanities, the creative foment it offers for disciplines that have traditionally taken questions of meaning to be essential to the human vocation.

What then takes the place of meaning, which has traditionally been at the center of humanistic inquiry? Properly to assess this question requires rethinking nonconscious cognition not only in complex technical systems, but within the human cognitive system as well. Research in cognitive neuroscience has, for the last two decades, demonstrated that nonconscious cognition operates in human neurology as well. The human cognitive nonconscious is capable of processing information at a much faster rate than consciousness, and it can discern patterns too complex for consciousness to see. It is also capable of learning, drawing inferences, and reaching conclusions about ambiguous or conflicting information. Indeed, without the cognitive nonconscious, consciousness would not be able to operate at all, for it would be overwhelmed with the information flooding in from sensory perceptions, somatic markers, and other chemical and electrical signals from other parts of the body, as well as from the outside world. The cognitive nonconscious serves the crucial functions of either passing on its conclusions to consciousness or, if not relevant to the immediate context, suppressing them.

This research suggests that such activities are indispensible components of meaning-making, for they are the foundation upon which higher level consciousness depends. Recognizing this fact would enable us to see the activities of text-analytic algorithms and other processing done in computational media not as mere calculation, or even more extremely, as a threat to the traditional values of humanistic inquiry (beliefs sometimes articulated by those who oppose the digital humanities), but rather as closely akin to the kind of information processing that precedes and enables meaning-making as consciousness understands it. This insight strongly suggests that there is a synergy between conscious meaning and the kinds of interpretive activities of nonconscious cognition, whether occurring in digital media or the human brain. Like external memory storage (a development that has been going on since the invention of writing), automated cognition re-creates in artificial media many of the processes, strategies, and tactics of internal human cognition. In my view, this need not imply that the brain operates like a computer. The brain has its own methods of generating cognition, and neurons are not binary switches but embodied and embedded processes that create complex recursive loops between brain, world, and body. That said, automated cognition increasingly operates within deep technical infrastructures that are enmeshed and enwebbed with human cognition and human life. To rise to the challenges, opportunities, and problems this situation creates, the humanities must broaden their traditional concepts of meaning to include nonconscious cognition, with all of implications and consequences nonconscious cognition poses. There is no better place to begin than with the digital humanities.

REFERENCES

#Alt-Academy. "MediaCommons." http://mediacommons.futureofthebook.org/alt-ac/

#transformDH. "This Is the Digital Humanities." *#transformDH*. 5 March 2012. Web. http://transformdh.tumblr.com/archive

4. Humanities. 2013. "Mission." 4Humanities. http://4humanities.org/mission/

Aarseth, Espen J. 1997. *Cybertext: Perspectives on Ergodic Literature*. Baltimore: Johns Hopkins University Press.

Abell, John C. 2009. "South Korean 'Prophet of Doom' Blogger Acquitted." *Wired*. April 20. http://www.wired.com/2009/04/south-korean-pr/.

"About HUMLab." HUMLab (website). http://admin.humlab.umu.se/about.

Abu-Lughod, Janet. 1989. *Before European Hegemony: The World System A.D. 1250-1350*. Oxford: Oxford University Press.

Addad, Edde. 2011. "charNG: A Character N-Gram Generator." eddeaddad.net. Last modified July 24. http://www.eddeaddad.net/charNG/

Adorno, Theodor. 1973. *Negative Dialectics*, trans. by E. B. Ashton. New York: Continuum.

Agamben, Giorgio. 2000. "The Face," in *Means without End: Notes on Politics*, 91–102. Translated by Vincenzo Binetti and Cesare Casarino. Minneapolis: University of Minnesota Press.

Agamben, Giorgio. 1993. "Without Classes." In *The Coming Community*, trans. by Michael Hardt. Minneapolis: University of Minnesota Press.

Akrich, Madeleine. 1992. "The De-Scription of Technical Objects." In *Shaping Technology, Building Society: Studies in Sociotechnical Change*, ed. by Wiebe Bijker and J. Law, 205–24. Cambridge: MIT Press.

Aldenderfer, Mark S. 1998. "Quantitative Methods in Archaeology: A Review of Recent Trends and Developments." *Journal of Archaeological Research* 6 (2): 91–120.

Alexander, Jonathan, and Elizabeth Losh. 2010. "Whose Literacy Is It Anyway? Examining a First-Year Approach to Gaming across Curricula." *Currents in Electronic Literacy*. http://currents.dwrl.utexas.edu/FIP/intro.html

Alexander Street Press. 2012. "The Digital Library of Classic Protestant Texts." Alexanderstreet.com. Last modified July. http://alexanderstreet.com/products/digital-library-classic-protestant-texts.

Allman, Mark, Vern Paxson, and Jeff Terrell. 2007. "A Brief History of Scanning." In *Proceedings of the 7th ACM SIGCOMM Conference on Internet Measurement*, pp. 77–82. ACM.

Alvarado, Rafael. 2012. "The Digital Humanities Situation." *Debates in the Digital Humanities*, ed. by Matthew K. Gold, 50–55. Minneapolis: University of Minnesota Press.

Amad, Paula. 2010. *Counter-Archive: Film, the Everyday, and Albert Kahn's Archives de la Planète*. New York: Columbia University Press.

Anthony, Laurence. 2006. "Developing a Freeware, Multiplatform Corpus Analysis Toolkit for the Technical Writing Classroom." *Professional Communication, IEEE Transactions on* 49 (3): 275–86.

Anthony, Lawrence. 2012. "AntConc." Lawrence Anthony's Website. http://www.antlab.sci.waseda.ac.jp/antconc_index.html

Apostolidès, Jean-Marie. 1981. *Le roi-machine: spectacle et politique au temps de Louis XIV*. Paris: Editions de Minuit.

Appuhn, Karl. 2009. *A Forest on the Sea: Environmental Expertise in Renaissance Venice*. Baltimore: Johns Hopkins University Press.

"ArcGIS." 1995–2013. Esri. http://www.esri.com/software/arcgis

Arminen, Ilkka. 2006. "Social Functions of Location in Mobile Telephony." *Personal and Ubiquitous Computing* 10 (5): 319–23.

Armstrong, Robert Plant. 1975. *Wellspring*. Berkeley: University of California Press.

Atkins, Daniel. 2003. "Revolutionizing Science and Engineering through Cyberinfrastructure: Report of the National Science Foundation Blue-Ribbon Advisory Panel on Cyberinfrastructure." National Science Foundation.

Austen, Jane. [1814] 2003. *Mansfield Park*, ed. by James Kinsley. Reprint, Oxford: Oxford University Press.

Bagnall, Kate, and Tim Sherratt. 2013. "The Real Face of White Australia. Invisible Australians: Living under the White Australia Policy." Last modified 19 June. http://invisibleaustralians.org/faces/

Baily, M. Z. 2011. "All the Digital Humanities Are White, All the Nerds Are Men, but Some of Us Are Brave." *Journal of Digital Humanities* 1 (1): 120–21.

Ballantyne, Tony. 2002. Empire, Knowledge, and Culture: From Proto-Globalization to Modern Globalization. In *Globalization in World History*, ed. A. G. Hopkins, 116–140. New York: Norton.

Balsamo, Anne. 2009. "Videos and Frameworks for 'Tinkering' in a Digital Age." *Spotlight on Digital Media and Learning.* January 30. http://spotlight.macfound.org/blog/entry/anne-balsamo-tinkering-videos/

Baneyx, A. 2008. "'Publish or Perish' as Citation Metrics Used to Analyze Scientific Output in the Humanities: International Case Studies in Economics, Geography, Social Sciences, Philosophy, and History." *Archivum Immunologiae et Therapiae Experimentalis* 56 (6): 363–71.

Banner, Stuart. 2007. *Possessing the Pacific: Land, Settlers, and Indigenous People from Australia to Alaska*. Cambridge: Harvard University Press.

Barbetti, Claire. 2009. *Ekphrastic Medieval Visions: A New Discussion in Ekphrasis and Interarts Theory.* Proquest dissertations and theses. Ann Arbor: ProQuest.

Bar-Ilan, J. 2007. "Google Bombing from a Time Perspective." *Journal of Computer-Mediated Communication* 12 (3): 910–38.

Barnett, G. A., and E. Sung. 2005. "Culture and the Structure of the International Hyperlink Network." *Journal of Computer-Mediated Communication* 11 (1): 217–38.

Barthes, Roland. [1968] 1978. "The Death of the Author." In *Image-Music-Text*, trans. by Stephen Heath. New York: Hill and Wang.

Bateson, Gregory. 1972. *Steps to an Ecology of Mind: A Revolutionary Approach to Man's Understanding of Himself.* San Francisco: Chandler Publishing House.

Bath, John, Richard Cunningham, Alan Galey, Brent Nelson, and Paul Werstine eds. *ArchBook.* 2009–2013. "Implementing New Knowledge Environments." University of Toronto iSchool. http://drc.usask.ca/projects/archbook/.

Baucom, Ian. 2005. *Specters of the Atlantic: Finance Capital, Slavery, and the Philosophy of History*. Durham: Duke University Press.

Jean, Baudrillard. 1994. *Simulacra and Simulation*, trans. by Sheila Faria Glaser. Ann Arbor: University of Michigan Press.

Bayly, C. A. 2002. 'Archaic' and 'Modern' Globalization in the Eurasian and African Arena, ca. 1750–1850. In *Globalization in World History*, ed. A. G. Hopkins, 45–72. New York: Norton.

BBC. 2005. "SMS Bible Launched in Australia." BBC News. October 7. http://news.bbc.co.uk/2/hi/4318750.stm

Beacham, R., H. Denard, and F. Niccolucci. 2006. "An Introduction to the London Charter." In *The e-volution of Information Communication and Technology in Cultural Heritage*, ed. by Marinos Ioannides, D. Arnold, F. Niccolucci, and K. Mania, 263–69. Budapest: Archaeolingua.

Behdad, Ali. 2005. "On Globalization, Again!" In *Postcolonial Studies and Beyond,* ed. by Ania Loomba, Suvir Kaul, Matti Bunzi, Antoinette Burton, and Jed Esty, 62–79. Durham: Duke University Press.

Bell, G. 2006. "No More SMS from Jesus: Ubicomp, Religion and Techno-spiritual Practices." In *Ubicomp 2006:Ubiquitous Computing*, ed. by Paul Dourish and Adrian Friday, 141–58. Berlin: Springer-Verlag.

Benedict, X. V. I. 2012. "Silence and Word: Path of Evangelization." Vatican.va. May 20. http://www.vatican.va/holy_father/benedict_xvi/messages/communications/documents/hf_ben-xvi_mes_20120124_46th-world-communications-day_en.html

Benjamin, Walter. 1968. "Theses on the Philosophy of History." In *Illuminations*, trans. by Harry Zohn, 253–64. New York: Schocken.

Benkler, Yochai. 2012. "Seven Lessons from SOPA/PIPA/Megaupload and Four Proposals on Where We Go from Here." January 25. TechPresident. http://techpresident.com/news/21680/seven-lessons-sopapipamegauplaod-and-four-proposals-where-we-go-here

Bennison, Amira. 2002. Muslim Universalism and Western Globalization. In *Globalization in World History*, ed. A. G. Hopkins, 73–98. New York: Norton.

Anna, Bentkowska-Kafel, Denard Hugh, and Drew Baker, eds. 2012. *Paradata and Transparency in Virtual Heritage*. London: Ashgate Publishing Company.

Berger, John. 1977. *Ways of Seeing*. New York: Penguin.

Berger, John. 2003. "Ways of Seeing." *The Feminism and Visual Culture Reader*, 37–39. New York: Penguin.

Bergvall, Caroline. 2005. "Via." In *Fig*. Cambridge: Salt Publishing.

Berry, David M. 2011. "The Computational Turn: Thinking about the Digital Humanities." *Culture Machine* 12:1–22.

Berry, David M. 2012. "Introduction: Understanding the Digital Humanities." In *Understanding Digital Humanities*, ed. by David M. Berry, 1–20. New York: Palgrave Macmillan.

Best, Stephen. 2011. "Neither Lost nor Found: Slavery and the Visual Archive." *Representations* 113 (1): 150–63.

Bialkowski, Voytek, Rebecca Niles, and Alan Galey. 2011. "The Digital Humanities Summer Institute and Extra-Institutional Modes of Engagement." *Faculty of Information Quarterly* 3 (4): 19–29.

Bianco, Jamie. "Skye." 2012. "This Digital Humanities Which Is Not One." In *Debates in the Digital Humanities*, ed. by Matthew K. Gold, 96–112. Minneapolis: University of Minnesota Press.

Bible League. 2013. "Akses Digital Bible Library." Bibleleague.org. http://www.bibleleague.org/what-we-do/how-we-serve/digital-bible-library.

Binford, Sally R., and Lewis R. Binford. 1968. *New Perspectives in Archaeology*. Chicago: Aldine Transaction.

Bhargava, Pushpa Mitra, and Chandana Chakrabarti. 2003. *The Saga of Indian Science since Independence . . . In a Nutshell*. Hyderabad, India: Sangam Books.

Black, Ira B. 2004. "Plasticity: Introduction." In *The New Cognitive Neurosciences III*, ed. by Michael S. Gazzaniga, 105–108. Cambridge: MIT Press.

Blackwell, Christopher, and David Neel Smith. 2009. "Homer Multitext: Nine Year Update." *Digital Humanities 2009 Conference Abstracts*. June 6–8. http://mith.umd.edu/dh09/wp-content/uploads/dh09_conferencepreceedings_final.pdf.

Blanchot, M. 1995. *The Work of Fire*, trans. by C. Mandell. Stanford: Stanford University Press.

Bloch, Ernst. 2000. *The Spirit of Utopia*, trans. by Anthony Nassar. Stanford: Stanford University Press.

Bodard, Gabriel, and Samuel Mahony, eds. 2011. *Digital Research in the Study of Classical Antiquity*. Farnham, UK: Ashgate.

Bodenhamer, David J. 2008. "History and GIS: Implications for the Discipline." *Placing History: How Maps, Spatial Data, and GIS Are Changing Historical Scholarship*, 219–33. Redlands: ESRI.

Bogost, Ian. 2006. *Unit Operations: An Approach to Video Game Criticism*. Cambridge: MIT Press.

Bogost, Ian. 2008. "Palin Generator, Chatterbot." Bogost.com. October 2. http://www.bogost.com/watercoolergames/archives/palin_generator.shtml

Bogost, Ian. 2012a. *Alien Phenomenology, or, What It's Like to Be a Thing*. Minneapolis: University of Minnesota Press.

Bogost, Ian. 2012b. "The Turtlenecked Hairshirt." In *Debates in the Digital Humanities*, ed. by Matthew K. Gold, 241. Minneapolis: University of Minnesota Press.

Bogost, Ian, ed. 2012c. "Home." *Platform Studies*. http://platformstudies.com/

Bogost, Ian, and Nick Montfort. 2009. "Platform Studies: Frequently Questioned Answers." Bogost.com. December 12–15. http://www.bogost.com/downloads/bogost_montfort_dac_2009.pdf

Borgman, Christine L. 2007. *Scholarship in the Digital Age: Information, Infrastructure, and the Internet*. Cambridge: MIT Press.

Borgman, Christine L. 2009. "The Digital Future Is Now: A Call to Action for the Humanities." *DHQ: Digital Humanities Quarterly* 3 (4) http://www.digitalhumanities.org/dhq/vol/3/4/000077/000077.html.

Bourdieu, Pierre. 1981. "Men and Machines." In *Advances in Social Theory and Methodology: Toward an Integration of Micro- and Macro-Sociologies*, ed. by Karin Knorr-Cetina and Aaron Victor Cicourel, 304–17. Boston: Routledge and Kegan Paul.

Bourdieu, Pierre, Jean-Claude Chamboredon, and Jean-Claude Passeron. 1991. *The Craft of Sociology: Epistemological Preliminaries*, trans. by Richard Nice. New York: Walter de Gruyter.

Bowker, Geoffrey C. 2005. *Memory Practices in the Sciences*. Cambridge: MIT Press.

Bozak, Nadia. 2011. *The Cinematic Footprint: Lights, Camera, Natural Resources*. New Brunswick: Rutgers University Press.

Brandt, Marisa. 2012. "From Ultimate Display to Ultimate Skinner Box: Virtual Reality and the Future of Psychotherapy." In *Media Studies Futures*, vol. 6, ed. by Kelly Gates. *The International Encyclopedia of Media Studies.* London: Blackwell.

Brassier, Ray. 2010. *Nihil Unbound: Enlightenment and Extinction*. London: Palgrave Macmillan.

Brennan, Timothy. 2008. Postcolonial Studies and Globalization Theory. In *The Postcolonial and the Global*, ed. Revathi Krishnaswarmy and John C. Hawley, 37–53. Minneapolis: University of Minnesota Press.

Brenneis, Don. 2009. "Anthropology in and of the Academy: Globalization, Assessment and Our Field's Future." *Social Anthropology* 17(3): 261–75.

Brewer, J. 2010. "Microhistory and the Histories of Everyday Life." *Cultural and Social History* 7 (1): 87–109.

Brewster, Bill, and Frank Broughton. 2000. *Last Night a DJ Saved My Life: The History of the Disc Jockey*. New York: Grove Press.

Brown, Bill. 2001. "Thing Theory." *Critical Inquiry* 28 (1): 1–22.

Brown, Susan, Patricia Clements, Isobel Grundy, Stan Ruecker, Jeffery Antoniuk, and Sharon Balazs. 2009. "Published yet Never Done: The Tension between Projection and Completion in Digital Humanities Research." *Digital Humanities Quarterly* 3 (2). http://www.digitalhumanities.org/dhq/vol/3/2/000040.html

Bruns, Gerald. 2011. *On Ceasing to be Human*. Stanford: Stanford University Press.

Buck-Morss, Susan. 1977. *The Origin of Negative Dialectics*. New York: Free Press.

Burdick, Anne, Johanna Drucker, Peter Lunenfeld, Todd Presner, and Jeffrey Schnapp. 2012. *Digital_Humanities*. Cambridge: MIT Press.

Bureau of Labor Statistics, U.S. Department of Labor. 2012. "Line Installers and Repairers." *Occupational Outlook Handbook, 2012–13 Edition*. March 29. www.bls.gov/ooh/installation-maintenance-and-repair/line-installers-and-repairers.htm

Burgess, J. E. 2007. "Vernacular Creativity and New Media." PhD dissertation. Queensland University of Technology. Qut.edu.au. http://eprints.qut.edu.au/16378/

Burkhalter, Byron. 1999. "Reading Race Online: Discovering Racial Identity in Usenet Discussions." In *Communities in Cyberspace*, ed. by Marc A. Smith and Peter Kollock. New York: Routledge.

Burman, L. 1999. "Det enkla valet: Konsekvenserna av en oproblematisk textsituation." In *Vid texternas vägskäl: Textkritiska uppsatser*, ed. by L. Burman and B. Ståhle Sjönell, 83–99. Stockholm, Sweden: Svenska vitterhetssamfundet.

Burroughs, William S. [1961] 2003. "The Cut-up Method of Brion Gysin." In *The New Media Reader*, ed. by Noah Wardrip-Fruin and Nick Montfort, 89–91. Cambridge: MIT Press.

Burson, Nancy. Nancyburson.com. http://nancyburson.com/human-race-machine/

Busa, Roberto A. 2004 "Foreword: Perspectives on the Digital Humanities." In Schreibman, Susan, Ray Siemens, and John Unsworth, eds. *A Companion to Digital Humanities*. Oxford: Blackwell.

Bush, Vannevar. 2003. "As We May Think." In *The New Media Reader*, ed. by Noah Wardrip-Fruin and Nick Montfort, 35–47. Cambridge: MIT Press.

Butler, Judith. 1990. *Gender Trouble*. New York: Routledge.

Butler, Judith. 2004. *Precarious Life: The Powers of Mourning and Violence*. London: Verso.

Buzzetti, Dino, and Jerome McGann. 2006. "Electronic Textual Editing: Critical Editing in a Digital Horizon." In *Electronic Textual Editing*, ed. by Lou Burnard, Katherine O'Brien O'Keeffe, and John Unsworth. New York: Modern Language Association of America. http://www.tei-c.org/About/Archive_new/ETE/Preview/mcgann.xml

Bynum, Caroline Walker. 2011. *Christian Materiality. An Essay on Religion in Late Medieval Europe.* New York: Zone Books.

Callebaut, D., D. Pletinckx, and Neil A. Silberman. "Why Multimedia Matter in Cultural Heritage: The Use of New Technologies in the Ename 974 Project." In *Multimedia Communication for Cultural Heritage*, ed. by Franco Niccolucci and Sorin Hermon, 65–72. Budapest: Archaeolingua.

Campbell, Heidi. A. 2010. *When Religion Meets New Media*. Abingdon, Oxon: Routledge.

Canguilhem, Georges. 1991. *The Normal and the Pathological*, trans. by Carolyn Fawcett. New York: Zone Books.

Cane, Greg. 2004. "Classics and the Computer: An End of the History." In *A Companion to Digital Humanities*, ed. by Susan Schreibman, Ray Siemens, and John Unsworth, 47–55. Oxford: Blackwell.

Canetti, Elias. 1966. *Crowds and Power*, trans. by Carol Stewart. New York: Viking Press.

Canning, Kathleen. 1994. "Feminist History after the Linguistic Turn: Historicizing Discourse and Experience." *Signs* 19 (2): 368–404.

Cantoni, L., E. Rapetti, S. Tardini, S. Vannini, and D. Arasa. 2012. "PICTURE: The Adoption of ICT by Catholic Priests." In *Digital Religion, Social Media and Culture: Perspectives, Practices and Futures*, ed. by P. Cheong, P. Fischer-Nielsen, S. Gelfgren, and C. Ess, 131–50. Oxford: Peter Lang Publishing.

Carlquist, J. 2011 "Jungfru Maria som förebild i det medeltida Sverige." *Research Database*, Umeå Universitet. http://www.umu.se/forskning/forskningsprojekt/forskningsdatabasen/visa-projekt/?code=573.

"Carriers at Work." 1903. Youtube video, 2:15. American Mutoscope and Biograph Company. Early Motion Picture collection of the Library of Congress. Filmed August 22. Posted April 15, 2010. http://www.youtube.com/watch?v=cNzTEveUURA

Carruthers, Mary. 1998. *The Craft of Thought. Meditation, Rhetoric, and the Making of Images, 400–1200*. Cambridge, UK: Cambridge University Press.

Carruthers, Mary. 2006. "On Affliction and Reading, Weeping and Argument: Chaucer's Lachrymose Troilus in Context." *Representations* 93(1): 1–21, esp., p. 17n. 26.

Carruthers, Mary. 2010. "The Concept of Ductus, or Journeying through a Work of Art." In *Rhetoric beyond Words. Delight and Persuasion in the Arts of the Middle Ages*, ed. by Mary Carruthers, 190–213. Cambridge, UK: Cambridge University Press.

Carter, Paul. 1987. *The Road to botany Bay: An Essay in Spatial History*. London: Faber.

Cartwright, Lisa, and Morana Alač. 2007. "Imagination, Multimodality and Embodied Interaction: A Discussion of Sound and Movement in Two Cases of Laboratory and Clinical Magnetic Resonance Imaging." In *Science Images and Popular Images of the Sciences*, ed. by Bernd Hüppauf and Peter Weingart, 199–223. New York: Routledge.

Cecire, Natalia. 2011. "When DH Was in Vogue; or, THATCamp Theory." *Works Cited* (blog). October 19. http://nataliacecire.blogspot.com/2011/10/when-dh-was-in-vogue-or-thatcamp-theory.html.

Chakrabarty, Dipesh. 2009. "The Climate of History: Four Theses." *Critical Inquiry* 35 (2): 197–222.

Chakrabarty, Dipesh. 2000. *Provincializing Europe: Postcolonial Thought and Historical Difference*. Princeton: Princeton University Press.

Champion, Erik, and Bharat Dave. 2002. "Where Is This Place?" In *Proceedings of ACADIA 2002: Thresholds between Physical and Virtual*, 87–97.

Chen, Kuan-Hsing. 2010. *Asia as Method: Toward Deimperialization*. Durham: Duke University Press.

Cheong, P., A. Halavais, and K. Kwon. 2008. "The Chronicles of Me: Understanding Blogging as a Religious Practice." *Journal of Media and Religion* 7 (3): 107–31.

Chesher, Chris. 2012. "Between Image and Information: The iPhone Camera in the History of Photography." In *Studying Mobile Media: Cultural Technologies, Mobile Communication, and the iPhone*, ed. by Larissa Hjorth, Jean Burgess, and Ingrid Richardson, 98–117. New York: Routledge.

Choudhury, Suparna, and Max Stadler. 2011. "CFP: Neuro-Reality Check Scrutinizing the 'neuro-turn' in the humanities and natural sciences." Call for Papers for a Workshop at the Max-Planck-Institute for the History of Science, Berlin, December 1–3.

Christen, Kim, and Chris Cooney. 2006. "Digital Dynamics across Cultures." *Vectors* 2 (1). http://vectors.usc.edu/projects/index.php?project=67

Christian, David. 2004. *Maps of Time: An Introduction to Big History. California World History Library 2*. Berkeley: University of California Press.

Chun, Wendy. 2006. *Control and Freedom: Power and Paranoia in the Age of Fiber Optics*. Cambridge: MIT Press.

Chun, Wendy. 2011. *Programmed Visions: Software and Memory*. Cambridge: MIT Press.

Claburn, Thomas. 2007. "Migration from MySpace to Facebook Shows Class Divide." *InformationWeek*. June 26. http://www.informationweek.com/migration-from-myspace-to-facebook-shows-class-divide/d/d-id/1056521.

Clarke, Arthur. 1953. *Childhood's End*. New York: Ballantine Books.

Clark, Andy. 2003. *Natural Born Cyborgs: Minds, Technologies, and the Future of Human Intelligence*. New York: Oxford University Press.

Clement, T. E. 2008. "A Thing Not Beginning and Not Ending: Using Digital Tools to Distant-read Gertrude Stein's The Making of Americans." *Literary and Linguistic Computing* 23 (3): 361–81.

Clement, Tanya. 2011. "Knowledge Representation and Digital Scholarly Editions in Theory and Practice." *Journal of the Text Encoding Initiative* 1 (June). http://jtei.revues.org/203.

Cohan, Peter. 2012. "To Boost Post-college Prospects, Cut Humanities Departments." *Forbes* 29 (May). http://www.forbes.com/sites/petercohan/2012/05/29/to-boost-post-college-prospects-cut-humanities-departments.

Cohen, Daniel. J. 2011. "Scholarly Communication, the Web Way." Paper presented at the NYU Humanities Initiative, New York, April 7.

Cohen, Daniel J., and Tom Scheinfeldt. 2013. *Hacking the Academy*. Roy Rosenzweig Center for History and New Media. http://hackingtheacademy.org/

Cohen, Yoel. 2011. "Haredim and the Internet: A Hate-Love Affair." In *Mediating Faiths: Religion and Socio-Cultural Change in the Twenty-First Century*, ed. by Michael Bailey and Guy Redden, 63–71. Farnham: Ashgate.

Cole, Jonathan R. 2009. *The Great American University*. New York: Public Affairs.

Cong-Huyen, Anne. 2011. "Toward an Asian American Digital Humanities." *Anne Cong-Huyen* (blog). January 10. http://anitaconchita.wordpress.com/2011/01/10/an-asian-american-digital-humanities-or-digital-asian-american-criticism/.

Connor, Steve. 2012. "Eureka! Cern Announces Discovery of Higgs BBoson 'God Particle.'" *The Independent*. July 5. http://www.independent.co.uk/news/science/eureka-cern-announces-discovery-of-higgs-boson-god-particle-7907677.html.

Conway, Paul. 2013. Preserving Imperfection: Assessing the Incidence of Digitization Error in HathiTrust. *Preservation, Digital Technology and Culture* 42 (1): 17–30.

Cook, Harold J. 2007. *Matters of Exchange: Commerce, Medicine and Science in the Dutch Golden Age*. New Haven: Yale University Press.

Coole, Diana, and Samantha Frost. 2010. *The New Materialisms: Ontology, Agency, Politics*. Durham: Duke University Press.

Corbett, John. 1994. *Extended Play: Sounding off from John Cage to Dr. Funkenstein*. Durham: Duke University Press.

Coyle, Karen. 2006. "Mass Digitization of Books." *Journal of Academic Librarianship* 32 (6): 641–45.

Cronon, William. 1983. *Changes in the Land: Indians, Colonists and the Ecology of New England*. London: Macmillan.

Cummings, Jonathon, and Sara Kiesler. 2003. "Coordination and Success in Multidisciplinary Scientific Collaborations." *ICIS 2003 Proceedings*. Paper 25. http://aisel.aisnet.org/icis2003/25/

Dahlström, Mats. 2010. "Critical Editing and Critical Digitization." In *Text Comparison and Digital Creativity: The Production of Presence and Meaning in Digital Text Scholarship*, ed. by Wido Van Peursen, Ernst D. Thoutenhoofd, and Adriaan Van Der Weel, 79–97. Amsterdam: Brill Academic Publisher.

Dahlström, Mats. 2011. "Editing Libraries." In *Bibliothek und Wissenschaft*, edited by C. Fritze, F. Fischer, P. Sahle, and M. Rehbein (Hrsgg.) *Digitale Edition und Forschungsbibliothek* 44: 91–106. Harrassowitz Verlag.

Daniel, Sharon, and Erik Loyer. 2007. "Public Secrets." *Vectors* 2 no. 2 (Winter). http://vectors.usc.edu/projects/index.php?project=57

Daniel, Sharon, and Erik Loyer. 2012. "Blood Sugar." *Vectors* 3, no. 2. http://vectors.usc.edu/projects/index.php?project=95

Darnton, Robert. 2009. *The Case for Books: Past, Present, and Future*. New York: PublicAffairs.

Darnton, Robert. 2010. *Poetry and the Police: Communication Networks in Eighteenth-century Paris*. Cambridge, MA: Belknap Press.

Daston, Lorraine J., and Peter Galison. 2007. *Objectivity.* Cambridge: Zone Books/MIT Press.

"data center facebook." 2011. Youtube video, 3:34. Posted by "Handaru ds." June 7. http://www.youtube.com/watch?v=-DRxqHrPrFw&feature=related

Davidson, Cathy N. 1999. "What If Scholars in the Humanities Worked Together, in a Lab." *Chronicle of Higher Education*: B4–B5. http://chronicle.com/article/Collaborative-Learning-for-the/128789/.

Davidson, Cathy N. 2011a. "Collaborative Learning for the Digital Age." *Chronicle of Higher Education,* August 26. http://chronicle.com/article/Collaborative-Learning-for-the/128789/.

Davidson, Cathy N. 2011b. *Now You See It: How the Brain Science of Attention Will Transform the Way We Live, Work, and Learn.* New York: Viking Adult.

Davidson, Cathy. 2011c. "Strangers on a Train: A Chance Encounter Provides a Lesson in Complicity and the Never-Ending Crisis in the Humanities." *Academe Online.* September–October. http://www.aaup.org/article/strangers-train#.Uh0kp0KPAVs

Davidson, Cathy N. 2012. "Assessing the Impact of Technology-Aided Participation and Mentoring on Transformative Interdisciplinary Research: A Data-Based Study of the Incentives and Success of an Exemplar Academic Network" National Science Foundation Grant Proposal 192089. August 2012. http://nsf.gov/awardsearch/showAward?AWD_ID=1243622.

Davidson, Cathy N., and David Theo Goldberg. 2007. "Cathy Davidson and David Theo Goldberg on digital media" (Video). MacArthur Foundation. August 7. http://www.macfound.org/videos/295/

Davidson, Cathy N., and David Theo Goldberg. 2011. *The Future of Thinking: Learning Institutions in a Digital Age.* Cambridge: MIT Press.

Davidson, Cathy N., and Mark Surman. 2012. "Why Web Literacy Should Be Part of Every Education." *Fast Co-Exist* (blog), August 8. http://www.fastcoexist.com/1680264/why-web-literacy-should-be-part-of-every-education

Davis, Juliet, Sue Gollifer, Alec MacLeod, Gwyan Rhabyt, Cynthia Beth Rubin, Gail Rubini, and Annette Weintraub. 2007. "Guidelines for Faculty Teaching in New-Media Arts." College Art Association. Last modified. *October* 28. http://www.collegeart.org/guidelines/newmedia07.

Davis, Natalie Zemon. 1983. *The Return of Martin Guerre.* Cambridge: Harvard University Press.

Dean, Jodi. 2002. *Publicity's Secret: How Technoculture Capitalizes on Democracy.* Ithaca: Cornell University Press.

de Certeau, Michel. 1984. *The Practice of Everyday Life.* Berkeley: University of California Press.

De Rond, M., and A. N. Miller. 2005. "Publish or Perish Bane or Boon of Academic Life?" *Journal of Management Inquiry* 14 (4): 321–29.

Deegan, Marilyn, and Kathryn Sutherland. 2009. *Transferred Illusions: Digital Technology and the Forms of Print*. Aldershot, England: Ashgate.

Deleuze, Gilles. 1988. *Foucault*, trans. by Seán Hand. Minneapolis: University of Minnesota Press.

Deleuze, Gilles. 1990. *The Logic of Sense*, ed. by Constantin V. Boundas; trans. by Mark Lester and Charles Stivale. London: Athlone Press.

Deleuze, Gilles. 1994. *Difference and Repetition*, trans. by Paul Patton. New York: Columbia University Press.

Deleuze, Gilles, and Félix Guattari. 1987. *A Thousand Plateaus: Capitalism and Schizophrenia*, trans. by Brian Massumi. Minneapolis: University of Minnesota Press.

Deleuze, Gilles, and Rosalind Krause. 1983. "Plato and the Simulacrum." *October* 27 (Winter): 45–56.

Derrida, Jacques. 1998. *Archive Fever: A Freudian Impression*, trans. by Eric Prenowitz. Chicago: University of Chicago Press.

Derrida, Jacques. 2002. "The University without Condition." In *Without Alibi*, trans. by Peggy Kamuf, 202–37. Stanford: Stanford University Press.

de Souza e Silva, Adriana and Jordan Frith. 2012. *Mobile Interfaces in Public Spaces: Locational Privacy, Control, and Urban Sociability.* New York: Routledge.

de Souza e Silva, Adriana, and Larissa Hjorth. 2009. "Playful Urban Spaces: A Historical Approach to Mobile Games." *Simulation and Gaming* 40 (5): 602–25.

Dewdney, A. K. 1985. "Computer Recreations." *Scientific American* 260: 122–25.

DHQ. 2012. "About DHQ." *Digital Humanities Quarterly*. http://www.digitalhumanities.org/dhq/about/about.html

"Digital Transformations in the Arts and Humanities: Big Data Research. Call for proposals." 2013. Arts and Humanities Research Council. July. http://www.ahrc.ac.uk/Funding-Opportunities/Documents/Big-Data-projects-call-document.pdf

digital.humanities@oxford. "What Are the Digital Humanities?" http://digital.humanities.ox.ac.uk/Support/whatarethedh.aspx

Dimock, Wai-Chee. 2006. *Through Other Continents: American Literature across Deep Time*. Princeton: Princeton University Press.

DiSalvo, Betsy, and Amy Bruckman. 2011. "From Interests to Values." *Communications of the ACM* 54 (8): 27–29.

The Dish (short film). 2008. Journeyman Pictures.

Donald, Merlin. 1991. *Origins of the Modern Mind: Three Stages in the Evolution of Culture and Cognition*. Cambridge: Harvard University Press.

Dosse, François. 2010. *Gilles Deleuze and Félix Guattari: Intersecting Lives*, trans. by Deborah Glassman. New York: Columbia University Press.

Dourish, Paul, and Genevieve Bell. 2011. *Divining a Digital Future: Mess and Mythology in Ubiquitous Computing*. Cambridge: MIT Press.

Downie, J. Stephen. 2008. "The Music Information Retrieval Exchange (2005–2007): A Window into Music Information Retrieval Research." *Acoustical Science and Technology* 29 (4): 247–55. doi:10.1250/ast.29.247.

Drayton, Richard. 2000. *Nature's Government: Science, Imperial Britain, and the "Improvement" of the World*. New Haven: Yale University Press.

Drescher, E. 2012. "Pixels Perpetual Shine: The Mediation of Illness, Dying and Death in the Digital Age." *Cross Currents* 62 (2): 204–18.

Drucker, Johanna. 2009a. "Blind Spots: Humanists Must Plan Their Digital Future." *Chronicle of Higher Education*: April 3. http://chronicle.com/article/Collaborative-Learning-for-the/128789/.

Drucker, Johanna. 2009b. *SpecLab: Digital Aesthetics and Projects in Speculative Computing*. Chicago: Chicago University Press.

Drucker, Johanna. 2012. "Humanistic Theory and Digital Scholarship." In *Debates in the Digital Humanities*, ed. by Matthew K. Gold, 85–95. Minneapolis: University of Minnesota Press.

Drucker, Johanna, and Bethany Nowviskie. 2004. "Speculative Computing: Aesthetic Provocations in Humanities Computing." In *A Companion to Digital Humanities*, ed. by Susan Schreibman, Ray Siemens, and John Unsworth. Oxford: Blackwell; http://www.digitalhumanities.org/companion/view?docId=blackwell/9781405103213/9781405103213.xml&chunk.id=ss1-4-10.

Du Bois, W.E.B. [1903] 1999. *The Souls of Black Folk: Authoritative Text, Contexts, Criticism*. New York: Norton.

Duguid, P. 2007. "Inheritance and loss? A brief survey of Google Books." *First Monday* 12 (8): 6. http://firstmonday.org/ojs/index.php/fm/article/view/1972/1847.

Duke University. 2012. "HASTAC Wins NSF Grant to Study Its Own Social Network." *Duke Today.* August 17. http://m.today.duke.edu/2012/08/hastacnsf

Dunn, Stuart. 2011. "Space as an Artefact: A Perspective on Neogeography from the Digital Humanities." In *Digital Research in the Study of Classical Antiquity*, ed. by Gabriel Bodard and Samuel Mahony, 53–69. Farnham: Ashgate.

During, Simon. 2014. "Stop Defending the Humanities." *Public Books.* March 1. http://www.publicbooks.org//nonfiction/stop-defending-the-humanities

Dyck, Paul. 2008. "A New Kind of Printing": Cutting and Pasting a Book for a King at Little Gidding. *Library* 9 (3): 306–33.

Dyer, J. 2012. "Powerful, Secure Software For Closed Countries . . . And You!" *Don't Eat the Fruit* (blog). February 28. http://donteatthefruit.com/2012/02/powerful-secure-bible-software-for-closed-countries-and-you/

Editions, Dynamic Variorum. "About Us." Dynamic Variorum Editions. http://sites.tufts.edu/dynamicvariorum/home

Ede, Lisa, and Andrea A. Lunsford. 2001. "Collaboration and Concepts of Authorship." *Publications of the Modern Language Association of America*: 354–69.

Education for Change. 2012. "Researchers of Tomorrow: The research behaviour of Generation Y doctoral students," *JISC.* June 28. http://www.jisc.ac.uk/publications/reports/2012/researchers-of-tomorrow.aspx

Electronic Literature Organization. 2006 and 2011. *Electronic Literature Collection,* vols. 1 and 2. http://collection.eliterature.org

Eliot, T. S. 1920. "Tradition and the Individual Talent." In *The Sacred Wood.* London: Methuen.

Eliot, T. S. [1922] 2011. *The Waste Land* (iPad App). London: Faber and Faber, and Touch Press. http://www.touchpress.com/titles/thewasteland/

Eliot, T. S. 1971. *Four Quartets.* New York: Harcourt Brace Jovanovich.

Eisenstein, S.M. [1922] 1988. "Beyond the Shot", *Selected Works.* Vol. 1: *Writings, 1922–34*, ed. and trans. by Richard Taylor. London: BFI Publishing.

Elkin, Larry M. 2012. "Publishers Perish: Ending Unjustified Subsidies for The University Press." *Palisades Hudson Financial Group LLC* (blog). August 2. http://www.palisadeshudson.com/2012/08/publishers-perish-ending-unjustified-for-the-university-press/

Ellison, Nicole B., Charles Steinfield, and Cliff Lampe. 2007. "The Benefits of Facebook Friends: Social Capital and College Students' Use of Online Social Network Sites." *Journal of Computer-Mediated Communication* 12 (4): 1143–68.

"Ename Charter 2008." 2008. Principles of Seville: International Principles of Virtual Archaeology. http://www.arqueologiavirtual.com/carta/?page_id=330

Ernesto. 2011. "Google's Anti-piracy Filter is Quite Effective." Torrent Freak. http://torrentfreak.com/googles-anti-piracy-filter-110712/

Ernst, Wolfgang. 2011. "Media Archaeography: Method and Machine versus History and Narrative of Media." In *Media Archaeology: Approaches, Applications and Implications*, ed. by Erkki Huhtamo and Jussi Parikka, 239–55. Berkeley: University of California Press.

Ernst, Wolfgang. 2013. *Digital Memory and the Archive*, ed. by Jussi Parikka. Minneapolis: University of Minnesota Press.

Esfandiari, Golnaz. 2012. "Nothing Comes between Iranians and Their Satellite Dishes: Not Even the Police." *Radio Free Europe*. "Persian Letters." March 13. www.rferl.org/content/persian_letters_satellite_dishes_iran_police/24514665.html

Ess, C. 2004. "Revolution? What Revolution? Successes and Limits of Computing Technologies in Philosophy and Religion." In *A Companion to Digital Humanities*, ed. by Susan Schreibman, Ray Siemens, and John Unsworth. Oxford: Blackwell.

"The European Landscape." European Commission. http://ec.europa.eu/research/infrastructures/index_en.cfm?pg=landscape

Evans, L., and S. Rees. 2012. "An Interpretation of Digital Humanities." In *Understanding Digital Humanities*, ed. by David M. Berry. New York: Palgrave Macmillan.

Fanon, Franz. 1967. *Black Skin, White Masks*, trans. by Charles Lam Marmann. New York: Grove.

Farman, J. 2010. "Mapping the Digital Empire: Google Earth and the Process of Postmodern Cartography." *New Media and Society* 12 (6): 869–88.

Farman, J. 2011. *Mobile Interface Theory*. London: Routledge.

Faulkner, W. 2000. "The Power *and* the Pleasure? A Research Agenda for 'Making Gender Stick' to Engineers." *Science, Technology and Human Values* 25 (1): 87–119.

Febvre, Lucien, and Henri-Jean Martin. 1979. *The Coming of the Book: The Impact of Printing 1450–1800*, ed. by Geoffrey Nowel-Smith and David Wootton; trans. by David Gerard. London: Verso.

Fernandez, Maria. 2002. "Cyberfeminism, Racism, Embodiment." In *Domain Errors! Cyberfeminist Practices*, ed. by Maria Fernandez, Faith Wilding, and Michelle M. Wright. Brooklyn: Autonomedia.

Field, Clive. 2012. "Resonate Online Panel." *British Religion in Numbers*. July 28. http://www.brin.ac.uk/news/2012/resonate-online-panel/

Finch, Dame Janet and the Working Group on Expanding Access to Published Research Findings. 2012. "Finch Report: Accessibility, sustainability, excellence: how to expand access to research publications." Research Information Network. http://www.researchinfonet.org/publish/finch/

Fish, Stanley. 2008. "Will the Humanities Save Us?" *The New York Times*. January 6. http://opinionator.blogs.nytimes.com/2008/01/06/will-the-humanities-save-us

Fitterman, Rob, and Vanessa Place. 2009. *Notes on Conceptualisms*. Brooklyn: Ugly Duckling Presse.

Fitzpatrick, Kathleen. 2006. "On the Future of Peer Review in Electronic Scholarly Publishing." *if:book—A Project of the Institute for the Future of the Book*. June 28. http://futureofthebook.org/blog/archives/2006/06/on_the_future_of_peer_review_i.html.

Fitzpatrick, Kathleen. 2010. "Reporting from the Digital Humanities 2010 Conference." *The Chronicle of Higher Education's ProfHacker*. July 13. http://chronicle.com/blogs/profhacker/reporting-from-the-digital-humanities-2010-conference/25473

Fitzpatrick, Kathleen. 2011. *Planned Obsolescence: Publishing, Technology, and the Future of the Academy*. New York: New York University Press.

Fitzpatrick, Kathleen, and Avi Santo. 2012. "Open Review: A Study of Contexts and Practices." *MediaCommons Press*. January 29. http://mcpress.media-commons.org/open-review/.

Flanders, Julia. 2005. "Electronic Textual Editing: The Women Writers Project: A Digital Anthology." In *Electronic Textual Editing*, ed. by John Unsworth, Lou Burnard, and Katherine O'Brien O'Keeffe. http://www.tei-c.org/About/Archive_new/ETE/Preview/flanders.xml

Fleck, Ludwig. 1979. *Genesis and Development of a Scientific Fact*, trans. by Fred Bradley and Thaddeus Trenn. Chicago: University of Chicago Press.

Flynn, William T. 2010. "Ductus figuratus et subtilis." In *Rhetoric beyond Words. Delight and Persuasion in the Arts of the Middle Ages*, ed. by Mary Carruthers, 250–80. Cambridge: Cambridge University Press.

Fossett, Judith Jackson. 2006. "Slavery's Ephemera." *Vectors* 2 no. 1 (Fall). http://vectors.usc.edu/projects/index.php?project=56

Foucault, Michel. 1972. *The Archaeology of Knowledge and the Discourse on Language*, trans. by A. M. Sheridan Smith. New York: Pantheon Books.

Foucault, Michel. 1979. "What Is an Author." In *Textual Strategies: Perspectives in Post-Structuralist Criticism*, ed. and trans. by Josué V Harari. Ithaca: Cornell University Press.

Foroutan, Hamid. 2011. "Photos: Iranian Police Destroying Satellite Dishes." *Payvand Iran News*. August 14. http://www.payvand.com/news/11/aug/1121.html

Forte, M., and A. Siliotti, eds. 1997. *Virtual Archaeology: Great Discoveries Brought to Life through Virtual Reality*. London: Thames and Hudson.

Forte, M. 2000. "About Virtual Archaeology: Disorders, Cognitive Interactions and Virtuality." In *Virtual Reality in Archaeology*, ed. by J. Barcelo, M. Forte, and D. Sanders, 247–263. Oxford: BAR International Series.

Forte, M. 2010. "Introduction to Cyber Archaeology." In *CyberArchaeology*, ed. by Maurizio Forte. Oxford: BAR International Series.

Forte, M., and N. Dell'Unto. 2010. "Embodied Communities, Second Life and Cyber Archaeology." In *Heritage in the Digital Era*, ed. by Mario Santana Quintero, 181–94. Hockley, Essex, UK: Multi-Science Publishing.

Forte M., N. Dell'Unto, J. Issavi, N. Lercari, and L. Onsurez. 2012. "3d Archaeology at Çatalhöyük." *International Journal on Digital Heritage* 3.

Forte, M., and G. Kurillo. 2010. "Cyberarchaeology Experimenting Teleimmersive Archaeology." 16th International Conference on Virtual Systems and Multimedia. October 20–23, Seoul, South Korea.

Forte, M., and S. Pescarin. 2012. "Behaviours, Interactions and Affordance in Virtual Archaeology." In *Paradata and Transparency in Virtual Heritage*, ed. by Anna Bentkowska-Kafel and Hugh Denard. Ashgate.

Forte, M., and E. Pietroni. 2009. "3D Collaborative Envirionments in Archaeology: Experiencing the Reconstruction of the Past." *International Journal of Architectural Computing Issue* 1 (7): 57–75.

Forte, M., and A. Siliotti, eds. 1997. *Virtual Archaeology: Great Discoveries Brought to Life through Virtual Reality*. London: Thames and Hudson.

Fox, Vanessa. 2010. "Can Google Tell Us What Men and Women Are REALLY Thinking?" Search Engine Land. January 19. http://searchengineland.com/can-google-tell-us-what-men-and-women-are-really-thinking-33691

Frohmann, B. 2004. *Deflating Information: From Science Studies to Documentation*. Toronto: University of Toronto Press.

Frischer, B., F. Niccolucci, N. S. Ryan, and J. A. Barceló. 2002. "From CVR to CVRO: the Past, Present and Future of Cultural Virtual Reality." In *Virtual archaeology—Proceedings of VAST 2000*, edited by Niccolucci F., 7–18. BAR International Series: Oxford.

Fuller, Matthew. 2008. *Software Studies: A Lexicon*. Cambridge: MIT Press.

Fuller, Matthew. 2009. "Evil Media: Making Good Use of Weights, Chains and Ranks," iTunesU podcast, 94:00, from the Animating Archives Conference, Brown University on December 4. https://itunes.apple.com/us/podcast/not-just-open-source/id381080290?i=84641746&mt=2

Gabler, Neal. 2011. "The Elusive Big Idea." *The New York Times*. August 14. http://www.nytimes.com/glogin?URI=http://www.nytimes.com/2011/08/14/opinion/sunday/the-elusive-big-idea.html&OQ=_rQ3D2Q26pagewantedQ3DallQ26&OP=89172651Q2FGQ51Q3CFGJdFGvvvGAFZ3GfQ51Q5D-lQ51Q51FNGNsyyGs0GyQ26GQ51Q3C(J(Q51JG-BJfQ23dGFAY5Y3B-(8Y5)(M5(fYQ23Q3AAFZ3.

Gabrys, Jennifer. 2011. *Digital Rubbish: A Natural History of Electronics*. Ann Arbor: University of Michigan Press.

Galina, Isabel, and Ernesto Priani. "Is There Anybody Out There? Discovering New DH Practitioners in Other Countries." *Proceedings of the Digital Humanities 2011 Conference*. DH 2011. June 19–22. http://dh2011abstracts.stanford.edu/xtf/view?docId=tei/ab124.xml;query=;brand=default

Galison, Peter. 1997. *Image and Logic: A Material Culture of Microphysics*. Chicago: University of Chicago Press.

Gavalli-Björkman, Görel. 2002. *Face to Face: Portraits from Five Centuries*. Stockholm: National Museum.

Geraci, R. 2010. *Apocalyptic AI: Visions of Heaven in Robotics, Artificial Intelligence, and Virtual Reality*. Oxford: Oxford University Press.

Gibbs, Frederick W., and Daniel J. Cohen. 2011. "A Conversation with Data: Prospecting Victorian Words and Ideas." *Victorian Studies* 54 (1): 69–77.

Gibson, J. 1950. *The Perception of the Visual World*. Westport, CT: Greenwood Press.

Ginzburg, Carlo. 1980. *The Cheese and the Worms: The Cosmos of a Sixteenth-Century Miller.* Trans. by John and Anne Tedeschi. London: Routledge.

Gitelman, Lisa. 2006. *Always Already New: Media, History and the Data of Culture*. Cambridge: MIT Press.

Gitelman, Lisa and Virginia Jackson. 2013. "Introduction." In*"Raw Data" Is an Oxymoron.* Cambridge: MIT Press, 1–14.

Goddard, Lisa. 2012. "What Is the Sound of Text?" TAPoR 2.0. February 8. http://www.tapor.ca/?id=11

Goggin, G., and L. Hjorth. 2009. "Waiting for Participate: An Introduction." *Communication, Politics and Culture* 42 (2): 1–5.

Gold, Matthew K., ed. 2012. *Debates in the Digital Humanities.* Minneapolis: University of Minnesota Press.

Gold, Matthew K. 2012. "Looking for Whitman: A Multi-Campus Experiment in Digital Pedagogy." In *Digital Humanities Pedagogy: Practices, Principles and Politics,* vol. 3, ed. by Brett D. Hirsch. Cambridge, UK: Open Book Publishers.

Gold, Matthew K. 2012. "Whose Revolution? Towards a More Equitable Digital Humanities." The Lapland Chronicles (blog). January 10. http://blog.mkgold.net/2012/01/10/whose-revolution-toward-a-more-equitable-digital-humanities/

Gold, Matthew K. 2012. "The Digital Humanities Moment." In *Debates in the Digital Humanities*, ed. by Matthew K. Gold. Minneapolis: University of Minnesota Press.

Goldberg, David Theo, Richard Marciano, and Chien-Yi Hou. 2012. "T-RACES: Testbed for the Redlining Archives of California's Exclusionary Spaces. *Vectors* 3, no. 2 (Summer). http://vectors.usc.edu/projects/index.php?project=93

Goldberg, David Theo, Stefka Hristova, and Erik Loyer. 2007. "Blue Velvet." *Vectors* 3, no. 1 (Fall). http://vectors.usc.edu/projects/index.php?project=82

Goldsmith, Kenneth. 2003. *Day*. Great Barrington, MA: The Figures.

Goldstein, Claire. 2008. *Vaux and Versailles: The Appropriations, Erasures, and Accidents That Made Modern France*. Philadelphia: University of Pennsylvania Press.

Gonzalez, V. V., and R. M. Rodriguez. 2003. "Filipina.com: Wives, Workers, and Whores on the Cyberfrontier." In *AsianAmerica.Net. Ethnicity, nationalism, and cyberspace*, ed. by R. C. Lee and S. C. Wong, 215–34. London: Routledge.

Google Inside Search. 2012. https://support.google.com/websearch/answer/106230?hl=en.

Gordon, E. and de Souza e Silva, A. 2011. *Net Locality.* Hoboken, NJ: Wiley.

Graduate School of Library Information Science University of Illinois at Urbana-Champaign. 2010. "Music Information Retrieval eXchange." imirsel. http://www.music-ir.org/?q=node/13

Graduate School of Library Information Science University of Illinois at Urbana-Champaign. 2010. "Networked Environment for Music Analysis. http://www.music-ir.org/?q=node/12

Graduate School of Library Information Science University of Illinois at Urbana-Champaign. 2010. "Structured Analysis of Large Amounts of Music Information." http://nema.lis.uiuc.edu/drupal/?q=node/14

Granka, L. A. 2010. "The Politics of Search: A Decade Retrospective." *Information Society* 26 (5): 364–74.

Grau, Oliver. 2003. *Virtual Art.* Cambridge: MIT Press.

Green, Rachel. 2000. "Web Work: A History of Internet Art." *Artforum* 9: 162–67, 190.

Greetham, D. 1992. "Textual scholarship." In *Introduction to scholarship in modern languages and literatures*, ed. by J. Gibaldi, 103–37. New York: MLA.

Greetham, D. 1999. *Theories of the Text.* Oxford: Oxford University Press.

Greetham, David. 1999. "Who's In, Who's Out: The Cultural Politics of Archival Exclusion." *Studies in the Literary Imagination* 32 (1): 1–28.

Grendler, Paul F. 2004. "The Universities of the Renaissance and Reformation." *Renaissance Quarterly* 57 (1): 1–42.

Grenier, Robert. 1978. *Sentences.* www.whalecloth.org/grenier/sentences_.htm

Grigar, Dene. 2012. "Electronic Literature Selected by Undergraduate Students for 'Future Writers,'" MLA 2012 Electronic Literature Exhibit. HASTAC. December 5. http://hastac.org/blogs/dgrigar/2011/12/05/electronic-literature-selected-undergraduate-students-future-writers-mla-20

Grimmelmann, J. 2007. "The Structure of Search Engine Law." *93 IOWA Law Review* 1 (18): 1–51.

Grosz, Elizabeth. 2001. *Architecture from the Outside: Essays on Virtual and Real Space*. Cambridge: MIT Press.

Guiliano, Jennifer. 2013. "Why You Shouldn't Be a Digital Humanist," Just another Day of DH site. April 8. http://dayofdh2013.matrix.msu.edu/jenguiliano/2013/04/08/why-you-shouldnt-be-a-digital-humanist/

Guillory, John. 2005. "Evaluating Scholarship in the Humanities: Principles and Procedures." *ADE Bulletin* 137: 18–33.

Guillory, John. 2010. "Genesis of the Media Concept." *Critical Inquiry* 36: 321–62.

Guldi, Jo. 2010. "Spatial Humanities: A Project of the Institute for Enabling Geospatial Scholarship. Scholars' Lab." http://spatial.scholarslab.org/

Guldi, Jo. 2012. *Roads to Power: Britain Invents the Infrastructure State*. Cambridge: Harvard University Press.

Gupta, S. P. 2002. *Report of the Committee on India Vision 2020*. New Delhi: Planning Commission of India.

Gutmann, M. 2012. "Questions without Borders: Why Future Research and Teaching Will Be Interdisciplinary" (lecture). University of Minnesota, Institute for Advanced Study forum: Minneapolis, February 13.

Gye, L. 2007. "Picture This: The Impact of Mobile Camera Phones on Personal Photographic Practices." *Continuum (Perth)* 21 (2): 279–88.

Hacking, Ian. 1998. *Mad Travelers: Reflections on the Reality of Transient Mental Illnesses*. Charlottesville: University Press of Virginia.

Hall, C. 1991. "Politics, Post-structuralism and Feminist History." *Gender and History* 3 (2): 204–10.

Hall, Stuart. 1981. "Notes on Deconstructing the Popular." In *People's History and Socialist Theory*, ed. by Raphael Samuel, 227–40. Boston: Routledge and Kegan Paul.

Hall, Stuart. 1998. "Subjects in History: Making Diasporic Identities."In *The House That Race Built*, ed. by Wahneema Lubiano, 289–99. New York: Vintage.

Naficy, Hamid. 2012. *A Social History of Iranian Cinema*, vol. 4. Durham: Duke University Press.

Hamilton, S. 2011. "The Ambiguity of Landscape." In *Evolutionary and Interpretive Archaeologies*, ed. by E. E. Cochrane and A. Gardner. Walnut Creek, CA: Left Coast Press.

Hannaway, Owen. 1986. "Laboratory Design and the Aim of Science: Andreas Libavius versus Tycho Brahe." *Isis* 77 (4): 584–610.

Hansen, Mark and Ben Rubin. 2009. "The Listening Post." YouTube video, 4:05. Posted by "MediaArtTube." April 19. http://www.youtube.com/watch?v=dD36IajCz6A

Hansen, Mark. 2004a. "Digitizing the Racialized Body; or, the Politics of Universal Address." *SubStance* 33: 107–33.

Hansen, Mark. 2004b. *New Philosophy for New Media*. Cambridge: MIT Press.

Hansen, Mark. 2006. *Bodies in Code: Interfaces with Digital Media*. New York: Routledge.

Hansson, Sven Ove. 2007. "Konsten att vara vetenskaplig." Unpublished manuscript. http://home.abe.kth.se/~soh/konstenatt.pdf

Haraway, Donna. [1991] 2004. "A Cyborg Manifesto: Science, Technology, and Socialist-Feminism in the 1980s." In *The Haraway Reader*, 7–46. New York: Routledge.

Harbin, Duane. 1998. "Fiat Lux: The Electronic Word." In *Formatting the Word of God: the Charles Caldwell Ryrie collection*, ed. by Valerie R. Hotchkiss and Charles C. Ryrie. Dallas: Bridwell Library.

Hargittai, E. 2007. "The Social, Political, Economic, and Cultural Dimensions of Search Engines: An Introduction." *Journal of Computer-Mediated Communication* 12 (3): 769–77.

Hargreaves, Steve. 2011. "The Internet: One Bit Power Suck." *CNNMoney*. May 9. http://money.cnn.com/2011/05/03/technology/internet_electricity/index.htm

Harley, Diane. 2012. "Comment on 'Open Review: A Study of Contexts and Practices.'" Media-Commons Press. June 18. http://mcpress.media-commons.org/open-review/contextualizing-questions/who-else-is-exploring-these-issues/.

Harrell, D. Fox. 2008. "GRIOT's Tales of Haints and Seraphs: A Computational Narrative Generation System." *Electronic Book Review*. February 19. http://www.electronicbookreview.com/thread/firstperson/generational

Harrell, Fox. 2009. "Computational and Cognitive Infrastructures of Stigma: Empowering Identity in Social Computing and Gaming." *Proceedings of the 7th ACM conference on Creativity and Cognition*. New York: ACM, 49–58.

Harris, Jonathan, and Sep Kamvar. 2006. *We Feel Fine: An Exploration of Human Emotion in Six Acts*. May. http://www.wefeelfine.org/

Harrison, Carol, and Ann Johnson. 2009. *National Identity: The Role of Science and Technology. Isis 24*. Chicago: University of Chicago Press.

Harrison, Steve, and Paul Dourish. 1996. "Re-place-ing Space: The Roles of Place and Space in Collaborative Systems." In *Proceedings of the 1996 ACM Conference on Computer Supported Cooperative Work*, 67–76. New York: ACM.

Hartwell, Robert. 1962. "A Revolution in the Chinese Iron and Coal Industries during the Northern Sung, 960–1126 A.D." *Journal of Asian Studies* 21 (2): 153–62.

Hartwell, Robert. 1967. "A Cycle of Economic Change in Imperial China: Coal and Iron in Northeast China, 750–1350." *Journal of Economic and Social History of the Orient* 10 (1): 102–59.

Harvey, David. 1990. *The Condition of Postmodernity*. Oxford: Blackwell.

Harwood, Graham. 1997. "Colour Separation." Mongrel. www.mongrel.org.uk/colourseparation

HASTAC. 2011. "Critical Code Studies Hastac Scholars Forum." *Hastac.org*. http://www.hastac.org/forums/hastac-scholars-discussions/critical-code-studies/.

Hastie, Amelie. 2007. *Cupboards of Curiosity: Women, Recollection, and Film History*. Durham: Duke University Press.

Hayles, N. Katherine. 2007a. "Electronic Literature: What Is It?" The Electronic Literature Organization. January 2. http://eliterature.org/pad/elp.html

Hayles, N. Katherine. 2007b. "Hyper and Deep Attention: The Generational Divide in Cognitive Modes." *Profession* 2007: 187–99.

Hayles, N. Katherine. 2012. *How We Think: Digital Media and the Contemporary Technogenesis*. Chicago: University of Chicago Press.

Hebdige, Dick. 1979. *Subculture, The Meaning of Style*. London: Methuen.

Heidegger, Martin. 1976. "Interview." *Der Spiegel*. 31 May: 193–219.

Heim, T., and T. Birdsong. 2010. *StickyJesus: How To Live Out Your Faith Online*. Nashville: Abingdon Press.

Heise, Ursula. 2011. "Lost Dogs, Last Birds, and Listed Species: Cultures of Extinction." *Configurations* 18: 39–62.

Heng, Geraldine. 2004. "Global Interconnections: Imagining the World 500–1500." *Medieval Academy Newsletter*. September.

Heng, Geraldine. 2007. "An Experiment in Collaborative Humanities: Imagining the World 500–1500." *ADFL Bulletin. Modern Language Association of America* 38 (3): 20–28.

Heng, Geraldine. 2009. "The Global Middle Ages." Special Issue on Experimental Literary Education. *ELN* 47 (1): 205–16.

Hertz, Garnet. 2012. *Critical Making—Handmade "Zine."*

Hertz, Garnet, and Jussi Parikka. 2012. "Zombie Media: Circuit Bending Media Archaeology into an Art Method." *Leonardo* 45 (5): 424–30.

Herzfeld, N. 2009. *Technology and Religion: Remaining Human in a Co-created World.* West Conshohocken: Templeton Press.

Higgin, Tanner. 2010. "Cultural Politics and the Digital Humanities." *Gaming the System* (blog). May 25. http://www.tannerhiggin.com/cultural-politics-critique-and-the-digital-humanities/

Hillier, Amy (director). 2013. "Mapping the Du Bois Philadelphia Negro." http://www.mappingdubois.org/

Hipps, S. 2009. *Flickering Pixels: How Technology Shapes Your Faith*. Grand Rapids: Zondervan.

Hjorth, L. 2005. "Locating Mobility: Practices of Co-presence and the Persistence of the Postal Metaphor in SMS/MMS Mobile Phone Customization in Melbourne." *Fibreculture Journal* 6. http://six.fibreculturejournal.org/fcj-035-locating-mobility-practices-of-co-presence-and-the-persistence-of-the-postal-metaphor-in-sms-mms-mobile-phone-customization-in-melbourne/.

Hjorth, L. 2007. "Snapshots of Almost Contact." *Continuum* 21 (2): 227–38.

Hjorth, L. 2011. "Locating the Online: Creativity and User-Created Content in Seoul." *Media International Australia* 141:118–27.

Hjorth, L. 2011a. "Mobile Spectres of Intimacy: The Gendered Role of Mobile Technologies in Love—Past, Present and Future." In *The Mobile Communication Research Series.* Vol. 3: *Mobile Communication: Bringing Us Together or Tearing Us Apart?* ed. by R. Ling and S. Campbell, 37–60. Edison, NJ: Transaction Books.

Hjorth, L., and K. Gu. 2012. "Placing, Emplacing and Embodied Visualities: A Case Study of Smartphone Visuality and Location-Based Social Media in Shanghai, China." *Continuum* 26 (5): 699–713.

Hjorth, L., and M. Arnold. 2013. *Online@AsiaPacific*. New York: Routledge.

Hjorth, L., R. Wilken, and K. Gu. 2012. "Ambient Intimacy: A Case Study of the iPhone, Presence, and Location-based Social Networking in Shanghai, China." In *Studying Mobile Media: Cultural*

Technologies, Mobile Communication, and the iPhone, ed. by L. Hjorth, J. Burgess, and I. Richardson, 43–62. London: Routledge.

Hodder, I. 2008. "Multivocality and Social Archaeology." In *Evaluating Multiple Narratives: Beyond Nationalist, Colonialist, Imperialist Archaeologies*, ed. by Junko Habu, Clare Fawcett, and John M. Matsunaga. New York: Springer.

Hofstadter, Douglas. 1979. *Gödel, Escher, Bach: An Eternal Golden Braid*. New York: Basic Books.

Holland, Dorothy, Gretchen Fox, and Vinci Daro. 2008. "Social Movements and Collective Identity: A Decentered, Dialogic View." *Anthropological Quarterly* 81 (1): 95–126.

Holland, John. H. 1995. *Hidden Order: How Adaptation Builds Complexity*. Reading MA: Addison -Wesley.

Horkheimer, Max. [1937] 2002. "Traditional and Critical Theory." In *Critical Theory: Selected Essays*, trans. by Matthew J. O'Connell, 188–243. New York: Continuum.

Horswill, Ian. 2008. "What Is Computation?" November 1. Northwestern Electrical Engineering and Computer Science. http://www.cs.northwestern.edu/~ian/What%20is%20computation .pdf

Horwatt, Eli. 2009. "A Taxonomy of Digital Video Remixing: Contemporary Found Footage Practice on the Internet." In *Cultural Borrowings: Appropriation, Reworking, Transformation*, ed. by Iain Robert Smith *Scope: An Online Journal of Film and Television Studies*: 76–91.

Hota, S. R., S. Argamon, M. Koppel, and I. Zigdon. 2006. "Performing Gender: Automatic Stylistic Analysis of Shakespeare's Characters." *Proceedings of Digital Humanities* 2006: 100–104. http://www.csdl.tamu.edu/~furuta/courses/06c_689dh/dh06readings/DH06-082-088.pdf.

"How do you define Humanities Computing/Digital Humanities?" 2011. TAPoR at Alberta Wiki. Last modified March 17. http://www.artsrn.ualberta.ca/taporwiki.

Howard, Jennifer. 2011. "Hard Times Sharpen the Lens on Labour and the Humanities." *Chronicle of Higher Education* 9 (January). http://chronicle.com/article/Hard-Times-Sharpen -the-MLAs/125905/.

Howe, Mark. 2012. "An Experiment in Online Communion." *Ship of Fools*. http://www.ship-of fools.com/features/2012/experiment_in_online_communion.html.

Huettel, Scott A., et al. 2009. *Functional Magnetic Resonance Imaging*. Sunderland, MA: Sinauer Associates.

Huggett, Jeremy. 2012. "Core or Periphery? Digital Humanities from an Archaeological Perspective." *Historical Social Research. Historische Sozialforschung* 37 (3): 86–105.

Huhtamo, Erkki. 2000. "T(h)inkering with Media: On the Art of Paul DeMarinis." In *Paul DeMarinis:Buried in Noise,*ed by Ingrid Beirer, Sabine Himmelsbach, and Carsten Seiffairth, 33–46. Heidelberg: Kehrer Verlag.

Huhtamo, Erkki. 2008. "Twin-Touch-Test-Redux: Media Archaeological Approach to Art, Interactivity, and Tactility." In *MediaArtHistories*, ed. by Oliver Grau. Cambridge: MIT Press.

Huizinga. Johan. 1951. *Homo Ludens.* Paris: Gallimard.

"The Human Microbiome: Me, Myself, Us." 2012. *The Economist.* August 18. http://www.economist.com/node/21560523

Hüppauf, Bernd, and Peter Weingart. 2008. "Images in and of Science." In *Science Images and Popular Images of the Sciences*, ed. by Bernd Hüppauf and Peter Weingart, 3–31. New York: Routledge.

Hutcheon, Linda, and Michael Hutcheon. 2001. "A Convenience of Marriage: Collaboration and Interdisciplinarity." *Publication of the Modern Language Association* 116 (5): 1364–76.

Hutchings, T. 2011. "Contemporary Religious Community and the Online Church." *Information Communication and Society* 14 (8): 1118–35.

Hypercities. 2011. "Hypercities." Beyond WebCT. Last modified November 9. http://beyondwebct.wikifoundry.com/page/Hypercities?t=anon.

ICOMOS International Scientific Committee on Interpretation and Presentation. 2007. "The Initiative." Last modified April 7. http://www.enamecharter.org/initiative_3.html

ICOMOS International Scientific Committee on Interpretation and Presentation of Cultural Heritage Sites. 2008. "ICOMOS Charter for the Interpretation and Presentation of Cultural Heritage Sites." October 4. http://www.international.icomos.org/charters/interpretation_e.pdf

Imagining America: Artists and Scholars in Public Life. http://imaginingamerica.org/

Digital, Immersion. 2012. "Glo: An Interactive Bible." Glo Bible. www.globible.com/aninteractivebible

Ingold, T. 2008. "Bindings against Boundaries: Entanglements of Life in an Open World."*Environment and Planning A* 40: 1796–1810.

Innis, Harold. 1991. *The Bias of Communication*. Toronto: University of Toronto Press.

"Inside a Google Data Center." 2009. Youtube video, 5:47. Posted by DataCenterVideos. April 2. http://www.youtube.com/watch?v=bs3Et540-_s

"Instruction of Paper Prototyping." http://issuu.com/hertz/docs/instruction_of_paper_prototyping

Introna, L., and H. Nissenbaum. 2000a. "Defining the Web: The Politics of Search Engines." *Computer* 33 (1): 54–62.

Introna, L., and H. Nissenbaum. 2000b. "Shaping the Web: Why the Politics of Search Engines Matters." *Information Society* 16 (3): 169–85.

Invisible Committee. The. 2011. "The Coming Insurrection." Self-published.

"Iran Police Confiscate 6,000 Satellite Dishes in Mazandaran Province." 2011. *Mohabat News.* October 20. http://mohabatnews.com/index.php?option=com_content&view=article&id=3143:iran-police-confiscate-6000-satellite-dishes-in-mazandaran-province-&catid=35:inside-iran&Itemid=278

Ito, Mizuko. 2003. "Mobiles and the Appropriation of Place." *Receiver* 8. Education, Values and Society (course website). http://academic.evergreen.edu/curricular/evs/readings/itoShort.pdf

Ito, Mizuko. 2009. *Hanging out, Messing around, and Geeking Out: Kids Living and Learning with New Media.* Cambridge: MIT Press.

Ito, Mizuko, and Daisuke Okabe. 2003. "Camera Phones Changing the Definition of Picture-Worthy." *Japan Media Review.* http://www.ojr.org/japan/wireless/1062208524.php (site discontinued)

Ito, Mizuko, and Daisuke Okabe. 2005. "Intimate Visual Co-Presence." Ubicomp, Takanawa Prince Hotel, Tokyo, Japan, September 11–14. http://www.itofisher.com/mito/archives/ito.ubicomp05.pdf

Ito, Mizuko, and Daisuke Okabe. 2006. "Everyday Contexts Of Camera Phone Use: Steps Towards Technosocial Ethnographic Frameworks." In *Mobile Communication In Everyday Life: An Ethnographic View*, ed. by J. Höflich and M. Hartmann, 79–102. Berlin: Frank and Timme.

Jackson, L. A., K. S. Ervin, P. D. Gardner, and N. Schmitt. 2001. "Gender and the Internet: Women Communicating and Men Searching." *Sex Roles* 44 (5–6): 363–79.

Jain, Sarah S. Lochlann. 2006. *Injury: the Politics of Product Design and Safety Law in the United States.* Princeton: Princeton University Press.

Jainpedia. 2011. "About." Jainpedia (blog). blog.jainpedia.org/about.

Jakubowicz, Andrew. 2003. "Ethnic Diversity, 'Race,' and the Cultural Political Economy of Cyberspace." In *Democracy and New Media*, ed. by Henry Jenkins and David Thornburn, 203–22. Cambridge: MIT Press.

James, William. [1907] 2000. *Pragmatism and Other Writings*, ed. by Giles B. Gunn. Reprint, Penguin.

"Japan Tohuko Earthquake." 2011. A CrisisCommons collaboration with GISCorps. MapShare. http://gis.ats.ucla.edu/japan/

Jay, Martin. 1973. *The Dialectical Imagination: A History of the Frankfurt School of Social Research, 1923–1950*. Boston: Little, Brown.

Jenkins, Henry. 2003. "Quentin Tarantino's Star Wars? Digital Cinema, Media Convergence, and Participatory Culture." In *Rethinking Media Change: The Aesthetics of Transition*, ed. by David Thorburn and Henry Jenkins, 281–314. Cambridge: MIT Press.

Jenkins, Henry. 2009. *Confronting the Challenges of Participatory Culture: Media Education for the 21st Century*. Cambridge: MIT Press.

Jenkins, Henry. 2006. *Convergence Culture: Where Old and New Media Collide*. New York: New York University Press.

Jobs, Steve. 2011. "Steve Jobs: Technology Alone Is Not Enough." Youtube video, 1:47. Filmed March 2 at iPad 2 unveiling, Yerba Buena Center for the Arts, San Francisco, CA. Posted by "Retorbapi," March 3, 2011. http://www.youtube.com/watch?v=sUCpuaqlISQ&playnext=1&list=PLC7BEB75B89776F7E&feature=results_main&noredirect=1

Johnson, Barbara. 1995. "Teaching Deconstructively." In *Writing and Reading Differently: Deconstruction and the Teaching of Composition and Literature*, ed. by Douglas Atkins and Michael L. Johnson. Lawrence, KS: Kansas University Press.

Joseph, Chris, Christine Wilks, and Randy Adams. 2013. R3/\/\1X\/\/0RX [remixworx] (blog). Last modified August 12. http://runran.net/remix_runran

Juhasz, Alexandra. 2008. "Learning the Five Lessons of YouTube: After Trying to Teach There, I Don't Believe the Hype." *Cinema Journal* 48 (2): 145–50.

Juhasz, Alexandra, and Craig Dietrich. 2011. *Learning from YouTube* (video-book). *Vectors*. http://vectors.usc.edu/projects/learningfromyoutube/

Juola, Patrick. 2008. "Killer Applications in the Digital Humanities." *Literary and Linguistic Computing* 23 (1): 73–83.

Kant, Immanuel. [1798] 1992. *The Conflict of the Faculties*. Trans. by Mary J. Gregor. Omaha: University of Nebraska Press: 53, 65.

Kanzog, K. 1970. *Prolegomena zu einer historisch-kritischen Ausgabe der Werke Heinrich von Kleists: Theorie und Praxis einer modernen Klassiker-Edition*. München: Hanser.

Katiyar, S. S. 2003. "2002 Presidential Address of the Indian Science Congress Association Health Care, Education and Information Technology." In *The Shaping of Indian Science*. Indian Science Congress Presidential Addresses, vol. 3. Universities Press (India): 2014.

Keen, Suzanne. 2006. "A Theory of Narrative Empathy." *Narrative* 14 (October): 2–31.

Kester, Grant H. 2011. *The One and the Many: Contemporary Collaborative Art in a Global Context*. Durham: Duke University Press.

Kinder, Marsha. 2003. "Designing a Database Cinema." In *Future Cinema: The Cinematic Imaginary after Film*, ed. by Jeffrey Shaw and Peter Weibel Cambridge, 346–53. Cambridge: MIT Press.

Kirschenbaum, Matthew. 2008. *Mechanisms: New Media and the Forensic Imagination*. Cambridge: MIT Press.

Kirschenbaum, Matthew. 2010. "What Is Digital Humanities and What's It Doing in English Departments." *ADE Bulletin* 150: 55–61.

Kirschenbaum, Matthew. 2012. "Digital Humanities As/Is a Tactical Term." In *Debates in the Digital Humanities*, ed. by Matthew K. Gold, 415–28. Minneapolis: University of Minnesota Press.

Kittler, Friedrich. 1999. *Gramophone-Film-Typewriter*, trans. by Geoffrey Winthrop-Young and Michael Wutz. Stanford: Stanford University Press.

Kjellman, U. 2006. *Från kungaporträtt till läsketikett: En domänanalytisk studie över Kungl. bibliotekets bildsamling med särskild inriktning mot katalogiserings- och indexeringsfrågor*. Uppsala: Uppsala University.

Klayman, Alison. director. 2012. *Ai Weiwei: Never Sorry*. DVD. Orland Park, Illinois: MPI Home Video.

Klein, Lauren. 2013. "The Image of Absence: Archival Silence, Data Visualization, and James Hemings." *American Literature* 85 (4): 661–88.

Knobel, Michele, and Colin Lankshear. 2008. "Remix: The Art and Craft of Endless Hybridization." *Journal of Adolescent and Adult Literacy* 52 (1): 22–33.

Knowles, Anne Kelly. W. Roush, C. Abshere, L. Farrell, A. Feinberg, and T. Humber. 2008. "What Could Lee See at Gettysburg." *Placing History: How Maps, Spatial Data, and GIS Are Changing Historical Scholarship*: 235–66. Redlands, CA: ESRI Press.

Knowles, Anne Kelly. 2000. "Introduction." Special Issue: Historical GIS: The Spatial Turn in Social Science History. *Social Science History* 24 (3): 451–70.

Koeberl, Christian. 2009. *The Late Eocene Earth: Hothouse, Icehouse, and Impacts.* Special paper 452. Boulder: Geological Society of America.

Koe, Adeline, and Roopika Risam. 2013. "Open Thread: The Digital Humanities as a Historical 'Refuge' from Race/Class/Gender/Sexuality/Disability?" Postcolonial Digital Humanities. May 10. http://dhpoco.org/blog/2013/05/10/open-thread-the-digital-humanities-as-a-historical-refuge-from-raceclassgendersexualitydisability/

Koh, Adeline, and Roopika Risam. "Mission Statement." Postcolonial Digital Humanities. http://dhpoco.org/mission-statement-postcolonial-digital-humanities/

Konvitz, Josef W. 1978. *Cities and the Sea: Port City Planning in Early Modern Europe*. Baltimore: Johns Hopkins University Press.

Koomey, Jonathan. 2011. "Growth in Data Center Electricity Use 2005 to 2010." Analytics Press. August 1. http://www.analyticspress.com/datacenters.html

Koskinen, I. 2007. "Managing Banality in Mobile Multimedia." In *The Social Construction and Usage of Communication Technologies: European and Asian Experiences*, ed. by R. Pertierra, 48–60. Singapore: Singapore University Press.

Krajewski, Markus. 2011. *Paper Machines: About Cards and Catalogs, 1548–1929*. Cambridge: MIT Press.

Kramnick, Jonathan. 2011. "Against Literary Darwinism." *Critical Inquiry* 37 (Winter): 315–47.

Kraus, Kari. 2012. "Introduction to Rough Cuts: Media and Design in Process." The New Everyday: A MediaCommons Project. MediaCommons. July 28. http://mediacommons.futureofthebook.org/tne/pieces/introduction

Langins, Janice. 2004. *Conserving the Enlightenment: French Military Engineering from Vauban to the Revolution*. Cambridge: MIT Press.

Latour, Bruno. 1987. *Science in Action*. Cambridge: Harvard University Press.

Lancashire, Ian, John Bradley, Willard McCarty, Michael Stairs, and T. R. Wooldridge. 1996. *Using TACT with Electronic Texts: A Guide to Text-Analysis Computing Tools: Version 2.1 for MS-DOS and PC DOS*. New York: MLA.

Laurence, Anthony. 2006. "Developing a Freeware, Multiplatform Corpus Analysis Toolkit for the Technical Writing Classroom Tutorial." *IEEE Transactions on Professional Communication* 49 (3): 275–86. doi:10.1109/TPC.2006.880753.

Lavagnino, John. 2009. Access. In: Literary and Linguistic Computing 24, 1, PP. 63–76

Lazzarato, Maurizio. 2012. *The Making of the Indebted Man: Essay on the Neoliberal Condition.* Cambridge: MIT Press.

Lee, Dong-Hoo. 2005. "Women's Creation of Camera Phone Culture." *Fibreculture Journal* 6. http://journal.fibreculture.org/issue6/issue6_donghoo_print.html.

Lee, Dong-Hoo. 2009. "Re-imaging Urban Space: Mobility, Connectivity, and a Sense of Place." In *Mobile Technologies*, ed. by G. Goggin and L. Hjorth, 235–51. London: Routledge.

Lee, Dong-Hoo. 2009a. "Mobile Snapshots and Private/Public Boundaries." *Knowledge, Technology and Policy* 22 (3): 161–71.

Lee, K. S. 2011. "Interrogating 'Digital Korea': Mobile Phone Tracking and the Spatial Expansion of Labour Control." *Media International Australia* 141: 107–17.

Lenssen, Phillip. 2007. "Google stops 'Did You Mean: He Invented.'" Google Blogoscoped. May 24. http://blogoscoped.com/archive/2007-05-24-n36.html

Lessig, Lawrence. 2008. *Remix: Making Art and Commerce Thrive in the Hybrid Economy*. New York: Penguin Press.

Levan, Golan, Kamal Nigam, and Jonathan Fein. 2006. "The Dumpster." FLONG. http://www.flong.com/projects/dumpster/

Levinas, Emmanuel. 1969. *Totality and Infinity: An Essay on Exterority*, trans. by Alphonse Lingus. Pittsburgh: Duquesne University Press.

Levinas, Emmanuel. 1989. "Ethics as First Philosophy." In *The Levinas Reader*, ed. by Seán Hand. Oxford: Blackwell.

Lindquist, Martin A. 2008. The Statistical Analysis of fMRI Data. *Statistical Science* 23 (4): 439–64.

Liu, Alan. 2004. "The Humanities: A Technical Profession." *ACLS Occasional Paper* 63: 13–22.

Liu, Alan. 2008. "Literature+." *Currents in Electronic Literacy*. Digital Writing and Research Lab, University of Texas at Austin. http://currents.cwrl.utexas.edu/Spring08/Liu

Liu, Alan. 2011. "The University in the Digital Age: The Big Questions." March 10, 2011. Texas Institute for Literary and Textual Studies 2010–2011. *"The Digital and the Human(ities)," Symposium II: Teaching and Learning*. Keynote Lecture. http://tilts.dwrl.utexas.edu/content/media#Liu.

Liu, Alan. 2012a. "The State of the Digital Humanities: A Report and A Critique." *Arts and Humanities in Higher Education* 11 (8): 8–41.

Liu, Alan. 2012b. "Where Is Cultural Criticism in the Digital Humanities?" In *Debates in the Digital Humanities*, ed. by Matthew K. Gold, 490–510. Minneapolis: University of Minnesota Press.

Lodge, David. 1995. *Small World*. New York: Penguin.

Lohan, M. 2000. "Constructive Tensions in Feminist Technology Studies." *Social Studies of Science* 30 (6): 895–916.

The London Charter for the Computer-based Visualisation of Cultural Heritage. 2009. Version 2.1. February. http://www.londoncharter.org

Lopez-Menchero, Victor, and Alfredo Grande. 2010. "The Seville Charter." http://cipa.icomos.org/fileadmin/template/doc/PRAGUE/096.pdf

Lough, John. 1985. *France Observed in the Seventeenth Century*. Boston: Oriel Press.

Lorimer, Rowland, Johanne Provençal, Brian Owen, Rea Devakos, David Phipps, and Richard Smith. 2011. *Digital Technology Innovation in Scholarly Communication and University Engagement*. Vancouver: Canadian Centre for Studies in Publishing Press.

Losh, Elizabeth. 2005. "Virtualpolitik: Obstacles to Building Virtual Communities in Traditional Institutions of Knowledge." June 1. eScholarship.http://escholarship.org/uc/item/9m44x5tf?query=losh

Losh, Elizabeth. 2012a. "Bodies in Classrooms: Feminist Dialogues on Technology, Part I." DMLcentral. August 6. http://dmlcentral.net/blog/liz-losh/bodies-classrooms-feminist-dialogues-technology-part-i

Losh, Elizabeth. 2012b. "Hacktivism and the Humanities: Programming Protest in the Era of the Digital University." In *Debates in the Digital Humanities*, ed. by Matthew K. Gold, 161–86. Minneapolis: University of Minnesota Press.

Losh, Elizabeth. 2012c. "Including Underrepresented Students in Hands-on Creative Work." In *Education in Action*, ed. by Diane Forbes Berthoud, Jim Lin, and Elizabeth Losh. http://sixth.ucsd.edu/experiential-learning-conference/#more

Losh, Elizabeth. 2012d. "Learning from Failure: Feminist Dialogues on Technology, Part II." DMLcentral. August 9. http://dmlcentral.net/blog/liz-losh/learning-failure-feminist-dialogues-technology-part-ii

Losh, Elizabeth. 2012e. "Play, Things, Rules, and Information: Hybridizing Learning in the Digital University." *Leonardo Electronic Almanac, DAC09: After Media: Embodiment and Context* 17 (2, June 20): 86–102.

Lorimer, Rowland, Johanne Provençal, Brian Owen, Rea Devakos, David Phipps, and Richard Smith. 2011. "Digital Technology Innovation in Scholarly Communication and University Engagement" (white paper). http://tkbr.ccsp.sfu.ca/files/2010/04/Lorimer-DigInnovationScholCmn.pdf (page discontinued)

Lothian, Alexis. 2011. "Marked Bodies, Transformative Scholarship, and the Question of Theory in the Digital Humanities." *Journal of the Digital Humanities* 1 (1). http://journalofdigitalhumanities.org/1-1/marked-bodies-transformative-scholarship-and-the-question-of-theory-in-digital-humanities-by-alexis-lothian/.

Lothian, Alexis, and Amanda Phillips. 2013. "Can Digital Humanities Mean Transformative Critique?" *Journal of e-Media Studies* 3, no. 1. DOI:10.1349/PS1.1938-6060.A.425

Lovink, Geert. 2005. "Tactical Media: The Second Decade," geertlovink.org. October 19. http://geertlovink.org/texts/tactical-media-the-second-decade/

Lowood, Henry. 2008. "Found Technology: Players as Innovators in the Making of Machinima." In *Digital Youth, Innovation, and the Unexpected*, ed. by Tara McPherson, 165–96. Cambridge: MIT Press.

Lyotard, Jean-Francois. 1991. *The Postmodern Condition: A Report on Knowledge*, trans. by Geoff Bennington and Brian Massumi. Minneapolis: University of Minnesota Press.

Macdonald, Kevin. dir. 2011a. *Life in a Day* (film). http://movies.nationalgeographic.com/movies/life-in-a-day/

Macdonald, Kevin. dir. 2011b. *Kevin Macdonald on Life In A Day*. http://www.youtube.com/watch?v=C_4uii96xqM

Mahendran, Delan. 2007. "The Facticity of Blackness, A Non-conceptual Approach to the Study of Race and Racism in Fanon's and Merleau–Ponty's Phenomenology." *Human Architecture: Journal of the Sociology of Self-Knowledge* 5: 191–204.

Malloy, Judy. 2013. *Authoring Software*. Last modified Autumn. http://www.well.com/user/jmalloy/elit/elit_software.html

Mandel, Laura. 2012. "Does It Work? Where Theory and Technology Collide." IDHMC Comment-press. http://idhmc.tamu.edu/commentpress/does-it-work-where-theory-and-technology-collude/

Manovich, Lev. 2001. *The Language of New Media*. Cambridge: MIT Press.

Manovich, Lev. 2003a. "New Media from Borges to HTML." In *The New Media Reader*, ed. by Noah Wardrip-Fruin and Nick Montfort, 13–25. Cambridge: MIT Press.

Manovich, Lev. 2003b. "The Paradoxes of Digital Photography." In *The Photography Reader*, ed. by Liz Wells, 240–49. London: Routledge.

Manovich, Lev. 2011. "Inside Photoshop." *Computational Culture: A Journal of Software Studies* 1 (November). http://computationalculture.net/article/inside-photoshop

Manovich, Lev. 2012. "Media Visualization: Visual Techniques for Exploring Large Media Collections." In *Media Studies Futures*, ed. by Kelly Gates. Oxford: Blackwell.

Manovich, Lev, Chanda Carey, and Xiado Wang. 2011. "Mondrian vs. Rothko: Footprints of Evolution in Style Space." *Software Studies Initiative*. June 29. http://lab.softwarestudies.com/2011/06/mondrian-vs-rothko-footprints-and.html

Magnet, Shoshana, and Kelly Gates. 2009. *New Media of Surveillance*. London: Routledge.

Duchamp, Marcel. 1951. "Bicycle Wheel" (collection listing). MOMA. http://www.moma.org/collection/object.php?object_id=81631

Marcuse, Herbert. [1937] 1969. "Philosophy and Critical Theory." In *Negations: Essays in Critical Theory*, ed. by Steffon G. Bohm; trans. by Jeremy Shapiro, 134–58. Boston: Beacon Press.

Marino, Mark. 2006. "Critical Code Studies," *Electronic Book Review*. December 4. http://www.electronicbookreview.com/thread/electropoetics/codology

Marriot, David. 2007. *Haunted Life: Visual Culture and Black Modernity*. New Brunswick: Rutgers University Press.

Massey, D. 2005. *For Space*. London: Sage.

Massumi, Brian. 2002. *Parables for the Virtual: Movement, Affect, Sensation*. Durham: Duke University Press.

Masters, Roger D. 1998. *Fortune Is a River: Leonardo da Vinci and Niccolò Machiavelli's Magnificent Dream to Change the Course of Florentine History*. New York: Free Press.

Mateas, Michael. 2005. "Procedural Literacy: Educating the New Media Practitioner." *On the Horizon*. Special issue on games in education 13 (2): 101–11.

Mateas, Michael, and Andrew Stern. 2003. "Fa√ßade: An Experiment in Building a Fully-Realized Interactive Drama." (conference presentation, *Game Developers Conference*, San Jose, CA. March 4–8). Façade: a one-act interactive drama. March. http://www.interactivestory.net/papers/MateasSternGDC03.pdf

Mathews, Harry, and Alistair Brotchie. 1998. *Oulipo Compendium*. London: Atlas Press.

Mattern, Shannon. 2013. "Methodolatry and the Art of Measure: The New Wave of Urban Data Science." *Places*. http://places.designobserver.com/feature/methodolatry-in-urban-data-science/38174/

Maturana, H., and F. Varela. 1980. "Autopoiesis and Cognition: The Realization of the Living." In *Boston Studies in the Philosophy of Science*, vol. 42. ed. by Robert S. Cohen. Robert S. and Max W. Wartofsky. Dordrecht: Reidel.

Mazlish, Bruce. 1999. "Review: Big Questions? Big History?" *History and Theory* 38 (2): 232–48.

McCarty, Willard, and Harold Short. 2002. "Mapping the Field." ALLC Meeting Report: Pisa, April.

McGann, Jerome. 1983. *The Romantic Ideology: A Critical Investigation*. Chicago: University of Chicago.

McGann, Jerome. 2001. *Radiant Textuality: Literature after the World Wide Web*. New York: Palgrave.

McGann, Jerome. 2013. *The Complete Writings and Pictures of Dante Gabriel Rossetti: A Hypermedia Archive*. Institute for Advanced Technology in the Humanities and NINES. http://www.rossettiarchive.org/index.html

McIntosh, Jonathan. 2009. "Buffy vs Edward: Twilight Remixed." *Blip TV* 19. YouTube. http://www.youtube.com/watch?v=RZwM3GvaTRM

McPherson, Tara. 2009a. "Introduction: Media Studies and the Digital Humanities." *Cinema Journal* 48 (2): 119–23.

McPherson, Tara. 2009b. Media Studies and the Digital Humanities." *Cinema Journal* 48 (2): 119–23.

McPherson, Tara. 2010. "Scaling Vectors: Thoughts on the Future of Scholarly Communication." *Journal of Electronic Publishing* 13 (2). DOI: http://dx.doi.org/10.3998/3336451.0013.208

McPherson, Tara. 2011. "U.S. Operating Systems At Mid-century: The Intertwining Of Race and Unix." In *Race after the Internet*, ed. by Lisa Nakamura and Peter A. Chow-White. New York: Routledge.

McPherson, Tara. 2012. "Why Are the Digital Humanities So White?" In *Debates in the Digital Humanities*, ed. by Matthew K. Gold, 490–510. Minneapolis: University of Minnesota Press.

Mead, George Herbert. 1962. *Mind, Self and Society*. Chicago: University of Chicago Press.

Michel, J. B., Y. K. Shen, A. P. Aiden, A. Veres, M. K. Gray, J. P. Pickett, D. Hoiberg, et al. 2011. "Quantitative Analysis of Culture Using Millions of Digitized Books." *Science* 331 (6014): 176–82.

"Microbes maketh man." 2012. *The Economist*. August 18. http://www.economist.com/node/21560559

Seminar, Midwest Faculty. 2013. "What Are the Digital Humanities" MFS. November 14–16. http://mfs.uchicago.edu/?conference.html.

Mignolo, Walter D. 2003. "Globalization and the Geopolitics of Knowledge: The Role of the Humanities in the Corporate University." *Nepantla, Views from South* 4 (1): 97–119.

Mindpirates. 2012. "Mindpirates Film Jockeys" (event description). ArtConnectBerlin Beta. July 7. Berlin. http://www.artconnectberlin.com/mindpirates/events/2608.

Mirzoeff, Nicholas. 2012. *We Are All Children of Algeria: Visuality and Countervisuality 1954–2011*. Last modified April 4. Scalar. http://scalar.usc.edu/nehvectors/mirzoeff/index

Mitchell, Alex P., and Nick Montfort. 2009. "Shaping Stories and Building Worlds on Interactive Fiction Platforms." *Proceedings of the Digital Arts and Culture Conference 2009: After Media: Embodiment and Context*. eScholarship. December 12–15. http://escholarship.org/uc/item/6pk7s4n6

Mittell, Jason. 2008. "Remix Video." *Media Technology 2008*. March 4. http://sites.middlebury.edu/middmedia/assignments/remix-video/.

Modern Language Association. 2012. "Guidelines for Evaluating Work in Digital Humanities and Digital Media." MLA.org. Last Modified January. http://www.mla.org/guidelines_evaluation_digital

Moner, William J. 2011. "Undercompensated Labor in *Life in a Day*." *Flow* 14, no. 1. June 9. http://flowtv.org/2011/06/crowdsourced-labor-in-life-in-a-day/?utm_source=Flow+TV+Journal+List&utm_campaign=aa45b6e369-New_Issue_of_Flow_Volume_14_Issue_016_12_2011&utm_medium=email

Montfort, Nick. 2003. *Twisty Little Passages: An Approach to Interactive Fiction*. Cambridge: MIT Press.

Montfort, Nick. 2007. "Riddle Machines: The History and Nature of Interactive Fiction." In *A Companion to Digital Literary Studies*, edited by Susan Schreibman and Ray Siemens, 267–282. Oxford: Blackwell, 2008. http://www.digitalhumanities.org/companion/view?docId=blackwell/9781405148641/9781405148641.xml&chunk.id=ss1-5-8&toc.depth=1&toc.id=ss1-5-8&brand=9781405148641_brand

Montfort, Nick. 2009. "Taroko Gorge." *nickm.com*. January 8. Last modified June 21, 2012. http://nickm.com/poems/taroko_gorge.html

Montfort, Nick. 2012. "ppg256 series: Perl Poetry Generators in 256 characters." *nickm.com*. June 8, 2:20 am. http://nickm.com/poems/ppg256.html

Montfort, Nick, Patsy Baudoin, John Bell, Ian Bogost, Jeremy Douglass, Mark C. Marino, Michael Mateas, Casey Reas, Mark Sample, and Noah Vawter. 2012. *10 PRINT CHR$(205.5+RND(1)); GOTO 10*. Cambridge: MIT Press.

Montfort, Nick, Dan Shiovitz, and Emily Short, and the MHTO Occupation Force. 2005. "Mystery House Taken Over." http://www.turbulence.org/Works/mystery/

Moretti, Franco. 2004. "Studying Literature by the Numbers." *The New York Times*. January 10.

Moretti, Franco. 2005. *Graphs, Maps, Trees: Abstract Models for a Literary History*. New York: Verso.

Mørk Petersen, S. 2009. "Common Banality: The Affective Character of Photo Sharing, Everyday Life and Produsage Cultures." PhD thesis, ITU Copenhagen.

Morozov, Evgeny. 2009. "Iran: Downside to the 'Twitter Revolution.'" *Dissent* 56 (4): 10–14.

Morozov, Evgeny. 2011. *The Net Delusion: The Dark Side of Internet Freedom*. New York: PublicAffairs.

Morris, Sean Michael. 2013. "Queequeg's Coffin: A Sermon for the Digital Human." *Hybrid Pedagogy: A Digital Journal of Learning, Pedagogy, and Teaching*. November 10. http://www.hybridpedagogy.com/Journal/queequegs-coffin-a-sermon-for-the-digital-human/.

Moscati P. 1996. "Archeologia e Calcolatori." *Archeologia Quantitativa: nascita, sviluppo e "crisi"* 7: 579–90.

Muir, Lynette R., and John White, eds. 1996. *Materials for the Life of Nicholas Ferrar*. Leeds: Leeds Philosophical and Literary Society Ltd.

Mukerji, Chandra. 1987. *Territorial Ambitions and the Gardens of Versailles*. Cambridge, UK: Cambridge University Press.

Mukerji, Chandra. 1997. *Territorial Ambitions and the Gardens of Versailles*. Cambridge, UK: Cambridge University Press.

Mukerji, Chandra. 2009. *Impossible Engineering*. Princeton: Princeton University Press.

Mukerji, Chandra. 2012. "Space and Political Pedagogy at the Gardens of Versailles." *Public Culture* 68 (3): 509–34.

Mulvey, Laura. 1975. "Visual Pleasure and Narrative Cinema." *Screen* 16 (3): 6–18.

Mumford, Lewis. 1934. *Technics and Civilization*. New York: Harcourt, Brace.

Mungiu-Pippidi, Alina, and Igor Munteanu. 2009. "Moldova's 'Twitter Revolution.'" *Journal of Democracy* 20 (3): 136–42.

Murch, Walter. 2005. *The Cutting Edge: The Magic of Movie Editing*. Dir. Wendy Apple. New York: Warner Home Video.

Nakamura, Lisa. 2002. *Cybertypes*. New York: Routledge.

Nakamura, Lisa. 2007. *Digitizing Race: Visual Cultures on the Internet*. Minneapolis: University of Minnesota Press.

Narine, Sandy. 2006. "Introduction." In *The Portrait Now*. London: National Portrait Gallery.

National Research Council. 2003. *Beyond Productivity: Information, Technology, Innovation, and Creativity*. Washington, DC: National Academies Press.

Navas, Eduardo. 2010. "Regressive and Reflexive Mashups in Sampling Culture". In *Mashup Cultures*, ed. by Stefan, Sonvilla-Weiss. Mörlenbach: Springer-Verlag/Wien.

Neatline: Plot Your Course in Space and Time. The Scholar's Lab. University of Virginia Library. http://neatline.org/

Nelson, Robert. 2000. "The Slide Lecture, or the Work of Art History in the Age of Mechanical Reproduction." *Critical Inquiry* 26: 414–34.

Nerad, Julie Cary. 2003. "Slippery Language and False Dilemmas: The Passing Novels of Child, Howells, and Harper." *American Literature* 75:813–41.

Neressian, Nancy J. 2006. "The Cognitive-Cutural Systems of the Research Laboratory." *Organization Studies* 27 (2): 125–45.

Newfield, Christopher. 2008. *Unmaking the Public University: The Forty-Year Assault on the Middle Class*. Cambridge: Harvard University Press.

Newfield, Christopher. 2003. *Ivy and Industry: Business and the Making of the American University, 1880–1980*. Durham: Duke University Press.

Nishime, LeiLani. 2005. "The Mulatto Cyborg: Imagining a Multiracial Future." *Cinema Journal* 44: 34–49.

Nitrogram. 2014. "Instagram Statistics." January 16. http://analytics.nitrogr.am/instagram-statistics/.

Noble, D. 1997. *The Religion of Technology: The Divinity of Man and the Spirit of Invention*. New York: Knopf.

Nowviskie, Bethany. 2011a. "It Starts on Day One." Bethany Nowviskie (blog). November 12. http://nowviskie.org/2011/it-starts-on-day-one/

Nowviskie, Bethany. 2011b. "Where Credit Is Due." Bethany Nowviskie (blog). May 21. http://nowviskie.org/2011/where-credit-is-due/

Nowviskie, Bethany. 2011c. "Where Credit Is Due: Preconditions for the Evaluation of Collaborative Digital Scholarship." *Profession* 1: 169–81.

Nowviskie, Bethany. 2012. "It Starts on Day One," *ProfHacker, The Chronicle of Higher Education*. January 12. http://chronicle.com/blogs/profhacker/it-starts-on-day-one/37893

Obadike, Keith. 2001. *Keith Obadike's Blackness*. Obadike.tripod.com. obadike.tripod.com/ebay.html.

Official Guide Book of the New York World's Fair 1939. 1939. New York: Exposition Publications.

Omi, Michael, and Howard Winant. 1994. *Racial Formation in the United States: From the 1960s to the 1990s*. New York: Routledge.

"One, Million Monkeys Typing." A Collaborative Writing Project. http://cqcounter.com/site/1000000monkeys.com.html (site discontinued)

Open Journal Systems. Public Knowledge Project. Simon Fraser University. https://pkp.sfu.ca/.

Otter, Malcolm, and Hilary Johnson. 2000. "Lost in Hyperspace: Metrics and Mental Models." *Interacting with Computers* 13 (1): 1–40.

Pérez, Rafael Pérez Ý., and Mike Sharples. 2001. "MEXICA: A Computer Model of a Cognitive Account of Creative Writing." *Journal of Experimental and Theoretical Artificial Intelligence* 13 (2): 119–39.

Palmer, D. 2012. "iPhone Photography: Mediating Visions of Social Space." In *Studying Mobile Media: Cultural Technologies, Mobile Communication, and the iPhone*, ed. by L. Hjorth, J. Burgess, and I. Richardson, 85–97. London: Routledge.

Palmquist. Stephen, R. 2006. "Philosophers in the Public Square," *Kant and the New Philosophy of Religion*, ed. by Chris L. Firestone and Stephen R. Palmquist, 235. Indiana University Press.

Palumbo-Liu, David. 2011. "Why the Humanities Are Indispensable." David Palumbo-Liu (blog). August 19. http://palumboliu.tumblr.com/post/9124363788/why-the-humanities-are-indispensable

Pannapacker, William. 2011. "Pannapacker at MLA: Digital Humanities Triumphant?" Brainstorm (blog). *Chronicle of Higher Education*, January 8. http://chronicle.com/blogs/brainstorm/pannapacker-at-mla-digital-humanities-triumphant/30915.

Parikka, Jussi. 2012. *What Is Media Archaeology?* Malden: Polity Press.

Parker, Matthew. 2009. *Lineman* (Yearbook). Los Angeles Trade-Technical College (Spring): 21.

Parks, Lisa. 2012. "Technostruggles and the Satellite Dish: A Populist Approach to Infrastructure." In *Cultural Technologies: The Shaping of Culture in Media and Society*, ed. by Göran Bolan, 64–84. London: Routledge.

Pärna, K. 2006. "Believe in the Net: The Construction of the Sacred in Utopian Tales of the Internet." *Implicit Religion* 9 (2): 180–204.

Pasolini, Pier Paolo. [1967] 1980. "Observations on the Long Take". Trans. by Norman MacAfee and Craig Owens. *October* 13:3–6.

Paterson, Mark. 2007. *The Sense of Touch: Haptics, Affects and Technologies*. Oxford: Berg.

Pattuelli, M. Cristina. 2012. "FOAF in the Archive: Linking Networks of Information with Networks of People: Final Report to OCLC." OCLC.org. March 13. http://www.oclc.org/url/?404;http://www.oclc.org/research/grants/reports/2012/pattuelli2012.pdf.

Pattuelli, M. Christina dir. 2013. "Linked Jazz: Revealing the Relationships of the Jazz Community." Linked Jazz. http://linkedjazz.org/

Peirce, Charles S. 1955. *Philosophical Writings of Peirce*. New York: Dover Publications.

Perloff, Marjorie. 2010. *UnOriginal Genius: Poetry by Other Means in the 21st Century*. Chicago: University of Chicago Press.

Perrault, Charles. 1697. *Histoires ou contes du temps passée*. Paris: Claude Barbin.

Perrault, Charles. 1982. *Le Labyrinthe de Versailles 1677*. Paris: Editions de Moniteur.

Peters, John Durham. 2003. "Space, Time, and Communication Theory." *Canadian Journal of Communication* 28 (4): 397–411.

Phillips, Amanda. 2011. "#transformDH—A Call to Action Following ASA 2011." HASTAC Scholar (blog). HASTAC. October 26. http://www.hastac.org/blogs/amanda-phillips/2011/10/26/transformdh-call-action-following-asa-2011.

Pickering, Andrew. 1995. *The Mangle of Practice: Time. Agency, and Science*. Chicago: University of Chicago Press.

Pinch, Trevor, and Frank Trocco. 2002. *Analog Days: The Invention and Impact of the Moog Synthesizer*. Cambridge: Harvard University Press.

Piper, Andrew. 2009. *Dreaming in Books: The Making of the Bibliographic Imagination in the Romantic Age*. Chicago: University of Chicago Press.

Pink, S. 2011. "Sensory Digital Photography: Re-thinking 'Moving' and the Image." *Visual Studies* 26 (1): 4–13.

Pink, S. 2012. *Situating Everyday Life*. London: Sage.

Plant, S. 1998. *Zeroes and Ones: Digital Women and the New Technoculture*. London: Harpercollins.

Poe, Edgar Allan. 1846. "The Philosophy of Composition." *Graham's Magazine*, April. http://xroads.virginia.edu/~HYPER/poe/composition.html

Pontifical Council for Social Communications. 2002. "The Church and Internet." Vatican.va. http://www.vatican.va/roman_curia/pontifical_councils/pccs/documents/rc_pc_pccs_doc_20020228_church-internet_en.html

Poschardt, Ulf. 1998. *DJ Culture*. London: Quartet Books.

Posner, Miriam. 2012. "Digital Humanities 2012 Notes" (Google document). https://docs.google.com/document/d/1mXgt5HGq50jGncZ_63zYjb5bB70YSZLxrz-uq77ctKM/edit

Poster, M. 2001. "Cyberdemocracy: The Internet and the Public Sphere." In *Reading Digital Culture*, ed. by D. Trend, 259–71. Oxford: Blackwell.

Potter, Rosanne G., ed. 1989. "Introduction." *Literary Computing and Literary Criticism: Theoretical and Practical Essays on Theme and Rhetoric*. Philadelphia: University of Pennsylvania Press.

Prelinger, Rick, and Raegan Kelly. 2006. "Panorama Ephemera." *Vectors* 2, no. 1 (Fall). http://vectorsjournal.org/projects/index.php?project=58

Prescott, Andrew. 2012. "Making the Digital Human: Anxieties, Posibilities, Challenges." Digital Riffs: Extemporizations, Excursion and Explorations in the Digital Humanities. July 5. http://digitalriffs.blogspot.com/2012/07/making-digital-human-anxieties.html

Presenceon, S., and P. Dourish. 1996. "Re-place-ing Space: The Roles of Place and Space in Collaborative Systems." *Proceedings of the ACM 1996 Conference on Computer Supported Cooperative Work*, Boston.

Presner, Todd. 2009. "The City in the Ages of New Media: From Ruttmann's Berlin: Die Sinfonie Der Grossstadt to Hypermedia Berlin." In *After the Digital Divide? German Aesthetic Theory in the Age of New Media*, ed. by Lutz Koepnick, 229–51. Camden House.

Presner, Todd. 2010. "HyperCities: A Case Study for the Future of Scholarly Publishing." In *The Shape of Things to Come*, ed. by Jerome McGann, 251–71. Houston: Rice University Press; http://cnx.org/content/m34318/latest/.

Presner, Todd, Yoh Kawano, and David Shepard. 2011. "HyperCities Now." http://egypt.hypercities.com/.

Presner, Todd, Jeffrey Schnapp, Peter Lunenfeld, et al. "The Digital Humanities Manifesto 2.0." *Humanities Blast*, http://www.humanitiesblast.com/manifesto/Manifesto_V2.pdf

Price, Kenneth. 2009. "Edition, Project, Database, Archive, Thematic Research Collection: What's in a Name?" *DHQ: Digital Humanities Quarterly* 3 (3). http://www.digitalhumanities.org/dhq/vol/3/3/000053/000053.html

"Programme Digital Humanities 2012." 2012. dh 2012. http://www.dh2012.uni-hamburg.de/conference/programme/.

Project MUSE. Johns Hopkins University. http://muse.jhu.edu

Quintilian. 1774. *Institutes of the Orator in Twelve Books*, vol. 2, Rollin (ed.). London: B. Law.

Rahman, A. 1995. India. In *Science, Technology and Development*, ed. by A. Rahman. Hoboken, NJ: Wiley.

Raina, D. 2009. *A Social Epistemology of Science and Society*. http://cscs.res.in/dataarchive/textfiles/social-epistemology-of--science-and-society-.pdf

Rajadhyaksha, Ashish. 2011. *The Last Cultural Mile: An Inquiry into Technology and Governance in India*. Bangalore: Centre for Internet and Society; http://cis-india.org/raw/histories-of-the-internet/last-cultural-mile.pdf.

Raley, Rita. 2009. *Tactical Media*. Minneapolis: University of Minnesota Press.

Ramsay, Stephen. 2001. *Reading Machines: Toward an Algorithmic Criticism*. Urbana: University of Illinois Press.

Ramsay, Stephen. 2004. "Databases." In *A Companion to Digital Humanities*, ed. by Susan Schreibman, Ray Siemens, and John Unsworth. Oxford: Blackwell; http://www.digitalhumanities.org/companion/view?docId=blackwell/9781405103213/9781405103213.xml&chunk.id=ss1-3-3&toc.depth=1&toc.id=ss1-3-3&brand=default.

Ramsay, Stephen. 2007. "An Approach to Videogame Criticism: Ian Bogost." *Literary and Linguistic Computing* 22 (3): 369–70. http://llc.oxfordjournals.org/content/22/3/369.citation.

Ramsay, Stephen. 2010. "The Hermeneutics of Screwing around; or What You Do with a Million Books." Unpublished book chapter. Playing with Technology in History. April 17. http://www.playingwithhistory.com/wpcontent/uploads/2010/04/hermeneutics.pdf

Ramsay, Stephen. 2011. "On Building." *Stephen Ramsay*. http://stephenramsay.us/text/2011/01/11/on-building/

Ramsay, Stephen, and Geoffrey Rockwell. 2012. "Developing Things." In *Debates in Digital Humanities*, ed. by Matthew K. Gold, 75–84. Minneapolis: University of Minnesota Press.

Rao, C. N. R. 2003. "1988 Presidential Address of the Indian Science Congress Association Frontiers in Science and Technology" in *The Shaping of Indian Science—Indian Science Congress Presidential Addresses,* vol. 3. Universities Press India: 671.

Raskin, Richard. 1998. "Five Explanations for the Jump Cuts in Godard's *Breathless*." *POV* 6 (December). http://pov.imv.au.dk/Issue_06/section_1/artc10.html

Ray, Benjamin. 2002. "Salem Witch Trials Documentary Archive and Transcription Project." University of Virginia. http://salem.lib.virginia.edu/home.html

Raymond, Eric Steven. 2000. "The Cathedral and the Bazaar" (version 3.0). *catb.org*. http://www.catb.org/~esr/writings/cathedral-bazaar/cathedral-bazaar/index.html

Readings, Bill. 1997. *The University in Ruins*. Cambridge: Harvard University Press.

Reagal, John. 2010. *Good Faith Collaboration: The Culture of Wikipedia*. Cambridge: MIT Press.

Reichelt, Leisa. 2007. "Ambient Intimacy." Disambiguity. March 1 http://www.disambiguity.com/ambient-intimacy/.

Reilly, P. 1990. "Towards a Virtual Archaeology." In *Computer Applications in Archaeology*, edited by K. Lockyear and S. Rahtz. BAR International Series 565, Oxford: Archaeopress: 133–39.

Renear, A. 2001. "Literal Transcription: Can the Text Ontologist Help?" In *New Media and the Humanities: Research and Applications*, ed. by D. Fiormonte and J. Usher, 23–30. Oxford: Oxford University.

Renfrew, C., and C. Scarre. 1999. *Cognition and Material Culture: The Archaeology of Symbolic Storage*. Cambridge: McDonald Institute.

"Research Infrastructures in the Digital Humanities." 2011. European Science Foundation, Science Policy Briefing 42. September. http://www.esf.org/fileadmin/Public_documents/Publications/spb42_RI_DigitalHumanities.pdf

Resnick, Mitchel, John Maloney, Andrés Monroy-Hernández, Natalie Rusk, Evelyn Eastmond, Karen Brennan, Amon Millner, et al. 2009. "Scratch: Programming for All." *Communications of the ACM* 52 (11): 60–67.

Rhoten, D. 2004. "Interdisciplinary Research: Trend or Transition?" *Items and Issues* 5 (1–2): 6–11.

Richards, Thomas. 1993. *The Imperial Archive: Knowledge and the Fantasy of Empire*. London, New York: Verso.

Richardson, I., and R. Wilken. 2013. "Parerga of the Third Screen: Mobile Media, Place, and Presence." In *Mobile Technology and Place*, ed. by R. Wilken and G. Goggin, 198–212. New York: Routledge.

Rieder, B., and T. Röhle. 2012. "Digital Methods: Five Challenges." In *Understanding Digital Humanities*, ed. by D. Berry, 67–84. New York: Palgrave Macmillan.

Riggs, M. 1986. *Ethnic notions (film)*. San Francisco: California Newsreel.

Riskin, Jessica. 2003. "The Defecating Duck, or, the Ambiguous Origins of Artificial Life." *Critical Inquiry* 29 (4): 599–633.

Riskin, Jessica, ed. 2007. *Genesis Redux: Essays in the History and Philosophy of Artificial Life*. Chicago: University of Chicago Press.

Riskin, Jessica. 2008. "Mechanical Christs, Hydraulic Brutes, and the Invention of Consciousness." Paper given at the conference, Technology and the Formations of Power. Unversity of California, San Diego.

Roberts-Smith, Jennifer, and Stan Ruecker. 2013. "Visualising Theatre Historiography: Judith Thompson's White Biting Dog (1984 and 2011) in the Simulated Environment for Theatre (SET)." *Digital Studies/Le champ numérique* 3, no. 2.

Robinson, Peter. 2000. "The One Text and the Many Texts." *Literary and Linguistic Computing* 15 (1): 5–14.

Rockwell, Geoffrey. 2003. "What Is Text Analysis, Really?" *Literary and Linguistic Computing* 18 (2): 209–19.

Rockwell, Geoffrey. 2010. "As Transparent as Infrastructure: On the Research of Cyberinfrastructure in the Humanities." *Online Humanities Scholarship: The Shape of Things to Come*, ed. by Jerome McGann. http://cnx.org/content/m34315/1.2/

Rockwell, Geoffrey, Stéfan Sinclair, and James Chartrand. 2005. "TAPoR: Five Views through a Text Analysis Portal." *ACH/ALLC 2005 Conference Abstracts*. 15–18 June. Victoria, BC: University of Victoria Humanities Computing and Media Center.

Rommel, Thomas. 2004. "Literary Studies." In *A Companion to Digital Humanities*, ed. by Susan Schreibman, Ray Siemens, and John Unsworth. Oxford: Blackwell; http://www.digitalhumanities.org/companion/view?docId=blackwell/9781405103213/9781405103213.xml&chunk.id=ss1-2-8.

Rose, Mark. 1995. *Authors and Owners: The Invention of Copyright*. Cambridge: Harvard University Press.

Rosenau, Helen. 1972. *The Ideal City, Its Architectural Evolution*. New York: Harper Row.

Rosental, Claude. 2004. *Une Sociology des formes de démonstration*. Paris: Mémoire pour l'Habilitation à diriger des Recherches.

Ross, Andrew. 1990. "Hacking away at the Counter-Culture." *Postmodern Culture* 1, no. 1. https://muse.jhu.edu/journals/postmodern_culture/v001/1.1ross.html

Rowe, Katherine, ed. 2010. *Shakespeare Quarterly* Open Review: "Shakespeare and New Media." MediaCommons. http://mcpress.media-commons.org/ShakespeareQuarterly_NewMedia/.

Roset, Rafel. 2012. *Personal Interview*. Barcelona: Institut Cartogràfic de Catalunya.

Ross, Andrew. 2010. "Digital Humanities Sessions at the 2011 MLA." *HASTAC*. December 13. http://www.hastac.org/blogs/marksample/digital-humanities-sessions-2011-mla.

The Royal Society. "History." The Royal Society. https://royalsociety.org/about-us/history/.

Rubenstein, D. 2005. "Cameraphone Photography: The Death of the Camera and the Arrival of Visible Speech." *Issues in Contemporary Culture and Aesthetics* 1:113–118.

Rubinstein, D., and K. Sluis. 2008. "A Life More Photographic: Mapping the Networked Image." *Photographies* 1 (1): 9–28.

Rushkoff, Douglas. 2011. *Program or Be Programmed: Ten Commands for a Digital Age*. Berkeley: Soft Skull Press.

Russell, Lynette, ed. 2001. *Colonial Frontiers: Indigenous-European Encounters in Settler Societies.* Manchester: Manchester University Press.

Ryan, N. 1996. "Computer-based Visualization of the Past: Technical 'Realism' and Historical Credibility." In *Imaging the Past: Electronic Imaging and Computer Graphics in Museums and Archaeology*, edited by Peter Main, Tony Higgins and Janet Lang. British Museum Occasional Papers 114. London: British Museum: 95–108.

Ryan, N. 2001. "Documenting and Validating Virtual Archaeology." *Archeologia e Calcolatori* 12 (November): 245–73.

Salvucci, Dario, and Joseph Goldberg. 2000. "Identifying Fixations and Saccades in Eye-Tracking Protocols." *Proceedings in the Eye Tracking Research and Applications Symposium*. New York: ACM Press: 71–78.

Sample, Mark. 2010. "I'm Mark, and Welcome to the Circus." Comment on HASTAC blog entry "I'm Chris. Where am I wrong?" September 10. http://www.hastac.org/blogs/cforster/im-chris-where-am-i-wrong.

Sample, Mark. 2012. "Scholarly Lies and the Deformative Humanities." Sample Reality (blog). May 17. http://www.samplereality.com/2012/05/17/scholarly-lies-and-the-deformative-humanities/

Samuels, Lisa, and Jerome McGann. 1999. "Deformance and Interpretation." *New Literary History* 30 (1): 25–56.

Sanderson, J., and P. Cheong. 2010. "Tweeting Prayers and Communicating Grief over Michael Jackson Online." *Bulletin of Science, Technology and Society* 30 (5): 328–40.

Sandoz, Devin. 2003. "Simulation / Simulacrum" (blog entry). The Chicago School of Media Theory. Winter. http://lucian.uchicago.edu/blogs/mediatheory/keywords/simulation-simulacrum/

Santo, Avi, and Christopher Lucas. 2009. "Engaging Academic and Nonacademic Communities through Online Scholarly Work." *Cinema Journal* 48 (2): 129–38.

Saxenian, AnnaLee. 1994. *Regional Advantage: Culture and Competition in Silicon Valley and Route 128*. Cambridge: Harvard University Press.

Sayers, Jentery. 2011. "Tinker-Centric Pedagogy in Literature and Language Classrooms." In *Collaborative Approaches to the Digital in English Studies*, ed. by Laura McGrath, 279–300. Logan: Utah State University Press; http://ccdigitalpress.org/cad/Ch10_Sayers.pdf.

Sayers, Jentery. J. James Bono, Curtis Hisayasu, and Matthew W. Wilson. 2012. "Standards in the Making: Composing with Metadata in Mind." *The New Work of Composing*, ed. by Debra Journet, Cheryl E. Ball, and Ryan Trauman. Logan: Utah State University Press. http://ccdigitalpress.org/nwc/chapters/wilson-et-al/index.html

Schaeffer, Pamela. 1999. "Scripture in Multimedia." *National Catholic Reporter Online*. May 14. http://natcath.org/NCR_Online/archives2/1999b/051499/051499a.htm

Scheinfeldt, Tom, and Dan Cohen eds. 2010. *Hacking the Academy*. Roy Rosenzweig Center for History and New Media. May 21–28. http://hackingtheacademy.org

Schiller, Jakob. 2012. "Photo Project Aims to Capture a Day in the Life of the World." *Wired*. April 13. http://www.wired.com/2012/04/photo-project-aims-to-capture-a-day-in-the-life-of-the-world/.

Schmidt, Ben. 2012. "Visualizing Ocean Shipping." *Sapping Attention* (blog). April 9. http://sappingattention.blogspot.com/2012/04/visualizing-ocean-shipping.html

Scholz, Trebor, ed. 2012. *Digital Labor: The Internet as Playground and Factory*. New York: Routledge.

Schreibman, Susan, Ray Siemens, and John Unsworth. 2004. "The Digital Humanities and Humanities Computing." In *A Companion to Digital Humanities*, ed. by Susan Schreibman, Ray Siemens, and John Unsworth, xxiii–xxvii. Oxford: Blackwell; http://www.digitalhumanities.org/companion/view?docId=blackwell/9781405103213/9781405103213.xml&chunk.id=ss1-1-3&toc.depth=1&toc.id=ss1-1-3&brand=default.

Schwartz, Joan M., and Terry Cook. 2002. "Archives, Records, and Power: The Making of Modern Memory." *Archival Science* 2: 1–19.

Scott, James. 1998. *Seeing Like a State: How Certain Schemes to Improve the Human Condition Have Failed*. New Haven: Yale University Press.

Scott-Railton, John. "Jan25 Voices" (twitter feed). https://twitter.com/Jan25voices

Scott-Railton, John. "Feb 17 voices" (twitter feed). https://twitter.com/feb17voices

Scott-Railton, John. 2011. "Jan25 Voices." Audioboo.fm. recorded phone calls. http://audioboo.fm/Jan25voices

Scott-Railton, John. 2012. "The Voices Feeds." John Scott-Railton. Last modified July 27. http://johnscottrailton.com/the-voices-feeds/

Seed, Patricia. 2012. "Preserving the Integrity of the Original When Georeferencing Historical Maps." Seventh International Workshop on Digital Cartography. Barcelona, Catalunya, Spain.

Sewell, William Hamilton. 2005. *Logics of History: Social Theory and Social Transformation. Chicago Studies in Practices of Meaning*. Chicago: University of Chicago Press.

Sewell, William Hamilton. 1980. *Work and Revolution in France: The Language of Labor from the Old Regime to 1848*. Cambridge, UK: Cambridge University Press.

Shah, Nishant. 2012. "UCHRI's Perspectives March 2012 with Nishant Shah." (Video). Vimeo. March. http://vimeo.com/38445362

Shah, Nishant. 2012. "Resisting Revolutions: Questioning the Radical Potential of Citizen Action." *Development* 55 (2): 173–80. http://www.academia.edu/1600603/Resisting_Revolutions_Questioning_the_radical_potential_of_citizen_action.

Shakespeare Quarterly Open Review. Folger Library. http://shakespearequarterly.folger.edu/openreview/

Shields, David. 2011. *Reality Hunger: A Manifesto*. New York: Vintage Books.

Shinji, Nishimoto, An. T. Vu, Thomas Naselaris, Yuval Benjamini, Bin Yu, Jack L. Gallant. 2011. "Reconstructing Visual Experiences from Brain Activity Evoked by Natural Movies." *Current Biology* 21 (19, October 11): 1641–46.

Shipman, Matt. 2012. "Creating an Online Portal into the Medieval World." The Abstract. June 13. http://web.ncsu.edu/abstract/technology/wms-medieval-online/

Siemens, Lynne, Wendy Duff, Richard Cunningham, and Claire Warwick. 2009. "It challenges members to think of their work through another kind of specialist's eyes": Exploration of the Benefits and Challenges of Diversity in Digital Project Teams. *Proceedings of the American Society for Information Science and Technology* 46 (1): 1–14. http://www.asis.org/Conferences/AM09/open-proceedings/openpage.html.

Siemens, Lynne, and Ray Siemens. 2012. "Notes from the Collaboratory: An Informal Study of an Academic Lab in Transition" (presentation). https://lecture2go.uni-hamburg.de/konferenzen/-/k/13921.

Siemens, Ray, Meagan Timney, Cara Leitch, Corina Koolen, and Alex Garnett, and with the ETCL, INKE, and PKP Research Groups. 2012. "Toward Modeling the Social Edition: An Approach to Understanding the Electronic Scholarly Edition in the Context of New and Emerging Social Media." *Literary and Linguistic Computing* 27 (4): 445–61. http://web.uvic.ca/~siemens/pub/2011-SocialEdition.pdf

Siemens, Ray, and Christian Vandendorpe. 2006. Canadian Humanities Computing and Emerging Mind Technologies. In *Mind Technologies: Humanities Computing and the Canadian Academic Community*, ed. by Ray Siemens and David Moorman. Calgary: University of Calgary Press.

Siemens, Ray, Claire Warwick, Richard Cunningham, Teresa Dobson, Alan Galey, Stan Ruecker, and Susan Schreibman, and the INKE Research Group. 2009. "Codex Ultor: Toward a Conceptual and Theoretical Foundation for New Research on Books and Knowledge Environments." *Digital Studies / Le Champ Numérique* 1 (2). http://www.digitalstudies.org/ojs/index.php/digital_studies/article/view/177/220

Siemens, Ray, Karin Armstrong, and Constance Crompton, and the Devonshire Manuscript Editorial Group eds. 2012. *A Social Edition of the Devonshire Manuscript BL MS Add 17, 492*. Wikibooks. http://en.wikibooks.org/wiki/The_Devonshire_Manuscript

Siemens, Ray, Meagan Timney, Cara Leitch, Corina Koolen, and Alex Garnett, and with the ETCL, INKE, and PKP Research Groups. 2012. "Toward Modeling the Social Edition: An Approach to Understanding the Electronic Scholarly Edition in the Context of New and Emerging Social Media." *Literary and Linguistic Computing* 27 (4): 445–61. http://web.uvic.ca/~siemens/pub/2011-SocialEdition.pdf

Silk Road Project. The Silk Road Project, Inc. http://www.silkroadproject.org.

Slack, Jennifer Daryl. 1996. The Theory and Method of Articulation in Cultural Studies. In *Stuart Hall: Critical Dialogues*, ed. by David Morley and Kuan-Hsing Chen, 113–29. New York: Routledge.

Slack, Jennifer Daryl, and J. Macgregor Wise. 2006. *Culture + Technology: A Primer*. New York: Peter Lang.

Smith, Gil R. 1997. *Architectural Diplomacy: Rome and Paris in the Late Baroque*. Cambridge: MIT Press.

Smith, Martha Nell. 2007. "The Human Touch Software of the Highest Order: Revisiting Editing as Interpretation." *Textual Cultures: Texts, Contexts, Interpretation* 2 (1): 1–15.

Sniderman, Zachary. 2010. "Boticelli's [sic] Venus Reimagined to Raise Breast Cancer Awareness." *Mashable* (October 15). http://mashable.com/2010/10/15/venus-samsung-breast-cancer

Snow, C. P. [1959] 1993. *The Two Cultures and the Scientific Revolution*. Cambridge, UK: Cambridge University Press.

Sonesson, Göran. 1998. "The Concept of Text in Cultural Semiotics." In *Sign System Studies 26*, ed. by Peeter Torop, Michail Lotman, and Kalevi Kull, 88–114. Tartu: Tartu University Press.

Spiro, Lisa. 2011. "Computing and Commmunicating Knowledge: Collaborative Approaches to Digital Humanities Projects." In *Collaborative Approaches to the Digital in English Studies*, ed. by

Laura McGrath, 44–82. Logan: Utah State University Press; http://ccdigitalpress.org/cad/Ch2_Spiro.pdf.

Spiro, Lisa. 2013. "Exploring the Signifiance of Digital Humanities for Philosophy." Digital Scholarship in the Humanities (blog). February 26. http://digitalscholarship.wordpress.com/2013/02/26/exploring-the-significance-of-digital-humanities-for-philosophy/

Srinivasan, Ramesh. 2011. "Starting with Culture before Technology: My Work from Egypt." Ramesh Srinivasan. August 30. http://rameshsrinivasan.org/2011/08/30/starting-with-culture-before-technology-my-work-from-egypt/

Srinivasan, Ramesh. 2012. "Digital Dissent and People's Power: Ramesh Srinivasan at TEDxSanJoaquin." TEDxSanJoaquin. October 10. San Joaquin, CA.

Srinivasan, Ramesh, R. Boast, J. Furner, and K. Bevar. 2009. "Digital Museums and Diverse Cultural Knowledges: Moving Past the Traditional Catalog." *Information Society* 25 (4): 265–78.

Star, Susan Leigh, and James R. Griesemer. 1989. "Institutional Ecology, Translations' and Boundary Objects: Amateurs and Professionals in Berkeley Museum of Vertebrate Zoology, 1907–39." *Social Studies of Science* 19 (3): 387–420.

Stafford, Barbara Maria. 1993. "Presuming Images and Consuming Words: The Visualization of Knowledge from the Enlightenment to Post-Modernism." In *Consumption and the World of Goods*, ed. by J. Brewer and R. Porter, 462–77; 473. New York: Routledge.

Steedman, Carolyn. 2001. "Something She Called a Fever: Michelet, Derrida and Dust." *American Historical Review* 106 (4): 1159–80.

Steedman, Carolyn. 2002. *Dust: The Archive and Cultural History*. Rutgers: Rutgers University Press.

Stephens, Greg J., Lauren J. Silbert, and Uri Hasson. 2010. "Speaker–Listener Neural Coupling Underlies Successful Communication." *Proceedings of the National Academy of Sciences of the United States of America* 107 (32): 14425–30.

Sterne, Jonathan. 2005. "Digital Media and Disciplinarity." *Information Society* 21: 249–56.

Sterne, Jonathan. 2007. "Out with the Trash: On the Future of New Media." *Residual Media*: 16–31.

Sterne, Jonathan. 2015. "Analog." In *Digital Keywords*, ed. by Ben Peters (forthcoming). Princeton: Princeton University Press.

Sonvilla-Weiss, Stefan. 2010. "Introduction: Mashups, Remix Practices and the Recombination of Existing Digital Content." *Mashup Cultures*: 8–23.

Stiegler, Bernard. 1998. *Technics and Time, 1: The Fault of Epimetheus*. Stanford: Stanford University Press.

Stoddart, E. 2011. *Theological Perspectives on a Surveillance Society: Watching and Being Watched*. Farnham: Ashgate.

Stojanov, Krassimir. 2012. "The Concept of Bildung and Its Moral Implications." In *Becoming Oneself*, 75–88. Wiesbaden: Springer Fachmedien Verlag für Sozialwissenschaften.

Stoler, Ann Laura. 2010. *Along the Archival Grain: Epistemic Anxieties and Colonial Common Sense*. Princeton: Princeton University Press.

Stone, Allucquere Rosanne. 1991. "Will the Real Body Please Stand Up?" *Cyberspace: First Steps*: 81–118.

Stone, Allucquere Rosanne. 1996. *The War of Desire and Technology at the Close of the Mechanical Age*. Cambridge: MIT Press.

Starosielski, Nicole. "Underwater Flow." 2011. *Flow* 15 (October 16).

Straw, Will. 2007. "Embedded Memories." *Residual Media*: 3–15.

Svedkauskaite, Asta. 2004. "MonoConc Pro 2.0 and the Corpus of Spoken Professional American English: Resources from Athelstan." *Style* 38 (1): 127–33.

Svensson, Patrik. 2010. "The Landscape of Digital Humanities." *Digital Humanities Quarterly* 4 (1). http://digitalhumanities.org/dhq/vol/5/1/000090/000090.html.

Svensson, Patrik. 2011. "From Optical Fiber to Conceptual Cyberinfrastructure." *Digital Humanities Quarterly* 5 (1) http://digitalhumanities.org/dhq/vol/5/1/000090/000090.html.

Svensson, Patrik. 2012. "Envisioning the Digital Humanities." *Digital Humanities Quarterly* 6 (1). http://www.digitalhumanities.org/dhq/vol/6/1/000112/000112.html.

Swedish Research Council. 2012. "The Swedish Research Council's Guide to Infrastructures 2012."Swedish Research Council Reports 3. http://vr.se/download/18.48d441ad1363c4c099af9/Swedish+Research+Council%C2%B4s+guide+to+infrastructures_rapport+3_2012.pdf

Sweet, L. 2012. *Viral: How Social Networking Is Poised to Ignite Revival*. Colorado Springs: Waterbrook Press.

Szerszynski, B. 2005. *Nature, Technology and the Sacred*. Oxford: Wiley-Blackwell.

"Table of Contents." *Critical Code Studies.* http://criticalcodestudies.com/wordpress/

Tanselle, G. Thomas. 1989. "Reproductions and Scholarship." *Studies in Bibliography* 42: 25–54.

Taylor, A., and R. Harper. 2003. "The Gift Of Gab? A Design Oriented Sociology of Young People's Use of Mobiles." *Computer Supported Cooperative Work* 12: 267–96.

Taylor, Diana. 2003. *The Archive and the Repertoire: Performing Cultural Memory in the Americas.* Durham: Duke University Press.

Taylor, Diana. 2009. "The Digital as Anti-Archive. Evil Media: Making Good Use of Weights, Chains and Ranks." iTunesU podcast, 94:00, from the Animating Archives Conference, Brown University, December 4. https://itunes.apple.com/us/podcast/keynote/id381080290?i=84641744&mt=2

Tene, Omer. 2008. "What Google Knows: Privacy and Internet Search Engines." Unpublished paper. ExpressO. http://works.bepress.com/omer_tene/2

Teusner, P. 2010. "Emerging Church Bloggers in Australia: Prophets, Priests and Rulers in God's Virtual World." PhD thesis. RMIT University, Melbourne.

Teusner, P., and R. Torma. 2010. "iReligion." *Studies in World Christianity* 17 (2): 137–55.

Thacker, Christopher. 1979. *The History of Gardens.* Berkeley: University of California Press.

Thomas, William. 2004. "Computing and the Historical Imagination." In *A Companion to Digital Humanities*, ed. by Susan Schreibman, Ray Siemens, and John Unsworth, xxiii–xxvii. Oxford: Blackwell; http://nora.lis.uiuc.edu:3030/companion/view?docId=blackwell/9781405103213/9781405103213.xml&chunk.id=ss1-2-5&toc.id=0&brand=9781405103213_brand.

Thomas III, William G., and Edward L. Ayers. 2003. "The Differences Slavery Made: A Close Analysis of Two American Communities." http://www2.vcdh.virginia.edu/AHR/

Thompson, Clive. 2004. "Art Mobs." *Slate*. Unz.org. July 21. http://www.unz.org/Pub/Slate-2004jul-00145

Thornton, Dora. 1997. *The Scholar in His Study: Ownership and Experience in Renaissance Italy.* New Haven: Yale University Press.

"Throwing Mail into Bags." 1903. Youtube video, 1:13. American Mutoscope and Biograph Company. Early Motion Picture collection of the Library of Congress. August 7. Posted April 15, 2010.http://www.youtube.com/watch?v=XKZQy0PtfSw

"Tidskrifter mister viktigt stöd." 2013. *Svenska Dagbladet*. November 15. Swedish newspaper article. http://www.svd.se/kultur/tidskrifter-mister-viktigt-stod_8729014.svd

Tilley, Christopher. 2008. "Phenomenological Approaches to Landscape Archaeology." *Handbook of landscape archaeology*, ed. by B. David and J. Thomas: 271–76. Walnut Creek, CA: Left Coast Press.

Trettien, Whitney Anne. 2009. "Computers, Cut-ups, and Combinatory Volvelles: An Archaeology of Text-Generating Mechanisms." Master's thesis. MIT. Whitneyannetrettien.com. http://whitneyannetrettien.com/thesis/

Trow, M. 2001. "From Mass Higher Education to Universal Access: The American Advantage." In *Defense of American Higher Education*, ed. by Philip Altbach, Patricia Gumpor, and Donald Bruce Johnstone. Baltimore: John Hopkins University Press.

Tryon, Chuck. 2009. *Reinventing Cinema: Movies in the Age of Media Convergence*. Piscataway: Rutgers University Press.

Tufekci, Zeynep. 2008. "Grooming, Gossip, Facebook, and MySpace: What Can We Learn about These Sites from Those Who Won't Assimilate?" *Information Communication and Society* 11: 544–64.

Turner, Fred. 2006. *From Counterculture to Cyberculture: Stewart Brand, the Whole Earth Network, and the Rise of Digital Utopianism*. Chicago: Chicago University Press.

Turner, Fred. 2012. "*The Family of Man* and the Politics of Attention in Cold War America." *Public Culture* 24 (1): 55–84.

Turner, Mark. 1996. *The Literary Mind: The Origins of Thought and Language*. New York: Oxford University Press.

University Committee on Planning and Budget. 2010. "The Choices Report." Academic Senate. March. http://senate.universityofcalifornia.edu/ucpb.choices.pdf.

Unsworth, John (chair), and the American Council of Learned Societies Commission on Cyberinfrastructure for the Humanities and Social Sciences. 2006. "Our Cultural Commonwealth: The Report of the American Council of Learned Societies Commission on Cyberinfrastructure for the Humanities and Social Sciences." acls.org. http://www.acls.org/programs/Default.aspx?id=644

Unsworth, John. 2012. "Interview with John Unsworth, April 2011, carried out and transcribed by Charlotte Tupman." *Collaborative Research in the Digital Humanities (Epub)*: 231.

Urban Simulation Team, University of California Los Angeles. 2013. "The World's Columbian Exposition of 1893." http://www.ust.ucla.edu/ustweb/Projects/columbian_expo.htm

Vaidhyanathan, S. 2011. *The Googlization of Everything (and Why We Should Worry).* Berkeley: University of California Press.

Vandendorpe, Christian. 2009. *From Papyrus to Hypertext: Toward the Universal Digital Library.* Champaign: University of Illinois Press.

van Oost, E. 2000. "Making the Computer Masculine. The Historical Roots of Gendered Representations." In *Women, Work and Computerization: Charting a Course to the Future*, ed. by E. Balka and R. Smith, 9–16. Dordrecht: Kluwer Academic.

Vaughan, L., and Y. Zhang. 2007. "Equal Representation by Search Engines? A Comparison of Websites across Countries and Domains." *Journal of Computer-Mediated Communication* 12 (3): 888–909.

Vee, Annette, Alexandra Lockett, Elizabeth Losh, David Rieder, Mark Sample, and Karl Stolley. 2012. "The Role of Computational Literacy in Computers and Writing." *Enculturation*. October 10. http://www.enculturation.net/node/5272.

Venegas, Cristina. 2010. "'Liberating' the Self: The Biopolitics of Cuban Blogging." *Journal of International Communication* 16 (2): 43–54.

Vérin, Hélène. 1993. *La Gloire des ingénieurs: l'intelligence technique du XVIe au XVIIIe siècle.* Paris: Albin Michel.

Veracini, Lorenzo. 2007. "Settler Colonialism and Decolonization." *Borderlands e-Journal* 6 (2). http://researchbank.swinburne.edu.au/vital/access/manager/Repository/swin:10838.

Vermeule, Blakey. 2011. *Why Do We Care about Literary Characters?* Baltimore: Johns Hopkins University Press.

Vernon, James. 2000. "'For Some Queer Reason': The Trials and Tribulations of Colonel Barker's Masquerade in Interwar Britain." *Signs* 26 (1, October 1): 37–62.

Villi, M. 2010. "Visual Mobile Communication: Camera Phone Photo Messages as Ritual Communication and Mediated Presence." PhD dissertation. Aalto University School of Art and Design, Helsinki.

Virilio, Paul. 1993. "The Third Interval: A Critical Transition." In *Rethinking Technologies*, ed. by Verena Conley, 3–12. Minneapolis: University of Minnesota Press.

von Thünen, Johann Heinrich. 1966. *Isolated State: An English Edition of Der isolierte Staat*. Pergamon Press.

Wagner, Rachel. 2012a. "First-Person Shooter Religion: Algorithmic Culture and Inter-Religious Encounter." *Cross Currents* 62 (2): 181–203.

Wagner, Rachel. 2012b. *GodWired: Religion, Ritual and Virtual Reality*. Abingdon: Routledge.

Wajcman, J. 2004. *Technofeminism*. Malden, MA: Polity Press.

Wallace, R. 2012. "Free Speech Fight over Arrest of Seoul Satirist." *The Weekend Australian*. (April 7–8): 9.

Wallerstein, Immanuel. 1976. *The Modern World-System I: Capitalist Agriculture and the Origins of the EuropeanWorld Economy in the Sixteenth Century*. New York: Academic Press.

Wallerstein, Immanuel. 1993. World System versus World-Systems. In *The World System: Five Hundred Years or Five Thousand?* ed. Andre Gunder Frank and Barry K. Gills, 292–296. New York: Routledge.

Wallerstein, Immanuel. 2004. *World-Systems Analysis: An Introduction*. Durham: Duke University Press.

Wallis, C. 2011. "Mobile Mobility: Marginal Youth and Mobile Phones in Beijing." In *Mobile Communication: Bringing Us Together and Tearing Us Apart*, ed. by R. Ling and S. W. Campbell, 61–81. New Brunswick: Transaction Books.

War Crimes Studies Center. 2013. "Virtual Tribunal." WCSC University of California Berkeley. http://wcsc.berkeley.edu/virtual-tribunal/

Ward, Brian. 2012. "Religion Starts 2012 as Facebook's Top Engaging Trend," *AllFacebook* (blog). January 3. http://allfacebook.com/facebook-engagement-2_b72349

Wardrip-Fruin, Noah. 2009. *Expressive Processing: Digital Fictions, Computer Games, and Software Studies*. Cambridge: MIT Press.

Wark, McKenzie. 2006. "The Weird Global Media Event and the Tactical Intellectual [version 3.0]." In *New Media, Old Media: A History and Theory Reader*, ed. by Wendy Hui Kyong Chun and Thomas Keenan, 265–76. New York: Routledge.

Wark, McKenzie. 2007. *Gamer Theory*. Cambridge: Harvard University Press.

Watt, R. J. C. 2013. *Concordance*. http://www.concordancesoftware.co.uk/

Wellander, Janna. 2012. "SPARC Europe's Response to the Finch Report." July 11. SPARC Europe. http://sparceurope.org/sparc-europe-response-to-the-finch-report/

Wellman, B. 2001. "Computer Networks as Social Networks." *Science* 293 (5537): 2031–34.

Wellman, B., Quan-Haase, A., Boase, J., and Chen, W., Hampton, Keith, Isla de Diaz, Isabel, and Miyata, Kakuko. 2003. "The Social Affordances of Networked Individualism." *Journal of Computer-Mediated Communication* 8 (3).

Werner, Sarah, ed. 2011. "Shakespeare and Performance." *Shakespeare Quarterly* Open Review. MediaCommons. http://mcpress.media-commons.org/shakespearequarterly performance/.

Wertheim, M. 1999. *The Pearly Gates of Cyberspace: A History of Space from Dante to the Internet.* London: Virago Press.

"What Are RIs?" European Commission. http://ec.europa.eu/research/infrastructures/index_en.cfm?pg=what

White, Hayden. 1978. *Tropics of Discourse: Essays in Cultural Criticism*. Baltimore: Johns Hopkins University Press.

White, Richard. 2010. "What Is Spatial History?" Spatial History Lab: Working paper http://www.stanford.edu/group/spatialhistory/cgi-bin/site/pub.php

Wikipedia. 2012. "Imagine a World without Free Knowledge." Wikimedia Commons. January 18. http://commons.wikimedia.org/wiki/File:History_Wikipedia_English_SOPA_2012_Blackout2.jpg

Wikipedia. 2013. "Protests against SOPA and PIPA." Wikipedia. Last modified September 30. http://en.wikipedia.org/wiki/Protests_against_SOPA_and_PIPA

Wilkinson, David. 2006. Globalizations: The First Ten, Hundred, Five Thousand and Million Years. In *Globalization and Global History*, ed. Barry K. Gills and William R. Thompson, 68–78. New York: Routledge.

Winthrop-Young, Geoffrey. 2012. "Kittler's Siren Recursions." Unpublished talk at the Phenomenology, Minds and Media Mellon Sawyer Seminar, Duke University, February 9. Forthcoming in *Kittler Now*, ed. by Stephen Sale and Laura Salisbury. Malden: Polity Press.

Wirth, W., T. Böcking, V. Karnowski, and T. von Pape. 2007. "Heuristic and Systematic Use of Search Engines." *Journal of Computer-Mediated Communication* 12 (3): 778–800.

Wolfram, Stephen. 2002. *A New Kind of Science*. Champaign, IL: Wolfram Media.

Wolin, Richard. 2006. *The Frankfurt School Revisited and Other Essays on Politics and Society*. New York: Routledge.

Women Writers Project. Brown University. 2013. "Women Writers Project." http://www.wwp.brown.edu/

Wu, Jean Yu-Wen Shen, and Thomas Chen, eds. 2010. "*Asian American Studies Now: A Critical Reader*." New Brunswick: Rutgers University Press.

Wyche, S., and R. Grinter. 2009. "Extraordinary Computing: Using Religion as a Lens for Reconsidering the Home." *Proceedings ACM SIGCHI Conference on Human Factors in Computing Systems (CHI '09)*. Boston, 749–758.

Xárene. 2009–2010. "Tehran Election Protests. Hypercities." June 6–February 11. http://hypercities.ats.ucla.edu/?collections/13549

Yamamura, Hiromi, Yasuhito Sawahata, Miyuki Yamamoto, and Yukiyasu Kamitani. 2009. "Neural art appraisal of painter: Dali or Picasso?" *Neuroreport* 20 (18): 1630–33.

Yashpal. 2003. "1990 Presidential Address of the Indian Science Congress Association "Frontiers in Science and Technology." In *The Shaping of Indian Science—Indian Science Congress Presidential Addresses,* vol. 3. Himayatnager, Hyderabad, India: Universities Press: 1711.

YouVersion. "About YouVersion." Bible.com. http://www.youversion.com/.

Zittrain, Jonathan. 2008. *The Future of the Internet and How to Stop It*. New Haven: Yale University Press.

INDEX